Numerical Techniques for Chemical and Biological Engineers Using MATLAB®

Numerical Techniques for Chemical and Biological Engineers Using MATLAB®

A Simple Bifurcation Approach

Said Elnashaie
Frank Uhlig

with the assistance of
Chadia Affane

 Springer

Professor Said
  S.E.H. Elnashaie
Pennsylvania State University
  at Harrisburg
Room TL 176 Capital College
777 W. Harrisburg Pike
Middletown, PA 17057-4898
sse10@psu.edu

Professor Frank Uhlig
Department of Mathematics
  and Statistics
Auburn University
312 Parker Hall
Auburn, AL 36849
uhligfd@auburn.edu

Chadia Affane
Department of Mathematics
  and Statistics
Auburn University
Auburn, AL 36849
affanac@auburn.edu

MATLAB® is a trademark of The MathWorks, Inc. and is used with permission. The MathWorks does not warrant the accuracy of the text or exercises in this book. This book's use or discussion of MATLAB® software or related products does not constitute endorsement or sponsorship by The MathWorks of a particular pedagogical approach or particular use of the MATLAB® software.

Additional material to this book can be downloaded from http://extras.springer.com.

ISBN 978-1-4939-5054-6   ISBN 978-0-387-68167-2 (eBook)
DOI 10.1007/978-3-387-68167-2
© 2007 Springer Science+Business Media, LLC
Softcover reprint of the hardcover 1st edition 2007

9 8 7 6 5 4 3 2 1

springer.com

SAID ELNASAIE

Born 1947, Cairo/Egypt; grew up in Egypt; married, two children,
three grandchildren.

Chemical Engineering student at Cairo University, University of
Waterloo, and University of Edinburgh.

B. S., Cairo University, 1968; M.S., U of Waterloo, Canada, 1970;
Ph.D., U Edinburgh, UK, 1973.

Postdoc, University of Toronto and McGill U, 1973 – 1974.

Professor, Cairo University, 1974 – 1992; King Saud University, Saudi
Arabia, 1986 – 1996; University of British Columbia,
Canada, 1990; University Putra Malaysia, 1996/97; King
Fahad University, Saudi Arabia, 1999; Auburn University,
1999 – 2005; Visiting Professor, U British Columbia,
Vancouver, 2004; U British Columbia, Vancouver, 2005 –

Vice President, Environmental Energy Systems and Services (EESS),
Egypt, 1996 – 1998.

Research Areas: Modeling, Simulation and Optimization of Chemical
and Biological Processes, Clean Fuels (Hydrogen, Biodiesel
and Ethanol), Fixed and Fluidized Bed Catalytic Reactors,
Nonlinear Dynamics, Bifurcation and Chaos,
Clean Technology, Utilization of Renewable Materials,
Sustainable Engineering.

300+ papers, 3+ books.

nashaie@chml.ubc.ca

FRANK UHLIG

Born 1945, Mägdesprung, Germany; grew up in Mülheim/Ruhr,
Germany; married, two sons.

Mathematics student at University of Cologne, California Institute
of Technology.

Ph.D. CalTech 1972; Assistant, University of Würzburg, RWTH
Aachen, Germany, 1972 – 1982.

Two Habilitations (Mathematics), University of Würzburg 1977,
RWTH Aachen 1978.

Visiting Professor, Oregon State University 1979/1980;

Professor of Mathematics, Auburn University 1982 –

Two Fulbright Grants; (Co-)organizer of eight research conferences.

Research Areas: Linear Algebra, Matrix Theory, Numerical Analysis,
Numerical Algebra, Geometry, Krein Spaces, Graph
Theory, Mechanics, Inverse Problems, Mathematical
Education, Applied Mathematics, Geometric Computing.

50+ papers, 3+ books.

uhligfd@auburn.edu   www.auburn.edu/~uhligfd

CHADIA AFFANE

Born 1968, Fes/Morocco; grew up in Morocco; married,
       two children.
Engineering student at École Supérieure de Technologie, E.S.T.,
       Fes, Morocco, 1989 – 1991.
Chemical engineering student at Texas A&M University, 1997 – 1999;
       B.S. Texas A&M University, 1999.
MS in Applied Mathematics, Auburn University, 2003.
Ph. D. student, Mathematics, Auburn University, 2004 –
Research Areas: Numerical Analysis, Applied Mathematics.

affanac@auburn.edu

Preface

This book has come about by chance.

The first author, Said Elnashaie, and his wife, Shadia Elshishini, moved next door to the second author, Frank Uhlig, and his family in 2000. The two families became good neighbors and friends. Their chats covered the usual topics and occasionally included random teaching, departmental, and university matters.
One summer day in 2003, Said showed Frank a numerical engineering book that he had been asked to review. Neither of them liked what they saw. Frank eventually brought over his "Numerical Algorithms" book and Said liked it. Then Said brought over his latest Modeling book and Frank liked it, too. And almost immediately this Numerical Chemical and Biological Engineering book project started to take shape.

Said had always felt more mathematically inclined in his work on modeling problems and bifurcation and chaos in chemical/biological engineering;
Frank had lately turned more numerical in his perception and efforts as a mathematician.

This book is the outcome of Said's move to Auburn University and his chance moving in next door to Frank. It was born by a wonderful coincidence!

Said and Frank's long evening walks through Cary Woods contributed considerably towards most of the new ideas and the educational approach in this book. We have both learned much about numerics, chemical/biological engineering, book writing, and thinking in our effort to present undergraduates with state of the art chemical/biological engineering models *and* state of the art numerics for modern chemical/biological engineering problems.

Chadia is a chemical engineer who has turned towards applied mathematics in her graduate studies at Auburn University and has helped us bridge the gap between our individual perspectives.

The result is an interdisciplinary, totally modern book, in contents, treatment, and spirit. We hope that the readers and students will enjoy the book and benefit from it.

For help with our computers and computer software issues we are indebted to A. J., to Saad, and to Darrell.

Auburn and Vancouver, 2006

Contents

Appendix 1: Linear Algebra and Matrices   533

Appendix 2:
Bifurcation, Instability, and Chaos in Chemical and Biological Systems   547

The question mark

Introduction

This book is interdisciplinary, involving two relatively new fields of human endeavor. The two fields are: Chemical/Biological[1] Engineering **and** Numerical Mathematics.

How do these two disciplines meet? They meet through mathematical modeling.

Mathematical modeling is the science or art of transforming any macro-scale or micro-scale problem to mathematical equations. Mathematical modeling of chemical and biological systems and processes is based on chemistry, biochemistry, microbiology, mass diffusion, heat transfer, chemical, biochemical and biomedical catalytic or biocatalytic reactions, as well as noncatalytic reactions, material and energy balances, etc.

As soon as the chemical and biological processes are turned into equations, these equations must be solved efficiently in order to have practical value. Equations are usually solved numerically with the help of computers and suitable software.

Almost all problems faced by chemical and biological engineers are nonlinear. Most if not all of the models have no known closed form solutions. Thus the model equations generally require numerical techniques to solve them. One central task of chemical/biological engineers is to identify the chemical/biological processes that take place within the boundaries of a system and to put them intelligently into the form of equations by utilizing justifiable assumptions and physico-chemical and biological laws. The best and most modern classification of different processes is through system theory. The models can be formed of steady-state design equations used in the design (mainly sizing and optimization), or unsteady-state (dynamic) equations used in start-up, shutdown, and the design of control systems. Dynamic equations are also useful to investigate the bifurcation and stability characteristics of the processes.

The complexity of the mathematical model depends upon the degree of accuracy required and on the complexity of the interaction between the different processes taking place within the boundaries of the system and on the interaction between the system and its surrounding. It is an important art for chemical/biological engineers to reach an optimal degree of sophistication (complexity) for the system model. By "optimal degree of sophistication" we mean finding a model for the process, which is as simple as possible without sacrificing the required accuracy as dictated by the specific practical application

---

[1]Biological engineering comprises both biochemical and biomedical engineering

1

of the model.

After the chemical/biological engineer has developed a suitable mathematical model with an optimal degree of "sophistication" for the process, he/she is then faced with the problem of solving its equations numerically. This is where stable and efficient numerical methods become essential.

The classification of numerical solution techniques lends itself excellently to the system theory classification as well.

A large number of chemical/biological processes will be presented, modeled, and efficient numerical techniques will be developed and programmed using MATLAB® [2]. This is a sophisticated numerical software package. MATLAB is powerful numerically through its built-in functions and it allows us to easily develop and evaluate complicated numerical codes that fulfill very specialized tasks. Our solution techniques will be developed and discussed from both the chemical/biological point of view and the numerical point of view.

Hence the flow of each chapter of this book will lead from a description of specific chemical/biological processes and systems to the identification of the main state variables and processes occurring within the boundaries of the system, as well as the interaction between the system and its surrounding environment. The necessary system processes and interactions are then expressed mathematically in terms of state variables and parameters in the form of equations. These equations may most simply be algebraic or transcendental, or they may involve functional, differential, or matrix equations in finitely many variables.

The mathematical model specifies a set of equations that reflects the characteristics and behavior of the underlying chemical/biological system. The parameters of the model can be obtained from data in the literature or through well-designed experimentation. Any model solution should be checked first against known experimental and industrial data before relying on its numerical solution for new data. To use a model in the design and control of a system generally requires efficient solution methods for the model equations.

Some of the models are very simple and easy to solve, even by hand, but most require medium to high-powered numerical techniques.

From chapter to chapter, we introduce increasingly more complex chemical/biological processes and describe methods and develop MATLAB codes for their numerical solution. The problem of validating a solution and comparing between different algorithms can be tackled by testing different numerical techniques on the same problem and verifying and comparing their output against known experimental and industrial data.

In this interdisciplinary text we assume that the reader has a basic knowledge of the laws governing the rates of different chemical, biological and physical processes, as well as of material and energy balances. Junior and senior undergraduates majoring in chemical/biological engineering and graduate students in these areas should be able to follow the engineering related portions of the text, as well as its notations and scientific deductions easily. Regarding mathematics, students should be familiar with calculus, linear algebra and matrices, as well as differential equations, all on the first and second

---

[2]MATLAB is a registered trade mark of The MathWorks, Inc., 3 Apple Hill Drive, Natick, MA 01760; http://www.mathworks.com .

year elementary undergraduate level. Having had a one semester course in numerical analysis is not required, but will definitely be helpful. We include an appendix on linear algebra and matrices at the end of the book since linear lgebra is often not required in chemical/biological engineering curricula.

In order to solve chemical/biological problems of differing levels we rely throughout on well tested numerical procedures for which we include MATLAB codes and test files. Moreover, a large part of this book is dedicated to explain the workings of our algorithms on an intuitive level and thereby we give a valuable introduction to the world of scientific computation and numerical analysis.

It is precisely the interdisciplinary bridge between chemical/biological engineering and numerical mathematics that this book addresses. A true melding of the two disciplines has been lacking up to now. We hope that this book is a step in the right direction.

Chemical/biological engineers need such a book in order to learn and be able to solve their continuously expanding problems (in size, complexity, and degree of nonlinearity). And numerical analysts can enrich their applied know-how from the civil, mechanical, and electrical engineering problem menu to include the richness of chemical/biological engineering problems.

An unusual aspect of this book addresses generality and special cases. What do we mean by that? We mean this in three different and distinct senses:

* The first regards reacting and nonreacting systems, that is, chemical/biological systems in which chemical and/or biological reactions take place within the boundaries of the system, and those with no reaction, such as encountered in nonreacting separating processes. Many texts treat nonreacting systems first for the obvious reason that they are simpler, followed by more complex reacting systems. Actually this is not optimal. From a pedagogical point of view and for a more efficient transfer of knowledge, it is advisable to do the opposite, i.e., to start with the more general reacting cases and then treat the nonreacting cases as special cases in which the rates of reactions are equal to zero.

* The second regards the number of states of a system (we will call them attractors for reasons explained inside the book). Almost all books by mistake treat systems as if they essentially have only one state (usually of one specific type which is another limitation), and consider systems with more than one state as if they were special or odd cases. Thus they rarely address cases having different numbers or types of states and attractors. In this book, and maybe for the first time in the engineering literature, we will address our systems in a more general and fundamental form. We will consider systems that have more than one state or attractor as the general case, with systems having one state being a special case thereof. We also address problems that have more than one type of steady state. This will necessitate a correction in terminology. We are afraid that this will be faced by some resistance. But this is to be expected when scientific research starts to change the way we treat our undergraduate curriculum and our terminology in a more fundamental manner. Consequently, the usual term "steady state" will have to be replaced by "Stationary Nonequilibrium State" (SNES) for example, to distinguish it from "Thermodynamic Equilibrium" which is also a steady state (no change with

time), but a dead one.

This SNES will also be called a "Fixed Point Attractor" (FPA) in order to distinguish it from oscillatory states which we will call "Periodic Attractors" (PA), and from quasi periodic states which we will call "Torus Attractors" (TA), and more complex states such as "Chaotic Attractors" (CA). Thus our more advanced approach will necessitate the definition of all basic principles of bifurcation theory early in the book, and it will make us introduce senior undergraduates early in our text to the generality and practical importance of these concepts. These modern engineering concepts will be reflected in the models' numerical solutions, specially with regard to initial guesses and numerical convergence criteria. Appendix 2 on bifurcation, multiplicity, and chaos in chemical and biological systems explains these natural phenomena in a simple manner.

Likewise, we treat numerical analysis from a modern and fresh perspective here.

* In the currently available (less than one handful) textbooks that treat numerical analysis and chemical/biological engineering problems, most of the numerics have the flavor of the 1960s. These texts labor long over Gaussian elimination to solve linear equations, over the intricacies of solving scalar equations given in standard form $f(x) = 0$ versus via fixed-point iterations $x_{i+1} = \phi(x_i)$ and so forth, and thus they build mostly on ancient and outdated methods and ideas. Worse, they generally fail to mention limitations of the proposed and often highly elementary algorithms. Many books in this area are not aware and do not make the readers aware of the instability of un-pivoted Gauss, of solving "unsolvable" linear systems in the least squares sense, or of the gross errors of standard zero finders in cases with multiple or repeated solutions, for example. In a nutshell, they do not convey the advances in numerical analysis over the last 40 or 50 years. This advance preeminently consists of a better understanding of numerical analysis and of the development of intricate verifiably stable codes for the efficient solution of standard numerical problems, as well as having learnt how to recognize ill-conditioned input data sets. The omission of these advances is very unfortunate and we intend to remedy this gap in the timely transfer of knowledge from one part of science to another. This book contains a plethora of new and stable algorithms that solve many chemical/biological engineering model problems efficiently. It does so by working with the stable codes inside MATLAB, rather than developing line by line original programs. There is no need to reinvent pivoted Gauss, or stable adaptive and efficient integration schemes, etc, over and over again.

All of our programs are available on the accompanying CD.

Our book is a *four pronged approach* at modernizing chemical/biological engineering education through numerical analysis.

The three prongs

(1) of addressing generalities versus what is a special case in chemical models,

(2) of treating bifurcation phenomena and multiple steady states as the norm and not the exception, and

(3) of dealing with stability and condition issues in modern numerical codes

are augmented by the fourth prong

(4) of color graphic visualization.

Throughout the book we emphasize graphic 2D and 3D representations of the solutions. All our MATLAB programs create color graphics. In the book, however, we can only use grayscale displays for economic reasons. Color is essential to fully understand many of the graphs. Therefore, we urge our readers to acquire a deeper understanding by accompanying a reading of the book with simultaneously recreating our figures in color on a computer. Sample calls for drawing the figures are given at the top of every MATLAB code on the accompanying CD.

We especially suggest to follow this advice and recreate the surface and multiple line graphs in color for Chapter 3, Chapter 4, Sections 4.1 and 4.3, Chapter 5, Sections 5.1 and 5.2, and Chapter 6, Section 6.4.

Through this colorful visual approach, students and readers will have a chance to experience the meaning, i.e., to "understand" for the first time (in the true definition of *to stand under*), what a solution means, how it behaves, looks, where it is strained, i.e., likely to be (nearly) unstable, where it is smooth, etc.

How to Use this Book

This book uses a modern systems approach. The problems that we study can be classified in three ways:

1. according to the system classification as closed, isolated or open systems;

2. according to the spatial variation as lumped or distributed systems; as well as

3. according to the number of phases as homogeneous or heterogeneous systems.

While we have made this book as self-contained as possible, it is beneficial if the student/reader has a background in chemical and biological modeling, calculus, matrix notation and possibly MATLAB before attempting to study this book.

As with any book on science and mathematics that contains mathematical equations, formulas and model derivations, it is important for our readers to take out *paper and pencil* and try to replicate the equations and their derivations from first principles of chemistry, physics, biology, and mathematics by him/herself.

Starting with Chapter 3, many relevant chemical/biological engineering problems are solved explicitly in the text. Each section of Chapters 3 to 7 contains its own unsolved exercises and each chapter contains further problems for the whole chapter at its end.

Students should first try to solve the worked examples inside each section on their own, with models of their own making and personally developed MATLAB codes for their solution. Then they should compare their results with those offered in the book. And finally they should try to tackle the unsolved section and chapter problems from the experience that they have gained.

The first chapter must be studied regardless of the particular background of the student or reader, but it need not be studied in class. Instead it can be assigned as reading and exploratory homework in the first days of class. Chapter 1 gives the reader an introduction to MATLAB and numerical analysis. It shows how to use MATLAB and explains how MATLAB treats many important numerical problems that will be encountered later in the book. As said before, the student/reader best studies Chapter 1 concretely on a MATLAB desktop displayed on his/her computer screen by mimicking the elementary examples from the book, varying them and thereby learning the fundamentals of MATLAB computation.

Chapter 2 introduces the essential principles of modeling and simulation and their relation to design from a systems point of view. It classifies systems based on system theory in a most general and compact form. This chapter also introduces the basic principles of nonlinearity and its associated multiplicity and bifurcation phenomena. More on this, the main subject of the book, is contained in Appendix 2 and the subsequent chapters.

One of the characteristics of the book is to treat the case with multiple steady states as the general case while the case with a unique steady state is considered a special case. When reading Chapter 2, our readers should notice the systems similarities between chemical and biological systems that make these two areas conjoint while uniquely disjoint from the other engineering disciplines.

We suggest to start a class based on our book with Chapter 3 and to assign or study the first two chapters as reading or reference material whenever the need arises in class.

Chapter 3 tries to give students the essential tools to solve lumped systems that are governed by scalar equations. It starts with the simplest continuous-start reactor, a CSTR in the adiabatic case. The first section should be studied carefully since it represents the basis of what follows. Our students should write their own codes by studying and eventually rewriting the codes that are given in the book. These personal codes should be run and tested before the codes on the CD are actually used to solve the unsolved problems in the book. Section 3.2 treats the nonadiabatic case.
Section 3.3 is devoted to biochemical enzyme systems in which the biocatalyst enzyme does not change during the progressing reaction. Biological systems whose biocatalysts change with time are presented later in the book.
In Section 3.4 we study several systems that have no multiple steady states and we introduce several transcendental and algebraic equations of chemical and biological engineering import. As always, the students and readers should find their own MATLAB codes for the various problems first before relying on those that are supplied and before solving the included exercises.

Chapter 4 studies problems that involve change over time or location and that therefore are modeled by differential equations. Specifically, we study initial value problems (IVP) here, including

1. one-dimensional distributed steady-state plug flow systems that are characterized by the complete absence of multiplicity and bifurcation phenomena for adiabatic

systems with cocurrent cooling or heating; and

2. unsteady states of lumped systems. These systems usually have multiple steady states when at least one of the processes depends nonmonotonically on one of the state variables. The unsteady-state trajectory that describes the dynamic behavior will, however, be unique since IVP cannot have multiple solutions. But different initial conditions can lead to different steady states.

The readers should study the four given examples very carefully and perform the associated exercises. Students should solve the solved-in-detail problems of this chapter first independently of the book and then solve the exercises of each section and of the chapter as a whole.

Chapter 4.1 deals with an important industrial problem, the vapor-phase cracking of acetone. Here the material- and energy-balance design equations are developed. We advise the students to try and develop the design equations independently before consulting the book's derivations. Numerical solutions and MATLAB codes are developed and explained for this problem and sample results are given that need to be checked against those of the students' codes.

The remaining sections of Chapter 4 treat an anaerobic digester, heterogeneous fluidized catalytic-bed reactors, and a biomedical problem of the neurocylce enzyme system.

Chapter 5 introduces a more difficult differential equations problem, namely boundary value problems (BVP). Such problems are very common in chemical and biological engineering but are unfortunately often given the least bit of attention in undergraduate training.

Two point boundary value problems of chemical/biological engineering typically arise from some "feedback" of information. They can result from any of the following sources.

1. diffusion;

2. conduction;

3. countercurrent operation; or

4. recycle or circulation.

Two point boundary value problems are much more difficult to solve and more demanding than initial value problems for differential equations. One of the strengths of MATLAB is that it has very good and efficient subroutines for solving both IVPs and BVPs.

First we introduce the reader to the principles of such problems and their solution in Sections 5.1.2 and 5.1.2. As an educational tool we use the classical axial dispersion model for finding the steady state of one-dimensional tubular reactors. The model is formulated for the isothermal case with linear kinetics. This case lends itself to an otherwise rare analytical solution that is given in the book. From this example our students can understand many characteristics of such systems.

Students and readers should be very familiar with the nature of the isothermal case before embarking on the nonlinear case and its numerical solution in Section 5.1.3. The students should study the nonlinear BVP carefully, as well as their associated bifurcation phenomena. MATLAB codes are again provided, explained and tested. Our readers should train themselves to become fluent in solving BVPs via MATLAB's collocation method. The exercises are a good training ground for this.

Section 5.2 is principally similar to Section 5.1. But now we analyze the porous catalyst pellet via boundary value problems for differential equations. Here the dimensionality of the problem doubles and the resulting singular BVP is easily handled in MATLAB due to its superbly capable matrices and vector implementation of code.

Chapter 6 deals with multiphase and multistage systems. Section 6.1 concentrates on heterogeneous systems, both lumped and distributed heterogeneous systems, as well as on isothermal and nonisothermal ones. Here the students should study the extension of the modeling principles for homogeneous systems to those for heterogeneous systems and become aware of the different numerical properties of the resulting model or design equations.

While many solved examples and complete MATLAB codes are given in the book and on the CD, the students should try to excel with his/her codes by personal initiative and by designing his/her own codes. These should be tested by solving the exercises at the end of the section.

Section 6.2 deals with high-dimensional lumped nonreacting systems, with special emphasis on multitray absorption.

Section 6.3 treats distributed nonreacting systems and specifically packed bed absorption, while Section 6.4 studies a battery of nonisothermal CSTRs and its dynamic behavior.

Chapter 7 is the climax of the book: Here the educated student is asked to apply all that he/she has learned thus far to deal with many common practical industrial units.

In Chapter 7 we start with a simple illustrative example in Section 7.1 and introduce five important industrial processes, namely fluid catalytic cracking in FCC units in Section 7.2, the UNIPOL® process in Section 7.3, industrial steam reformers and methanators in Section 7.4, the production of styrene in Section 7.5, and the production of bioethanol in Section 7.6.

However, here we leave all numerical procedures and MATLAB coding as exercises to the students and readers. For each problem, all the necessary modeling and data is included, as well as samples of numerical results in the form of tables and graphs. Our readers should now be able to use the models and the given parameters to develop their own MATLAB codes along the lines of what has been practiced before. Then the students should be try to solve the exercises given at the end of each section and finally the general exercises at the end of the chapter.

This, we think, is the best preparation of our students for a successful career as a chemical or biological engineer in the practical modern industrial world.

Manueline ceiling

Chapter 1

Numerical Computations and MATLAB

[This chapter should be read and studied by our readers early on, with fingers on the keyboard of a computer, in order to gain a first working knowledge of MATLAB, and also later throughout the book as we introduce new numerical techniques and codes.]

The history of human mathematical computations goes back for several millenia. The need for numerical computations has increased since the age of enlightenment and the industrial revolution three centuries ago. For the last 50 years, the human race has become more and more dependent on numerical computations and digital computers. Computational techniques have developed from early hand computations, through table look up, mechanical adding and multiplying devices, the slide rule etc, to programmable electronic computers, mainframes, PCs, laptops, and notebooks.

The earliest electronic computers were programmed in machine language. Later, programming languages such as ALGOL, FORTRAN, C, etc. were developed. Over the last decade, many of our serious mathematical computations have begun to be performed via **software** rather than individual line-by-line coded programs. Software allows for a simpler command structure and an easier interface, and it offers ready graphics and ready coding error detection among its many advantages over computer language coding.

One ideally suited software for engineering and numerical computations is MATLAB® [1]. This acronym stands for "***Mat**rix **Lab**oratory*". Its operating units and principle are vectors and matrices. By their very nature, matrices express linear maps. And in all modern and practical numerical computations, the methods and algorithms generally rely on some form of linear approximation for nonlinear problems, equations, and phenomena. Nowadays all numerical computations are therefore carried out in linear, or in matrix and vector form. Thus MATLAB fits our task perfectly in the modern sense.

---

[1]MATLAB is a registered trade mark of The MathWorks, Inc., 3 Apple Hill Drive, Natick, MA 01760; http://www.mathworks.com

MATLAB contains a library of built-in state of the art functions and functionality for many standard numerical and graphics purposes, such as solving most any equation, whether scalar, functional, or differential, and plotting multidimensional surfaces and objects. It is a software that allows its users to adapt the built-in functions of MATLAB easily and intuitively to any problem specific needs with a few lines of code. Typically, users build their own codes using problem specific MATLAB functions, intertwined with command lines of their own design and for their own purpose. One of the most wonderful attributes of numerical and graphics software packages such as MATLAB is the ease with which they can display chemico-physical problems and their solutions readily by spatial surfaces and objects. This gives new means and meaning to our "understanding" and it is the source of great pleasure as well.

MATLAB comes as one main body of built-in functions and codes, and there are many additional specialized MATLAB "toolboxes" for various applications. As this book is primarily directed towards undergraduate and beginning graduate students, we have restricted ourselves deliberately to using the main body of MATLAB only in our codes and none of its many toolboxes.

1.1   MATLAB as a Software and a Programming Language

All what follows in this chapter and the rest of the book assumes that the user has easy access to MATLAB on a computer. This access may be provided by the college, university, or department in the form of a computer with MATLAB already installed, or through special licensing from The Mathworks, or by purchasing an individual copy of MATLAB or its student edition for a personal computer. The program codes of this book were begun under MATLAB version 6.5 and finished under version 7.1. Since MATLAB is designed to be backward compatible, all our codes should be able to run in any MATLAB version from 6.5 on up. We simply start our discourse assuming that there is a MATLAB desktop and command window on the computer screen with its >> command prompt in front of the reader. MATLAB m file code lines will henceforth always be displayed between two thick black horizontal lines.

Due to the special structure of MATLAB, readers should be familiar with the mathematical concepts pertaining to matrices, such as systems of linear equations, Gaussian elimination, size and rank of a matrix, matrix eigenvalues, basis change in n-dimensional space, matrix transpose, etc. For those who need a refresher on these topics there is a concise Appendix on linear algebra and matrices at the end of the book.

1.1.1   The Basics of MATLAB

The basic unit of MATLAB is a matrix.

If the **matrix** A contains m rows and n columns of entries, we call A an m by n matrix (*"**rows before columns**"*) and write $A_{m \times n}$ or $A_{m,n}$ in short mathematical notation. Column **vectors** of length k can be viewed as k by 1 matrices and row vectors as 1 by n matrices if they contain n entries.

Real or complex **scalars** c thus can be viewed as 1 by 1 matrices if we wish.

To represent or to generate and store the 3 by 4 matrix

$$A = \begin{pmatrix} 1 & 2 & 3 & 4 \\ 5 & 6 & 7 & 8 \\ 9 & 10 & 11 & 12 \end{pmatrix}$$

in MATLAB, we type the following line after the >> MATLAB prompt.
```
>> A = [1 2 3 4;5 6 7 8;9,10,11,12]
```
After pressing the return key, the MATLAB command window will display A as follows:
```
>> A = [1 2 3 4;5 6 7 8;9,10,11,12]
A =
    1     2     3     4
    5     6     7     8
    9    10    11    12
```
Note that entry-wise defined matrices and vectors are delimited by square brackets [and] in MATLAB commands.

Note further that the entries of one row of a matrix (or of a row vector) can be entered into MATLAB's workspace either separated by a blank space or by a comma, while a semicolon ; indicates the start of a new row. A comma or a blank at the end of a command line will cause screen display of the object that has just been defined, while a semicolon ; after a command will not.

Here is another sequence of commands and their screen output:
```
>> A = [1 2 3 4;5 6 7 8;9,10,11,12];
>> B = zeros(5)
B =
    0     0     0     0     0
    0     0     0     0     0
    0     0     0     0     0
    0     0     0     0     0
    0     0     0     0     0
```
The first command generates the matrix A. A is stored in MATLAB's **workspace**, but it is not displayed on screen due to the ; following the command. With the second command, the matrix B is generated, stored, and displayed on screen because there is a blank and no ; following the second command. Note that a comma , after a MATLAB command has the same effect as a blank as regards not suppressing screen output. In MATLAB, one can put as many comma or semicolon delimited commands on one line as one wishes and will fit.

One can alter a stored matrix $A = (a_{ij})$ by reassigning one element, say $a_{22} = 12$, as follows:
```
>> A(2,2) = 12
A =
    1     2     3     4
    5    12     7     8
    9    10    11    12
```

One can reassign a whole row (or similarly a whole column) of a matrix A to obtain an updated version of A with a new third row, for example, by using the "all entries" MATLAB symbol : appropriately.

```
>> A(3,:)=[1 0 1 0]
A =
     1     2     3     4
     5    12     7     8
     1     0     1     0
```

Here $A(3, :)$ denotes all entries of row 3 of A, while $A(:, 2)$ would designate column 2 of A.

One can update a whole block of A at one time as well, for example, the block

$$A(2:3, 2:4) = \begin{pmatrix} a_{22} & a_{23} & a_{24} \\ a_{32} & a_{33} & a_{34} \end{pmatrix}$$

as follows:

```
>> A(2:3,2:4)=[1 2 3;0 9 7]
A =
     1     2     3     4
     5     1     2     3
     1     0     9     7
```

Here $A(2:3, 2:4)$ describes the submatrix of A comprised of the rows 2 and 3 and the columns 2, 3, and 4, i.e., the lower right 2 by 3 block of A.

One can alter the shape of A by adding one or several rows or columns of compatible size, or by deleting one or several rows or columns of A by setting some rows or columns equal to the **empty row** or column [ ], respectively.

```
>> A = [A; A(3,:)]
A =
     1     2     3     4
     5     1     2     3
     1     0     9     7
     1     0     9     7
>> A(:,4) = []
A =
     1     2     3
     5     1     2
     1     0     9
     1     0     9
>> A(3,:)=[]
A =
     1     2     3
     5     1     2
     1     0     9
```

The first command makes A into a square 4 by 4 matrix by replicating the third row in row four of the new A.

The second command then deletes the last column of A, and the third command takes off the last column and creates a square 3 by 3 matrix with the same name A.

Note that any change prescribed in MATLAB for a named object such as our matrix A saves the changed object in the original object's place. Hence our final matrix A does not resemble the original A at all.

Once stored in the workspace, MATLAB can operate on these matrices A and B. A useful MATLAB task is to find the size of a stored matrix, i.e., the numbers of rows and columns of the matrix.

```
>> size(A), size(B), length(A), length(B), result = [size(A'), length(A')]
ans =
    3    4
ans =
    5    5
ans =
    4
ans =
    5
result =
    4    3    4
```

Let us look at the above sequence of commands and the output. If

$$A = \begin{pmatrix} 1 & 2 & 3 & 4 \\ 5 & 6 & 7 & 8 \\ 9 & 10 & 11 & 12 \end{pmatrix}$$

in MATLAB's workspace, then A has 3 rows and 4 columns. Its size vector is (3 4) in MATLAB (remember "rows before columns"), while that of the zero matrix B defined earlier is (5 5). **length** measures the maximal dimension of a matrix.

The commas after our MATLAB commands above are necessary as **delimiters** to be able to place several commands onto one line of code. They create screen output, whereas semicolons ; would have suppressed it. The final entry **result = [size(A'), length(A')]** on the command line creates the vector (4, 3, 4), called "result" by us. It contains the number of rows (4) of A^T, followed by A^T's number of columns (3) and its "length" or maximal dimension (4).

Here the matrix **A'** denotes the **transpose** of A, namely

$$A^T = \begin{pmatrix} 1 & 5 & 9 \\ 2 & 6 & 10 \\ 3 & 7 & 11 \\ 4 & 8 & 12 \end{pmatrix}.$$

In A^T the rows of A appear as columns, and the columns of A as rows in their natural order.

If an item is entered in MATLAB *without* a designation such as x = ... and is not followed by a ; , such an item will always be designated as **ans** on screen. This is short for "answer". Such an object will be stored as **ans** in the workspace. Note that the contents of **ans** is freely and frequently overwritten. Please compare with the on-screen output of the first four size and length commands above that are comma delimited. If an item is "named" in MATLAB code, such as A, B, or **result** above are, it will carry that name throughout the computations (until reassigned) and be displayed on screen only if followed by a , or by a blank.

For the above defined matrix $A_{3,4}$ we can form the matrix product $D_{3\times 5} = A_{3\times 4}{\cdot}C_{4\times 5}$, but not $F = C_{4\times 5} \cdot A_{3\times 4}$ for any 4 by 5 **matrix** C, such as the **random matrix C = randn(4,5)** that contains normally distributed random entries.

This is due to the nature of **matrix multiplication**:

Matrix multiplication is based on the **dot product** defined for any two vectors of equal size. The product matrix P of two conformally sized matrices K and L is a matrix of size "number of rows of K" by "number of columns of L". To be compatible for multiplication, the rows of K and the columns of L must have the same length. If this is so then the entries p_{ij} of the matrix product $P = K \cdot L = (p_{ij})$ are computed as the dot products of row i of the first matrix factor K with the column j of the second matrix factor L. We refer the reader to the annex on matrices.

```
>> C = randn(4,5); D = A*C, F = C*A
D =
        6.2593        -1.5887         1.2617         8.7846        -1.9003
       11.444         -2.2827        -1.0572        22.73          -2.6528
       16.628         -2.9768        -3.376         36.676         -3.4053
??? Error using ==> *
Inner matrix dimensions must agree.
```

Here the MATLAB on-screen term `Inner matrix dimensions` refers to the underlined "inner" size numbers of the matrix factors of D such as depicted by our underlining in $D_{3\times 5} = A_{3\times \underline{4}} \cdot C_{\underline{4}\times 5}$. These inner dimensions are both equal to 4 for the matrix product $D = A \cdot C$.

However, the inner dimension numbers do not agree when we try to form $F = C_{4\times \underline{5}} \cdot A_{\underline{3}\times 4}$, one being 5, the other 3. Therefore, F cannot be formed and does not exist. Note that the matrix product $D = A \cdot C$ inherits the (not underlined) "outer dimensions" 3 by 5 from its factors $A_{3,4}$ and $C_{4,5}$, or $D = A \cdot C$ is a 3 by 5 matrix.

Solving **linear equations** $Ax = b$ is done in MATLAB via the backslash $A\backslash b$ command. For example, let us consider a system of 4 linear equations in 4 unknowns x_1, x_2, x_3, x_4 such as

$$
\begin{aligned}
x_1 + 3x_2 - 5x_3 - x_4 &= 2 \\
-x_1 + 17x_2 - x_3 - 5x_4 &= 12 \\
5x_1 - x_3 + 31x_4 &= 0 \\
x_1 + x_2 + x_3 &= -11
\end{aligned}
\tag{1.1}
$$

In order to solve such systems of linear equations on a computer, it helps to realize that only the coefficients of the system (1.1) play a role. For efficient computer use, such systems should be rewritten in matrix form $Ax = b$ by extracting the coefficient matrix A on the left hand side of (1.1) and the vector b on the right hand side of (1.1). Here

$$
A = \begin{pmatrix} 1 & 3 & -5 & -1 \\ -1 & 17 & -1 & -5 \\ 5 & 0 & -1 & 31 \\ 1 & 1 & 1 & 0 \end{pmatrix} \text{ and } b = \begin{pmatrix} 2 \\ 12 \\ 0 \\ -11 \end{pmatrix}.
$$

Note that we have to place zeros in the positions of A that correspond to unused variables in any equation in (1.1). Below are the corresponding MATLAB commands that let us find the solution vector

$$x = \begin{pmatrix} x_1 \\ x_2 \\ x_3 \\ x_4 \end{pmatrix}$$

for the system of linear equations

$$Ax = \begin{pmatrix} 1 & 3 & -5 & -1 \\ -1 & 17 & -1 & -5 \\ 5 & 0 & -1 & 31 \\ 1 & 1 & 1 & 0 \end{pmatrix} \begin{pmatrix} x_1 \\ x_2 \\ x_3 \\ x_4 \end{pmatrix} = \begin{pmatrix} 2 \\ 12 \\ 0 \\ -11 \end{pmatrix} = b \,.$$

Our two command lines below first generate the coefficient matrix A and the right hand side vector b for (1.1), followed by the MATLAB backslash linear equations solver that computes the solution vector x. This is followed by a simple verification of the error inherent in the **residual vector** $A \cdot x - b$ for our numerical solution x. This error is nearly zero since in MATLAB the number $-1.3323e-15$ describes the real number $-1.3323 \cdot 10^{-15}$.

```
>> A = [1 3 -5 -1;-1 17 -1 -5;5 0 -1 31;1 1 1 0],
b = [2;12;0;-11], x=A\b, error = A*x-b,
A =
    1     3    -5    -1
   -1    17    -1    -5
    5     0    -1    31
    1     1     1     0
b =
    2
   12
    0
  -11
x =
      -9.1947
       0.44732
      -2.2526
       1.4104
error =
   -1.3323e-15
   -1.7764e-15
    0
    0
```

The MATLAB backslash command solves all linear systems of equations, with rectangular or square matrices alike. If we instead want to solve the linear system $Ax = b$ for our earlier matrix

$$A = \begin{pmatrix} 1 & 2 & 3 & 4 \\ 5 & 6 & 7 & 8 \\ 9 & 10 & 11 & 12 \end{pmatrix} \quad \text{and} \quad b = \begin{pmatrix} -3 \\ 2 \\ 0 \end{pmatrix},$$

we could effect this in MATLAB via the following commands.

```
>> A = [1 2 3 4;5 6 7 8;9,10,11,12]; b = [-3;2;0], x = A\b, error = A*x - b,

b =
    -3
     2
     0
Warning: Rank deficient, rank = 2  tol =     1.3293e-14.
ans =
       1.1111
            0
            0
      -0.73611
error =
       1.1667
      -2.3333
       1.1667
```

Here A is suppressed on screen and only the right hand side b is displayed since it was followed by a comma , on the command line instead of the suppressing ; . Next MATLAB issues a warning that our chosen 3 by 4 matrix A has deficient **rank** two instead of three, and then it gives the **least squares solution**

$$x = \begin{pmatrix} 1.1111 \\ 0 \\ 0 \\ -0.73611 \end{pmatrix}$$

of the linear system of equations $Ax = b$, followed by the (now sizable) error vector $Ax - b$. Recall that the **rank** of a matrix A refers to the number of linearly independent rows or columns in A, or equivalently to the number of pivots in a row echelon form of A; refer to Appendix 1 on Linear Algebra and Matrices for more details.

The **least squares** solution x of an **unsolvable linear system** $Ax = b$ such as our system is the vector x that minimizes the error $\|Ax - b\|$ in the euclidean **vector norm** $\|x\|$ defined by $\|x\| = \sqrt{x_1^2 + \ldots + x_n^2}$ when the vector x has n real entries x_i.

We also note that a *multiplication* $*$ *symbol* is required in MATLAB whenever a multiplication is to be performed. Our students should compare the output of the two commands $\gg$ `2 pi` and $\gg$ `2*pi` to learn the difference.

The latter command will give the answer as 6.2832 in `format short g`, while the screen output will be `6.283185307179586e+00` in `format long e`. Powers of the base 10 are always displayed in MATLAB via the e-extension as $be \pm a$ in `format e`. Thus $400 =$ `4e+02` and $1/1000 =$ `1e-03` in MATLAB's **exponential output and screen display**. Format statements can be entered at any prompt such as $\gg$ `format short`, or inside MATLAB code where desired. Commanding $\gg$ `format` toggles the output format back to the previous format setting.

No matter what format is specified, MATLAB always computes in **double precision**, i.e., it carries about 16 digits in all of its computations. For detailed numerical analysis we prefer the extended digit output of `format long e`, while for most engineering output `format short g` will give sufficient information. `format short g` limits the output to 9 digits and writes numbers $0.0001 \le |x| \le 9.99999899 \cdot 10^8$ in standard decimal form and numbers outside of this range in exponential form $be{\pm}a$.

In this book we shall use MATLAB codes and explain more involved features of MAT-LAB as we encounter them. MATLAB has a built-in help menu; typing `help format` at the >> prompt, or `help \`, for example, will show the syntax and variations of these two commands "format" and 'backslash'. Whenever a student encounters a MATLAB command that is not self explanatory, we suggest using this built-in help function of MATLAB.

There are many MATLAB tutorials available on the web; please enter "MATLAB tutorial" into your favorite internet search engine (such as google, hotbot, infoseek, ly-cos, yahoo, ... , etc) and follow the offered links. Unfortunately, we cannot include a full fledged MATLAB tutorial in this book due to space and time concerns.

This chapter continues with a description of a few basic numerical methods and their underlying principles. However, a solid first course in numerical analysis cannot be replaced by the concise intuitive explanations of numerical methods and phenomena that follow below.

1.2 Numerical Methods and MATLAB Techniques for Chemical and Biological Engineering Problems

In this section we give an overview of numerical analysis in general, and of the aspects of numerical analysis that are needed for problems encountered specifically in chemical and biological engineering[2]. This overview will, by necessity, be rather brief and it cannot substitute for a full semester course on Numerical Analysis. It is meant as a refresher only, or as a grain-of-salt type introduction to the theory and practice of mathematical computation. Many of the key terms that we introduce will remain only rather loosely defined due to space and time constraints. We hope that the unfamiliar reader will consult a numerical analysis textbook on the side; see our Resources appendix at the end of the book for specific recommendations. This we recommend highly to anyone, teacher or student, who does not feel firm in the concepts of numerical analysis and in its fundamentals.

In this book we aim to place general and specific chemical/biological problems in the context of standard numerical techniques and we try to exhibit and explain special numerical considerations that are needed to solve these chemical/biological problems using MATLAB. These concerns are due in part to the special nonlinearities that occur in chemical/biological engineering. They are often caused by the Arrhenius[3] dependence (2.1) with its exponential nonlinearity of reacting systems, or by other subject specific nonlinearities in both chemical and biological systems.

We hope that this interdisciplinary book will open the door for more interdisciplinary research so that industrial chemical/biochemical/biomedical plants can be run more eco-

---

[2]Biological engineering comprises both biochemical and biomedical engineering
[3]Svante August Arrhenius, Swedish chemist, 1859 − 1927

nomically and environmentally safer through efficient numerical simulation and the optimal utilization of numerical analysis and digital computers.

1.2.1   Solving Scalar Equations

Scalar equations have played a fundamental role in the history and development of Numerical Analysis. But they represent only the simplest problems and have little bearing on the advanced chemical/biological engineering problems that we want to solve numerically. Our very first "numerics for chemical/biological engineering problems chapter", Chapter 3, will deal with scalar equations rather quickly and easily using MATLAB. Since the current subsection is of little importance to the rest of the book, it can be skipped at first and it is best read later with a grain-of-salt looseness of mind, if only to study the MATLAB codes and learn from them. Nothing much or deep depends on scalar equations nowadays as they can be solved readily by software.

A **scalar equation** is an equation of the form $f(x) = 0$ for a function f that depends on one variable x. In mathematical notation, we express functions f that map one real variable x to the real value $f(x)$ symbolically by writing $f : \mathbb{R} \to \mathbb{R}$, where $\mathbb{R}$ denotes the set of all real numbers.

What mathematicians call "functions" is often referred to as "state variables" by engineers. These are parallel languages. If t is the independent time variable and $T(t)$ represents the temperature at time t, then in "engineering language", $T(t)$ is a dependent state variable in the independent variable or parameter t, while mathematically speaking, $T(t)$ is the temperature function dependent on time t. Use of the **function terminology** is more recent and allows for treating multi-variable, multi-output functions such as

$$F(t, x, y) = \begin{pmatrix} F_1(t, x, y) \\ \vdots \\ F_n(t, x, y) \end{pmatrix} : \mathbb{R}^3 \to \mathbb{R}^n$$

where each **component function** $F_i(t, x, y) : \mathbb{R} \to \mathbb{R}$ for $i = 1, 2, ..., n$ is a state variable itself depending on the three independent variables t, x, and y by the same 'function formalism'. F as defined above is not a "state variable", because it combines several states in one subsuming function F. The formal ease when speaking of "functions" rather than "state variables" gives this mathematical concept a clear advantage in modeling complex problems. Engineers should be aware of the limitations of the "state variable" concept and try to embrace the more versatile "function" notation.

Polynomial equations such as $x^3 - 2x^2 + 4 = 0$, for example, have been studied for many centuries. Over the last hundred years, there have been over 4,000 research publications and many books written on how to solve the general polynomial equation $p(x) = 0$, or how to find n roots x_i with $p(x_i) = 0$ for $i = 1, .., n$, when

$$p(x) = a_n x^n + a_{n-1} x^{n-1} + ... + a_1 x + a_0$$

is a **polynomial** of **degree** n with real or complex coefficients a_k, provided that $a_n \neq 0$. Note that if $a_n = 0$ in p, then $p(x)$ does not have degree n but a lower degree. It is clear that a polynomial in one variable x is a special type of function $p : \mathbb{R} \to \mathbb{R}$, given in polynomial form.

Let us look at the process of finding polynomial roots as an input to output process:

$$\boxed{\textbf{INPUT}} \text{ polynomial } p \text{ of degree } n \; \longrightarrow \; \boxed{\boxed{\text{root finder}}} \; \longrightarrow \; \boxed{\textbf{OUTPUT}} \, n \text{ roots } x_i$$

Here the "root finder" can be any one of the algorithms from the vast literature. Assuming that the chosen algorithm works properly, it takes the n real or complex coefficients a_i for $i = 0$ to $i = n - 1$ of p as input and produces n real or complex numbers x_i, the roots of p, as output. To do so, we have tacitly assumed that the leading coefficient a_n of p is normalized to equal 1 since dividing p of degree n by a constant ($a_n \neq 0$) does not alter its roots.

We start with a **short history** of the polynomial-root finding problem that will explain the eminent role of matrices for numerical computations by example.

The study of polynomial roots literally lies at the origin of modern **Numerical Analysis**.

In 1824, Abel[4] proved that it is the impossible to solve a general polynomial equation of degree five or higher by radicals, such as the quadratic formula

$$x_{1,2} = \frac{b \pm \sqrt{b^2 - 4ac}}{2a}$$

does for second-degree polynomials $p(x) = ax^2 + bx + c = 0$, for example.

Since there is no direct mathematical way to write down general formulas for the roots of general polynomials of degree larger than 4, the roots of such higher degree polynomials can only be computed iteratively by numerical procedures, giving both birth and need to Numerical Analysis.

About 20 years after Abel's discovery, matrices were invented and their theory developed. One of the fundamental uses of matrices lies in their ability to model processes and phenomena of many branches of engineering and applied science. One of the most pressing needs at the time was to understand and find the periodical and oscillatory behavior of mechanical and other engineering systems. This quest involves the study of scalars λ with $Ax = \lambda x$ for a given model matrix $A_{n,n}$ and any nonzero vector x. If, for example, there is one such value λ with $|\lambda| > 1$, then the iterates $A^n x = \lambda^n x$ necessarily become arbitrarily large, i.e., they "blow up", predicting a dangerous disaster for the plant or system. For this purpose mathematicians study the question for which scalars λ the corresponding homogeneous matrix equation $Ax - \lambda x$ has a nonzero solution vector x. In matrix terms, see our Appendix 1, this question is equivalent to finding those values of λ for which the matrix $A - \lambda I$ is **singular** or **noninvertible**, where I denotes the n by n identity matrix

---

[4]Niels Abel, Norwegian mathematician, 1802 - 1829

with ones on its main diagonal and zeros elsewhere, i.e.,

$$I = \begin{pmatrix} 1 & & 0 \\ & \ddots & \\ 0 & & 1 \end{pmatrix}_{n,n}.$$

For almost another century, many mathematicians tried to solve the related determinantal equation $\det(A - \lambda I) = 0$, in order to find these particular values λ of matrices numerically. Incidentally $\det(A - \lambda I) = 0$ is a polynomial of degree n in λ.

This equation is called the **characteristic equation** of the matrix A; it is an nth degree polynomial in λ and its roots λ_1, ..., λ_n are commonly called the **eigenvalues** of A. The eigenvalues λ_i of A describe the growth, decay, and oscillatory behavior of the states of the underlying model in detail. Such models and, apparently, reliable polynomial-root finders for them became a dire necessity much more recently when, for example, trying to break the sound barrier in flight by attenuating wing oscillations sufficiently to avoid break-up caused when the real parts of the eigenvalues of the corresponding model matrix lay in the right half of the **complex plane** $\mathbb{C} = \mathbb{R} + i\mathbb{R} = \{a + bi \mid a,\ b \in \mathbb{R},\ i = \sqrt{-1}\}$. In chemical engineering the computation of the eigenvalues of linearized model equations were at the basis of the discovery of periodic (oscillatory) behavior of chemical/biological reactors and enzyme membranes. This work has been developed further during the last three decades leading to the discovery of chaos in these systems. Such practical applications have driven polynomial-root algorithms to the fore during the first 50 years of the last century. Unfortunately, none of the algorithms of the early half of the 1900s has proved to be much good.

Things changed for the better when matrix means were further developed in the latter half of the 20th century with the advent of electronic and digital computers. The breakthrough for the matrix eigenvalue problem came from judicious matrix factorizations such as the LR factorization $A =: A_0 = L_0 \cdot R_0$. The LR matrix factorization expresses the standard Gaussian[5] elimination, or equivalently the row echelon form reduction of a matrix A as the product of a lower triangular matrix L and an upper triangular one called R. It took a stroke of genius by Rutishauser[6] to re-multiply the LR matrix factors of A_0 in reverse order as $A_1 = R_0 \cdot L_0$, then to factor A_1 as $A_1 = L_1 \cdot R_1$ again and to form A_2 by reverse order multiplication, to re-factor and to reverse order multiply $A_2 := R_1 \cdot L_1 = L_2 \cdot R_2$, and so forth. This is only meaningful since matrix products $L \cdot R \neq R \cdot L$ in general, i.e., matrix products generally do not commute. Finally Francis'[7] **QR algorithm** extended Rutishauser's LR algorithm by using the **QR factorization** of matrices instead in which the first factor Q is orthogonal, and the second factor R is upper triangular. The QR algorithm finds general matrix eigenvalues more quickly and accurately than ever before.

And the original modeling problem of oscillatory engineering phenomena was not only best modeled via matrices, but it was also best solved by matrix methods themselves.

---

[5] Johann Carl Friedrich Gauß, German mathematician, 1777 – 1855
[6] Heinz Rutishauser, Swiss mathematician, 1918 - 1970
[7] John Guy Feggis Francis, English systems engineer, Oct. 10, 1934 –

Nowadays, numerical analysts give the following global ready advice when any applied problem leads to trying to find roots of a polynomial: "*to go back to before the problem was expressed in polynomial form*". There was likely a linear model at an earlier stage of the problem development whose eigen information could be extracted much more reliably from the respective matrix than the roots can be from the unfortunate polynomial.

More specifically in most chemical/biological engineering problems, the system is described by nonlinear differential equations (DEs). When these are linearized in a neighborhood of certain steady states they lead to linear DEs whose characteristics can be determined by analyzing the corresponding matrix eigenvalues.

The expense and accuracy distinguishes between all known polynomial-root finding algorithms. The most economical known polynomial-root finder `pzero.m`, see the folder chap1.2m on the accompanying CD, performs $O(n^2)$ operations on the n input data a_i of p of degree n to produce its n roots. It is based on a tridiagonal matrix representation of p and can account for moderate repeats of the roots of p. MATLAB in turn uses an $O(n^3)$ algorithm that forms the companion matrix (see the definition below) for p and then finds the eigenvalues of the companion matrix via the QR algorithm. The **companion matrix** of a normalized polynomial $p(x) = x^n + a_{n-1}x^{n-1} + ... + a_1x + a_0$ is the n by n matrix $C = C(p)$ whose first row contains the negative coefficients $-a_{n-1}, -a_{n-2}, ..., -a_1, -a_o$ of p in the positions $(1,1)$, $(1,2)$, ..., $(1,n-1)$, $(1,n)$, respectively, it contains ones on the first subdiagonal and zeros everywhere else, i.e.,

$$C = C(p) = \begin{pmatrix} -a_{n-1} & -a_{n-2} & \cdots & -a_1 & -a_0 \\ 1 & 0 & \cdots & \cdots & 0 \\ 0 & \ddots & \ddots & & \vdots \\ \vdots & & \ddots & \ddots & \vdots \\ 0 & \cdots & 0 & 1 & 0 \end{pmatrix}_{n,n} . \qquad (1.2)$$

MATLAB's polynomial-roots finder `roots` does not handle repeated or clustered roots very well, but otherwise it is the best $O(n^3)$ root finder available. **Note** that an **operations count** of $O(n^j)$ for an algorithm signifies that the algorithm performs $K \cdot n^j$ additions and multiplications (for some algorithm specific constants K and j, but depending on n) to obtain its output from n input data. Most of the polynomial-root finders of the last century unfortunately were even slower $O(n^4)$ algorithms and all in all much too slow and inaccurate.

To illustrate we first verify the identical behavior of the MATLAB QR based polynomial-root finder `roots` and MATLAB's QR based matrix eigenvalue finder `eig` for p's companion matrix $P = C(p)$: First we define p by its coefficient vector in MATLAB's workspace, then we invoke the MATLAB polynomial-root finder `roots`, followed by its matrix eigenvalue finder `eig` on the companion matrix of p. Finally we display the companion matrix P of p. As an example we use $p(x) = x^3 - 2x^2 + 4$ here and represent p by its coefficient vector `[1 -2 0 4]` in the following line of MATLAB commands.

```
>> p = [1 -2 0 4]; [roots(p), eig(compan(p))], P = compan(p)
```

```
ans =
      1.5652 +      1.0434i      1.5652 +      1.0434i
      1.5652 -      1.0434i      1.5652 -      1.0434i
     -1.1304                    -1.1304
P =
      2      0     -4
      1      0      0
      0      1      0
```

[We advise our students to follow our text and to always copy our MATLAB command lines onto their MATLAB desktop as they read through this chapter. Our students and readers should observe the execution of these commands, and we encourage them to alter these commands slightly to learn more about MATLAB.]

Note that the output of `roots(p)` and `eig(compan(p))` each is a complex column vector of length three, i.e., each output lies in $\mathbb{C}^3$; and the two solution vectors are identical. Our trial polynomial $p(x) = x^3 - 2x^2 + 4$ has one pair of complex conjugate roots $1.5652 \pm 1.0434 \cdot i$ and one real root -1.1304. The (first row) **companion matrix** P of a **normalized nth degree polynomial** p (normalized, so that the coefficient a_n of x^n in p is 1) is the sparse n by n matrix $P = C(p)$ as described in formula (1.2). Note that our chosen p is normalized and has zero as its coefficient a_1 for $x = x^1$, i.e., the $(1,2)$ entry in P is zero. For readers familiar with determinants and Laplace expansion, it should be clear that expanding $\det(P - xI)$ along row 1 establishes that our polynomial $p(x)$ is the characteristic polynomial of P. Hence P's eigenvalues are precisely the roots of the given polynomial p.

Polynomial equations are not rare in chemical/biological engineering problems, they are typically met in most local stability problems. However, nonalgebraic or **transcendental equations** are more common in our subject. A function in one variable x is called transcendental, if it is not merely a polynomial in x, nor a ratio of polynomials (called **rational function**), but contains nonalgebraic transcendental expressions in x, such as

$$e^x - x^2, \ \sqrt[5]{x^3 - 1/x^{0.5}}, \ \frac{x^2 e^{2x}}{1 - \sin(x)e^{-2x}}, \ \sin(x) + \log(x), \ \log(x^2) - 17x^{-3/2} + 2x^3 - x^2 + 4,$$

etc.

Historically speaking, most polynomial and transcendental one variable equations $f(x) = 0$, have been solved by one of two methods: by **Newton**[8] **iteration** or by **bisection/inclusion** (or by a judicious mixture of the two).

Newton iteration starts with an approximation x_0 of a root x^* of f and goes along the tangent line to the curve at $(x_0, f(x_0))$ until it intersects the x axis. This intersection is labeled x_1. It is an improved iterative approximation for the actual root, and the process continues leading from x_1 to x_2 via the tangent to f at x_1, etc. until the difference between successive iterates becomes negligible, see Figure 1.1 for our trial polynomial equation $p(x) = x^3 - 2x^2 + 4 = 0$ and the start $x_0 = -2$.

In Figure 1.1 we have marked our iteration start x_0, and the first and second iterates x_1, x_2 that are the respective intersections of the tangents to the curve at $(x_0, f(x_0))$ and

[8]Isaac Newton, British physicist and mathematician, 1643-1727

$(x_1, f(x_1))$ with the x axis, as well the solution $x^* = -1.1304$ towards which the iterates x_i converge. Clearly the solution x^* near -1 of $p(x) = 0$ is the point of intersection of the graph of p with the x axis where $p = 0$.

Newton's method leads to the formal iteration rule

$$x_{i+1} \;=\; x_i - \frac{f(x_i)}{f'(x_i)} \;. \tag{1.3}$$

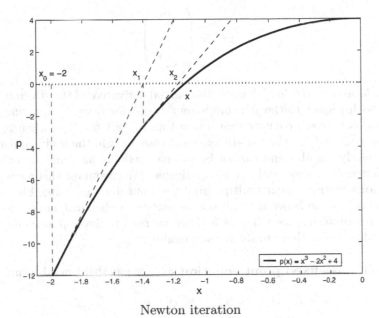

Newton iteration

Figure 1.1

Equation (1.3) follows readily from the **point-slope equation** of a line, applied to the tangent at $(x_i, f(x_i))$, namely to

$$y - f(x_i) \;=\; f'(x_i) \cdot (x - x_i) \;. \tag{1.4}$$

Solving (1.4) for x and setting $x_{i+1} = x$ as well as $y = 0$ gives rise to the iteration (1.3).

Newton's method is the basis of many root-finding algorithms. It can be extended to several, i.e., n real variables

$$z = \begin{pmatrix} z_1 \\ \vdots \\ z_n \end{pmatrix} \quad \text{and } n \text{ variable functions } f(z) = \begin{pmatrix} f_1(z) \\ \vdots \\ f_n(z) \end{pmatrix} : \mathbb{R}^n \to \mathbb{R}^n \;.$$

The division by $f'(x_i)$ in (1.3) is replaced in the n dimensional case by multiplication with the matrix inverse $\left(Df(z^{(i)})\right)^{-1}$ of the **Jacobian**[9] n by n matrix $Df = (\partial f_\ell / \partial z_j)$

[9]Carl Jacobi, German mathematician, 1801-1851

of f:

$$z^{(i+1)} = z^{(i)} - \left(Df\left(z^{(i)}\right)\right)^{-1} \cdot f\left(z^{(i)}\right). \tag{1.5}$$

In order to distinguish between the n components of each iterate, marked by subscripts, we mark the iterates $z^{(i)}$ in $\mathbb{R}^n$ themselves by superscripts. Thus fully written out the ith Newton iterate $z^{(i)}$ in $\mathbb{R}^n$ is the vector

$$z^{(i)} = \begin{pmatrix} z_1^{(i)} \\ \vdots \\ z_n^{(i)} \end{pmatrix}$$

when $n > 1$. Moreover, $\partial f_i / \partial z_j$ denotes the partial derivative of the ith component function f_i of f with respect to the jth component z_j of z for $i, j = 1, ..., n$, and each iterate $z^{(k)} \in \mathbb{R}^n$ is a real vector. We note that if $n = 1$ and $f'(x_i) \approx 0$ (i.e., when x_i is close to a near-multiple root of f) in the one-dimensional case, or if in the multi-dimensional case, $Df(z^{(k)})$ is nearly singular and cannot be inverted reliably as a matrix for some iterate $z^{(k)}$, then Newton's method will not be applicable. We encounter this phenomenon with scalar equations routinely near multiple roots, see our discourse after Figure 1.2 for details, as well as in some Newton iterations in particular chemical/biological engineering boundary value problems, see Chapter 5. Then we need to develop new and more appropriate methods to solve these problems successfully.

Formula (1.3) is a **fixed point equation** since it has the general form

$$x_{i+1} = \phi(x_i)$$

with $\phi(x) = x - f(x)/f'(x)$ for Newton's method. The word "fixed" derives from the fact that if $x_i = x^*$, i.e., if we are at the solution x^* of $f(x) = 0$, then by (1.3), $x_{i+1} = x^* - f(x^*)/f'(x^*) = x^*$ as well. And the iteration has come to a stand-still, or it has become "fixed" at the solution x^*. Finally we note that any scalar function equation of the form $h(x) = k(x)$ can be readily transformed into the **standard form** $f(x) = 0$ (with zero on the right hand side) by setting $f(x) = h(x) - k(x)$.

Another method to solve scalar equations in one real variable x uses **inclusion** and **bisection**. Assume that for a given one variable continuous function $f : \mathbb{R} \to \mathbb{R}$ we know of two points $x_\ell < x_{up} \in \mathbb{R}$ with $f(x_\ell) \cdot f(x_{up}) < 0$, i.e., f has opposite signs at x_ℓ and x_{up}. Then by the intermediate value theorem for continuous functions, there must be at least one value x^* *included* in the open interval (x_ℓ, x_{up}) with $f(x^*) = 0$. The art of inclusion/bisection root finders is to make judicious choices for the location of the root $x^* \in (x_\ell, x_{up})$ from the previously evaluated f values and thereby to *bisect* the interval of inclusion $[x_\ell, x_{up}]$ to find closer values $v < u \in [x_\ell, x_{up}]$ with $|v - u| < |x_\ell - x_{up}|$ and $f(v) \cdot f(u) < 0$, thereby closing in on the actual root. Inclusion and bisection methods are very efficient if there is a clear intersection of the graph of f with the x axis, but for slanted, near-multiple root situations, both Newton's method and the inclusion/bisection method are often inadequate. In particular, for the dynamic chemical/biological transcendental equations of Chapter 3 we shall develop a variant of the **level curve method** that helps solve our equations where the standard Newton or bisection/inclusion methods fail. For details, see Sections 3.1 and 3.2.

MATLAB has a built-in root finder for scalar equations $f(x) = 0$ in one real variable x that are in standard form. The built-in MATLAB function is `fzero`. The use of `fzero` hinges on a user-defined function, such as the "function" f inside the following `fzero` tester, called `fzerotry1`, that we apply to our previously studied third degree polynomial.

We note that `fzerotry1` is a MATLAB **function m file** that is stored in its folder with the extension .m as `fzerotry1.m`. Our program code is annotated with comments following the % symbol. Anything that follows after a % symbol on a line of code is not executed in MATLAB. More on MATLAB files, their storage, creation, etc. is given in Section 1.2.5. Specific built-in MATLAB functions and their use, such as `fzero` in the code below, should always be scrutinized by our students for their input/output syntax etc. using the built-in help MATLAB command >> `help fzero` for example.

```
   function fzerotry1                    % function m file
%  Sample call: fzerotry1
%  Input : None
%  Output: Newton iterate and number of iterations used
%  Computes the zero of the polynomial f(x) = x^3 - 2*x^2 + 4 near x = -2 first
%  and then within the interval [-2,-1].
%  Method used : MATLAB's fzero built-in root finding function

format long g, format compact            % specify long format
[Newt,fval,exitflag,output] = fzero(@f,-2), %  call fzero for f from x = -2
iterations = output.iterations;
From_minus_2 = [Newt, iterations],        % give screen output
[incl,fval,exitflag,output] = fzero(@f,[-2 -1]); % call fzero for f on [-2,-1]
iterations = output.iterations;
From_interval = [incl, iterations], format  % give screen output

function [y] = f(x)                       % function m file for
y = x.^3 - 2*x.^2 + 4;                    %    evaluating f(x)
```

Note that `fzerotry1.m` uses no input from and creates no output to the workspace. Calling `fzerotry1` creates the following screen output:

```
>> fzerotry1
From_minus_2 =
         -1.13039543476728                  24
From_interval =
         -1.13039543476728                   9
```

The first call of `fzero` inside `fzerotry1` takes 24 iterations to arrive at the real root $x^* = -1.1304$ of our trial polynomial $p(x) = x^3 - 2x^2 + 4$ when starting at $x_0 = -2$, while the second call converges after 9 iterations when looking for real roots of p inside the interval $[-2, -1]$. Please look up >> `help fzero` to learn more about this MATLAB function and how it was used.

We now give a simple example of an equation with a root of high multiplicity. This illustrates the *limits of even the best standard root finders*.
We investigate the 9th degree polynomial

$$p(x) = x^9 - 18 \cdot x^8 + 144 \cdot x^7 - 672 \cdot x^6 + 2016 \cdot x^5 - 4032 \cdot x^4 + 5376 \cdot x^3 - 4608 \cdot x^2 + 2304 \cdot x - 512 \ . \quad (1.6)$$

p can be rewritten more simply as $p(x) = (x-2)^9$. In the plots of Figure 1.2 we plot both p and its derivative

$$p'(x) = 9 \cdot (x-2)^8 = 9x^8 - 144x^7 + 1008x^6 - 4032x^5 + 10080x^4 - 16128x^3 + 16128x^2 - 9216x + 2304$$

from their "naive" extended polynomial form above (plotted as small circles in Figure 1.2) and from their simple, i.e., "powers of $(x-2)$" form (plotted as a curve in Figure 1.2) using MATLAB.

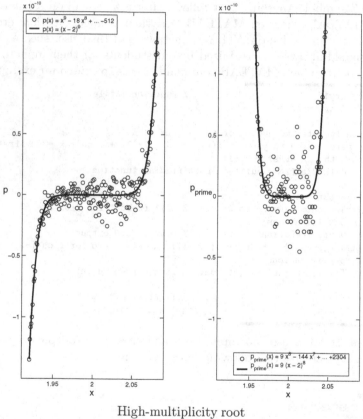

High-multiplicity root

Figure 1.2

In our **script m file** `poly9.m` below, we evaluate both p and p' from their extended polynomial form, rather than use the more stable and slightly more accurate **Horner scheme**, which would have us rewrite p as

$$p(x) = ((...(((x - 18)x + 144)x - 672)x + ... - 4608)x + 2304)x - 512$$

for example and thus evaluate $p(x)$ from the inside out.

A script m file in MATLAB does not start with a function ... declaration as the function m file `fzerotry1.m` does. A script m file rather consists only of MATLAB commands. See Section 1.2.5 for more details on these two types of m files in MATLAB.

```
                     % poly9 script m file
%   Sample call:  poly9
% Plots the graph of
%     p(t) = (t-2)^9 = t.^9 - 18*t.^8 + 144*t.^7 - 672*t.^6 + 2016*t.^5 - ...
%                 4032 *t.^4 + 5376*t.^3 - 4608*t.^2 + 2304*t - 512
% in one subplot, both in closed form (curve) and in extended form (o)
% near t = 2.
% And repeats the same for the derivative
%     p'(t) = 9*(t-2)^8 = 9*t^8 + ... + 9*256
% of p in both forms, closed in a curve, and extended (o) in a second subplot.

t = linspace(1.92,2.08,200);          % form a partition t of the interval
                                      %    [1.92, 2.08] of 200 points
f = t.^9-18*t.^8+ 144*t.^7 -672*t.^6 + 2016*t.^5 -4032 *t.^4 ...
   + 5376*t.^3 -4608*t.^2+ 2304*t -512; % naive formula for evaluating f = p
f1= 9*(t.^8 -16*t.^7 + 112 *t.^6 -448*t.^5 + 1120 *t.^4 -1792*t.^3 ...
   + 1792*t.^2 -1024*t + 256);         % naive formula for evaluating f' = p'

subplot(1,2,1), plot(t,f,'ok'),        % plot (t,f(t)) on first subplot
   axis([1.91 2.09 -1.5*10^ 10 1.5*10^-10]), hold on, % plot more on same plot
   plot(t,(t-2).^9,'-k','LineWidth',2),          % plot (t,(t-2)^9)
   xlabel('x','FontSize',14),
   ylabel('p','Rotation',0,'FontSize',14)   % label the axes
   legend('p(x) = x^9 - 18 x^8 + ... -512',...
          'p(x) = (x - 2)^9' ,2)        % include legend
          hold off                      % clear left graphics window

subplot(1,2,2), plot(t,f1,'ok'),       % plot (t,f'(t)) on second subplot
   axis([1.91 2.09 -1.5*10^-10 1.5*10^-10]), hold on, % plot more on same plot
   plot(t,9*(t-2).^8,'-k','LineWidth',2),   % plot the derivative (t,9*(t-2)^8)
   xlabel('x','FontSize',14), ylabel('p_{prime}','Rotation',0,...
          'FontSize',14)                % label the axes
   legend('p_{prime}(x) = 9 x^8 - 144 x^7 + ... +2304',...
          'p_{prime}(x) = 9 (x - 2)^8' ,0) % include legend
          hold off                      % clear right graphics window
```

We urge the interested reader to look up >> `help linspace`, >> `help plot`, >> `help subplot`, >> `help axis`, >> `help xlabel`, >> `help ylabel`, and >> `help legend` to fully understand the code of `poly9.m`. However, over time many students will find these and other MATLAB commands quite obvious from their use and from the generated output. MATLAB programming is very simple and intuitive. Built-in MATLAB functions can be knitted into personal codes that perform particular tasks and can compute or plot most anything in engineering.

In Figure 1.2 we note that for x within around 3% of the multiple root $x^* = 2$ of p, the values of $p(x)$ and $p'(x)$ in extended polynomial form (both plotted as dots in Figure 1.2) are small random numbers of magnitudes up to around $\pm 5 \cdot 10^{-11}$. Hence any bisection/inclusion algorithm performed on p in extended form (1.6) will fail to get closer to the actual root $x^* = 2$ than by about 0.06, i.e., every such algorithm will have an error of around 3% when trying to find x^*. Likewise for Newton iterations, the iteration process turns into a nonconverging random hit and miss exercise near $x^* = 2$ as we shall see.

The following experiments validate our assessment of troubles with Newton or bisection root finders for multiple roots. First we use the bisection method based MATLAB root finder `fzero`, followed by a simple Newton iteration code, both times using the chosen polynomial $p(x)$ of degree 9 in its extended form (1.6).

```
    function fzerotry2          % function m file
% Sample call: fzerotry2
% Input : None
% Output: Newton iterate and number of iterations used
% Computes the zero of the polynomial
%      f(x) =  x^9 - 18*x^8 + 144*x^7 - 672*x^6 + 2016*x^5 - 4032*x^4 ...
%             + 5376*x^3 - 4608*x^2 + 2304*x - 512
% near x = 1.5 first and then within the interval [1.91, 2.1].
% Method used : MATLAB's fzero built-in root finding function

format long g, format compact
[Newt,fval,exitflag,output] = fzero(@f,1.5);        % use fzero from x_0 = 1.5
iterations = output.iterations;
From_1point5 = [Newt, iterations],                  % screen output
[incl,fval,exitflag,output] = fzero(@f,[1.91 2.1]); % fzero on interval
iterations = output.iterations;
From_interval = [incl, iterations], format          % screen output

function [y] = f(x)          % long extended form of polynomial f(x) = (x-2)^9:
y = x^9-18*x^8+ 144*x^7 -672*x^6 + 2016*x^5 -4032*x^4 ...
    + 5376*x^3 -4608*x^2+ 2304*x -512;  % function evaluation subfunction f
```

The output of `fzerotry2` is as follows:
```
>> fzerotry2
From_1point5 =
            2.0542835318567                      35
From_interval =
            1.96209266952178                     6
```

Note that MATLAB's `fzero` algorithm stops after 35 iterations when it is about 2.7% away from the true root 2 with an approximate start at 1.5. From the inclusion interval [1.91, 2.1] it also ends rather prematurely and off target as well.

The Newton method fares no better here; see the code below and its output on the next page.

```
    function X = newtonpoly9(xstart,k)      % function m file
%    Sample call :  newtonpoly9(3,42);
% Input: xstart = start of newton iteration
%        k = number of iterations
% Output: (on screen)
%         list of Newton iterates and iteration indices in three columns
% Tries to find the roots of p(x) = (x-2)^9 given in extended form
%    p(x) = x^9 - 18*x^8 + 144*x^7 - 672*x^6 + 2016*x^5 ...
```

```
%                - 4032*x^4  + 5376*x^3 - 4608*x^2 + 2304*x - 512
% via Newton's method starting at xstart, and performing  k iterations

i = 0; X = xstart;                    % start at xstart
while i < k+2                         % do k Newton steps
  x = xstart;                         % x := x - p(x)/p'(x)
  xstart = xstart - (x^9-18*x^8+ 144*x^7 -672*x^6 + 2016*x^5 ...
      -4032*x^4  + 5376*x^3 -4608*x^2+ 2304*x -512)/(9*(x^8 ...
      -16*x^7 + 112 *x^6 -448*x^5 + 1120 *x^4  -1792*x^3 ...
      + 1792*x^2 -1024*x + 256));
  X = [X;xstart]; i = i+1; end        % Accumulate the Newton iterates
                                      % display on screen :
disp(['iteration counter      x(i)         counter     x(i)    '...
     ,         counter    x(i)']), format short g
disp([[0:13]',X(1:14),[14:27]',X(15:28),[28:41]',X(29:42)]), format
```

The code above is a MATLAB **function m file** with two inputs **xstart** and **k** and one output **X**. Its first code line

```
        function X = newtonpoly9(xstart,k)        % newton poly 9
```

contains the function name declared to be **newtonpoly9**. Once this m file is stored in the current working directory of the MATLAB desktop, its name **newtonpoly9** becomes a valid MATLAB command and this command needs two inputs such as **xstart,k** to generate its output **X**. In order to find and be able to execute the commands of the function **newtonpoly9** from the desktop, MATLAB must find an m file named "newtonpoly9.m" in its current working directory. The name of the function **newtonpoly9** is followed by its two inputs (**xstart,k**). And its output, the vector **X** containing the starting point **xstart**, followed by k 1 Newton iterates precedes it in front of the = sign on the first line of code.

In order to call >> **newtonpoly(start,n)** implicitly on a command line with success, the two implicit inputs "start" and "n" must have been declared previously and must be available in the current workspace. Of course, one can also call **newtonpoly9** explicitly by entering >> **newtonpoly(21,42)**, for example, on the command line if one wants to see the list of 41 Newton iterates for the same polynomial-root problem, starting from $start = x_0 = 21$.

Our MATLAB programs generally carry many comments after a % sign so that the reader can better understand what we do when and why. This will help our users to adapt our printed programs for other purposes.

Here is the output for the Newton iterates 0 - 41 starting with the initial guess $x_0 = 3$.

```
>> newtonpoly9(3,42);
iteration counter      x(i)         counter     x(i)         counter     x(i)
                0         3         14     2.1922          28     2.0606
                1    2.8889         15     2.1709          29     2.0655
                2    2.7901         16     2.1519          30     2.0524
                3    2.7023         17      2.135          31      1.977
                4    2.6243         18      2.12           32     4.1992
                5    2.5549         19     2.1067          33     3.9548
                6    2.4933         20     2.0949          34     3.7376
                7    2.4385         21     2.0845          35     3.5446
```

8	2.3897	22	2.0756	36	3.3729
9	2.3464	23	2.0665	37	3.2204
10	2.3079	24	2.0563	38	3.0848
11	2.2737	25	2.0482	39	2.9643
12	2.2433	26	2.0764	40	2.8571
13	2.2163	27	2.0678	41	2.7619

Note how slowly the Newton iterates decrease from our start at $x_0 = 3$ to 2.0524 in 30 steps. Then the iterates slip below the root $x^* = 2$ to 1.977 in step 31 and the 32nd iterate jumps far away from its aim $x^* = 2$ to 4.1992. At $x = 1.977...$, obviously the "randomness" in evaluating p and p' near $x = 2$ takes the Newton iterates completely off target to 4.199..., from where the Newton iterates come back slowly again to within about 3% of $x^* = 2$. Then they may converge or get pushed off target once more, etc.

The students should try other starting values and observe the inherent limits of both bisection and Newton's method for finding multiple roots or finding nearly flat and slanted intersections of the graph of f with the x axis.

MATLAB's $O(n^3)$ polynomial-root finder **roots**, used for the same polynomial p, encounters different problems and computes 4 complex conjugate root pairs instead. These lie on a small radius circle around the ninefold root 2. As input for **roots**, we represent our polynomial $(x - 2)^9$ of degree 9 in extended form by its coefficient vector [1 -18 ... 2304 -512].

```
>> roots([1 -18 144 -672 2016 -4032 5376 -4608 2304 -512])
ans =
     2.0609
     2.0458 +   0.039668i
     2.0458 -   0.039668i
     2.0089 +   0.059025i
     2.0089 -   0.059025i
     1.9696 +   0.050203i
     1.9696 -   0.050203i
     1.9453 +   0.019399i
     1.9453 -   0.019399i
```

roots is based on the MATLAB matrix eigenvalue finder **eig**. This matrix eigenvalue finder tries to find all roots of p from the associated companion matrix $C(p)$ by using the QR algorithm. The computed roots are quite inaccurate and most are accidentally computed as complex numbers.

Only the earlier mentioned faster $O(n^2)$ polynomial-root finder **pzero** discovers the ninefold real root 2 of p correctly; see the Resources appendix for a quote of the literature for **pzero** and the folder pzero on our CD for the actual MATLAB code of **pzero**.

```
>> pzero([1 -18 144 -672 2016 -4032 5376 -4608 2304 -512])
ans =
     2
     2
     2
     2
     2
     2
     2
     2
     2
```

Unfortunately, in chemical and biological engineering problems multiple roots or near multiple roots occur often; for example, look at Figure 3.2 on p. 71 and the slanted intersections marked by (1), (2), and (3). These force us to devise specific better numerical methods for such problems in Sections 3.1 and 3.2.

Exercises

1. Use the m file `fzerotry1.m` on p. 27 as a template for finding the roots of other functions f of your own choice. Vary the initial guesses as fits your chosen f.

   (It is best to use a plot routine for f when looking for likely root candidates; see the next problem.)

2. Develop a plotting code for Problem 1 that is based on the plotting commands in `poly9.m` on p. 27. (You will probably not need to use the `subplot(...)` command.)

3. Adapt `fzerotry2.m` on p. 30 to various other polynomials of your choice. Use polynomials of degrees less than 7 that have some multiple roots, as well as no multiple roots. What happens to the complex roots of a polynomial under `fzero`?

4. Repeat Problem 3 with `newtonpoly9.m` on p. 30.

5. Modify the MATLAB function m file `newtonpoly9.m` on p. 30 to print out three columns of iterates with their iteration indices for all values of k.
   [Hint: the first output column should contain the iterates indexed from 0 to $\left\lceil \frac{k}{3} \right\rceil - 1$, the second one those from $\left\lceil \frac{k}{3} \right\rceil$ to $2 \cdot \left\lceil \frac{k}{3} \right\rceil - 1$ and the last from $2 \cdot \left\lceil \frac{k}{3} \right\rceil$ to k, padded by blanks if needed. To achieve this, you need to adjust/replace the last three lines of code only.]

6. Learn to plot complex numbers on the complex plane by calling `>> x = roots([1 -18 144 -672 2016 -4032 5376 -4608 2304 -512])`, for example, first and then plotting the real and imaginary parts of the output vector x, as well as $x^* = 2$. Use the * symbol to plot the entries of x and the + symbol for $x^* = 2$. Where do the computed roots lie in the complex plane? How can you describe their location? Also evaluate `abs(x-2)`; what do these numbers mean and signify?

In closing we remark that not all scalar equations are solvable. For example, equations such as $\sin(x) = 7$, $e^{\sin(x)} = 0$, $e^{\cos(x)} = 20$, or $|x|^2 + |x|^4 = -2$ are all unsolvable since none of the values chosen for the right hand sides lie within the range of the respective functions on the left. In chemical/biological and any branch of engineering, however, correctly posed problems will result in equations that always have solutions.

However, many equations that we shall encounter in this book will have more than one solution for a given set of the parameters, indicating that the modeled chemical/biological process can actually be in one of several steady states, depending on its recent history or start-up and on the internal dynamics of the system as described by differential equations. *Such behavior signifies bifurcation and this is where the fun begins.*

1.2.2   Differential Equations; the Basic Reduction to First Order Systems

Scalar equations, as studied earlier, depend on one or several real variables x, y, t, T, ...
Differential equations instead link various derivatives of one or more functions $f(...)$, $g(...)$, $y(...)$, ..., each with any number of variables. These mathematical "functions" describe "state variables" in engineering parlance. Differential equations are equations in one or more variables **and** in one or more functions of the variables and in their derivatives. They involve independent variables such as space and time, and dependent, so called "state variables" or functions and their derivatives. Many physico-chemical processes are governed by differential equations or systems thereof, that involve unknown functions f, g, ... in various variables and various of their derivatives f', g', ...; f'', g'', ... etc.

The task of modeling is to obtain valid scalar, differential, or other type of equations (integral, integro-differential, etc) that describe a given physical system accurately and efficiently in mathematical terms. Numerical analysis and computations then lead us to the solution values or to the solution function(s) themselves from the model equations.

When trying to solve differential equations, we look for functions as supplying us with the "solution" to our problem. Differential equations problems generally include additional information on the solution function in their formulation, such as knowledge of some initial conditions for the solution, knowledge of some function behavior at the boundaries of the variable range, or a mixture thereof. Hence differential equations naturally fall into the categories of **IVP**s or of **BVP**s, signifying **initial value problems** or **boundary value problems**, respectively.

A different classification scheme for **DE**s, short for **differential equations**, separates those DEs with a single independent variable dependence, such as only time or only 1-dimensional position, from those depending on several variables, such as time and spatial position. DEs involving a single independent variable are routinely called **ODE**s, or **ordinary differential equations**. DEs involving several independent variables such as space *and* time are called **PDE**s, or **partial differential equations** because they involve partial derivatives.

As this book is mainly intended for undergraduates, we only treat those chemical/biological processes and problems that can be modeled with ODEs. PDEs are significantly more complicated to understand and solve. However, we will often have to solve systems of ODEs, rather than one single ODE. Such ODE systems contain several DEs in one and the same independent variable, but they generally involve several functions or "state variables" in their formulation. In fact, systems of ODEs (and matrix DEs) occur quite naturally in chemical/biological engineering problems.

A single nth order ODE for one unknown function $f(x)$ is an equation of the form

$$f^{(n)}(x) + a_{n-1}(x)f^{(n-1)}(x) + ... + a_1(x)f'(x) + f(x) = g(x) \qquad (1.7)$$

which involves $f(x)$ and the derivatives $f'(x)$, $f''(x)$, ..., $f^{(n)}(x)$ of f up to and including order n. Here we have normalized the coefficient function of $f^{(n)}(x)$ to be 1. In (1.7) each of the coefficient functions $a_i(x)$ can be a constant function $a_i(x) = a_i \in \mathbb{R}$ for all x, or a function $a_i(x) : \mathbb{R} \to \mathbb{R}$ in the variable x. Analytical descriptions of the solution f in form of a "formula" for f are desirable, but unfortunately almost always impossible to find. On the other hand, numerical solutions can routinely be found for (1.7) by converting the nth order DE first into a system of n first-order DEs involving n auxiliary functions $y_i(x)$ as follows:

$$
\begin{aligned}
\text{Set} \quad y_1(x) &:= f(x), \\
y_2(x) &:= f'(x), \\
&\ \ \vdots \\
y_{n-1}(x) &:= f^{(n-2)}(x), \text{ and} \\
y_n(x) &:= f^{(n-1)}(x).
\end{aligned}
$$

And define

$$
y := y(x) = \begin{pmatrix} y_1(x) \\ \vdots \\ y_n(x) \end{pmatrix} \in \mathbb{R}^n
$$

by its **component functions** $y_i : \mathbb{R} \to \mathbb{R}$. Then the function $y(x) : \mathbb{R} \to \mathbb{R}^n$ satisfies the following system of DEs for its components y_i:

$$
y'(x) = \begin{pmatrix} y'_1 \\ y'_2 \\ \vdots \\ y'_{n-1} \\ y'_n \end{pmatrix} = \begin{pmatrix} y_2 \\ y_3 \\ \vdots \\ y_n \\ g(x) - a_{n-1}(x)f^{(n-1)}(x) \ldots - a_1(x)f'(x) - f(x) \end{pmatrix} \tag{1.8}
$$

since $y'_1 = y_2 = f'$, $y'_2 = y_3 = f''$, ..., and $y'_{n-1} = y_n = f^{(n-1)}$ by definition and

$$
\begin{aligned}
y'_n(x) &= f^{(n)}(x) \\
&= g(x) - a_{n-1}(x)f^{(n-1)}(x) - \ldots - a_1(x)f'(x) - f(x) \\
&= g(x) - a_{n-1}(x)y_n(x) - \ldots - a_1(x)y_2(x) - y_1(x)
\end{aligned}
$$

according to (1.7).

We can rewrite (1.8) in matrix form:

$$
y'(x) = \begin{pmatrix} 0 & 1 & 0 & \cdots & 0 \\ 0 & 0 & 1 & & \vdots \\ \vdots & & \ddots & \ddots & 0 \\ 0 & \cdots & \cdots & 0 & 1 \\ -1 & -a_1(x) & \cdots & \cdots & -a_{n-1}(x) \end{pmatrix} \begin{pmatrix} y_1 \\ y_2 \\ \vdots \\ y_{n-1} \\ y_n \end{pmatrix} + \begin{pmatrix} 0 \\ 0 \\ \vdots \\ 0 \\ g(x) \end{pmatrix} = A(x)\,y(x) + \begin{pmatrix} 0 \\ 0 \\ \vdots \\ 0 \\ g(x) \end{pmatrix}
$$

$$\tag{1.9}$$

Here the DE system matrix $A(x)$ is a last row **companion matrix** whose entries in the last row may depend on x. More specifically, most of A's entries are zero, except for ones in its upper co-diagonal and for the negative coefficient functions -1, $-a_1(x)$, ..., $-a_{n-1}(x)$

of (1.7) in its last row. Note that the last row companion matrix $A(x)$ is **similar** to a first row companion matrix for polynomials p of the form $C(p)$ in (1.2) that we have encountered earlier with MATLAB's `roots` function. To see this, we simply need to reverse the order of the components y_i of y. This corresponds to a **basis change** $E^{-1} \cdot A(x) \cdot E$ that is affected by the counterdiagonal matrix

$$E = \begin{pmatrix} 0 & & 1 \\ & \cdot^{\cdot^{\cdot}} & \\ 1 & & 0 \end{pmatrix}_{n,n} ;$$

refer to the Appendix on linear algebra and matrices. E turns any vector

$$y = \begin{pmatrix} y_1 \\ \vdots \\ y_n \end{pmatrix} \quad \text{upside down to become} \quad Ey = \begin{pmatrix} y_n \\ \vdots \\ y_1 \end{pmatrix}$$

as can be readily verified by multiplying matrix E times vector y.

An nth order DE is called **homogeneous** if $g(x)$ in (1.7) is the zero function; and it is called **nonhomogeneous** if g is not the zero function.

Example: (Transforming a third order ODE into a system of three first-order ODEs)

Assume that $f = f(t)$ is a real valued real variable function in the independent time variable t. Assume that f satisfies the third order ODE

$$f'''(t) + a_1 f''(t) + a_2 f'(t) + a_3 f(t) = \frac{d^3 f}{dt^3} + a_1 \frac{d^2 f}{dt^2} + a_2 \frac{df}{dt} + a_3 f = F(t, f) , \quad (1.10)$$

where the coefficients a_i can be scalars or functions $a_i = a_i(t)$ of t and the right hand side is an arbitrary given function F of t and f.

If we set $y_1(t) = f(t)$, $y_2(t) = dy_1/dt = df/dt = f'(t)$, and $y_3(t) = dy_2/dt = d^2 y_1/dt^2 = d^2 f/dt^2 = f''(t)$, then $dy_3/dt = d^3 f/dt^3 = f'''(t)$. And equation (1.10) can be written as

$$\frac{dy_3}{dt} + a_1 y_3 + a_2 y_2 + a_3 y_1 = F(t, y_1) .$$

Thus the given third order ODE (1.10) is equivalent to the following system of three first-order ODEs.

$$\frac{dy_1}{dt} = 0 \cdot y_1 + y_2 + 0 \cdot y_3$$

$$\frac{dy_2}{dt} = 0 \cdot y_1 + 0 \cdot y_2 + y_3 \qquad\qquad (1.11)$$

$$\frac{dy_3}{dt} = -a_3 y_1 - a_2 y_2 - a_1 y_3 + F(t, y_1)$$

If we further combine the individual functions $y_i : \mathbb{R} \to \mathbb{R}$ into one vector valued function

$$y = \begin{pmatrix} y_1 \\ y_2 \\ y_3 \end{pmatrix} : \mathbb{R} \to \mathbb{R}^3 \text{ with } y' = \begin{pmatrix} y_1' \\ y_2' \\ y_3' \end{pmatrix} ,$$

then the system of three ODEs (1.11) becomes the following matrix ODE system

$$y'(t) = \begin{pmatrix} 0 & 1 & 0 \\ 0 & 0 & 1 \\ -a_3 & -a_2 & -a_1 \end{pmatrix} y(t) + \begin{pmatrix} 0 \\ 0 \\ F(t, y_1) \end{pmatrix} ,$$

exemplifying formula (1.9).

Numerical ODE solvers generally work for first-order systems

$$y' = F(x, y) \tag{1.12}$$

and $y : \mathbb{R} \to \mathbb{R}^n$ for arbitrary dimensions n. Here F is allowed to be much more general than in formula (1.9) In the next subsections we first deal with initial value problems where initial conditions such as $y(x_0) = y_0 \in \mathbb{R}^n$, $y'(x_0) = y_0' \in \mathbb{R}^n, \ldots$ for $y(x)$ are given, and then with solving boundary value problems where boundary conditions such as $y(a) = y_a \in \mathbb{R}^n$, $y(b) = y_b \in \mathbb{R}^n$, $y'(a) = y_a' \in \mathbb{R}^n$, $y'(b) = y_b' \in \mathbb{R}^n, \ldots$ are given instead for an interval $a \leq x < b$. All ODE algorithms and numerical codes can be adapted readily by "vectorizing the codes" to work with n functions

$$y(x) = \begin{pmatrix} y_1(x) \\ \vdots \\ y_n(x) \end{pmatrix}$$

and all higher order systems of ODEs can be transformed into first-order n-dimensional systems (1.12) $y' = F(x, y)$. To solve specific ODEs numerically, the user first has to rewrite the model DEs in first-order multi-component form (1.12) and thereafter he/she can use any appropriate vectorized MATLAB ODE solver. We shall practice this "rewriting" repeatedly whenever it comes up in our chemical/biological models.

1.2.3 Solving Initial Value Problems

Due to the preceding remarks, we can limit our considerations to the case of one first-order IVP ($n = 1$) in one function or state variable $y : \mathbb{R} \to \mathbb{R}$ and one independent variable x, starting at $x = a$ with the initial condition $y(a) = y_a \in \mathbb{R}$. We want to solve the given IVP for the unknown function $y(x)$ and the given right hand side $F(x, y)$. In other words, we seek the values of $y(x)$ for all values of x in the interval $[a, b]$. Expressed in mathematical notation, we thus seek to solve

$$y'(x) = F(x, y) \quad \text{with } x \in [a, b] \text{ and } y(a) = y_a \in \mathbb{R} . \tag{1.13}$$

Here F and y_a are known. The function $y(x) : \mathbb{R} \to \mathbb{R}$ is unknown on the interval $[a, \ b]$, except for its initial value $y(a) = y_a$, called the **initial condition**.

To start we consider a **partition** $a = x_0 < x_1 < x_2 < \ldots < x_N = b$ of the interval $[a, b]$. If we integrate (1.13) from x_i to x_{i+1}, then by (1.13)

$$\int_{x_i}^{x_{i+1}} y'(x)dx \ = \ \int_{x_i}^{x_{i+1}} F(x, y(x))dx \ .$$

The fundamental theorem of calculus allows us to rewrite the left hand side as $y(x_{i+1}) - y(x_i)$. Thus for each i we obtain

$$y(x_{i+1}) \ = \ y(x_i) + \int_{x_i}^{x_{i+1}} F(x, y(x))dx \ \left(= \int_{x_i}^{x_{i+1}} y'(x)dx \right) \ . \qquad (1.14)$$

Numerical ODE solvers use this equation to find approximations y_{i+1} for the value $y(x_{i+1})$ of y at x_{i+1} from F, y_i, and possibly other data. Sophisticated ODE solvers use several previously computed approximate values y_{i-1}, y_{i-2}, ... of $y(x_{i-1})$, $y(x_{i-2})$, ... To do so by approximately integrating the right hand side function F of (1.13) from x_i to x_{i+1} in (1.14) according to (1.14). To solve IVPs means to integrate $F = y'$ numerically in order to find y_{i+1}, y_{i+2}, ... and so forth from known values of $y_j \approx y(x_j)$ for $j \leq i$.

If we know x_i, y_i, and F, we can evaluate $F(x_i, y_i)$ $(\approx y'(x_i))$ to learn the direction that the unknown curve y takes from the point (x_i, y_i). This is depicted in Figure 1.3 for **Euler's**[10] **rule** and $i = 0, 1, 2, 3$. In Euler's rule the computed value of $y_i \approx y(x_i)$ is used to compute the next value $y_{i+1} \approx y(x_{i+1})$ of equation (1.14) by linear extrapolation using the point-slope formula of a line. This line is determined by the point (x_i, y_i) and the slope $F(x_i, y_i) \approx y'(x_i)$. Using this data, y_{i+1} is taken as the y-value on this line that is reached at x_{i+1}, or

$$y_{i+1} = y_i + F(x_i, y_i) \cdot (x_{i+1} - x_i) \quad \text{for} \ \ i = 0, \ldots, N-1 \ .$$

---

[10]Leonhard Euler, Swiss mathematician, 1707-1783

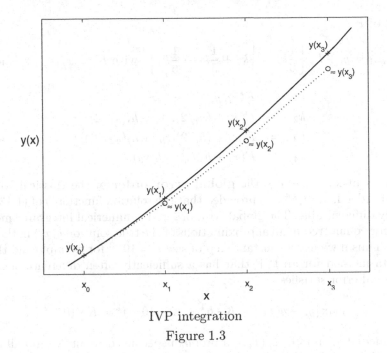

IVP integration

Figure 1.3

In Euler's rule we move from an approximate point (x_i, y_i) along the line with the slope $F(x_i, y_i) \approx y'(x_i)$ for $x_{i+1} - x_i$ units to reach the approximate value y_{i+1} of $y(x_{i+1})$ at x_{i+1}. Then we recompute the slope at x_{i+1} as $F(x_{i+1}, y_{i+1}) \approx y'(x_{i+1})$ and move along a line with this slope for $x_{i+2} - x_{i+1}$ units to reach the approximation y_{i+2} of $y(x_{i+2})$ etc., as depicted in Figure 1.3 in the dotted line segments. The Euler method is called a **one-step method** since y_{i+1} is computed from only one earlier value y_i and $F(x_i, y_i)$. The practical limits of this simple integrator are obvious from Figure 1.3. The true curve $y(x)$, drawn out continuously, diverges from our computed approximations, depicted by small circles o, and ever worse so: the farther away from the starting point $(x_0, y(x_0))$ we integrate and the larger the steps of integration are, the larger becomes the difference between the curves and the computational error.

There are more refined one-step integrators than Euler's method. **Multi-step integration methods** use more than just one previously computed y_i and F value and thereby they can better account for y's curvature and higher derivatives. Thus they follow the actual solution curve much closer. But in turn, they need more supporting F data computations.

Euler's formula has been improved manyfold over the last century or two. To better account for the "turning", or for the concavity of the solution curve $y(x)$, improved integration formulas involve several evaluations of F in the interval $[x_i, x_{i+1}]$ and then average. For example, the **classical Runge[11]-Kutta[12] integration formula** uses four

---

[11] Carle Runge, German mathematician, 1856-1927
[12] Martin Kutta, German mathematician, 1867-1944

evaluations of F:

$$y_{i+1} = y_i + h_i \left\{ \frac{1}{6}k_1 + \frac{1}{3}k_2 + \frac{1}{3}k_3 + \frac{1}{6}k_4 \right\} \text{ with } h_i := x_{i+1} - x_i \text{ and}$$

$$
\begin{aligned}
k_1 &= F(x_i, y_i), \\
k_2 &= F(x_i + (h_i/2), \, y_i + h_i(k_1/2)), \\
k_3 &= F(x_i + (h_i/2), \, y_i + h_i(k_2/2)), \\
k_4 &= F(x_i + h_i, \, y_i + h_i k_3).
\end{aligned}
$$

With $h_{max} := \max(x_{i+1} - x_i)$, the **global error order** of the classical Runge-Kutta method is of order 4, or $O(h_{max}^4)$, provided that the solution function y of (1.13) is 5 times continuously differentiable. The global error order of a numerical integrator measures the maximal error committed in all approximations of the true solution $y(x_i)$ in the computed y values y_i. Thus if we use a constant step of size $h = 10^{-3}$ for example and the classical Runge-Kutta method for an IVP that has a sufficiently often differentiable solution y, then our global error satisfies

$$\max |y_i - y(x_i)| \le K \cdot h^4 = K \cdot (10^{-3})^4 = K \cdot 10^{-12}$$

for all intermediate points $(x_i, y(x_i))$, a problem specific constant K, and all i. For comparison, the Euler method has global error order $O(h_{max})$. If applied to the same problem and step size, it will only compute $y(x_i)$ with global errors bounded by $C \cdot 10^{-3}$ for another constant C.

The price we pay for more accurate higher order methods such as the classical Runge-Kutta method has to be paid with the effort involved in their increased number of function evaluations. Many different Runge-Kutta type integration formulas exist in the literature for up to and including order 8, see the Resources appendix.

Ideally, an integrator should account for the variations in the solution $y(x)$. In regions of x where $y(x)$ has constant slope, for example, one can integrate y correctly over large intervals even with Euler's simple line method. When the graph of $y(x)$ twists and turns rapidly, the integrator should be able to recognize this and use higher order methods combined with smaller steps. This has lead to **adaptive integrators** with an internally varying step size control that depends on y's variation. Adaptive methods usually are built upon two integrators of differing error orders, in which the high order integrator reuses some of the function evaluations that the low-order integrator uses so that the total computational effort is minimized. The results of the two methods at x_{i+1} are compared and if they agree satisfactorily, the solution curve is deemed to be well matched, allowing for subsequently larger steps. If they disagree enough, then the step size is reduced to account for the relatively high variation of the solution. Such integrator pairs are called **embedding formulas**.

There are several adaptive integrators available in software form inside MATLAB. Their names are `ode45` for a Runge-Kutta embedding formula of orders 4 and 5, `ode23` of a Runge-Kutta embedding formula of orders 2 and 3, and `ode113` of the predictor-corrector formula of Adams-Bashford-Moulton [13] with global error order 6. In fact, Adams and Bashforth developed the predictor part of the predictor-corrector integration formula in the mid 1800s and Moulton added the corrector part early in the 20th century to make it truly adaptive.

All of the above mentioned MATLAB integrators are recommended for general use. Their differences are subtle and too tedious to explain here; please look at MATLAB's help. For stiff differential equations MATLAB offers the integrators `ode15s`, `ode23s`, `ode23t`, and `ode23tb`. Stiff ODEs are quite common in chemical/biological engineering IVPs and BVPs, but we will defer discussing stiffness until it appears naturally in Chapters 4 and 5 with specific initial and boundary value problems.

This is but a short introduction to numerical methods.
For more help to decide which ODE solver to use, type `>> more on, type ode...,` `more off`, rather than only `help ode....`

We shall subsequently use the MATLAB ODE solvers as black boxes, sometimes varying between the individual ones only for higher speed or better accuracy when warranted. Our students should experiment freely with using either of the above seven ODE solvers to learn which is more advantageous where. It only takes a different call of MATLAB to find out.

To make certain that a chosen ODE solver has solved the IVP correctly, we recommend to decrease the allowed error bounds repeatedly inside the respective MATLAB `ode...` call and to verify that the computed solution does not change significantly. See p. 201 in Chapter 4 for more details.

For those eager to see some MATLAB ODE examples, we advice to click on the MATLAB desktop "Help" button and to search for "Differential equations" on the left side "Navigator" under the "Search" option. Scrolling down, you will find six clickable items on ODEs and the "Examples" contain more on ODEs in MATLAB.

As told earlier for scalar equations, we must be well aware that there are **unsolvable IVPs** for which the solution may, for example, "blow up", i.e., have a pole before reaching the right endpoint of the target interval $[a, b]$. Likewise, an IVP may have multiple solutions, especially if the number of initial conditions is less than the order of the DE or the dimension of the first-degree system. But under mild mathematical assumptions of continuity and Lipschitz[14] boundedness $\|F(x, y) - F(x, \tilde{y})\| \leq L\|y - \tilde{y}\|$ of the function F in (1.12), every ODE system $y' = F(x, y)$ with $y = y(x) : \mathbb{R} \to \mathbb{R}^n$ and with n specified initial conditions $y(x_0)$, $y'(x_0)$, ... , $y^{(n-1)}(x_0) \in \mathbb{R}^n$ at $x_0 \in \mathbb{R}$ will have a **unique solution**; we refer the reader to the literature quoted in the Resources appendix at the end of this book.

---

[13] John Couch Adams, English astronomer, 1819-1892
Francis Bashforth, English astronomer, mathematician, and ballastician, 1819-1912
Forest Ray Moulton, US astronomer, 1872-1951
[14] Rudolf Lipschitz, German mathematician, 1832-1903

1.2.4   Solving Boundary Value Problems

For **boundary value problems** of ordinary differential equations the first step is to reduce the problem to a first-order system of DEs of dimension $n \geq 1$ as described in Section 1.2.2. Once this is done, one needs to solve the system of DEs

$$y'(x) \; = \; F(x, y) \tag{1.8}$$

for the unknown function $y : \mathbb{R} \to \mathbb{R}^n$ with n component functions

$$y(x) = \begin{pmatrix} y_1(x) \\ \vdots \\ y_n(x) \end{pmatrix}$$

and all $x \in [a, b]$. In BVPs certain **boundary conditions** are imposed on the solution y at the interval ends a and b. The initial IVP data of size n that specifies $y(a)$ and possibly $y'(a)$ etc. at the initial value $x = a$ of an IVP is distributed here among the two endpoints of the BVP in order to tie the solution $y(x)$ down at both ends $x = a$ and $x = b$. Since fewer than n initial conditions at a are specified in a BVP than in the corresponding IVP, one can use this freedom at $x = a$ to parameterize the non specified initial conditions so that the resulting IVP with the parameterized set of initial conditions reaches the right endpoint $y(b) = y_b$ as desired. The standard approach to solving a BVP is to vary the parameterized and unspecified initial conditions of the corresponding IVP at the start $x = a$ until the desired terminal condition(s) at $x = b$ are satisfied. This method is called the **shooting method**.

It goes back to the ancients:

If your arrow strikes above or behind the target, just lower the bow, i.e., decrease the slope or the derivative of the solution curve at the start, in order to hit the target. Vary this initial angle (the free parameter) slightly until the arrow hits just right.

This shooting and adjusting process itself can be modeled by a nonlinear equation that has to be solved in order to move along the trajectory y from a to b that satisfies the ODE system (1.12) *and* the boundary conditions. In a numerical shooting method, we solve an underdefined IVP first, with some unspecified free parameters, and then we try to solve the boundary value problem by adjusting the free parameters to satisfy the right endpoint boundary conditions. This may be possible to do, or the BVP may not be solvable at all. A BVP may have many, even infinitely many solutions. Finding the proper set of initial conditions involves solving a nonlinear system of equations iteratively after each parameter specified IVP has been solved. The IVP is generally solved by one of the methods described in the previous section, while the intermediate nonlinear equation is generally solved by a Newton iteration.

In practice it is advantageous to subdivide the interval $a = x_0 < x_1 < ... < x_N = b$ and to solve the N sub-BVPs as parametric IVPs on each subinterval from x_i to x_{i+1} for $i = 0, 1, ..., N - 1$ first. This is followed by solving the resulting nonlinear system in the parameters so that the values of adjacent sub-functions match at each node x_i in order to obtain a global solution $y(x)$ of the BVP from a to b. Such **multi-shooting methods** rely heavily on an initial guess of the shape of the solution y. The initial guess serves as a first approximation for each of the N sub-IVPs from x_i to x_{i+1}. The variable parameters at each node x_i are then adjusted by a Newton iteration of the form (1.5) for the corresponding nonlinear system of equations $f = 0$ in the free initial value parameters until the individual pieces of the solution y match at the nodes and the concatenated function y satisfies the ODE and the given endpoint boundary conditions.

MATLAB uses a **collocation method** in its BVP solver bvp4c that improves on the multi-shooting method in situations where the function value at the endpoint of an integration is very sensitive to variations of the free parameters at the starting point of the sub-IVPs and where ordinary shooting and multi-shooting methods generally fail. The numerical integrator used by bvp4c on each subinterval is a **Simpson rule**[15] of order 4; see the quoted literature in the Resources appendix. The subinterval solutions are represented in bvp4c as third degree polynomials $S_i(x)$. These are computed so that their derivatives S_i' agree not only at the respective endpoints x_i and x_{i+1} of each subinterval, but also so that they agree at the subinterval midpoints with F as defined in (1.12), i.e., $S'((x_i + x_{i+1})/2) = F((x_i + x_{i+1})/2, S((x_i + x_{i+1})/2))$. To make this matching possible with third degree local polynomials S_i, collocation methods internally subdivide the parameter intervals further when indicated.

The bvp4c MATLAB code can deal with singular ODEs and we shall explain its use and the necessary preparations in Chapter 5. In fact there we show how to modify the inner workings of the built-in MATLAB BVP code bvp4c so that it does not stop when an intermediate Newton iteration encounters a singular or near singular Jacobian matrix, but rather continues with the least squares solution. The modifications to bvp4c will be explained when there is need in Chapter 5.

Under the MATLAB "Help" button, when searching for "Differential Equations", a list of eight topics on BVPs will pop up for those who want to have a preview of MATLAB's BVP capabilities.

This short introduction has given only a broad overview of BVP solvers; for more information we refer the reader to the relevant items mentioned in our Resources appendix.

1.2.5 MATLAB m and Other Files and Built-in MATLAB Functions

In the previous sections we have used MATLAB commands that were typed in on the $\gg$ desktop command line, as well as MATLAB commands that invoke special stored MATLAB files, called **m files**. When starting to work with MATLAB, we advise our users to create a special folder for their personal MATLAB m files once and to continue to use this folder to store personal m files. To access the stored m files from the MATLAB desktop, the user has to point the small desktop window called "Current Directory" to the personal m file folder. Then all stored m files, as well as all built in MATLAB functions can be accessed, called upon, and used from the command line. All stored personal

---

[15]Thomas Simpson, English mathematician, 1710-1761

m files can be opened by clicking on the "open file" icon; they can be printed, edited, and saved by the customary icon clicks. And new m files can be created as well.

MATLAB m files can contain any number of MATLAB commands. They are handy for long sets of MATLAB instructions or data and afford easy repeated use and easy modification. There are two kinds of MATLAB m files: **function** and **script m files**. **Script m files** such as `poly9.m` on p. 28 use the data that is defined inside of them and any undefined data is taken from the data (with the proper name) that is currently active on the workspace; look at the top left window commonly displayed on the MATLAB desktop for a list of defined variables and their type. Script files overwrite the workspace data as instructed in their code.

Function m files such as `fzerotry1m` on p. 27 work in their own closed environment instead: all data used must be supplied as input or generated inside the function m file. Data is not exchanged with the workspace, except for the variables mentioned as "output" and "input" on a function m file's first line such as `function y = f(x) %` `evaluate y=f(x)` for y and x, for example. By a call of $\gg$ `y = f(x)` the input variable x is not altered in the workspace, but the output variable y is. Calling $\gg$ `x = f(x)` for a function m file named `f.m` is allowed and updates the workspace variable x to become equal to $f(x)$. Note that a **function m file** starts with the word `function`, followed by a blank space, and the name of the (optional) output variable(s) or vector(s) such as `y` or `[y,n,error]` for example, if output is desired. The output designation is followed by the equal sign = and the name of the function with its input(s) from the workspace, such as `function y = g(x,z,k)` if f allows three input variables x, z, and k and computes one output y for example. Another function called `funct` is defined on the following lines by specifying its output variables y and s from its input variables x, z, and k. This example function m file computes the kth root of $x^2 + z^2$ and produces the output

$$y = \sqrt[k]{x^2 + z^2} \, ,$$

as well as an "indicator" s (for possible later use in further computations), that is set equal to 1 if $x \cdot z \geq 0$ and equal to $s = -1$ if $x \cdot z < 0$. Here is the example program `funct`:

```
function [y,s] = funct(x,z,k) % compute kth-root(x^2+z^2) and a sign indicator s
y = (x^2 + z^2)^(1/k);
s = 1; if x*z < 0, s = -1; end
```

If this function m file is stored as `funct.m` in the current desktop working directory, it can be called and executed in MATLAB by typing $\gg$ `funct(3,5,17)` or $\gg$ `x=3; z=5;` `k=17; [yy,sign] = funct(x,z,k)` for example. These calls will create the following sequence of outputs.

```
>> funct(3,5,17)
ans =
        1.2305
>> x=3; z=5; k=17; [yy,sign] = funct(x,z,k)
yy =
        1.2305
sign =
     1
```

```
>> [why,ka] = funct(3,5,17); pause, why, pause, ka
why =
      1.2305
ka =
      1
```

In our first call above, we have not named an output and we have omitted the "no screen output" colon ; so that the first (unnamed) output variable y is displayed only as **ans**. In the second call of **funct** we have preceded the call of **funct** by assigning names to both outputs and have followed the command **funct(x,y,k)** with a blank space. This creates the screen output of both variables, called yy and **sign** by us. In the last call the command names the output **why** and **ka** and it is followed by a colon ; . Therefore, we have no output to the screen initially. Only after the **pause** command is ended by hitting the "return" key does the output called **why** appear on screen. After another "return" stroke to overcome the second **pause** command, the second output **ka** is displayed. Note that in the second command the computed values for yy and sign are both available in the workspace whether we use a blank, a comma, or a colon following the call of **funct** since the output is explicitly defined on the command line. Likewise for the third call of **funct** and **why** and **ka**. If on the first command line, however, we had finished the command **funct(3,5,17);** with a colon ; instead, then there would have been no output at all, not onto the screen, nor onto the workspace.

The reader should study our earlier MATLAB m files now, such as the script m file **poly9.m** on p. 28 and the function m file **fzerotry1.m** on p. 27 and look for their differences.

Data files can be saved from one session of MATLAB to another by the **save** command. Below is the screen output of commands that create a 3 by 3 matrix A, then save it (as **A.mat**) in the current working directory, and finally clear all variables in the workspace with the **clear all** command. Commanding only **clear A** would have sufficed to clear, i.e., delete A alone from the MATLAB workspace. We then check that A is in fact gone from the workspace; A now resides in the file **A.mat** in the current working directory of MATLAB. We can resurrect the data in A by using the **load** command for A and A will be again available on the MATLAB workspace. Just follow our diary below. Note that in MATLAB the n by n **identity matrix** $I_{n,n}$ with zeros everywhere except for ones on its main diagonal, can be generated by the command **eye(n)**.

```
>> A = randn(3)-20*eye(3)
A =
      -19.142      -1.441        0.69
       1.254      -19.429       0.81562
      -1.5937      -0.39989     -19.288
>> save A
>> A
A =
      -19.142      -1.441        0.69
       1.254      -19.429       0.81562
      -1.5937      -0.39989     -19.288
>> clear all
>> A
??? Undefined function or variable 'A'.
```

In our last call >> A, we have asked MATLAB to display the "variable" A that was present on the workspace before we cleared the whole workspace with the `clear all` command. Of course at that moment A is no longer available on the workspace, having been cleared from workspace storage. To view or use our stored copy of A again we simply need to load it with a `load A` command onto the current workspace and we can continue working with A as we please. The MATLAB `save` and `load` commands are very useful for storing and reusing data from session to session, over days or weeks. And often it helps MATLAB's efficiency to call `clear all` after long sessions in order to free up workspace memory. Here is what happens when we reload A and display it.

```
>> load A
>> A
A =
    -19.142      -1.441       0.69
      1.254     -19.429       0.81562
    -1.5937      -0.39989    -19.288
```

MATLAB can keep a diary of a session or parts thereof via the commands `diary` or `diary June12`, for example, which collects all screen output in a file called "diary" or "June12", respectively, in the current working directory. The command `diary off` stops the collection of the on-screen information. If the `diary on` command is given later in the same session, the subsequent screen output will be appended to the previous diary and saved. Please read up on `diary` by entering >> `help diag`.

Figures and plots can be generated easily via MATLAB. These can be transformed into and stored in many formats, such as MATLAB's generic `,,,.fig`, or as Jpeg, postscript, RAW, TIFF, etc. files as desired. Simply look at the dialogue box that pops up from the save icon of a MATLAB figure. For a quick introduction to MATLAB graphics, type >> `demo matlab graphics` and a "Help" window will open, that lets you choose several clickable graphics demonstrations. We defer further explanations to our relevant graphics codes in later chapters.

The last part of this section lists a few standard MATLAB operations, functions, and commands, collected into groups, together with short descriptions. This may help our readers to more easily find and use built-in MATLAB functions in their own MATLAB program codes. Please note that our MATLAB function descriptions below are very few and very short by necessity. The user should use the `help ...` command to find the full length MATLAB reference guide entry for each MATLAB function when the need arises. This will help our readers use the full power and functionality of MATLAB commands and will enable them to browse for and find related built-in MATLAB functions.
The printed MATLAB Function Reference Guide book consists of three volumes amounting to about 2,000 pages that list every one of about 1,000 built-in MATLAB functions.

MATLAB Desktop Utilities

clear Deletes items from the workspace. **clear all** clears the whole workspace. Recommended after long sessions on the same MATLAB desktop.

clf This command clears the current figure window and thereby avoids superimposed images if a **hold** command is still active on the window.

close Deletes the current figure, or a specified figure. **close all** closes all figure windows.

help Calling >> **help** ... at the MATLAB command line displays the comment lines on screen that explain the function ... .

quit This command exits the current MATLAB desktop. Upon calling >> **quit** MATLAB saves the previous command history, while simply closing the desktop via the mouse does not keep the history of commands in MATLAB's memory.

↑, ↓ ↑ entered on the keyboard brings up the previous command on the command line, and the command before the previous one when pressing ↑ again, etc. ↓ allows to go forward in the command history.

Arithmetic Operations in MATLAB

+, -, *, /, ^ Elementary mathematical operations : plus, minus, times, divide, and power or exponentiation.
If any of the arithmetic symbols is preceded by a dot . such as in **x.*y** for two row vectors x and y of the same length, for example, the indicated operation is performed element by element and results in the row vector $(x_1 y_1, ..., x_n y_n)$, while **x.^y** becomes $(x_1^{y_1}, ..., x_n^{y_n})$.

Special Characters in MATLAB

@ Creates a function handle, so that functions can be passed to MATLAB function m files as variables are.

... Three dots ... indicate a line break inside a MATLAB command. Useful to break long lines of contiguous code, such as with title text.

% The percentage sign makes all that follows on the same line into commentary to be skipped during execution of the file. See p. 232 for details on how to effectively (un)comment whole blocks of MATLAB code.

Elementary Functions in MATLAB

feval Evaluates a function for the supplied arguments. The function may be given by its "handle" as @func for a function m file named `func.m`.

clock Saves the current time as a date vector.

etime Evaluates elapsed time, often used in conjunction with `clock`.

exp Exponential function e^x for the Euler[16] number $e = 2.71...$.

log, log10 Logarithm for base e or base 10.

Elementary Matrix Generation in MATLAB

eye Creates the identity matrix.

ones Creates a matrix of all 1s.

rand, randn Creates a matrix with random entries; uniformly distributed or normally distributed, respectively.

zeros Creates a matrix with all entries equal to zero.

norm Computes the norm of a matrix.

diag For vector input, `diag` creates a matrix whose diagonal is as prescribed with zeros elsewhere. For a matrix input `diag` extracts the diagonal entries of the matrix in a vector.

Elementary Matrix Functions in MATLAB

size Computes the size vector (number of rows, number of columns) of a matrix.

length Computes the maximal size max(number of rows, number of columns) of a matrix.

linspace Creates a partition vector with linearly spaced nodes.

logspace Creates a partition vector with logarithmically spaced nodes.

**** Solves a linear system of equations.

lu The call of [L,U] = lu(A) computes a lower triangular matrix L and an upper triangular one U with $A = L \cdot U$ if possible, using Gaussian elimination.

---

[16]Leonhard Euler, Swiss mathematician, 1707-1783

qr The call [Q,R] = qr(A) computes a unitary matrix Q and an upper triangular matrix R with $A = Q \cdot R$ using Householder elimination.

eig Calling eig(A) computes the eigenvalues of a square matrix A.

svd Calling svd(A) computes the singular values of a matrix A.

Sorting, Finding, and Comparison Functions in MATLAB

sort Sorts a real vector by size of the entries; a complex vector by the magnitude of the entries.

abs Computes the componentwise absolute values of the entries of a vector or a matrix.

max, min Compute the maximal (or minimal) entry of each column of a real matrix, or of the entries of a real vector.
To find the absolute largest entry in a real matrix A, use max(max(A)).

find The call [i,j] = find(A ~= 0) for example returns the row and column indices of all nonzero entries in the matrix A.

fliplr, flipud Reverse the order of the entries of a vector or of the columns or rows of a matrix, either left to right or up-down.

end When used as an *index* inside a matrix or vector statement this denotes the "final" index of the vector/matrix. Not to be confused with the terminating "end" command of "for", "while" or "if" loops. Avoids finding the size of a vector or matrix first before working on the maximal index entry.

Logical Functions in MATLAB

==, ~=; <, <=, >, >= Equal sign in logic comparisons, not-equal sign; comparison signs.

&, | Logical "and" and "or" symbols.

for ... A "do loop" uses the syntax for "list", "action"; end such as in x = zeros(1,21); for i = 1:2:21, x(i) = -i; end which sets every second component in x equal to its negative index, i.e., $x = (-1, 0, -3, 0, ..., -19, 0, -21)$.

while ... A "while loop" uses the syntax while "condition", "action"; end such as in
i = 1; while i <= 21, x(i) = -i; i = i + 1; end which sets every component in $x \in \mathbb{R}^{21}$ equal to its negative index, i.e., $x = (-1, -2, -3, ..., -20, -21)$.

if ... else ... end The command if x > 2, A = 5; else, A = 20; end, for example, sets A to be 5 if $x > 2$, and equal to 20 otherwise.
The syntax of an "if" statement is as follows: if "condition", "action"; end,

`try ... catch ... end` The commands
`A = [0 1;1,2]; b = [0,10]';` `try, x = A/b;` `catch, x = 20; end, x,`
`lasterr,` for example, try to solve a linear system of equations $Ax = b$ for x. But the command `x = A/b;` uses the wrong direction slash due to a typo. If this command had been typed correctly as `x = A\b;`, then the above line of commands would create no error message and the program would continue with $x = A^{-1}b = \begin{pmatrix} 10 \\ 0 \end{pmatrix}$. As the command inside the original "try" segment fails, however, x is set to 20 inside the "catch" segment and the program continues with that value of x.

Check the output of our command line to verify this and then alter the forward slash typo into a backslash for the correct output of the solution $x = \begin{pmatrix} 10 \\ 0 \end{pmatrix}$ to $Ax = b$.

The syntax of a "try ... catch" statement is : `try, "action1"; catch, "action2"; end, .`

Special Input and Output Functions in MATLAB

`nargin, nargout` Number of input or output arguments of a MATLAB function. Used to specify certain input parameters in the preamble of a MATLAB program or at the end.

`disp` Causes screen output of its arguments.

`error` Displays an error message and terminates the program.

`num2str` Creates a string for a number to be used inside text strings, mainly used in plot or screen annotations.

Definite Integral Evaluation in MATLAB

`quad, quad8, quadl` Evaluate definite integrals via different integration formulas.

Ordinary Differential Equation Solvers in MATLAB

For Initial Value Problems:

`ode45,ode23 etc.` Initial value problem solvers for various types of stiff and nonstiff DEs.

For Boundary Value Problems:

`bvp4c` Boundary value problem solver for DEs.

Graphing and Plotting Utilities in MATLAB

figure Creates a figure graphics object. Commands such as figure(1) and figure(2) etc, preceding plot commands refer to different figure windows on the desktop.

subplot Divides the current figure window into multiple panes. Number of pane rows, columns and the location of the next plot need to be specified.

meshgrid Generates the x and y coordinate matrices for plotting 3D surfaces.

plot The command plot(x,y) for two equal sized vectors x and y plots the lines between each point (x_i, y_i) and the subsequent point $(x_{i+1}, y_{i+1}) \in \mathbb{R}^2$ in a linear 2-D plot in the plane. Line type, line width, and color etc. can be further specified.

plot3 The command plot3(x,y,z) for three equal sized vectors x, y, and z plots the lines between each point (x_i, y_i, z_i) and the subsequent point $(x_{i+1}, y_{i+1}, z_{i+1}) \in \mathbb{R}^3$ in a linear 3-D plot that projects the curve in $\mathbb{R}^3$ onto a planar image. Line type, line width, and color etc. can be further specified.

semilogx, semilogy The command semilogx(x,y) for two equal sized vectors x and y plots the lines connecting each point (x_i, y_i) and the subsequent point $(x_{i+1}, y_{i+1}) \in \mathbb{R}^2$ in a semi-logarithmic 2-D plot in the plane. Here the scale of the (horizontal) x axis is logarithmic, while the vertical y scale is linear. Line type, line width, and color etc. can be further specified.
semilogy(x,y) plots linearly in the horizontal scale and logarithmically on the y scale.

loglog The command loglog(x,y) for two equal sized vectors x and y plots the lines between each point (x_i, y_i) and the subsequent point $(x_{i+1}, y_{i+1}) \in \mathbb{R}^2$ in a two axes logarithmically scaled plot in the plane. Line type, line width, and color etc. can be further specified.

interp1 A call of Y = interp1(x,y,X) creates interpolating data X and Y from given data x and y of points $(x_i, y_i) \in \mathbb{R}^2$. Used to plot smooth curves that interpolate the (sparse) data x and y to obtain (usually) much larger data X and Y. Various methods of interpolation such as splines etc. can be specified.

interp2 Same as interp1, except for three dimensional data x, y, and z of surfaces.

contour Two dimensional contour plot of a surface. For three equal dimensioned matrices X, Y, and Z, contour(X,Y,Z) draws the surface $Z = z(x, y)$ over the rectangle with partitioned edges in X and Y. Level lines and types can be explicitly set, as well as colors.

contour3 Plots contour lines on a 3D surface plot.

surf, surfc For three equal dimensioned matrices X, Y, and Z, surf(X,Y,Z) creates a three dimensional color-shaded surface plot. And surfc(X,Y,Z) draws an additional contour plot below the surface.

surface For three equal dimensioned matrices X, Y, and Z, surface(X,Y,Z) plots the parametric surface of the data X, Y, and Z. Color can be specified.

Figure Window Manipulations

⊕ , ⊖ Figure zoom functions to magnify or reduce an area of a plot. Used via the mouse from the Figure window toolbar.

rotation icon When activated on the Figure window toolbar, the mouse can be used to rotate 3D plots for better visualization. Can be used in conjunction with ⊕ , ⊖ .

Color and Line Types in MATLAB

Color specifications: In MATLAB 'y' stands for "yellow", 'r' for red, 'g' for green, 'b' for blue, 'k' for "black", etc. Alternatively one can specify color as an RGB triple such as [1 0 0] for red, [0 0 0] for "black", [1 1 0] for "yellow", etc.
For surface and line color drawings it is advisable to define a colormap such as colormap(hsv(128)) that contains 128 colors, varying from "red" hues through "yellow", "green", "cyan", "blue", and "magenta" hues before returning back to "red" hues.

Line type specifications: Specifying the line type by – creates a solid line; – – creates a dashed line, : makes a dotted line etc. Plots can be marked by symbols such as + for the "plus" sign, or o, i.e., the lower case letter o, for small "circles", or s for squares, etc.
Line width can be specified as in plot(....,'LineWidth',2), for example, by a width given in "points" such as 1 for thin lines or 2 for wider ones in most of the MATLAB plotting functions.

Fontsize specifications: Plot annotating utilities such as xlabel, ylabel, text, and title allow fontsize settings for displayed text such as xlabel(...,'FontSize', 12). Here 12 indicates the font to be used. Fontsize 14 is rather large and fontsize 8 rather small.

Rotation specifications: Plot annotating utilities such as xlabel, ylabel, text, and title allow rotation of the text via text(....,'Rotation',90). Here 90 indicates a rotation of ninety degrees from horizontal positioning. A rotation by "0" will not turn the text at all.
Note that in MATLAB ylabel defaults to a rotation of the text by 90 degrees, i.e., it displays the y text information sideways, unless a rotation by 0° is specified.

Figure text special commands: Titles and axes labels can carry mixed alphabetical and numerical text. For this the string of words and numbers must be enclosed by a [...] bracket inside the title, xlabel etc. command. Consecutive words

are enclosed by ' ', followed by a comma , and `num2str(bt)` if we want the numerical value of β to appear amongst the plot annotations as in this example: `xlabel(['Example for standard data with \beta = ',num2str(bt),' for n = ',num2str(n)])`.
Note that MATLAB accepts LaTeX style alphanumeric input such as \beta above. It creates the greek symbol β as intended.

MATLAB is a very simple language for computations. Its lines of numerical code are often few and short. By comparison, the visual output and the proper texting of graphs created in MATLAB may take much more effort than the numerics.
The MATLAB codes printed in this book give our readers a varied and solid introduction into the above and many other built-in MATLAB functions, as well as into their functionality.

Students should always try to develop their own codes first for a problem before reading or using the codes of this book and the CD. Then we think it is best to study, mimic, and even imitate codes if they run smoother or better. Therefore, we encourage our readers to read and study our codes well, but only after having tried their own.

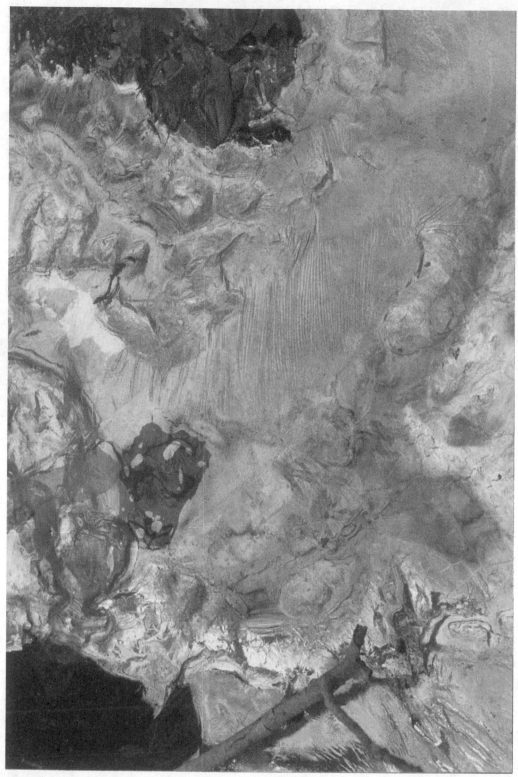

Environmental system

Chapter 2

Modeling, Simulation, and Design of Chemical and Biological Systems

After the introduction of the basic building blocks of MATLAB and the fundamentals of numerical analysis with an eye on solving scalar and differential equations in Chapter 1, we now introduce mathematical models in chemical and biological engineering[1]. Our subsequent chapters will combine these two areas by introducing models for specific chemical and biological processes and finding detailed numerical solutions via MATLAB. This chapter is quite short and condensed. Any reader or student who wants to gain a deeper insight into model formulation should consult the literature on math modeling in the Resources appendix.

2.1 System Theory and its Applications

This section introduces some of the basic concepts of system theory in relation to modeling. Our presentation is rather brief since our aim is to integrate known models for chemical/biological processes with numerical techniques to solve these models for simulation and design purposes, rather than to give a broad introduction to either system theory or modeling itself. For references on modeling, see the Resources appendix.

We first give an overview of system theory and modeling as it applies to our subject in the rest of this book.

2.1.1 Systems

What is a *system*?

---

[1] Biological engineering comprises both biochemical and biomedical engineering

55

- A system is a whole, composed of its parts or elements.

- The decision of what constitutes a system, a subsystem, or an element thereof is relative and depends upon the level of our analysis. For example, we can view the human body as a system, with the heart, the arms, the liver, etc. as its elements. Or we can consider these elements as subsystems in themselves and can analyze them further by their smaller constituents, such as the valves and chambers of the heart, or the bones, tendons, muscles, etc. of the arm, and so on.

- The parts or elements of a system can be parts in the physical sense of the word or they can be processes. In the strictly physical sense, the parts of a human body or those of a chair constitute a physical system. In our studies of chemical/biological equipment that performs certain chemico-physical functions, we must also consider the various chemico-physical processes that take place inside the system as elements thereof. These processes interact very often with each other to perform the task of the particular chemical plant, called the system. A simple example of this is a chemical reactor in which processes such as mixing, chemical reactions, heat evolution, heat transfer, etc. take place in a controlled way to achieve the task of the reactor, i.e., the change of the input reactants to the output products.

- The properties of the whole system are not the sum of the properties of its components or elements. However, the properties of the components affect the whole system. The system properties in chemical/biological engineering are usually the result of highly nonlinear interactions between the system's components (elements). For example, one of the "system qualities" of human beings is "consciousness". But this is not the property of any of its components or elements, in however much detail we may look at the human system. In our chemical/biological case, the mass transfer with chemical reactions has certain properties which are not the properties of the chemical reaction or of the mass transfer alone. An example for this is the possibility of multiple steady states (bifurcation), a phenomenon which occurs often in nature, as well as in industrial chemical, biochemical and biomedical equipment. This phenomenon shall comprise a large part of our studies in this book. See Appendix 2 for an overview of this important phenomenon of chemical and biological engineering systems.

The above list is a very elementary presentation of system theory. We will revisit this subject repeatedly and learn more and deeper facts.

Generally, a chemical/biological system has a boundary that distinguishes it from the environment. The interaction between the system and its environment determines the type of the system and its main characteristics as will be detailed later.

For simplicity, we shall refer to any chemical/biological system from now on simply as a *system*, omitting the qualifying adjectives, since we shall only consider such chemical/biological systems from now on.

2.1.2 Steady State, Unsteady State, and Thermodynamic Equilibrium

A stable system may be stationary with regards to time, or it may be changing with time. An open system which exchanges matter with the environment is usually changing with time until, if it is stable, it reaches a stationary state. This stable state is called the **steady state** in chemical/biological engineering. A more correct name for this stationary stable state would be: a **stationary nonequilibrium state**. In Chapters 4 through 7 of this book, we prefer to call the stable steady state of a system a *point attractor* to distinguish it from other types of attractors, such as *periodic* and *chaotic attractors*. Unstable states, such as saddles and unstable nodes are not attractors, but rather repellers. Closed and isolated systems which do not exchange matter with their surroundings also change with time till they reach a **stationary equilibrium state**. This is a dead stationary state, unlike the stationary nonequilibrium state for open systems which is well known in chemical/biological engineering as the steady state of continuous processes.

Generally, we consider three quantities associated with a system:

The state of the system:
The state of a system is rigorously defined through the state variables of the system. The state variables of any system are chosen according to the nature of the system. The state of a boiler, for example, can be described by temperature and pressure, that of a heat exchanger by temperature, the state of a nonisothermal reactor by the concentration of the different components and their temperature, and the state of a bioreactor by the substrate concentrations and pH.
Recall from Chapter 1 that mathematically speaking each "state variable" is a function of one or several variables, such as of time or location.

Input variables (or **parameters**):
These are not state variables of the system. They are external to the system and affect the system, or in other words they "work on the system". For example, the feed temperature and composition of the feed stream for a distillation column or a chemical reactor, or the feed temperature of a heat exchanger are input variables.

Initial conditions:
For systems that have not reached their stationary state (steady state or thermodynamic equilibrium), the behavior with regards to time cannot be determined without knowing the initial conditions, or the values of the state variables at the start, i.e., at *time* = 0. When the initial conditions are known, the behavior of the system is uniquely defined. Note that for chaotic systems, the system behavior has infinite sensitivity to the initial conditions; however, it is still uniquely defined. Moreover, the feed conditions of a distributed system can act as initial conditions for the variations along the length.

2.2   Basic Principles for Modeling Chemical and Biological Engineering Systems

The simplest definition for **modeling** is *"putting the physical reality into an appropriate mathematical form"*.
A model of a system is a mathematical representation in the form of equations and inequalities that relates the system behavior over time to its inputs and predicts the outputs.

Model Development

Every elementary procedure for model building generally includes the following steps:

1. Defining the system.

2. Defining the boundaries of the system.

3. Defining the type of system, whether it is open, closed, or isolated.
   Open systems exchange matter with the environment, closed and isolated systems do not. Details to follow.

4. Defining the state variables.

5. Defining the input variables (sometimes called input parameters).

6. Defining the design variables (or parameters).

7. Defining the nature and type of interaction between the system and the surrounding environment (if any).

8. Defining the physical and chemical processes that take place within the boundaries of the system.

9. Defining the rate of the different processes in terms of the state variables and physico-chemical parameters, and introducing equations that reflect the physical and chemical relations between the states at the different phases.

10. Writing mass-, heat-, energy-, and/or momentum-balance equations to obtain the **model equations** that relate the system input and output to the state variables and the physico-chemical parameters. These mathematical equations describe the state variables with respect to time and/or space.

Solution of the model equations

The equations of the developed model need to be solved for certain inputs, certain design objectives, and given physico-chemical parameters in order to predict the output and, for design purposes, the variation of the state variables within the boundaries of the system. In order to solve the model equations we have two tasks:

1. Determine the physico-chemical parameters of the model experimentally or obtain them from the literature if some or all of them are available there.

2. Develop a method that finds the solution of the mathematical model equations. The method may be analytical or numerical. Its complexity needs to be understood if we want to monitor a system continuously. Whether a specific model can be solved analytically or numerically and how, depends to a large degree upon the complexity of the system and on whether the model is linear or nonlinear.

Algorithmic and computational solutions for model (or design) equations, combined with chemical/biological modeling, are the main subjects of this book. We shall learn that the complexities for generally nonlinear chemical/biological systems force us to use mainly numerical techniques, rather than being able to find analytical solutions.

2.3 Classification of Chemical and Biological Engineering Systems

It is very useful to classify different basic system models now in relation to their mathematical description.

I. **Classification according to the variation or constancy of the state variables with time:**

  - **Steady-state models**: described by transcendental equations or ODEs (Ordinary Differential Equations) or PDEs (Partial Differential Equations).

  - **Unsteady-state models**: described by ODEs or PDEs.

II. **Classification according to the spatial variation of the state variables:**

  - **Lumped models** (usually called "lumped parameter models", which is wrong terminology since the state variables are lumped, not the input variables or parameters); described by transcendental equations for the steady state and ODEs for the unsteady state.

  - **Distributed models** (usually called "distributed parameter models", which is again wrong terminology since the state variables are distributed, not the parameters). These are models for systems for which the state variables are distributed along one or several spatial directions. They are described by ODEs or PDEs for the steady-state and by PDEs for the unsteady-state case.

III. **According to the functional dependence of the rate-governing laws upon the state variables:**

  - **Linear models**: described by linear equations (these can be constant or variable coefficient systems of linear equations, or linear ODEs or PDEs). These models can often be solved analytically, except for high dimensions.

- **Nonlinear models**: described by nonlinear equations (these can be transcendental systems of scalar equations, ODEs, or PDEs). These models can generally only be solved numerically.

IV. **According to the type of processes taking place within the boundaries of the system:**

- Mass transfer (example: isothermal absorption).

- Heat transfer (example: heat exchangers).

- Momentum transfer (example: pumps or compressors).

- Chemical reaction (example: homogeneous reactors).

- Combination of any two or more of the above processes (example: heterogeneous reactors).

V. **According to the number of phases in the system:**

- Homogeneous models.

- Heterogeneous models (more than one phase).

VI. **According to the number of stages in the system:**

- Single stage.

- Multi-stage.

VII. **According to the mode of operation of the system:**

- Batch (closed or isolated system).

- Fed-Batch.

- Continuous (open system).

VIII. **According to the system's thermal relation with the surroundings:**

- Adiabatic (neither heating nor cooling).

- Nonadiabatic cocurrent or countercurrent cooling or heating.

IX. **According to the thermal characteristics of the system:**

- Isothermal.

- Nonisothermal.

All of these classifications are naturally interrelated for a given chemical/biological engineering problem, and the best approach is to choose one main classification and then use the other problem's classifications as subdivisions. The most fundamental classification of systems is usually based upon their thermodynamical characteristics. This classification is the most general; it divides systems into open, closed, and isolated systems which are defined as follows:

Isolated Systems: Isolated systems exchange neither energy nor matter with the environment. The simplest example from chemical or biological engineering is the adiabatic batch reactor. Isolated systems naturally tend towards their thermodynamic equilibrium with time. This state is characterized by maximal entropy, or the highest possible degree of disorder.

Closed Systems: Closed systems exchange energy with their environment through their boundaries, but they do not exchange matter. The simplest example is the nonadiabatic batch reactor. These systems also tend towards a thermodynamic equilibrium with time, again characterized by maximal entropy, or the highest possible degree of disorder.

Open Systems: Open systems exchange both energy and matter with their environment through their boundaries. The most common example from chemical/biological engineering is the continuously stirred tank reactor (CSTR). Open systems do not tend towards a thermodynamic equilibrium with time, but rather towards a "stationary nonequilibrium state", characterized by minimal entropy production, i.e., the least degree of disorder. This state is usually called the "steady state" in chemical/biological engineering.

Efficient solution techniques (whether analytical or numerical) for solving chemical and biological engineering models will be presented throughout this book.

Mathematical Programs

For most chemical/biological engineering problems, the mathematical models are usually quite complex due to the generally high degree of complexity of both the physico-chemical processes and their nonlinear interactions. Thus the solution algorithms will generally need to use elaborate computer programs and software. The models can be solved in whatever programming language or computational environment one wants to use. However, we have decided to train our readers throughout this book to use **MATLAB**, as it is one of the most versatile computational softwares. MATLAB is extremely well suited for numerical analysis, while quick and very intuitive to learn and use.

2.4 Physico-Chemical Sources of Nonlinearity

The main reason that chemical/biological engineers need extensive use of numerical methods is twofold:

1. Most chemical and biological engineering problems involve highly nonlinear processes, specially in reacting systems.

2. The high dimensional models and representations that most chemical/biological engineering problems require cannot be solved analytically even when they are linear.

The nonlinearity of chemical/biological processes can be strong or weak. One strong nonlinearity is the ***Arrhenius dependence***[2] of the rate or reaction constant k upon temperature T, i.e.,

$$k = k_0 \cdot e^{-E/(R \cdot T)} .$$ (2.1)

On the other hand, most weak nonlinearities can be associated with the dependence of specific heats (C_p) upon temperature, or of diffusion coefficients upon concentrations, etc.

Needless to say, strong nonlinearities have stronger effects and pose more numerical problems than weak nonlinearities. Almost all nonlinear models can only be solved numerically, whether the system is described by scalar equations or differential equations. One way to solve a nonlinear problem analytically is to linearize it in a certain region of the variables to obtain an approximate solution in this region. Linearization is widely used in several simple process dynamics and control problems. It is well known that linearization generally does not give an accurate description of the system. However it may be useful to obtain some insight into the basic characteristics of the system. Linearization of nonlinear problems will be discussed later, where its usefulness and limitations will be highlighted.

For chemical/biological systems it is important to point out that even when the system is approximated by linear differential equations, such a system cannot generally be solved analytically when the dimension of the system is high. An ODE initial value system of 40 linear differential equations $y' = Ay$, for example, although solvable analytically, at least in theory via the Jordan normal form of the system matrix A, cannot be solved that way in reality since it is numerically impossible to compute the Jordan normal form of sizable matrices A correctly. Hence a proper understanding of the theory and of the numerics of matrices is essential for developing competent and efficient numerical algorithms. Such algorithms, for the most part, ***do not*** mimic mathematical or theoretical matrix theory such as using determinants, but are rather inspired by modern numerical analysis practices such as matrix factorizations.

Nonlinearities are always associated with the very nature of the process. Examples include the nonlinear dependence of the rate of reaction upon temperature, the non-monotonic dependence of the rate of reaction upon the reactants' concentrations, the dependence of specific heats and diffusion coefficients upon temperature and concentrations, etc. High dimensions, i.e., large numbers of state variables in our models, are generally associated with the complex chemical and biological equipment being used, such as the large number of trays in absorption or distillation columns, as well as with the large number of components that the chemical or biological equipment under study is handling.

In a nutshell, we state that almost all chemical/biological engineering systems are nonlinear and that the physico-chemical sources of the nonlinearities are associated with the chemical and biological reactions and with the nonlinear dependence of physical parameters on state variables. The most common (and also the strongest) nonlinearity is associated with the exponential Arrhenius dependence (2.1) of the rate of reaction

---

[2]Svante August Arrhenius, Swedish chemist, 1859 – 1927

upon temperature. However, other nonlinearities associated with reaction rate dependence upon concentrations are possible. For example:

1. **Simple reaction rates** r, for which the order of reaction is neither one nor zero, such as in

$$r = f(C_A) = k \cdot C_A^{1.3}$$

for example, where C_A is the concentration of the reactant and k is the rate constant. This is considered a simple weak nonlinearity and is graphed in Figure 2.1.

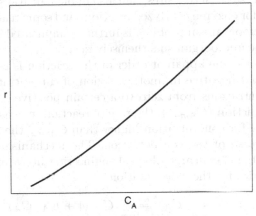

Weak nonlinearity

Figure 2.1

2. **Nonlinearity with saturation.** This type of nonlinearity is quite common in biochemical engineering and waste water treatment systems.

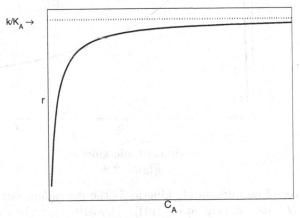

Nonlinearity with saturation

Figure 2.2

A typical rate equation of this type has the form

$$r = f(C_A) = k \cdot C_A/(1 + K_A \cdot C_A) \,.$$

The degree of nonlinearity here is higher than in the previous example. The corresponding curve is shown in Figure 2.2.

Nonmonotonic dependence:

All of the previously mentioned nonlinearities are actually monotonic. Nonmonotonic functions are very common in gas-solid catalytic reactions due to competition between two reactants for the same active sites, and also in biological systems, such as in substrate inhibited reactions for enzyme catalyzed reactions and some reactions catalyzed by microorganisms. The microorganism problem is further complicated in a nonlinear manner due to the growth of the microorganisms themselves.

For these reactions, the apparent order of the reaction is positive in a certain region of concentrations and negative in another region of concentrations. This means that in the region of concentrations from zero to a certain positive value corresponding to the maximum rate of reaction (C_{Amax}), the rate of reaction increases with an increase of the concentration, while for concentration higher than C_{Amax} the rate of reaction decreases with a further increase of the concentration. The mechanisms for this phenomenon is well established in the chemical/biological engineering literature. Figure 2.3 shows such nonmonotonic kinetics for the rate equation

$$r = r(C_A) = k \cdot C_A/(1 + K_A \cdot C_A)^2 \,.$$

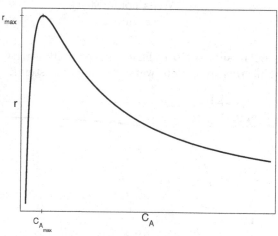

Nonmonotonic kinetics
Figure 2.3

Another source of nonmonotonic kinetics is the dependence of most biological reactions upon the H^+ ion concentration (pH). This situation gives rise to nonmonotonic kinetics since the dependence of the enzyme rate of reaction upon pH is nonmonotonic (bell shaped) as shown in Figure 2.4.

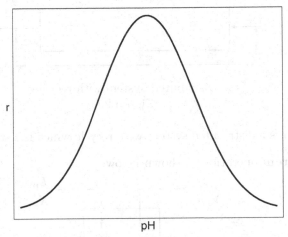

Nonmonotonic dependence of biological reaction rate on pH
Figure 2.4

The last and most common nonmonotonic rate of reaction occurs under nonisothermal operation where the rate of reaction, even for a first-order reaction, is highly nonlinear, namely

$$r = k_0 \cdot e^{-E/(R \cdot T)} \cdot C_A \ . \tag{2.2}$$

Here C_A is the concentration of the reactant A. If the reaction is exothermic, i.e., if the process generates heat, then the rate equation (2.2) is nonmonotonic with respect to conversion. As the reaction proceeds, the conversion increases, i.e. the concentrations of the reactants decrease (tending to cause a decrease in the rate of reaction), while the temperature increases (tending to cause an increase in the rate of reaction). These opposing effects obviously give nonmonotonic dependence upon the conversion or the concentration or the temperature. This case is investigated in more detail in Chapter 3. It is one of the natural causes of multiple steady states and bifurcation for reacting systems.

2.5   Sources of Multiplicity and Bifurcation

The multiplicity of steady states and associated bifurcation phenomena are associated with open systems which allow more than one stationary nonequilibrium state for the same set of parameters. This multiplicity of a steady state may result from many causes. The most important sources of multiplicity in chemical/biological engineering processes are:

1. **Nonmonotonic** dependence of one rate process (or several) on one or more state variables.

2. **Feedback** of information.
   A recycle in a tubular reactor or axial dispersion which will be discussed in Section 7.2.

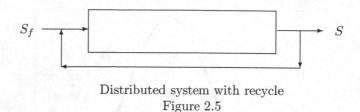

Distributed system with recycle
Figure 2.5

Figure 2.5 shows a distributed system with recycle which is a source of multiplicity.

3. **Countercurrent** operation as shown below.

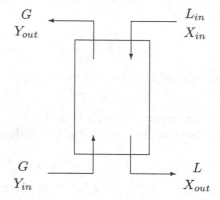

Countercurrent process
Figure 2.6

These sources of multiplicity are very frequent in actual chemical/biological processes. Chapter 7 highlights examples of multiplicity, bifurcation, and chaotic behavior for a number of experimental and industrial chemical/biochemical processes. We also include Appendix 2 on multiplicity and bifurcation at the end of the book.

Dr. Said, stirring Brunswick stew at Whatley farms

Chapter 3

Some Models with Scalar Equations, *with and without Bifurcation*

It is important to introduce the reader at an early stage to simple examples of nonlinear models. We will first present cases with bifurcation behavior as the more general case, followed by special cases without bifurcation. Note that this is deliberately the reverse of the opposite and more common approach. We take this path because it sets the important precedent of studying chemical and biological engineering systems first in light of their much more prevalent multiple steady states rather than from the rarer occurrence of a unique steady state.

3.1 Continuous Stirred Tank Reactor: The Adiabatic Case

One of the simplest practical examples is the homogeneous nonisothermal and adiabatic **continuous stirred tank reactor** (CSTR), whose steady state is described by nonlinear transcendental equations and whose unsteady state is described by nonlinear ordinary differential equations.

We will consider a very simple irreversible reaction

$$A \implies B$$

that takes place inside an adiabatic CSTR as shown in Figure 3.1. The rate of reaction is described by a simple first-order rate equation, namely by

$$r = k \cdot C_A \,,$$

and the dependence of k upon the temperature T is given by the well known Arrhenius form

$$k = k_0 \cdot e^{-E/(R \cdot T)} .$$

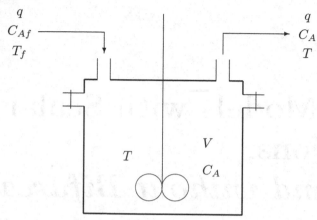

Schematic diagram of an adiabatic CSTR
Figure 3.1

Under the simple assumptions of constant volume V and volumetric flow rate q, the steady states of the system are described by the two equations

$$q \cdot C_A = q \cdot C_{Af} - V \cdot k_0 \cdot \left(e^{-E/(R \cdot T)} \right) \cdot C_A \qquad \text{(material balance design equation)}$$

and

$$q \cdot \rho \cdot C_p \cdot (T - T_f) = V \cdot k_0 \cdot \left(e^{-E/(R \cdot T)} \right) \cdot C_A \cdot (-\Delta H) . \qquad \text{(heat balance design equation)}$$

Here

q	=	volumetric flow rate, in l/min;
C_A	=	reactant concentration at exit and at every point in the CSTR, in $mole/l$;
ρ	=	mixture average (constant) density, in g/l;
C_p	=	mixture average (constant) specific heat, in $cal/(g \cdot K)$;
C_{Af}	=	reactant feed concentration, in mol/l;
V	=	active reactor volume; in l;
k_0	=	frequency factor for the reaction, in $1/min$;
E	=	reaction activation energy, in $J/mole$;
R	=	general gas constant, in $J/(mole \cdot K)$;
T	=	temperature at the exit and at every point in the CSTR, in K;
T_f	=	feed temperature, in K;
ΔH	=	heat of reaction, in $J/mole$.

These two balance equations can be reduced easily to one equation and a simple linear relation between T and C_A. This is done by multiplying the material balance design equation by $(-\Delta H)$ and by adding it to the heat balance design equation to obtain

$$(-\Delta H) \cdot q \cdot C_A + q \cdot \rho \cdot C_p \cdot (T - T_f) = (-\Delta H) \cdot q \cdot C_{Af} .$$

Rearrangement gives

$$C_A = C_{Af} + \frac{\rho \cdot C_p \cdot (T_f - T)}{(-\Delta H)} . \qquad (3.1)$$

Thus the two design equations can be rewritten in terms of the single variable T as

$$q \cdot \rho \cdot C_p \cdot (T - T_f) = V \cdot k_0 \cdot \left(e^{-E/(R \cdot T)} \right) \cdot \left\{ C_{Af} + \frac{\rho \cdot C_p \cdot (T_f - T)}{(-\Delta H)} \right\} \cdot (-\Delta H) \quad (3.2)$$

and relation (3.1).

   If we consider exothermic reactions, where $(-\Delta H)$ is positive, and think of the physical meaning of the two sides of equation (3.2), we realize that the left-hand side represents the affine heat removal function $R(T) = q \cdot \rho \cdot C_p \cdot (T - T_f)$ due to the flow, while the right-hand side represents the more complicated exponential heat generation function $G(T)$ of the reaction. And obviously steady states, or more accurately the stationary nonequilibrium states, occur when the heat removal equals the heat generation, i.e., when $R(T) = G(T)$. We plot the heat removal function $R(T)$ and the heat generation function $G(T)$ against T symbolically in Figure 3.2.

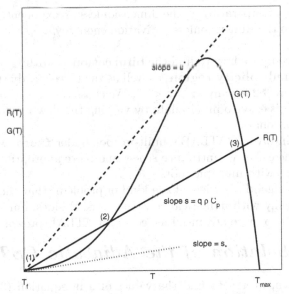

Heat generation and heat removal functions
Figure 3.2

It is clear that in a certain region of the parameter $s = q \cdot \rho \cdot C_p$, the adiabatic CSTR has three steady states with $R(T) = G(T)$, marked by (1), (2), (3) in Figure 3.2. From a steady-state analysis point of view, the two steady states (1) and (3) are stable, while the steady state (2) is unstable, as explained later; also refer to Appendix 2. Here a steady state is called a **stable steady state** if after a small disturbance of the system is introduced and removed shortly afterwards, the system returns to its original state. An **unstable steady state** is one that does not return to its original state once a small disturbance has occurred and been removed.

The varying behavior of the multiple steady states in Figure 3.2 is called **bifurcation**. The **bifurcation points** for the parameter s are determined by the tangent lines with extreme slopes s^* and s_* as depicted by the dashed and dotted lines in Figure 3.2. For any $s_* < s < s^*$ there are three steady states, while for any $s > s^*$ or for any $s < s_*$ there is only one steady-state solution of the system. And of course, for $s = s_*$ and $s = s^*$ there are precisely two steady states.

Equation (3.2) $R(T) = G(T)$ can be put into dimensionless form by dividing both sides by $q \cdot \rho \cdot C_p \cdot T_f$ to give us the equation

$$y - 1 \; = \; \alpha e^{-\gamma/y}(1 + \beta - y) \, . \tag{3.3}$$

Similarly relation (3.1) becomes $x_A = (1 + \beta - y)/\beta$ for $x_A = C_A/C_{Af}$, β, and y. Here we have set

$$y = \frac{T}{T_f}, \quad \alpha = \frac{V \, k_0}{q}, \quad \gamma = \frac{E}{R \, T_f}, \quad \text{and} \quad \beta = \frac{(-\Delta H) \, C_{Af}}{\rho \, C_p \, T_f} \, ,$$

for the dimensionless temperature y, the dimensionless preexponential factor α, the thermicity factor β, and the dimensionless activation energy γ.

We will now investigate how to find the **bifurcation points** s_* and s^*, which are the boundaries of the **multiplicity region**, as well as the three steady states marked by (1), (2), and (3) in Figure 3.2 for any $s_* < s < s^*$. We chose $\alpha = V \cdot k_0/q$ as the **bifurcation parameter** since α is easy to manipulate by varying the flow rate q. Note that α and q are inversely proportional.

Our first attempt involves MATLAB's built-in root finder `fzero`, which uses the bisection method and thereafter we introduce a new and more appropriate numerical method for solving equations with multiple roots.

This example gives a good overview of the kind of problems that chemical/biological engineers encounter daily with numerical computations. Besides, our numerical codes will introduce the reader to more advanced aspects of MATLAB programming and plotting.

Numerical Solution of the Adiabatic CSTR Problem

In our first approach, we try to find the values of y in equation (3.3) that correspond to the points labeled (1), (2), and (3) in Figure 3.2 for given values of α, β, and γ by solving (3.3) via a generic root finder such as the bisection method of `fzero` in MATLAB.

```
function [y,fy] = solveadiabxy(al,bt,ga)
%    solveadiabxy(al,bt,ga)
% Sample call    :  [y,fy] = solveadiabxy(50000,1,15)
% Sample output :   y =
```

```
% (on screen)          1.0202       1.2692       1.9595
%                  fy =
%                      4.5103e-17   1.1102e-16  -1.4433e-15ans =
% input : al, -1 <= bt <= 1, ga
% program finds the y values on all branches of the bifurcation curve.
% output : y and fy (=f(y)) "error" values (fy ~ 0)
% CAUTION: always check critical applications against a run of
%          [y,fy] = adiabNisographsol(al,bt.ga) and
%          [a1,a2] = adiabNisoauxf1(bt,ga,100) for confirmation.
%          (the solvadiabxy routine is slightly more reliable near the upper
%           bifurcation points than the lower one; the graphic solver is
%           reliable up to the bifurcation points and beyond.)
% local subroutines : adiab(x,al,bt,ga)  and Matlab's own "fzero" root finder.

warning off;                  % to satisfy the grandfather warnings re. fzero
ltol = 10^-14;                % local tolerance for fzero; change if critical
options = optimset('dis',[],'tolx',ltol);
x0 = fzero(@adiab,1,options,al,bt,ga), % call fzero from
                              % the y interval ends 1 and 1 + bt, with
                              % optimization parameters set
x1 = fzero(@adiab,1+bt,options,al,bt,ga);
if abs(x0 - x1) > 10^-6, % three roots (bifurcation) likely; check with graph
   try
        xmid = fzero(@adiab,[1.01*x0,0.99*x1],options,al,bt,ga);
                % we use the target interval [1.01*x0,0.99*x1] to find xmid
   catch, xmid = NaN; end    % dummy setting that does not plot if error
else
   xmid = NaN;                % dummy value for xmid if no bifurcation
   if x1 < 1+0.4*bt, x1 = NaN; end % taking care of the no bifurcation cases
   if x0 > 1+0.4*bt, x0 = NaN; end
end
y = [x0 xmid x1];  fy = adiab(y,al,bt,ga); % evaluate f at y (nearly zero)
warning on

function f = adiab(x,al,bt,ga)        % adiabatic function f(x) = 0 in (3.2)
% evaluates the adiabatic-non-iso function at x for given
%  al, bt, ga values. (vector version)
f = x - 1 - al*exp(-ga./x) .* (1 + bt - x);
```

Note that the above program works for both endothermic reactions ($\beta < 0$) and exothermic reactions ($\beta > 0$) and that only exothermic reactions can have multiple steady states. The built-in MATLAB root finder **fzero** finds the roots of a function f from a starting guess a if we call **fzero**($@f, a, \dots$), i.e., if we attach the "function handle" @ to f and follow this with the appropriate list of parameters in MATLAB.

Rather than work with (3.3) directly, the MATLAB m function **solveadiabxy.m** above evaluates the associated standard function

$$f(y) = y - 1 - \alpha e^{-\gamma/y}(1 + \beta - y) \qquad (3.4)$$

inside its subfunction named **adiab** at the end of the above code and searches for its zeros y with $f(y) = 0$ from judicious initial guesses. Note that the arithmetic operations inside **adiab** use the **vector MATLAB operations** ./ and .* so that **fy** can be evaluated in a single command line at the computed y vector of steady state locations on the last line of the main code.

Our initial guesses for the zeros of (3.4) are the extreme temperatures $y = 1$ and $y = 1 + \beta$ of the system, called x0 and x1 inside the program. If these two starting values for the MATLAB zero finder fzero give rise to two different solutions, then for our parameter set there must be a third solution, called xmid in between, i.e., x0 < xmid < x1. This middle solution is often hard, if not impossible to find via a bisection algorithm (or Newton's method) near the bifurcation limits. Therefore we have to be cautious with the output of the central line xmid = fzero(@adiab,[1.01*x0,0.99*x1],options,al,bt,ga) of the program. If fzero fails in its attempt to find xmid, then MATLAB sends out an error message that normally terminates the whole run. To avoid this early termination in one run of solveadiabxy.m for multiple α values in runsolveadiabxy.m, we have put the xmid = ... line of code inside a try ... catch ... end setting. This MATLAB programming tool allows us to continue computing after a command inside the try and catch bracket has failed. More specifically, if there is an error detected inside the try ... catch bracket, then the commands inside the secondary catch ... end bracket are executed instead. And specifically for our situation we simply set xmid = NaN, where NaN signifies "Not a Number" if fzero fails inside the try ... catch bracket. In MAT-LAB, NaN data entries are not plotted.

Figure 3.3 shows the typical limited range of useful output from the bisection method in the bifurcation range for this problem when $\beta = 1$ and $\gamma = 8.5$.

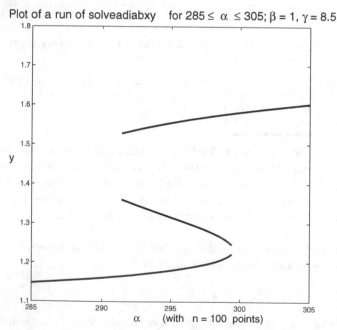

Limited bifurcation output from a bisection method

Figure 3.3

Note the large gaps in Figure 3.3 between the three branches where `fzero` fails to give us any useful output. Such data gaps near the bifurcation points are often sizable: too large to allow us to use the bisection method here. For a comparison with the graphical level-set method for this problem, see Figure 3.11.

We next discuss the MATLAB file `runsolveadiabxy.m`, which draws Figure 3.3 upon the command `runsolveadiabxy(285,305,1,8.5)`.

Note that almost all calling sequences of MATLAB function m files start with the function's name, such as `runsolveadiabxy` above, followed by a list of parameters in parentheses (...). Our particular call `runsolveadiabxy(285,305,1,8.5)` uses the interval limits 285 and 305 for α as its first two parameters, followed by the values of β and γ for a specific chemical reaction. In our m files the list of possible parameters is always explained in the first comment lines of code. Often one or several of the parameters are optional. If they are not specified in the calling sequence, they are internally set to default values inside the program, such as n and *anno* are here.

```
function Result = runsolveadiabxy(al1,al2,bt,ga,n,anno)
%     runsolveadiabxy(al1,al2,bt,ga,n)
% Sample call : runsolveadiabxy(285,305,1,8.5);
% Input : al1, al2 : the limits of alpha values to be used; al1 < al2 needed
%         bt, ga : the system parameters beta and gamma
%         n : number of linearly spaced intermediate points from al1 to al2
%         anno : if anno is set to 0, graph is not annotated/labeled;
%                default with annotation or anno = 1.

if nargin < 4, error('Not enough input data'), end
if al1 >= al2, error('alpha 2 is less than alpha 1; cancel run'), end
if nargin == 4, n = 100; anno = 1; end      % default values for n and anno
if nargin == 5, anno = 1; end
                                            % initial settings :
step = (al2-al1)/n; Result = zeros(n+1,3); AL = [al1:step:al2]';
for i=0:n,                                  % sweep through alpha values
   [y,fy] = solveadiabxy(al1+i*step,bt,ga);  % solve and store data
   Result(i+1,:) = y; end

plot(AL,Result(:,1),'k'), hold on, % plot
plot(AL,Result(:,2),'r'),plot(AL,Result(:,3),'g'), hold off, % Plot labels :
if anno == 1            % comment this title command for Fig 3.11
  title(['Plot of a run of solveadiabxy   for ',num2str(al1,'%10.5g'),...
     ' \leq \alpha \leq ',num2str(al2,'%10.5g'),'; \beta = ',...
     num2str(bt,'%10.5g'),', \gamma = ',num2str(ga,'%10.5g')],'FontSize',14)
  xlabel(['                    \alpha      (with n = ',num2str(n,'%10.5g'),...
     ' points)'],'FontSize',12),
  ylabel('y','Rotation',0,'FontSize',12), end
```

`runsolveadiabxy.m` takes the output from a sequence of runs of `solveadiabxy` for increasing values of α and stores it in an $n+1$ by 3 matrix called `Result`. Each row of `Result` contains three entries: that of the bottom branch of the solution to $f(y) = 0$ in equation (3.4) in column 1, followed by the middle branch solutions in column 2 and those of the top branch in column 3. Many of these data entries may be set equal to NaN inside `solveadiabxy.m` as warranted, and then these points will not be plotted. The (up to three) solution branches for equation (3.4) are then plotted in different colors in three separate plotting commands inside `runsolveadiabxy.m`.

The command `hold on` after the first `plot` command ensures that all plots appear in the same graph. The command `hold off` at the end of our multiple plottings makes sure that the picture of one `solveadiabxy` run is not preserved in the figure window and that subsequent `solveadiabxy` plots for different data are not appended to the earlier plots, but rather drawn by themselves.

If annotations are desired, then the last three commands, `title`, `xlabel`, and `ylabel`, write the relevant descriptions onto the margins of the graph, as seen in Figure 3.3. Note the `'FontSize'` specifications, as well as the way of associating variable names such as β with their numerical value via `num2str(bt,'%10.5g')` and the set of square brackets around the mixed text/numerical annotations in the `title`, `xlabel` and `ylabel` lines. Finally, note the use of LATEX language such as `\gamma` to generate the Greek γ.

Since ordinary zero finders fail us often in root-finding problems with multiple roots, we now set out to develop a more reliable **graphical level-set method** for finding all y values of the solutions to (3.4) for any range of α parameters.

The MATLAB function `adiabNisoplot.m` finds the zero crossings of the graph of f in (3.4) graphically. Graphs are generally very helpful when trying to solve equations since they help us visualize the points of intersection of the function f in question with the x-axis, that is, where $f = 0$.

```
function adiabNisoplot(N,al,bt,ga,book)
%    adiabNisoplot(N,al,bt,ga)
% Sample call  : adiabNisoplot(100,1300,.8,10)
% Input: N : number of intermediate points
%        al, bt, ga  are the system parameters
%        book : If "book" is set to 1, we plot for display in the textbook with
%               wider lines and larger fonts. Default is book = 0.
% Plots adiabatic transcendental function from N nodes and given al, bt, ga.
% Looking for zero crossings.

if nargin == 4, book = 0; end    % default setting: narrow lines, small fonts

y=1:(bt/N):1+bt; % create y nodes; then plot adiab (3.2) curve and horiz. axis
if book == 1,    % for use in textbook: wider lines, larger fonts
  plot(y,adiabNiso(y,al,bt,ga),'b','LineWidth',2), hold on, % adiab (3.2) curve
  plot([1,1+bt],[0,0],'-r','LineWidth',2),                  % horizontal axis
  xlabel([' y      ( N = ',num2str(N,'%9.4g'),' )'],'FontSize', 14);
  ylabel(' adiab function ','FontSize', 14);
  title([' Adiab CSTR function plot of (3.2) ;   \alpha = ',...
        num2str(al,'%9.4g'),' , \beta = ',num2str(bt,'%9.4g'),...
        ' , \gamma = ',num2str(ga,'%9.4g')],'FontSize', 14); hold off
else             % for ordinary screen display
  plot(y,adiabNiso(y,al,bt,ga),'b'), hold on, % adiab (3.2) curve
  plot([1,1+bt],[0,0],'-r'),                  % horizontal axis
  xlabel([' y      ( N = ',num2str(N,'%9.4g'),' )']);
  ylabel(' adiab function ');
  title([' Adiab CSTR function plot of (3.2) ;   \alpha = ',...
        num2str(al,'%9.4g'),' , \beta = ',num2str(bt,'%9.4g'),...
        ' , \gamma = ',num2str(ga,'%9.4g')]); hold off; end
```

The call `adiabNisoplot(100,1300,.8,10)` with $N = 100$ intermediate points and $\alpha = 1300$, $\beta = 0.8$, and $\gamma = 10$ for example gives rise to the plot in Figure 3.4.

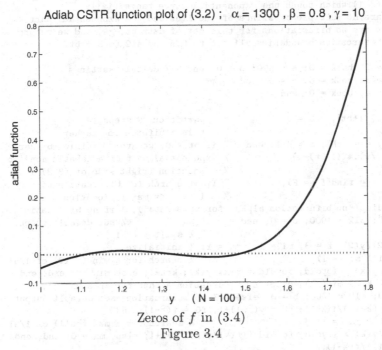

Zeros of f in (3.4)

Figure 3.4

Looking at the three shallow intersections of the horizontal axis with the graph of f in Figure 3.4, we are reminded of the problems encountered in Chapter 1 on p. 30 and 31 with both the bisection and the Newton root finder for polynomials with repeated roots. The common wisdom is that the shallower these intersections become, the worse the roots will be computed by standard root-finding methods (see the exercises below), and multiple roots will easily be missed.

Let us therefore proceed by a different route, using a **level-set method** approach that is ultimately more promising for this problem. We proceed in two stages.

First we determine the range of α for which there are multiple steady states. We do this semigraphically by using Figure 3.2 as our guide in the following code.

```
function [al1,al2] = adiabNisoauxf1(bt,ga,N,p,book)
%     [al1,al2] = adiabNisoauxf1(bt,ga,N,p)
% Sample call   :  [al1,al2] = adiabNisoauxf1(1,15,500,1)
% Auxiliary function for adiabatic Non-isothermal CSTR equation.
% Input : beta (bt), gamma (ga), # of steps in y direction = N, N <= 1000
%         typical input values: 0.1 <= bt <= 18; 5 <= ga <= 50
%         [larger values of N make the bifurcation start and end points al1
%          and al2 come out more precisely.]
%         p = 1: plot the exponential part of the curve; p ~= 1: do not plot
%         If book = 1, we plot with wide lines and big fonts for the
%         textbook. Default: book = 0, or unspecified.
```

```
% Output: limits of bifurcation for bt, ga: al1 and al2; N may be reduced.
% Here we evaluate the exponential part of the equation and find the
% min/max slopes 1 <= al1 < al2 <= 1+bt of 1/al(y-1) = exponential part of
% equation (3.1) which touch the exponential curve tangentially.
% If there are no multiple crossings of the line and the exponential part,
% then there are no bifurcations for this set of data bt, ga, and we return
% with the bifurcation boundaries al1 set to 0.5 and al2 to 3 + bt.

if nargin == 2, book = 0; N = 500; p = 0; end   % default settings
if nargin == 3, book = 0; p = 0; end
if nargin == 4, book = 0; end

y = [1:(bt/N):1+bt];    b = 1;          % y partition, N steps;
                                        %  b is a bifurcation marker
if bt < 0, y = y(1:N); N = N-1; end   % if bt < 0, we avoid division by zero
f = exp(-ga./y).*((1+bt)-y);          % exponential part of adiabatic non-iso
                                      %  equation (right side of (3.3) )
F = max(f); i = find(f == F);         % limit search for bifurcation to
                                      %  1 <= y <= max(f) location
if i == 1, disp('no bifurcation alphas for this data'), % if no bifurcation
   al1 = 50; al2 = 1000; b = 0; end                     % set default output
y = y - 1;                            % shift y by 1
foyold = f(2)/y(2); k = 3; mi = 1;  ma = 1; % initialize
while (k < i & mi == 1)               % search for a min (=al2) of 1/al
   if f(k)/y(k) < foyold, foyold = f(k)/y(k); k=k+1; else mi = 0; end, end,
if (k >= N | k >= min(i)), disp('no bifurcation alphas for this data'),
   al1 = 50; al2 = 1000; b = 0; else, % no bifurcation: set default output
   al2 = (y(k-1))/f(k-1); y0 = y(k-1)+1; f0 = f(k-1); end
while (k < i & ma == 1)               % search for a max (=al1) of 1/al
   if f(k)/y(k) > foyold, foyold = f(k)/y(k); k=k+1; else, ma = 0; end, end
al1 = (y(k-1))/f(k-1);
if p == 1,                            % if plot is desired
   if book == 1,
     plot(y+1,f,'-r','LineWidth',2), hold on,
     ylabel('(y-1) / \alpha          ','Rotation',0,'Fontsize',14),
   else
     plot(y+1,f,'-r'), hold on,
     ylabel('(y-1) / \alpha          ','Rotation',0,'Fontsize',14), end
   if b == 1,                         % in case of bifurcation:
     plot(y0,f0,'o'), hold on,        %   plot bottom tangent
     if book == 1,
     text(1+2*bt/3,2*f0*bt/(3*(y0-1)),' slope = 1/\alpha_2','Fontsize',12)
     plot([1 1+2*bt/3],[0 2*bt*f0/(3*(y0-1))],':','LineWidth',2)
     plot(y(k-1)+1,f(k-1),'+'),
     plot([1 1.2*y(k-1)+1],[0 1.2*f(k-1)],'--','LineWidth',2),
    text(.73*y(k-1)+1,f(k-1),'slope = 1/\alpha_1','Fontsize',12) % top tangent
     xlabel([' y  from 1 to ',num2str(1 + bt,'%10.5g'),...   % make note of
 ' ;  bifurcation,  N = ',num2str(N,'%6.5g')],'Fontsize',14), % bifurcation
     title({['       adiabNisoauxf1 : extreme \alpha :']},{['  \alpha_1 = ',...
           num2str(al1,'%10.5g'),',  \alpha_2 = ',num2str(al2,'%10.5g'),...
            ' ;  \beta = ',num2str(bt,'%10.5g'),',  \gamma = ',...
           num2str(ga,'%10.5g'),'here']},'Fontsize',14),
   else
     text(1+2*bt/3,2*f0*bt/(3*(y0-1)),' slope = 1/\alpha_2')
     plot([1 1+2*bt/3],[0 2*bt*f0/(3*(y0-1))],':')
     plot(y(k-1)+1,f(k-1),'+'),
```

```
      plot([1 1.2*y(k-1)+1],[0 1.2*f(k-1)],'--'), % plot top tangent spot ->
      text(.73*y(k-1)+1,f(k-1),'slope = 1/\alpha_1')
      xlabel([' y  from 1 to ',num2str(1 + bt,'%10.5g'),...   % make note of
  ' ;   bifurcation,  N = ',num2str(N,'%6.5g')],'Fontsize',14), % bifurcation
      title({'          adiabNisoauxf1 : extreme \alpha :',['  \alpha_1 = ',...
             num2str(al1,'%10.5g'),',  \alpha_2 = ',num2str(al2,'%10.5g'),...
             ' ;  \beta = ',num2str(bt,'%10.5g'),',  \gamma = ',...
             num2str(ga,'%10.5g')]},'Fontsize',14), end
else,
    if book == 1,
    xlabel([' y  from 1 to ',num2str(1 + bt,'%10.5g'),...   % or note no
  ' ;   No bifurcation,  N = ',num2str(N,'%6.5g')],'Fontsize',14) % bifurcation
      title(['       adiabNisoauxf1 :  no bifurcation \alpha values for  ',...
             ' \beta = ',num2str(bt,'%10.5g'),',  \gamma = ',...
             num2str(ga,'%10.5g')],'Fontsize',14),
    else
    xlabel([' y  from 1 to ',num2str(1 + bt,'%10.5g'),...   % or note no
  ' ;   No bifurcation,  N = ',num2str(N,'%6.5g')],'Fontsize',14) % bifurcation
      title(['       adiabNisoauxf1 :  no bifurcation \alpha values for  ',...
             ' \beta = ',num2str(bt,'%10.5g'),',  \gamma = ',...
             num2str(ga,'%10.5g')],'Fontsize',14), end
end, hold off, end
```

This code relies on a slight variant of equation (3.3), namely

$$\frac{1}{\alpha}(y-1) = e^{-\gamma/y}(1+\beta-y) \quad (=\tilde{f}(y)) . \tag{3.5}$$

The left-hand side of (3.5) describes a line with slope $1/\alpha$ in the variable y with the point $(y,\tilde{f}) = (1,0)$ on its graph. The right-hand side is an exponential function in y with $(y,\tilde{f}) = (1,e^{-\gamma} \cdot \beta)$ on its graph for $e^{-\gamma} \cdot \beta > 0$.

The algorithm of adiabNisoauxf1.m initially limits the search for tangents from $(1,0) \in \mathbb{R}^2$ to the graph of the exponential curve $\tilde{f}$ to y values below the maximum of the right-hand side of (3.5). It then slides backward along the curve, computing the maximal and minimal ratios of $\tilde{f}$ and y, i.e., it computes the two extreme slopes of tangent lines to the graph of $\tilde{f}$ that pass through the point $(1,0)$. The reciprocals of the extreme slopes then give us the extreme parameters al1 and al2 as the bifurcation points. If the input parameter $p = 1$ is specified as the last variable in a MATLAB call of [al1,al2] = adiabNisoauxf1(1,15,400,1), for example, then we draw the plot as in Figure 3.5, on which the bifurcation limits are displayed numerically on screen and in the plot's title line, as well as drawn out on the graph.

```
>> [al1,al2] = adiabNisoauxf1(1,15,500,1); extreme_alphas = [al1,al2]
extreme_alphas =
       15757         93777
```

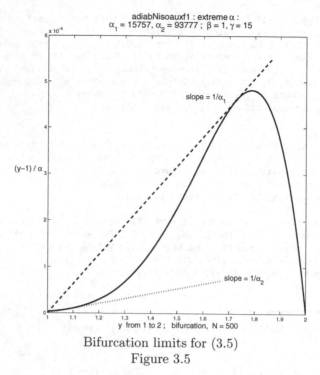

Bifurcation limits for (3.5)
Figure 3.5

If the input parameter p of `adiabNisoauxf1` is equal to 1, then `adiabNisoauxf1` outputs the bifurcation limits α_1 and α_2 only in the workspace and does not plot a figure.

Secondly, having found the bifurcation points α_1 and α_2, we then plot the 3D surface

$$z \;=\; F(\alpha, y) \;:=\; y - 1 - \alpha e^{-\gamma/y}(1 + \beta - y) \tag{3.6}$$

as a function of two variables α and y for $0.9 \cdot \alpha_1 \le \alpha \le 1.1 \cdot \alpha_2$ and $1 \le y \le 1 + \beta$. This includes the parameter range of α in which there is bifurcation. Equation (3.6) with $z = F(\alpha, y) = 0$ reinterprets our earlier equation $f(y) = 0$ in (3.4) and replaces our 1-dimensional root-finding attempts by a two-dimensional approach to root-finding in the two variables α and y.

The bifurcation curve relating y and α is the projection of the level curve $F(\alpha, y) = 0$ onto the α-y plane, where $z = 0$. Our 3D plot of the surface $z = F(\alpha, y)$ in `adiabNisosurf` contains this level curve marked in black on the surface, as well as a second, isolated plot of it below the surface on the ground plane: see Figure 3.6.

```
function adiabNisosurf(a1,a2,bt,ga,N)
%    adiabNisosurf(a1,a2,bt,ga,N)
% Sample call   : adiabNisosurf(15000,95000,1,15,100)
% Plots the adiab equation surface and the contour = 0 curve below it.
% Input : limiting values for alpha [al1, al2] (supplied by adiabNisoauxf1),
%         beta, gamma, N = # of nodes (N = 80 or 100 ok)

if nargin == 4, N = 100; end,
```

```
[y,al] = meshgrid([1:bt/N:1+bt],[0.9*a1:(1.1*a2-0.9*a1)/N:1.1*a2]);
                % make grids for y and alpha in relevant ranges for y and beta
z = adiabNiso(y,al,bt,ga);          % z = adiab Non-iso function value
h = surf(al,y,z); hold on;
shading interp; colormap(hsv(128)); colorbar,% plot surface
a = get(gca,'zlim'); zpos = a(1);   % Always put 0 contour below the plot
[cc,hh] = contour3(al,y,z,[0 0],'-k'); % draw zero contour data on surface

[ccc,hhh] = contour3(al,y,z,[0 0],'-b'); % put 0 contour data at the bottom
for i = 1:length(hhh)                % size zpos to match the data
    zzz = get(hhh(i),'Zdata');
    set(hhh(i),'Zdata',zpos*ones(size(zzz))); end

xlabel('\alpha','FontSize',14),         % annotate top 3D plot
ylabel('y','Rotation',0,'FontSize',14),
zlabel('adiabNiso(y,\alpha,\beta,\gamma )','FontSize',12),
title([' adiabNisosurf 3D plot :    \alpha_1 = ',num2str(a1,'%10.5g'),...
     ', \alpha_2 = ',num2str(a2,'%10.5g'),' ;   \beta = ',...
     num2str(bt,'%10.5g'),', \gamma = ', num2str(ga,'%10.5g')],...
     'FontSize',12), hold off
```

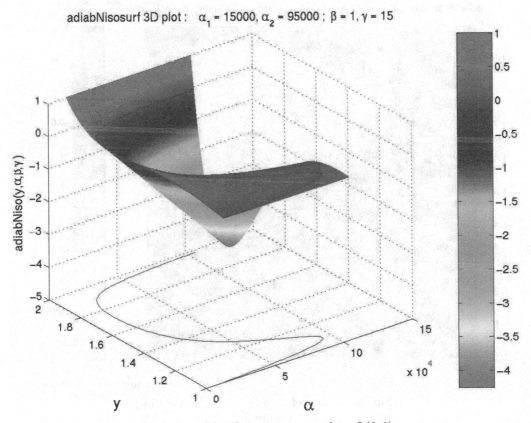

Surface and level-zero contour plot of (3.6)

Figure 3.6

Figure 3.6 depicts the 3D surface $z = F(\alpha, y) = y - 1 - \alpha \cdot e^{-\gamma/y}(1 + \beta - y)$ as introduced in equation (3.6). This surface depends on the two parameters α and y. It is colored[1] by our `adiabNisosurf.m` file as a geographical map would be. Namely, the actual color of any spot $(\alpha, y, z) \in \mathbb{R}^3$ on this surface depends on the height or the level of $z = y - 1 - \alpha e^{-\gamma/y}(1 + \beta - y)$ above the α-y plane. In this particular graph, the "height" of $z = F(\alpha, y)$ ranges between around -4 and $+1$ in the surface plot and in the associated colorbar on the right side of Figure 3.6. Additionally we draw the intersection of the level-zero plane $(\alpha, y, 0)$ of $\mathbb{R}^3$ with the surface $z = F(\alpha, y)$ in black on the surface itself. This curve is called the **level-zero contour** or **level-zero curve** of the surface. On the ground plane of our 3D surface plot we draw this curve separately in **blue**. The level-zero curve or contour depicts all the solutions to our original adiabatic CSTR equation (3.3) and thus it solves our original adiabatic CSTR problem in the contemplated parameter range.

Next we draw a grayscale contour map of this surface, as well as the level curve $F(\alpha, y) = 0$ drawn in black using the MATLAB function `adiabNisocolorcontour.m`. On a computer screen the same coloring scheme as in Figure 3.6 is used.

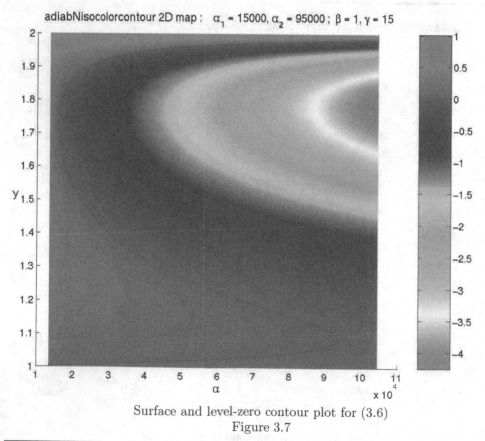

Surface and level-zero contour plot for (3.6)

Figure 3.7

---

[1]References to color always refer to a computer-generated color graph of the figure in question.

Figure 3.7 is drawn by calling `adiabNisocolorcontour(15000,95000,1,15,100)`. The code of the contour-plotting MATLAB m file `adiabNisocolorcontour.m` follows below.

```
function adiabNisocolorcontour(a1,a2,bt,ga,N)
%    adiabNisocolorcontour(a1,a2,bt,ga,N)
% Sample call   : adiabNisocolorcontour(15000,95000,1,15,100)
% Plots the adiab equation surface as a 2D contour map with the contour = 0
% curve marked in black.
% Input : limiting values for alpha [al1, al2] (supplied by adiabNisoauxf1),
%         beta, gamma, N = # of nodes (N = 80 or 100 ok)

if nargin == 4, N = 100; end,

[y,al] = meshgrid([1:bt/N:1+bt],[0.9*a1:(1.1*a2-0.9*a1)/N:1.1*a2]);
             % make grids for y and alpha in relevant ranges for y and beta
z = adiabNiso(y,al,bt,ga);          % z = adiab Non-iso function value
h = surface(al,y,z); hold on; shading interp; colormap(hsv(128)); % 2D map
colorbar
[cc,hh] = contour3(al,y,z,[0 0],'-k'); % put 0 contour curve onto map

xlabel('\alpha','FontSize',12),        % annotate 2D color map
ylabel('y','Rotation',0,'FontSize',12),
title([' adiabNisocolorcontour 2D map :   \alpha_1 = ',num2str(a1,'%10.5g'),...
      ', \alpha_2 = ',num2str(a2,'%10.5g'),' ; \beta = ',...
      num2str(bt,'%10.5g'),', \gamma = ', num2str(ga,'%10.5g')],...
  'FontSize',12), hold off
```

The last two codes contain many intricate and useful plotting and contour commands that are self-explanatory when one uses the MATLAB `help` ... function for the MATLAB graphics commands `meshgrid`, `surface`, `contour3`, `xlabel`, `ylabel`, `title`, `colormap`, etc. Students should study these graphics commands of MATLAB in order to be learn how to display the easily computed numerical data well. Please refer to MATLAB `help` ....

In Figures 3.6 and 3.7, note the steep slopes of the surface $z = F(\alpha, y)$ near $y = 1 + \beta = 2$ and the relatively shallow slopes near $y = 1$. Such slope disparities make great numerical obstacles and they ultimately contribute to the failure of simple root finders.

Figures 3.5, 3.6, and 3.7 are combined into a single figure with three plot windows via the m file `runadiabNiso.m` in Figure 3.8.

```
function runadiabNiso(bt,ga,N,p)
%    runadiabNiso(bt,ga,N,p)
% Sample call   : runadiabNiso(1,15,500,1)
% runs all three adiabNiso plots:
% Input : -1 < bt < 2; 2 < ga < 25; N (= 60, 100, or 500); p (= 0 or 1)
% first determine the bifurcation points a1 and a2, in adiabNisoauxf1.m;
% then the surface plot, and the bifurcation curve, if any,
% in adiabNiso3dplot.m.
% For standard values of bt and ga, set N = 60, 100, 500 or 1,000 for even
% greater accuracy of the bifurcation limits a1 and a2.
% Set p = 1 for plotting.

if nargin == 2, N = 100; p = 0; end        % default settings
if nargin == 3, p = 0; end
if bt <=-1,
```

```
    disp('ALERT : bt is not larger than -1, reject the data; NO computations'),
       return, end
   subplot(3,1,1);
   [a1, a2] = adiabNisoauxf1(bt,ga,N,p);                % plot adiab graph on top
   N = min(N,100); adiabNiso3dplot(a1,a2,bt,ga,N);      % limit partition lengths
          % plot surface and 0 level contour for a limited value of N (for  speed)
```

A call of `runadiabNiso(1,15,500,1)` gives numerical and graphical output as shown in Figure 3.8 for the associated CSTR problem.

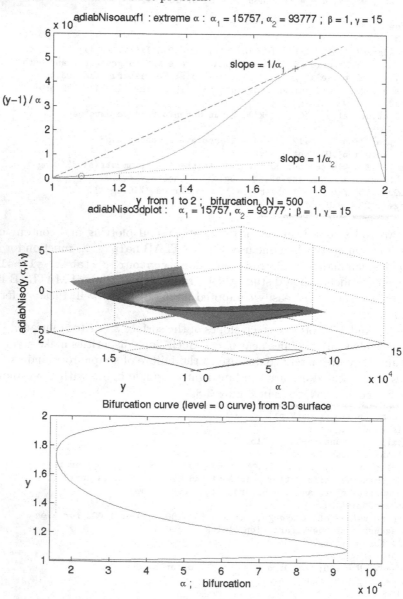

Combined function, surface, and level-zero contour plot for (3.6)

Figure 3.8

The m function `runadiabNiso` relies on `adiabNiso3dplot`, which combines our $z = F(\alpha, y)$ surface plot and a simplified version of the level-zero contour plot.

Our next code and plot draws only the bottom curve of Figure 3.8 by itself from the inputs α, β, and γ. It uses `adiabNisoauxf1` to find the actual bifurcation points of the curve, depicts them by the dotted vertical lines in the figure window, and then plots the bifurcation curve itself. This plot gives the most practical output in our view.

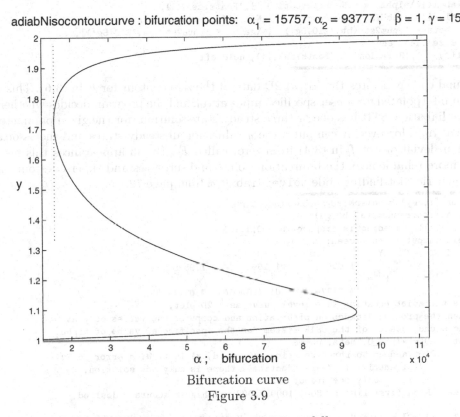

Bifurcation curve

Figure 3.9

The plot-generating MATLAB code for Figure 3.9 is as follows:

```
function adiabNisocontourcurve(bt,ga)
%    adiabNisocontourcurve(bt,ga)
% Sample call   : adiabNisocontourcurve(1,15)
% Plots the adiab equation zero level curve as a 2D plot
% Input : beta, gamma

N = 500; [a1,a2] = adiabNisoauxf1(bt,ga,N); % find bifurcation points a1, a2

[y,al] = meshgrid([1:bt/N:1+bt],[0.8*a1:(1.2*a2-0.8*a1)/N:1.2*a2]);
%[y,al] = meshgrid([1.1:0.6/N:1.7],[285:20/N:305]); % For Figure 3.11
           % make grids for y and alpha in relevant ranges for y and beta
z = adiabNiso(y,al,bt,ga);             % z = adiab Non-iso function value
contour(al,y,z,[0 0],'b'); hold on,    % draw 0 level curve
```

```
if a2 ~= 1000,                         % in case of bifurcation
   xlabel(['\alpha ;    bifurcation' ],'Fontsize',14),
   title([' adiabNisocontourcurve : bifurcation points:   \alpha_1 = ',...
          num2str(a1,'%10.5g'),', \alpha_2 = ',num2str(a2,'%10.5g'),...
          ' ;   \beta = ',num2str(bt,'%10.5g'),', \gamma = ',...
          num2str(ga,'%10.5g')],'Fontsize',14),
   plot([a1,a1],[1+bt/3,1+0.98*bt],':')   % at left bifurcation a1
   plot([a2,a2],[1.02,1+bt/2.4],':')      % at right bifurcation a2
else,                                  % in case of no bifurcation
   xlabel(['\alpha ;    No bifurcation '],'Fontsize',14),
   title([' adiabNisocontourcurve :   No  bifurcation points ;',...
 '   \beta = ',num2str(bt,'%10.5g'),', \gamma = ',num2str(ga,'%10.5g')],...
 'Fontsize',14), end
ylabel('y   ','Rotation',0,'Fontsize',14), hold off
```

Our final code generates the (α, y) 2D data of the zero contour for F in (3.6). This data is then interpolated for a user-specified input at α_0, and the program decides whether the given adiabatic CSTR has one or three steady state solutions for the given parameters β and γ at α_0. Moreover, it computes the y values for all steady states and the associated function deviations of f in (3.4) from zero, called Fy. By all appearances this method is far more reliable near the bifurcation points and surpasses and supersedes our initial more generic root-finding code solveadiabxy.m from page 72.

```
function [Y, Fy] = adiabNisographsol(a0,bt,ga)
%    adiabNisographsol(a0,bt,ga)
% Sample call   : adiabNisographsol(50000,1,15)
% Sample output :  on screen:
%              Y =
%                    1.0202        1.2692        1.9595
%              Fy =
%                    4.3927e-07   6.5708e-06  -6.6714e-05
% Plots the adiab equation zero level curve as a 2D plot,
%    then the program decides on bifurcation and computes the values of y at
%    the steady states of the CSTR system for the specified a0 value of alpha.
% Input : a0 = alpha_0, beta, gamma
% Output: Plot and up to three solutions x0, xmid, x1 in Y. With error in Fy.
%              (if entries in Y are identical, there is only one solution,
%               i.e., only one steady state for this input.)
% Increase N on first line to 800, 1000, 2000, if higher accuracy desired

N = 500; [a1,a2] = adiabNisoauxf1(bt,ga,N); % find bifurcation points a1, a2
a11 = min(a1,a0); a22 = max(a2,a0);         % extend alpha range to contain a0
three = 0;                                  % triple steady states ?
[y,al] = meshgrid([1:bt/N:1+bt],[0.8*a11:(1.2*a22-0.8*a11)/N:1.2*a22]);
              % make grids for y and alpha in relevant ranges for y and beta
z = adiabNiso(y,al,bt,ga);                  % z = adiab Non-iso function value
C = contour(al,y,z,[0 0],'b'); hold on,     % draw 0 level curve
if a1 < a2,
  yy = find(C(1,1:end-1)-C(1,2:end) > 0);
  if length(yy) == 0, itop = length(C); ibot = 1;
  else                                      % find left bifurcation point
     itop = yy(1); [bull,ibot] = min(C(1,itop:end));
     ibot=ibot+itop-1; end                  % add the front end length!
  if a1 <= a0 & a2 >= a0 & length(yy) > 0, three = 1; % in case of multiplicity
     x0 = interp1(C(1,1:itop),C(2,1:itop),a0,'spline'); % bottom solution (1)
```

```
       xmid = interp1(C(1,itop:ibot),C(2,itop:ibot),a0,'spline');  % middle (2)
       x1 = interp1(C(1,ibot:end),C(2,ibot:end),a0,'spline');      % top sol (3)
     elseif a0 < a1,
       x0 = interp1(C(1,1:itop),C(2,1:itop),a0,'spline'); % only one steady state
       xmid = x0; x1 = x0;
     else x0 = interp1(C(1,ibot:end),C(2,ibot:end),a0,'spline'); % one steady st.
       xmid = x0; x1 = x0; end,
   else x0 = interp1(C(1,1:end),C(2,1:end),a0,'spline');  % no bifurcation;
       xmid = x0; x1 = x0; end,                           % only one steady state
   Y = [x0 xmid x1]; Fy = adiab(Y,a0,bt,ga);              % prepare output
   if a2 ~= 1000,                         % in case of bifurcation
     plot(a0,1,'+r')                      % mark requested value for alpha_0
     plot([a1,a1],[1+bt/3,1+0.98*bt],':') % plot left bifurcation point a1
     plot([a2,a2],[1.02,1+bt/2.4],':')    % plot right bifurcation point a2
     if three == 1;                       % mark 3 steady states
         xlabel(['\alpha        (three steady states for \alpha_0 = ',...
             num2str(a0,'%12.9g'),' , \beta = ',num2str(bt,'%10.5g'),...
             ', \gamma = ',num2str(ga,'%10.5g'),')'],'FontSize',12),
         title([' adiabNiso graphical solution  :   y_1 = ',...
             num2str(x0,'%10.6g'),';   y_2 = ',num2str(xmid,'%10.6g'),...
             ' ;    y_3 = ',num2str(x1,'%10.6g')],'FontSize',12),
         plot(a0,x0,'+r'), plot(a0,xmid,'+r'), plot(a0,x1,'+r'),
     else                                 % mark single steady state
         xlabel(['\alpha        (single steady state for \alpha_0 = ',...
             num2str(a0,'%12.9g'),' , \beta = ',num2str(bt,'%10.5g'),...
             ', \gamma = ',num2str(ga,'%10.5g'),')'],'FontSize',12),
         title([' adiabNiso graphical solution  :    y_1 = ',...
             num2str(x0,'%10.6g')],'FontSize',12),
   plot(a0,x0,'+r'), end % mark single steady state
   else,                                  % in case of no bifurcation
       xlabel(['\alpha        (single steady state for \alpha_0 = ',...
             num2str(a0,'%12.9g'),' , \beta = ',num2str(bt,'%10.5g'),...
             ', \gamma = ',num2str(ga,'%10.5g'),')'],'FontSize',12),
       title([' adiabNiso graphical solution  :   y_1 = ',...
             num2str(x0,'%10.6g')],'FontSize',12),
     plot(a0,x0,'+r'),  plot(a0,1,'+r'), end  % mark single steady state
   ylabel('y','Rotation',0,'FontSize',12), hold off

   function f = adiab(x,al,bt,ga)       % adiabatic function f(x) = 0 in (3.2)
   % evaluates the adiabatic-non-iso function at x for given
   %  al, bt, ga values. (vector version)
   f = x - 1 - al*exp(-ga./x) .* (1 + bt - x);
```

The screen output and plot of calling `>> [y,fy]=adiabNisographsol(50000,1,15)` are
as follows:

```
>> [y, fy]=adiabNisographsol(50000,1,15)
y =
      1.0202       1.2692       1.9595
fy =
   4.3927e-07   6.5708e-06  -6.6714e-05
```

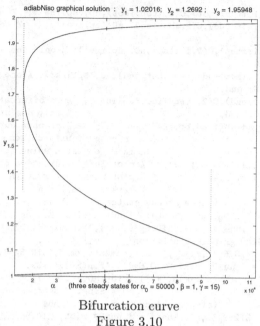

Bifurcation curve
Figure 3.10

Note that the graph itself includes the computed y values for the steady states marked by + signs thrice and that their respective y values are written out on the title line. The user-specified parameter values of α_0, β, and γ are given below the plot in Figure 3.10.

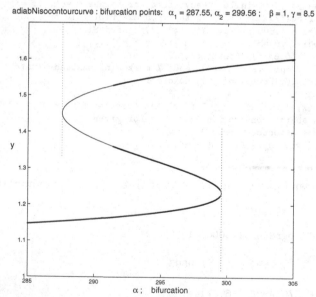

Bifurcation curve: graphical method output as a thin line; bisection output superimposed as a thick line
Figure 3.11

To finish this section, we have revisited the earlier bisection algorithm `solveadiabxy.m` in Figure 3.11. To illustrate we superimpose the curve data from Figure 3.3 as a thick curve on top of the level-set method's graphics output depicted by a thin line in Figure 3.11. This figure is drawn by the command sequence `adiabNisocontourcurve(1,8.5)`, `hold on`, `runsolveadiabxy(285,305,1,8.5)`; after the `meshgrid` line near the top of `adiabNisocontourcurve.m` has been altered for Figure 3.11 as indicated inside the program.

Figure 3.11 makes it obvious that the level-set method for equation (3.6) gives much more meaningful numerical results and clearer graphical representations of the multiple steady state solutions of the CSTR problem (3.3).

Figure 3.11 contains several independent validations of our graphical methods: The bifurcation points are determined graphically to be $\alpha_1 = 287.55$ and $\alpha_2 = 299.56$ using `adiabNisoauxf1.m` and equation (3.5). These limits of bifurcation are indicated in Figure 3.11 by the two dotted vertical lines. The thin part of the graph of the temperature function $y(\alpha)$ is determined in `adiabNisocontourcurve.m` as the level-zero curve for the surface $z = F(\alpha, y)$ as defined in equation (3.6) by interpolating the α-y-F 3D function data. A visual inspection shows a perfect match of the bifurcation points and the extremes of the function excursions. Moreover, the graphical data and the bisection algorithm data match perfectly on the part of the graph where bisection data is available.

In conclusion, the graphical output of the level-set method is far superior to the bifurcation data obtained via any of the more standard root-finding algorithms.

We shall use the same graphical approach again in the next section for the nonadiabatic CSTR problem.

Exercises for 3.1

1. Fix $\beta = 1$ and find the range of γ values for which there is bifurcation for the associated adiabatic CSTR problem by running `runadiabNiso.m` with N set to 400 and to 40, with and without plotting.

2. Find the maximal and minimal α values experimentally, correct to 5 digits, for which `solveadiabxy(al,1,15)` determines bifurcation correctly.

3. Repeat Problem 2 for `adiabNisographsol` and compare.

4. Test `adiabNisographsol.m` for $\beta = 1$, $\gamma = 8$, and many positive α_0. Watch the labels and title, as well as the red + mark(s) of the steady states.

5. Repeat Problem 4 for $\beta = 1$, $\gamma = 18$, and $\alpha_0 = 80,000$; $200,000$; and $8,000,000$. Watch all plot annotations and make sure you understand the label and title trees inside the program well.

6. View the surfaces defined in (3.6) via `adiabNisosurf.m` for $\beta = 1$ and $\gamma = 8$ or $\gamma = 18$ and various values of $a1$ and $a2$. How do the two surfaces differ for the same values of α? How do they change with changing α values?

7. Repeat the previous problem with `adiabNisocolorcontour.m` instead.

Conclusions

The kind of bifurcation that we have just encountered is called **static bifurcation** (SB). This behavior is very common in many chemical/biological processes. A limited number of industrial examples of this behavior include:

1. Fluid catalytic cracking (FCC) units for producing high-octane gasoline from gas-oil or naphta: see Section 7.2.

2. The UNIPOL® process for polyethylene and polypropylene production using the Ziegler-Natta[2] catalyst: see Section 7.3.

3. Catalytic oxidation of carbon monoxide CO in car exhaust reactors.

4. Ethanol production by fermentation at high substrate concentrations: see Section 7.6.

Understanding such static bifurcation behavior is very important for the design, optimization, startup, operation, and control of the system. The bifurcation points determine the boundaries of the region where multiple steady states exist. Even if the design and operating parameters of a system cause it to operate outside of the multiplicity region, detailed knowledge of its multiplicity region is very important for startup and control purposes. This is so because a system that operates in its unique steady state region but near its multiplicity region is not like a system in which bifurcation does not exist. A unique steady state system operating near the multiplicity region may be exposed to a disturbance during operation (or startup) that moves its operating parameters into the multiplicity region. This complicates the system's behavior and imposes the features of the multiplicity region on the system's design, operation, and control.

On the other hand, if the chosen design and operating conditions determine that the system is operating in the multiplicity region, then the system is completely affected according to the following underlying principles:

1. **Which steady state to choose**: the high-temperature one (with y near $1 + \beta$), the middle one, or the low-temperature one (with y near 1)?

   For a simple reaction

   $$A \Rightarrow B$$

   the high-temperature steady state is a likely choice from a process point of view (i.e., with respect to conversion and productivity). But the associated temperature may be too high for the construction material of the reaction vessel. In this case one might be forced to operate at the middle or the low-temperature steady states. But the low-temperature steady state may be at too low a temperature (quenched reaction) or gives too low a conversion to be economical. Thus one may be forced to operate at the middle steady state, which unfortunately is unstable and requires stabilization through control. For a more detailed analysis, we refer to our discussions on p. 117 following Figure 3.29 in Section 3.3.

---

[2]Karl Ziegler, German chemist, 1898-1973
Giulio Natta, Italian chemist, 1903-1979

2. **More complex reactions**, such as

$$A \Rightarrow B \Rightarrow C$$

where B is the desired product.

Here it is often the case that the high-temperature steady state uses a temperature that is too high and does not produce enough of the desired intermediate product B; in fact the system often overreacts all reactants directly to C, such as in fluid catalytic cracking. On the other hand, the low temperature steady state generally corresponds to too low a conversion. Thus the middle, unstable steady state is often the only productive choice, and it yields the maximal amount of product B. A typical case is that of fluid catalytic cracking (FCC) units for producing high-octane gasoline.

3. **Operating at the middle unstable steady state** requires using some means of control for the plant, such as a stabilizing controller or nonadiabatic operation with carefully chosen parameters to stabilize the saddle-point type of the unstable steady state.

4. **Startup and multiplicity**: When the operating parameters are chosen to be in the unique steady state region, then almost any startup policy will lead to the desired unique steady state. However, when the operation parameters place the reaction in the multiplicity region, then different startup policies may lead to different steady states. Here is an actual example: when a catalytic hydrogenation reactor was designed to give an 80% steady-state conversion and when the plant gave only a 7% conversion, many trials to check for catalyst deactivation and other classical approaches did not yield any useful remedy. Through bifurcation analysis, it was finally discovered that the reactor had three steady states under the chosen design conditions, a stable low temperature one at 7% conversion, an unstable middle saddle-type one at 35% conversion, and a high-temperature, high-conversion one at 80% conversion. Preheating at startup to get the system to the high-temperature steady state was sufficient to solve the problem and to operate the reactor at 80% conversion as desired.

The Roles of the Parameters α , β , and γ :

β : This dimensionless parameter is the thermicity factor; it is positive for exothermic reactions and negative for endothermic ones. Given a constant feed concentration and feed temperature for a CSTR, larger positive values of β in the reaction mean higher exothermicity (a higher value of the exothermic heat of reaction), while lower negative values of β signify a higher endothermicity of the reaction, i.e., a higher value of the endothermic heat of reaction.

γ : This dimensionless parameter describes the activation energy. For constant feed temperatures, a higher value of γ indicates a higher activation energy. In other words, the larger γ is, the higher the reaction's sensitivity will be to temperature.

α : This dimensionless parameter involves k_0, the factor preceding the exponential term of the Arrhenius reaction formula (2.1) or (2.2), called the preexponential factor or the frequency factor; q, the volumetric flow rate; and V, the volume of the reactor, which are related via the formula $\alpha = k_0 \cdot V/q$. Here k_0 is usually quite large. If we keep k_0 constant, then α is directly proportional to the volume V, and inversely proportional to the flow rate q. The values of α can reach very large magnitudes due to the nature of k_0 and its range of values. In general, α is the most widely varying parameter of the reaction. Therefore we usually investigate bifurcation with α chosen as the bifurcation parameter.

3.2 Continuous Stirred Tank Reactor: The Nonadiabatic Case

For the nonadiabatic case we consider the same simple irreversible reaction

$$A \Longrightarrow B ,$$

but now with a cooling jacket with constant temperature T_c surrounding the tank.

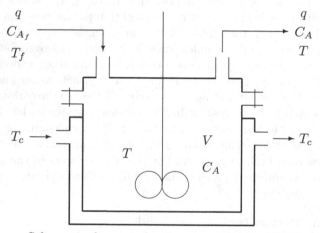

Schematic diagram for the nonadiabatic CSTR
Figure 3.12

Here the material balance design equation is the same as for the adiabatic case, namely

$$q \cdot C_{A_f} = q \cdot C_A + V \cdot k_0 \cdot e^{-E/(R \cdot T)} \cdot C_A . \tag{3.7}$$

In dimensionless form and with the variable settings identical to those below equation (3.3), equation (3.7) becomes

$$1 = x_A + \alpha e^{-\gamma/y} \cdot x_A . \tag{3.8}$$

The heat balance equation for the nonadiabatic case is

$$q \cdot \rho \cdot C_p \cdot (T - T_f) \;=\; V \cdot k_0 \cdot e^{-E/(R \cdot T)} \cdot C_A \cdot (-\Delta H) - U \cdot A \cdot (T - T_c) \,,$$

where U is the overall heat transfer coefficient between the jacket and the tank, measured in $J/(m^2 \cdot min \cdot K)$, A is the overall area of heat transfer between the tank and the cooling jacket, measured in m^2, and T_c is the constant temperature of the cooling jacket. We can write this equation in dimensionless form as

$$y - 1 \;=\; \alpha e^{-\gamma/y} \cdot x_A \cdot \beta - K_c(y - y_c) \,, \tag{3.9}$$

where the only new parameters and variables are $K_c = (U \cdot A)/(q \cdot \rho \cdot C_p)$, which measures the heat transfer between the tank and the jacket, and $y_c = T_c/T_f$, the ratio of cooling and feed temperatures.

The material balance equation (3.8) and the heat balance equation (3.9) can be reduced to a single equation by using different substitutions depending on the design and/or simulation investigation that is desired.

First form in terms of y:

Equation (3.8) is equivalent to

$$x_A = \frac{1}{1 + \alpha e^{-\gamma/y}} \,. \tag{3.10}$$

By inserting (3.10) into (3.9) we obtain

$$y - 1 \;=\; \frac{\alpha e^{-\gamma/y} \beta}{1 + \alpha e^{-\gamma/y}} - K_c(y - y_c) \,. \tag{3.11}$$

Rearrangement gives

$$(1 + K_c)y - (1 + K_c y_c) \;=\; \frac{\alpha e^{-\gamma/y} \beta}{1 + \alpha e^{-\gamma/y}} \,. \tag{3.12}$$

This form is preferred for studying the effect of varying K_c for a constant value of α.

Second form in terms of y:

Starting with equation (3.8), namely $x_A - 1 = -\alpha e^{-\gamma/y} \cdot x_A$, we multiply both sides by β and add this to (3.9) to obtain the following sequence of equations:

$$\beta(x_A - 1) \;=\; -\alpha \beta \, e^{\gamma/y} \cdot x_A \qquad\qquad [\beta \cdot (3.6)]$$

$$y - 1 + \beta x_A \;=\; \beta - K_c y + K_c y_c \,, \text{ or} \qquad [\beta \cdot (3.6) + (3.7)]$$

$$x_A \;=\; \frac{1 + \beta - y - K_c y + K_c y_c}{\beta} \,. \tag{3.13}$$

Inserting (3.13) into (3.9), we finally get

$$\frac{(1 + K_c)y}{\alpha} - \frac{(1 + K_c y_c)}{\alpha} \;=\; e^{-\gamma/y} \cdot (1 + \beta - (1 + K_c)y + K_c y_c) \,. \tag{3.14}$$

This is most suitable for studying the effect of varying α while holding K_c constant.

Third form in terms of x_A:

Sometimes it is desirable to express the nonadiabatic equation (3.8) in terms of the variable x_A alone, instead of in terms of y. To do so, we solve (3.13) for y to obtain

$$y = \frac{1 + K_c y_c + \beta(1 - x_A)}{1 + K_c} \ . \tag{3.15}$$

Plugging this expression for y into (3.8) gives us the following equivalent nonadiabatic nonisothermal CSTR equation, which depends solely on x_A:

$$1 - x_A = \alpha e^{-\gamma \cdot (1+K_c)/(1+K_c y_c + \beta(1-x_A))} \cdot x_A \ . \tag{3.16}$$

Numerical Solution of the Nonadiabatic CSTR Problem

As in Section 3.1 for the adiabatic CSTR problem, we again start with a generic MATLAB `fzero.m` based root finder to try to settle the issues of multiplicity in the nonadiabatic CSTR case. The MATLAB m file `solveNadiabxy.m` below finds the values for y (up to three values if α lies in the bifurcation region) that satisfy equation (3.12) for the given values of α, β, γ, K_c, and y_c using MATLAB's root finder `fzero`.

```
function [y,fx,x] = solveNadiabxy(al,bt,ga,kc,yc)
%    solveNadiabxy(al,bt,ga,kc,yc)
% Sample call    :  [y,fx,x] = solveNadiabxy(180000,1,15,1,1)
% Sample output :  y =
% (on screen)             1.0515        1.197        1.4
%                   fx =
%                        -1.5543e-15   3.8858e-16   1.1102e-16
%                    x =
%                         0.19995       0.60592      0.89707
% Input : al, bt, ga, kc, yc
% Program finds the xA and the y values on all branches of the curve.
% Output : y values at x for all steady states, error at each xA, xA
% CAUTION: always check critical applications against a run of
%          [Y, Fy,x] = NadiabNisoalgraphsol(al,bt,ga,kc,yc) and
%          [a11,a12,a21,a22,bif] = NadiabNisoauxfalxA(bt,ga,kc,yc,100,tol,w)
%          for confirmation.
%          NadiabNisoalgraph is more reliable near bifurcation limits than
%          solsolveNadiabxy.
% local subroutines : Nadiab(x,al,bt,ga,kc,yc)
%                     [differs from NadiabNiso which does the same for the
%                     vector variable y]
%                     And Matlab's own "fzero" root finder.

warning off; midtol = 10^-13; % to satisfy the grandfather warnings re. fzero
```

```
ltol = 10^-14; mid = 1;        % local tolerance for fzero; change if critical
options = optimset('dis',[],'tolx',ltol);
x0 = fzero(@Nadiab,0,options,al,bt,ga,kc,yc);
x1 = fzero(@Nadiab,1,options,al,bt,ga,kc,yc);
if abs(x0 - x1) > 10^-10, % three roots (bifurcation) likely; check with graph
    xmid = fzero(@Nadiab,[1.01*x0,0.99*x1],options,al,bt,ga,kc,yc);
else, xmid = (x0+x1)/2;  end, % dummy value for xmid if no bifurcation
x = [x0 xmid x1]; fx = Nadiab(x,al,bt,ga,kc,yc);
                                % compute error in x coordinates
y = sort((1+kc*yc + bt*(1-x))/(1+kc));        % transform from xA to y
warning on

function f = Nadiab(x,al,bt,ga,kc,yc)          % uses equation (3.14)
% evaluates the non-adiabatic-non-iso function at x = xA for given
% al, bt, ga, kc, yc values. (vector version)
f = 1 - x - al*exp(-ga*(1+kc)./(1+kc*yc + bt *(1 - x))) .* x;
```

Compared to `solveadiabxy.m` for the adiabatic CSTR case in Section 3.1, the above MATLAB function `solveNadiabxy.m` depends on the two extra parameters K_c and y_c that were defined following equation (3.9). It uses MATLAB's built-in root finder `fzero.m`. As explained in Section 3.1, such root-finding algorithms are not very reliable for finding multiple steady states near the borders of the multiplicity region. The reason – as pointed out earlier in Section 1.2 – is geometric; the points of intersection of the linear and exponential parts of equations such as (3.16) are very shallow, and their values are very hard to pin down via either a Newton or a bisection method, especially near the bifurcation points.

Instead, we shall rely again on graphical solvers and the level-set method, but with the added twist and complication of the two additional parameters K_c and y_c.

NOTE :   In the sequel, we shall no longer print out every function m file as we have done so far in this book. All our m files are stored on the accompanying CD and they can and, if desired, should be viewed and accessed from the CD. Check Appendix 3 with the contents list of our CD and look at the relevant *Readme* files first, please.

Whenever we explain m files in extended detail from now on, we will print them out in the book, but sometimes we just print their graphical output, or parts thereof or nothing at all, especially if the respective file is truly auxiliary or self-explanatory in nature. This helps to tighten our presentation and is ultimately helpful for our students by avoiding needless chatter and clutter.

The accompanying CD contains three plotting routines for the nonadiabatic CSTR problem under the file names `NadiabNisoplotfgxy.m`, `NadiabNisoplotfgxA.m`, and `NadiabNisoplotfgxAy.m` in the folder for Chapter 3. These plot graphs of different versions of the nonadiabatic CSTR equations. Figure 3.13 shows the result of calling `NadiabNisoplotfgxy(180000,1,15,1,1,100)`, for example.

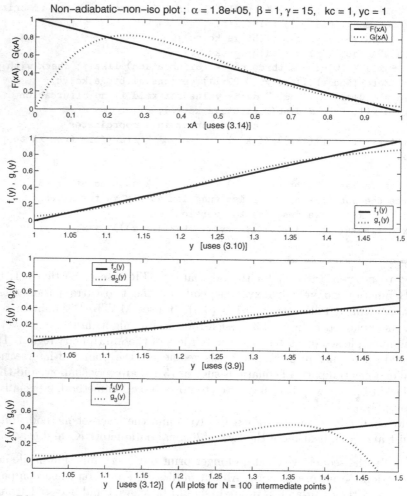

Plots for equations (3.16), (3.12), (3.11), and (3.14), respectively

Figure 3.13

Each of the four plots in Figure 3.13 contains two graphs: that of a solid line for the linear left-hand sides of equations (3.11) through (3.16) and that of the respective exponential function from the respective right-hand sides, in a dotted curve. The very shallow intersections in Figure 3.13 indicate the location of the multiple steady states of the given nonadiabatic CSTR system.

Similarly, `NadiabNisoplotfgxA.m` plots the linear and exponential parts separately, first of the two equations (3.16) in terms of x_A, and then of (3.11) in terms of y in more detail, while `NadiabNisoplotfgxAy.m` creates the following three graphs: first it repeats the linear and exponential parts plot of (3.16), followed by plotting the equation (3.16) converted to standard form, i.e., converted to an "f equal to zero" equation. And finally the same is done with equation (3.14). These differing plots are useful when one is trying to visualize the behavior of a nonadiabatic nonisothermal CSTR system for one specific set of inputs.

To gain further and broader insights into the bifurcation behavior of nonadiabatic, nonisothermal CSTR systems, we again use the level-set method for nonalgebraic surfaces such as $z = f(K_c, y)$. This particular surface is defined via equation (3.14) as follows for a given constant value of y_c with the bifurcation parameter K_c:

$$z = f(K_c, y) = (1+K_c)y - (1+K_c y_c) - \alpha \cdot e^{(-\gamma/y)} \cdot ((1+\beta+K_c y_c) - (1+K_c)y) \,. \quad (3.17)$$

To obtain this equation we have multiplied (3.14) by α and converted it into an equation in standard form $z = f(...) = 0$. We print out only the descriptive initial comment lines of runNadiabNisokc.m below.

```
function runNadiabNisokc(al,bt,ga,yc,N,tol,kcstart,kcend)
%   runNadiabNisokc(al,bt,ga,yc,N,tol,kcstart,kcend)
% Sample call   : runNadiabNisokc(180000,1,15,1,100,.001,.7,1.2)
% Runs NadiabNisokc double plot:    [uses equation (3.12) for variable kc]
%                   surface plot on top with bifurcation curve (variable kc)
%                   marked; and bifurcation curve on bottom.
% Input : al, -1 < bt, ga, yc, N (= number of points used),
%         tolerance tol (= 0.01, ..., 0.0001; optional),
%         kcstart ( >= 0 ), kcend (both optional)
%         [If unspecified, we plot the whole bifurcation region.]
% First we determine which range of kc1 < kc2 ensures bifurcations,
% then we plot the surface, followed by the y versus kc multiplicity region
% plot.
% [Note: if the given kcend or kcstart lies inside the bifurcation region
%  [kc1,kc2], the bifurcation ending dotted lines will take the kcstart/kcend
%  values rather than the true ones; simply widen the inputs [kcstart,kcend]
%  interval in this case.]
% For standard al, bt, ..., yc, set N = 80 or 100, for extreme data set
% N = 200 or 300.
% External subroutines : NadiabNisoauxfkcxA(al,bt,ga,yc,N,tol)
%                        [to find bifurcation limits]
%                        NadiabNiso(y,al,bt,ga,Kc,yc) [to evaluate function]
```

We run runNadiabNisokc.m to obtain a plot of the surface $z = f(K_c, y)$ with the zero-level curve depicted in black on the surface itself and also below on the K_c-y plane in blue, as well as a separate plot of the zero-level curve in a second window below. In particular, a call of runNadiabNisokc(180000,1,15,1,100,.001,.7,1.2) exhibits bifurcation for $0.77125 \le K_c \le 1.12452$ when $\alpha = 1.8 \cdot 10^8$, $\beta = 1$, $\gamma = 15$, and $y_c = 1$ as shown in Figure 3.14.

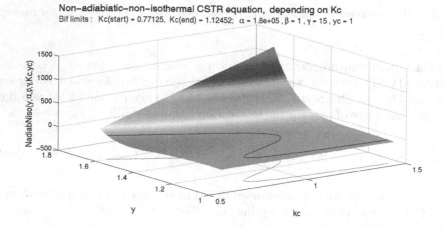

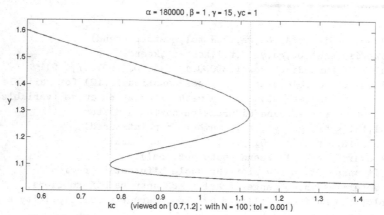

Nonadiabatic CSTR surface and bifurcation curve for K_c
Figure 3.14

Generally speaking, there are three possible outcomes of `runNadiabNisokc(...)` if the K_c viewing interval [`kcstart, kcend`] is specified in the call: there may be no bifurcation at all for the α, β, γ, and y_c input data, or there may be bifurcation: either somewhere other than near the desired viewing interval [`kcstart, kcend`], or the viewing interval [`kcstart, kcend`] may overlap the multiplicity region in case of bifurcation. The plot annotations will reflect this clearly for our users. If the K_c viewing region is not specified, defaults are set automatically and the program tries to plot the whole multiplicity region in case bifurcation is detected somewhere for the input data.

The `runNadiabNisokc.m` plotting function relies heavily on the auxiliary function `NadiabNisoauxfkcxA.m`, which decides whether there is any multiplicity for the given inputs α, β, γ, and y_c. The key to these decisions lies in its auxiliary function [`b,li,lj`] = `lowhighkc(i,j,heat)`, which we shall explain following the `NadiabNisoauxfkcxA.m` listing below.

```
function [kc11,kc22,bif] = NadiabNisoauxfkcxA(al,bt,ga,yc,N,tol,kc11,kclim)
```

```
%    NadiabNisoauxfkcxA(al,bt,ga,yc,N,tol,kc11,kclim)
% Sample call    : [kc11,kc22,bif] = NadiabNisoauxfkcxA(180000,1,15,1,100,.001)
% Sample output : [ The kc bifurcation limits are :
%                     kc11 (= lower bifurcation limit)
%                     and kc22 (= upper bifurcation limit),
%                     if bif = 1, i.e., if bifurcation has been detected.]
%    on screen : kc11 =
%                     0.77124 (= lower limit for Kc with bifurcation)
%                 kc22 =
%                     1.1245  (= upper bifurcation limit)
%                 bif =          ( bifurcation indicator)
%                    1
% searches for values of -1 <= kc11 <= kc <= kc22 with bifurcation,
% for a certain tolerance; tol = 10^-5 gives very stringent results;
% tol = 10^-3 is generally ok. [N = 60 or 100; tol = 0.01 or 0.001 will work.]
% Inputs kc11 ( >= 0 )and kclim are optional and demarcate the range of kc
% values in which we search for bifurcation, if they are specified.
% On output: bif = 1, if there is bifurcation, bif = 0, if there is not.
% Special subroutines used: external: NadiabfkcxA(x,al,bt,ga,kc11,yc)
%                           internal: lowhighkc(i,j)

if nargin == 6, kc11 = 0.00001; kclim = 20; end, kcstep = (kclim-kc11)/10;
x = 0:1/N:1; bif = 0; biff = 1;  % set starting values; adjust kclim to defaults
F1 = NadiabfkcxA(x,al,bt,ga,kc11,yc);
b1 = lowhighkc(find(F1 > 0),find(F1 <= 0));
if b1 == -1,
 disp(['W A R N I N G 1: no nonnegative values for kc lead to bifurcation,',...
       ' we give up']),    % return ?
    kc12 = kc11; kc21 = kclim; kc22 = kclim; biff = 0;   end % abandon program
if b1 == 0, kc12 = kc11; bif = 1; end
           % bifurcation : set up lower bifurcation limits kc11, kc12 first
if b1 == 1, kc12 = kc11; bif = 1; % bifurcation inside prescribed kc interval
    while b1 == 1 & kc12 < kclim  % search for its lower limit kc11
        kc11 = kc12; kc12 = kc12 + kcstep; F1 = NadiabfkcxA(x,al,bt,ga,kc12,yc);
        b1 = lowhighkc(find(F1 > 0),find(F1 <= 0));  end,
    if kc12 >= kclim,
      disp(['W A R N I N G 2: no apparent bifurcation for any kc <= kclim = ',...
            num2str(kclim,'%9.4g')]),        % return ?
        kc11 = 0; kc12 = 0; kc21 = kclim; kc22 = kclim;
        if biff == 0, bif = 0; end, end, end
if bif == 1,
 kc21 = kc12; F2 = NadiabfkcxA(x,al,bt,ga,kc21,yc);
 b2 = lowhighkc(find(F2 > 0),find(F2 <= 0)); % set upper bifurcation limits
 if b2 == -1, kc22 = kc21; kc21 = kc22 - kcstep; end
 if b2 == 0,
    while b2 == 0 & kc21 < kclim,  % search for upper bifurcation limit
        kc21 = kc21 + kcstep; F2 = NadiabfkcxA(x,al,bt,ga,kc21,yc);
        [b2,li,lj] = lowhighkc(find(F2 > 0),find(F2 <= 0)); end,
```

```
    if kc21 >= kclim,            % upper bif limit outside kc interval
      disp(['W A R N I N G 3: bifurcations for kc >= kclim = ',...
            num2str(kclim,'%9.4g')]), end,
  kc22 = kc21; kc21 = kc22 - kcstep;  end
  while kc12-kc11 > tol * kc12      % refine lower bifurcation limits
    k1 = (kc11+kc12)/2; Fk1 = NadiabfkcxA(x,al,bt,ga,k1,yc);
    b1 = lowhighkc(find(Fk1 > 0),find(Fk1 <= 0));
    if b1 == 1, kc11 = k1; else, kc12 = k1; end, end, k2 = kc21; K2 = k2;
  while kc21 > 0 & kc22-kc21 > tol * abs(kc21) & k2 > 0, % refine upper bif
    K2 = k2; k2 = (kc21+kc22)/2; Fk2 = NadiabfkcxA(x,al,bt,ga,k2,yc);
    [b2,li,lj] = lowhighkc(find(Fk2 > 0),find(Fk2 <= 0));
    if k2 > 0,
      if b2 == -1, kc22 = k2; else, kc21 = k2; end,
    else
      if b2 == -1, kc22 = K2; else, kc21 = K2; end, end, end, end
if bif == 1,
  kc11 = (kc11+kc12)/2;             % half the differences
  if kc21*kc22 <= 0, kc22 = K2; else, kc22 = (kc21+kc22)/2; end,
  if kc11 == kc22, bif = 0; end,
else, kc22 = kclim; end


function [b,li,lj] = lowhighkc(i,j)   % length comparison function for kc
% input : two strings of integers i, j
% output b : b = 0 if kc is good (bifurcation);
%             b = -1 if reducing kc leads to good kc;
%             b = 1 if increasing kc leads to good kc
li = length(i); lj = length(j); b = -1;           % going down default
if i(li) == li & j(1) == li+1, if li < lj, b = 1; end; % going up now
else, b = 0; end                                   % bifurcation
```

MATLAB code line pairs such as

```
F1 = NadiabfkcxA(x,al,bt,ga,kc11,yc);
b1 = lowhighkc(find(F1 > 0),find(F1 <= 0));
```

on lines 3 and 4 above appear six times inside the function m file `NadiabNisoauxfkcxA.m`. Such a line pair first evaluates the nonadiabatic nonisothermic system equation. In particular, this equation is the standard form $(f(...) = 0)$ of equation (3.16). This code line pair helps the program to decide whether the current value of K_c, called `kc11`, the partition x, and the constants α, β, γ, and y_c describe a system with multiplicity, denoted by `b1 = 0`, or whether reducing K_c will likely lead to the multiplicity region by setting `b1` equal to –1, or if increasing K_c will likely do so by setting `b1 = 1`.

Here is a plot of the three possibilities for multiplicity when solving equation (3.16) in its standard form

$$F1(x) = 1 - x - \alpha \cdot e^{-\gamma \cdot (1+kc)/(1+kc \cdot yc + \beta(1-x))} \cdot x = 0 \qquad (3.18)$$

for varying values of K_c and the same parameter set as has been used for Figure 3.14, namely $\alpha = 18 \cdot 10^4$, $\beta = 1$, $\gamma = 15$, and $y_c = 1$.

When viewing the graphs of Figure 3.15, please recall that multiple solutions of $F1(x) = 0$ for $x \in [0,1]$ signify bifurcation for the underlying CSTR system.

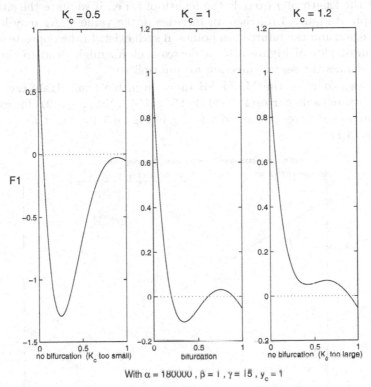

With $\alpha = 180000$, $\beta = 1$, $\gamma = 15$, $y_c = 1$

Bifurcation behavior of equation (3.18), depending on K_c
Figure 3.15

Figure 3.15 was drawn using the MATLAB m file kctriple.m on the CD that accompanies the book.

In the three example plots of Figure 3.15, the graph of $F1$ in (3.18) crosses the horizontal axis where $F1 = 0$, i.e., $F1$ changes sign for $x \in [0,1]$ and any value of K_c at least once, since $F1(0) = 1$ and $F1(1) < 0$ for all K_c. If K_c lies inside the multiplicity region, then $F1$ changes signs more than once on the interval $[0,1]$, and it does so an odd number of times. In order to decide which is the case for one particular K_c, we form two vectors for the index sets i and j of all points of the x partition where $F1 > 0$ in i and where $F1 \leq 0$ in j. Here $F1$ is computed by NadiabfkcxA(x,al,bt,ga,kc11,yc).

Inside the subprogram lowhighkc we look at the last index of i. If we are in the situation of the left- or rightmost graphs of Figure 3.15, then the last index of i must be equal to i's length, since there is only one contiguous x interval where the function $F1$ is positive. Moreover, in these two situations, the vector j of all indices of the partition x with negative $F1$ values must have its first entry precisely equal to one plus the length of the vector i, since the index vector j starts exactly after i has ended. Observe that in case of multiplicity, depicted in the center graph of Figure 3.15 for $F1$, neither of these two observations can hold, since i must contain the index sets of at least two disjoint x intervals with $F1(x) > 0$, and likewise for j.

Moreover, if the length of j exceeds the length of i, i.e., if we have the situation of the leftmost graph of Figure 3.15, then an increase in the value of K_c may lead us to the bifurcation region and the bifurcation points, if such exist. In the opposite case, depicted in the rightmost plot of Figure 3.15, a decrease of K_c might lead to the multiplicity region. This defines the search direction for multiplicity.

Our CD also contains the MATLAB function m file `runNadiabNisokccurve.m`. A call of `runNadiabNisokccurve(180000,1,15,1,100,.001,.7,1.2)`, for example, plots only the bifurcation curve with respect to K_c in Figure 3.16, i.e., it repeats the bottom plot in Figure 3.14.

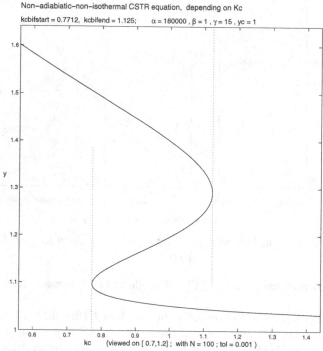

Nonadiabatic CSTR bifurcation curve for K_c as the bifurcation parameter
Figure 3.16

Equation (3.17) allows a different interpretation of the underlying system's bifurcation behavior by taking K_c and y_c as fixed and letting α vary, for example. We now study the bifurcation behavior of nonadiabatic and nonisothermal CSTR systems via their level-zero curves for the associated transcendental surface $z = g(\alpha, y)$. The surface is defined as before, except that here we treat K_c and y_c as constants and vary α and y in the 3D surface equation

$$z = g(\alpha, y) = (1 + K_c)y - (1 + K_c y_c) - \alpha \, e^{(-\gamma/y)} \cdot ((1 + \beta + K_c y_c) - (1 + K_c)y) \, . \quad (3.19)$$

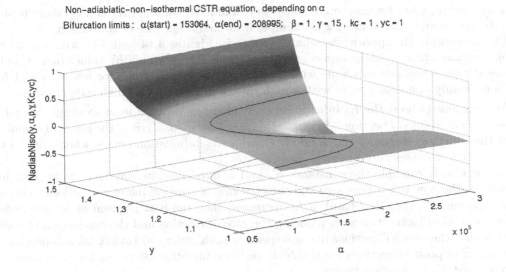

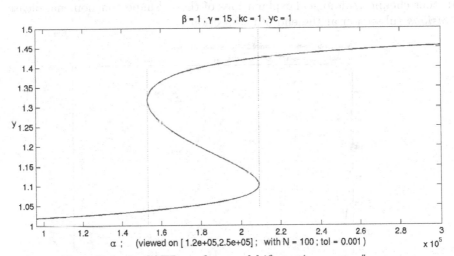

Nonadiabatic CSTR surface and bifurcation curve for α

Figure 3.17

If we run `runNadiabNisoal.m` (contained on the CD), we obtain a plot of the surface $z = g(\alpha, y)$ and a separate plot of its zero-level curve. In particular, a call of `runNadiabNisoal(1,15,1,1,100,.001,120000,250000)` exhibits bifurcation for 153,064 $\leq \alpha \leq$ 208,984 when $\beta = 1$, $\gamma = 15$, $k_c = 1$, and $y_c = 1$ as depicted in Figure 3.17.

Note the difference between the shapes of the K_c bifurcation curves in Figures 3.14 and 3.16 on the one hand and that of the α bifurcation curve in Figure 3.17 for the same equation (3.14) on the other hand: The α bifurcation curves have the general form of the letter $\mathcal{S}$, as seen in the bottom graph of Figure 3.17, while the curve in Figure 3.16 has the mirror image $\mathcal{S}$ shape. For very high values of α and fixed values for K_c and y_c, there is only one steady state at a high rate of conversion and consequently at a high temperature, while for very small values of α and fixed values of K_c and y_c, there is also only one steady state, but at a low temperature and a low rate of conversion.

This is completely opposite to what happens in Figures 3.14 and 3.16 with regard to bifurcation when K_c varies and α is kept constant. For very high K_c values there is only one steady state, but this time with a low conversion rate, while for low values of K_c there is only one steady state with a high rate of conversion and high temperature.

As mentioned above, the K_c bifurcation curves have an inverted letter-$\mathcal{S}$ shape. We refer to the conclusions of Section 3.2, and to Chapter 7 for an analysis of the physical meaning of the differing shapes of the K_c and α parameter bifurcation curves when applied to industrial processes and reactors.

In our CSTR example the constants K_c and α have opposite physical effects. If α increases, the flow rate q decreases and thus the rate of reaction increases, as does the heat of reaction. On the other hand, if K_c increases, then the heat removal by heat transfer to the cooling jacket increases, reducing the rate of reaction and the production of heat. Note that the search directions in our respective `lowhighkc` and `lowhighal` sub-programs point in opposite directions for the $\mathcal{S}$-shaped α bifurcation curves and for the inverted $\mathcal{S}$-shaped K_c bifurcation curves.

For further chemical/biological explanations of these "shape" phenomena, please see the *Conclusions* subsection at the end of this section.

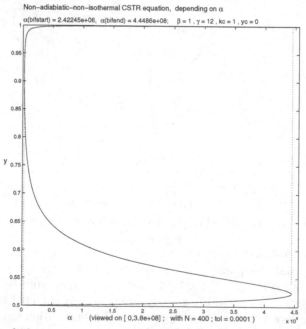

Nonadiabatic, nonisothermal CSTR bifurcation curve for α

Figure 3.18

Again we plot the bifurcation curve with respect to α as was done in Figure 3.17 separately. We use the command `runNadiabNisoalcurve(1,12,1,0,400,0.0001,0,3.8*10^8)` for Figure 3.18, for example, where the m file `runNadiabNisoalcurve.m` has been taken from the CD. This gives us the following plot.

Thus far we have explored the bifurcation behavior of equation (3.14) with respect to K_c via equation (3.17) in Figures 3.14 through 3.16, and with respect to α via (3.19) in Figures 3.17 and 3.18. Since different K_c and α values can lead to bifurcation behavior for the same nonadiabatic, nonisothermal CSTR system, it is of interest and advantageous to be able to plot the joint bifurcation region for the parameters K_c and α as well.

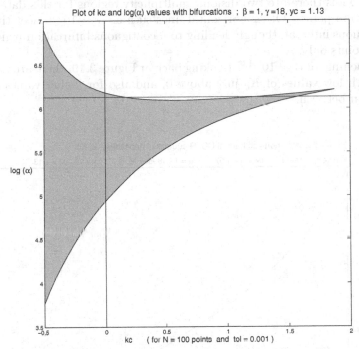

Nonadiabatic, nonisothermal CSTR joint bifurcation region for K_c and α
Figure 3.19

The gray area in Figure 3.19 is drawn by the call `NadiabNisoalkcplot(1,18,1.13,100,.001,-0.5,2)`. It describes the multiplicity area in terms of both α and K_c. The terminal gray points of any line parallel to the α axis denote the boundaries of the multiplicity region for α and the chosen fixed value of K_c, while the nonnegative points for any line parallel to the K_c axis denote the multiplicity interval(s) for K_c and the chosen fixed α. Note that the axis scales in Figure 3.19 are chosen so as to give a reasonable and instructive plot of the parameter ranges. The parameters K_c with bifurcation generally lie in the range $0 \leq K_c \leq 5$, while α's range of bifurcation often covers several powers of ten. Therefore we have chosen a logarithmic scale (to base 10) for the vertical axis in Figure 3.19.

Choosing $K_c \geq 0$ and α inside the gray region of Figure 3.19 gives the associated CSTR system multiple steady states, and conversely, choosing $K_c \geq 0$ and α outside the shaded bifurcation region ensures a unique steady state for this particular system with its given parameters $\beta = 1$, $\gamma = 18$, and $y_c = 1.13$.

Let us look in more detail at Figure 3.19: For any fixed value of K_c, we can read the bifurcation limits (in terms of α) of the system from the graph. The bifurcation region for each fixed K_c is a simple, connected interval since any vertical line in Figure 3.19 that meets the gray region meets it contiguously with only two bifurcation points.

The situation is quite different for fixed α: look, for example, at the black horizontal line drawn additionally in Figure 3.19 for $\alpha = 10^{6.13}$. It intersects the gray bifurcation region twice. Thus there are two disjoint multiplicity regions for this data with at least three bifurcation points. Other horizontal lines such as $\alpha = 10^{5.8}$ meet the gray region in one contiguous interval, though, leading to a contiguous bifurcation region and to two bifurcation points only.

What is happening for $\alpha = 10^{6.13}$? Looking back at Figure 3.19, there are multiple steady states for both low values of K_c just above 0, and also for higher values of K_c around 1.5, but not in between.

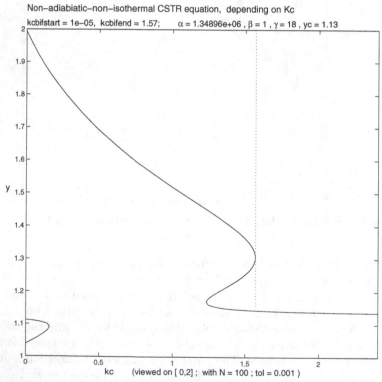

Nonadiabatic, nonisothermal CSTR system with disjoint multiplicity regions for K_c
Figure 3.20

Figure 3.20 uses `runNadiabNisokccurve.m` with $\alpha = 10^{6.13} = 1.34896 \cdot 10^6$, $\beta = 1$, $\gamma = 18$, $y_c = 1.13$ for $0 \le K_c \le 2$ and describes the bifurcation behavior with respect to K_c as represented by the intersection of the black horizontal line with the gray bifurcation region of Figure 3.19 for $K_c \ge 0$. We note the standard inverted $\mathcal{S}$-shape of the curve in Figure 3.20 from around $K_c = 1.2$ to about $K_c = 1.57$, as well as a little "dimple" for $0 \le K_c \le 0.2$ or thereabouts. This "dimple" corresponds to part of the black line and gray region intersection immediately to the right of the black vertical line $K_c = 0$ in Figure 3.19.

If we decrease α sufficiently for the same set of data, the two areas of multiplicity will eventually join into one. For $\alpha = 10^{6.092389299}$, for example, a call of `runNadiabNisokccurve` `(10^6.092389299,1,18,1.13,800,.0001,0,2)` produces the plot in Figure 3.22 with an almost contiguous multiplicity region where two of the bifurcation points come very close to each other.

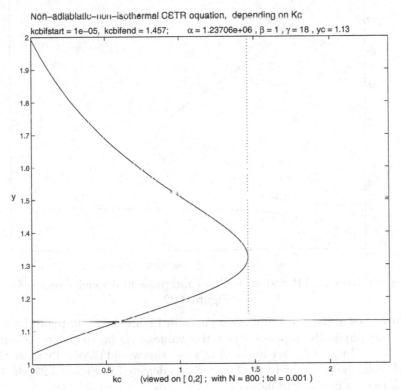

Nonadiabatic, nonisothermal CSTR system with an almost contiguous multiplicity region for K_c

Figure 3.21

Note the near-horizontal and slightly disconnected black line at $y \approx y_c = 1.13$ in Figure 3.21. For the chosen value of $\alpha = 1.23706 \cdot 10^6 = 10^{6.092389299}$ the bifurcation diagram becomes an imperfect pitchfork. A perfect pitchfork bifurcation diagram occurs when α is slightly decreased, so that the corresponding black horizontal line in Figure 3.19

becomes precisely tangent to the upper limiting curve of the gray joint bifurcation region of Figure 3.19. For this critical value of α, the middle steady state remains constant at the cooling jacket temperature $y_c = 1.13$ for all values of K_c.

When α is decreased below this critical value, the pitchfork phenomenon disappears. In particular, for $\alpha = 10^{6.08}$ we observe two unconnected branches for the three steady states of the system in Figure 3.22.

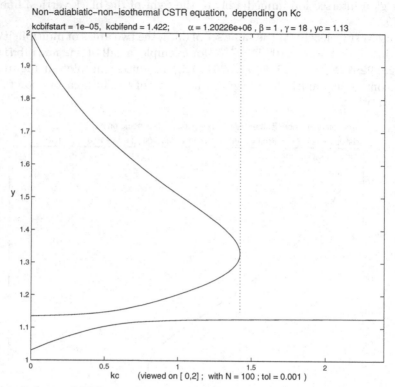

Nonadiabatic CSTR system with a contiguous multiplicity region for K_c
Figure 3.22

The two branches of the top bifurcation curve in Figure 3.22 will rejoin for some large-magnitude but physically impossible negative value of K_c far to the left of our window's edge. Negative values for K_c are impossible, since this would physically mean that heat is transferred from the cold part to the hot part. As depicted in Figure 3.22, the bifurcation curves look like an incomplete **isola**.

We are made aware of the lively change in bifurcation behavior here: just a third digit change in α can cause absolutely different bifurcation behavior of the associated CSTR system, as witnessed by Figures 3.20 to 3.22.

For this reason, only highly reliable models coupled with accurate numerical routines such as those presented here are useful for the professional chemical/biological engineer.

What has just been done for K_c and α can be repeated for y_c and α as well. Here is the joint multiplicity region plot for y_c and α, obtained by calling `NadiabNisoalycplot(1,15,1,100,.001,-.4,2)` with $\beta = 1$, $\gamma = 15$, and $K_c = 1$ and the imposed bounds $-0.4 \leq y_c \leq 2$.

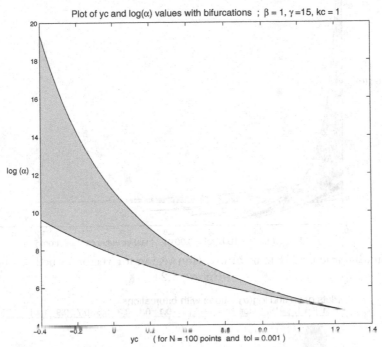

Nonadiabatic CSTR joint bifurcation region for y_c and α

Figure 3.23

We note that split bifurcation regions and pitchfork bifurcation such as depicted in Figures 3.20 and 3.21 for K_c were never encountered by us for any fixed α in terms of y_c.

Next we show the graphical output of two multiplot routines from our CD (in terms of several y_c values plotted with respect to K_c, or in terms of several K_c values plotted with respect to y_c). A call of `NadiabNisoalkcmultiplot(1,20,[1.5:.5:4],100,.001,-.4,2)` plots six superimposed K_c and α joint multiplicity regions for $y_c = 1.5$, 2, 2.5, 3, 3.5, and 4 in Figure 3.24.

The topmost drawn multiplicity region in Figure 3.24 represents the one for the last entry of the y_c vector, and the one for the first entry of the vector y_c is the bottom region plotted in the stack of regions in Figure 3.24. If we desire to depict the y_c multiplicity region for $y_c = 1.5$ on top in Figure 3.23, we should call `NadiabNisoalkcmultiplot(1,20,fliplr ([1.5:.5:4]), 100,.001,-.4,2)`, or equivalently `NadiabNisoalkcmultiplot(1,20, [4:-.5:1.5],100,.001,-.2,2)` instead.

Moreover, recall that negative values of K_c are physically meaningless here. But our numerical techniques are robust enough to plot even for unrealistic negative K_c values.

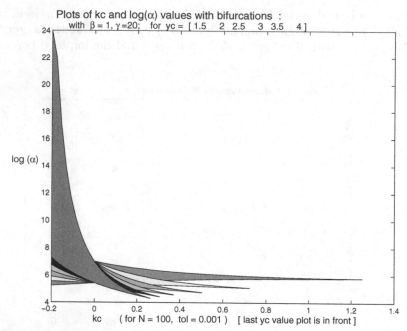

Nonadiabatic CSTR joint bifurcation region for vector valued y_c and α
Figure 3.24

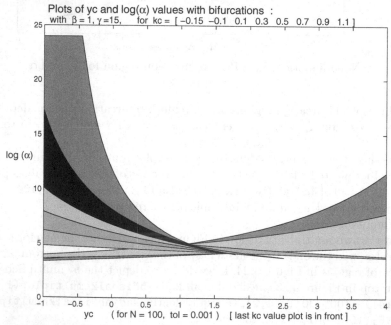

Nonadiabatic CSTR joint bifurcation region for vector valued K_c and α
Figure 3.25

Likewise, we can plot multiple joint bifurcation regions for multiple values of K_c and α. Calling `NadiabNisoalycmultiplot(1,15,[-.15,-.1,[.1:.2:1.1]],100,.001,-1,4)` with `NadiabNisoalycmultiplot.m` from our CD draws Figure 3.25 in color. We note that the range of α allowed in the auxiliary function `NadiabNisoauxfalxA.m` has been deliberately limited to below 10^{24}. Therefore there is an artificial "top shelf" at $\alpha = 10^{24}$ for the top purple bifurcation region drawn for the negative values of y_c when $K_c = 1.1$ in Figure 3.25.

Note how the multiplicity regions here become wider to the left for smaller values of y_c if $K_c > 0$ and how they open up to the right for larger y_c if $K_c < 0$. Of course, negative values for K_c have no physical meaning, since heat cannot be transferred from a cold part to a hot part; see our comments following Figure 3.22. These negative K_c value plots are included only to show the versatility and strength of our numerical methods that go well beyond the physically meaningful applications of our model. These methods may well turn out to be useful in other applications. The same comments apply to our earlier Figures 3.19 and 3.24, which also allowed for physically impossible negative values of K_c.

Our final MATLAB m file in this section rounds out our efforts just as Figure 3.10 did for the adiabatic CSTR problem. It uses the plotting routine for Figure 3.18 in conjunction with a MATLAB interpolator to mark and evaluate the (multiple) steady state(s) graphically for nonadiabatic, nonisothermal CSTR problems. Our function m file `NadiabNisoalgraphsol.m` from the N_adiab folder on the CD reads like this:

```
function [Y, Fy,x] = NadiabNisoalgraphsol(a0,bt,ga,kc,yc)
%    NadiabNisoalgraphsol(a0,bt,ga,kc,yc)
% Sample call   : [Y, Fy,x] = NadiabNisoalgraphsol(300000000,1,12,1,0)
% Sample output :  (on screen)
%           Y =
%           0.50821      0.55562       0.99973
%           Fy =
%               -0.0087544   1.5137e-05   2.4585e-06
%           x =
%           0.00054877      0.88877      0.98358
% Plots the (3.12) Nadiab equation zero level curve as a 2D plot,
%    then the program decides on bifurcation and computes the values of y at
%    the steady states of the CSTR system for the specified a0 value of alpha.
% Input : a0 = alpha_0, beta, gamma, kc, and yc
% Output: Plot of up to three steady state solutions y0, ymid, y1 in Y on the
%         bifurcation curve.
%         And on screen: Y values; function error in Fy, and converted xA
%         values in x (via equation (3.11))
%         (If entries in Y are identical, there is only one solution,
%              i.e., only one steady state for this input.)
% Increase N on first line to 800, 1000, 2000, and possibly decrease tol,
%         if higher accuracy desired.
```

```
N = 500; tol = 0.001; w = 0;
[a1,a12,a2,a22,bif] = NadiabNisoauxfalxA(bt,ga,kc,yc,N,tol,w);
a11 = min(a1,a0); a22 = max(a2,a0);        % extend alpha range to contain a0
three = 0;                                 % triple steady states ?
miny = (1+kc*yc)/(1+kc); maxy = (1+bt+kc*yc)/(1+kc); % prepare plot
[y,Al]=meshgrid([miny:(maxy-miny)/N:maxy],...
                [0.8*a11:(1.2*a22-0.8*a11)/N:1.2*a22]);
             % make grids for y and alpha in relevant ranges for y and beta
z = NadiabNiso(y,Al,bt,ga,kc,yc);          % z = adiab Non-iso function value
C = contour(Al,y,z,[0 0],'b'); hold on,    % draw 0 level curve
if a1 < a2,
  yy = find(C(1,1:end-1)-C(1,2:end )> 0);
  if length(yy) == 0, itop = length(C); ibot = 1;
  else                                     % find left bifurcation point
      itop = yy(1); [bull,ibot] = min(C(1,itop:end));
      ibot=ibot+itop-1;   end              % add the front end length!
  if a1 <= a0 & a2 >= a0 & length(yy) > 0, three = 1; % in case of multiplicity
   x0 = interp1(C(1,1:itop),C(2,1:itop),a0,'linear');  % bottom solution (1)
   xmid = interp1(C(1,itop:ibot),C(2,itop:ibot),a0,'linear'); % middle (2)
   x1 = interp1(C(1,ibot:end),C(2,ibot:end),a0,'linear');   % top sol (3)
  elseif a0 < a1,
     x0 = interp1(C(1,1:itop),C(2,1:itop),a0,'spline'); % only one steady state
     xmid = x0; x1 = x0;
  else x0 = interp1(C(1,ibot:end),C(2,ibot:end),a0,'spline'); % one steady st.
     xmid = x0; x1 = x0; end,
else x0 = interp1(C(1,1:end),C(2,1:end),a0,'spline'); % no bifurcation;
     xmid = x0; x1 = x0; end,              % only one steady state
Y = [x0 xmid x1]; x = sort((1+bt-Y-kc*(Y-yc))/bt);    % prepare output
Fy = Nadiab(x,a0,bt,ga,kc,yc);

if bif == 1,                               % in case of bifurcation
  plot(a0,miny,'+r'),                      % mark requested value for alpha_0
  plot([a2,a2],[maxy,miny+0.08*(maxy-miny)],':');
  plot([a1,a1],[miny,miny+.8*(maxy-miny)],':'); % plot bifurcation limits
  if three == 1;                           % mark 3 steady states
      xlabel(['\alpha       (three steady states for \alpha_0 = ',...
          num2str(a0,'%12.9g'),' , \beta = ',num2str(bt,'%10.5g'),...
          ', \gamma = ',num2str(ga,'%10.5g'),')'],'FontSize',12),
      title([' NadiabNiso graphical solution  :   y_1 = ',...
          num2str(x0,'%10.6g'),';   y_2 = ',num2str(xmid,'%10.6g'),...
          ' ;     y_3 = ',num2str(x1,'%10.6g')],'FontSize',12),
      plot(a0,x0,'+r'), plot(a0,xmid,'+r'), plot(a0,x1,'+r'),
  else                                     % mark single steady state
      xlabel(['\alpha       (single steady state for \alpha_0 = ',...
          num2str(a0,'%12.9g'),' , \beta = ',num2str(bt,'%10.5g'),...
          ', \gamma = ',num2str(ga,'%10.5g'),')'],'FontSize',12),
      title([' adiabNiso graphical solution  :   y_1 = ',...
      num2str(x0,'%10.6g')],'FontSize',12),
```

```
    plot(a0,x0,'+r'), end % mark single steady state
else,                                            % in case of no bifurcation
    xlabel(['\alpha        (single steady state for \alpha_0 = ',...
            num2str(a0,'%12.9g'),' , \beta = ',num2str(bt,'%10.5g'),...
            ', \gamma = ',num2str(ga,'%10.5g'),')'],'FontSize',12),
    title([' adiabNiso graphical solution  :   y_1 = ',...
           num2str(x0,'%10.6g')],'FontSize',12),
    plot(a0,x0,'+r'),  plot(a0,1,'+r'), end  % mark single steady state
ylabel('y','Rotation',0,'FontSize',12), hold off

function f = Nadiab(x,al,bt,ga,kc,yc) % uses xA parameter and equation (3.14)
% evaluates the non-adiabatic-non-iso function at x = xA for given
%  al, bt, ga, kc, yc values. (vector version)
f = 1 - x - al*exp(-ga*(1+kc)./(1+kc*yc + bt *(1 - x))) .* x;
```

This m file produces three steady-state values for y_c as marked when called by the MAT-LAB command `NadiabNisoalgraphsol(300000000,1,12,1,0)`, for example, in Figure 3.26.

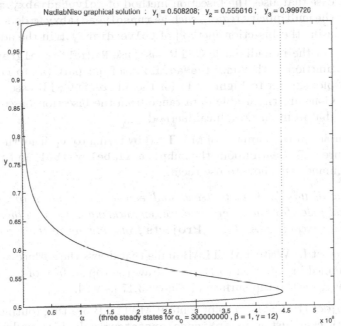

Bifurcation curve with multiple steady states marked by +
Figure 3.26

It is worthwhile to explore the differences between the results of `NadiabNisoalgraphsol` and `solveNadiabxy` from the beginning of this section. For the nonadiabatic case such an exploration will duplicate what we have already learned about the differences between our `fzero` based and our graphics based steady-state finders for the adiabatic CSTR case in Section 3.1; see the exercises below.

Exercises for 3.2

1. Compare the results obtained from `solveNadiabxy.m` with those from `NadiabNisoalgraphsol.m`, especially near the boundaries of the multiplicity regions for various sets of inputs.

2. Compare the graphical output from `NadiabNisoalkcplot.m` and `NadiabNisoalycplot.m` for the same values of K_c and y_c by measuring the endpoints of the respective α bifurcation regions with the help of a ruler. These should be the same. Are they?

3. Use `runNadiabNisokccurve.m` with judiciously adjusted kcstart and kcend values to find the lower limit of the bifurcation interval around $K_c = 1.5$ graphically for $\alpha = 10^{6.13}$, $\beta = 1$, $\gamma = 18$, and $y_c = 1.13$ as used in Figure 3.20. How can you likewise find the upper multiplicity limit for the bifurcation region around $K_c = 0$ for the same input data and cleverly chosen values for kcstart and kcend?

4. (a) Create a program called `runsolveNadiabxy.m` for the nonadiabatic CSTR case that uses the bisection method of `solveNadiabxy.m` to plot $y(\alpha)$ in the multiplicity region, such as `runsolveadiabxy.m` does in Figure 3.3 by using the bisection method of `solveadiabxy.m` in the adiabatic case.

    (b) For the nonadiabatic CSTR case, use `NadiabNisoalgraphsol.m` in conjunction with `runsolveNadiabxy.m` from part (a) to create an overlay plot similar to Figure 3.11 for the adiabatic CSTR case.
    Compare the usable data range from the bisection based algorithm with that from the graphical method.

5. Exercise your command of MATLAB by trying to replicate the three graphs in Figure 3.15. Learn about the `subplot`, `xlabel`, `ylabel`, and `title` MATLAB commands and how to use them.

One aim of this book is to teach and enable our readers to develop relevant numerical codes for chemical/biological engineering models on their own. For this purpose we include MATLAB **Projects** *from now on in the Exercise sets.*

6. **Project I:** Write a MATLAB m file that draws the surface $z = f(K_c, y)$ with f defined in equation (3.17), i.e., draw the top surface of Figure 3.14 alone. Repeat for the top surface of Figure 3.17 as well.

7. **Project II:** Create a color contour MATLAB plotting routine for the nonadiabatic case, just as `adiabNisocolorcontour.m` did for α and y in the adiabatic case in Figure 3.7. The aim is to create a color-contoured version of Figure 3.16.
    Repeat for K_c and y, i.e., add color contours to Figure 3.16.
    Finally, explore the K_c and y color contour plots for the disjoint multiplicity situations depicted in Figures 3.20 to 3.22.

8. **Project III:** Combine and adapt the m files `NadiabNisoalgraphsol.m`, `runNadiabNisoalcurve.m`, and `runNadiabNisokccurve.m` to create a new

multiplicity finder named `NadiabNisokcgraphsol.m` for y in terms of K_c, i.e., write a MATLAB m file that produces a plot like Figure 3.26, but with respect to K_c rather than with respect to α.

Compare your results from `NadiabNisoalgraphsol.m` and `NadiabNisokcgraphsol.m` analogously to Exercise 1 above.

Conclusions

The current section has covered numerical techniques and MATLAB codes for investigating the static bifurcation behavior of nonadiabatic lumped systems.

It is clear that the behavior of nonadiabatic systems is more complicated than that of the adiabatic ones that were treated in Section 3.1.

Starting with Sections 3.1 and 3.2, we are progressing to learn how to design and apply efficient numerical methods to investigate industrial chemical, biochemical, and biomedical systems with and without bifurcation.

For the nonadiabatic case we have demonstrated that the variables α and K_c have opposite effects on the system's behavior. This is consistent with the physical meaning of these two parameters. Increasing α increases the heat production, while increasing K_c increases the heat dissipation. Thus for a constant α and variable K_c, at low K_c values the heat dissipation is small, leading to a unique high-temperature steady state of the system, while at large values of K_c, when heat dissipation is high the system can have only one low-temperature steady state. And for intermediate K_c values there may be three steady states when α is kept constant. The opposite holds when K_c is kept constant and α varies. For low values of α the rates of reaction and heat production are low, giving rise to a unique low-temperature steady state for the system. And for large values of α with a high rate of heat production, there will be a unique high-temperature steady state, while intermediate α values may lead to three steady states.

Note that a battery of three CSTRs and its dynamic behavior are studied in Section 6.4.

3.3   A Biochemical Enzyme Reactor

Similar behavior to that of the nonisothermal CSTR system will be observed in an isothermal bioreactor with nonmonotonic enzyme reaction, called a continuous stirred tank enzyme reactor (**Enzyme CSTR**). Figure 3.27 gives a diagram.

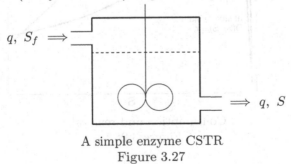

A simple enzyme CSTR

Figure 3.27

For simplicity, we assume that the enzymes are immobilized inside the reactor, i.e., there is no outflow of enzymes nor any washout. At the same time we assume that there is no mass transfer resistance. Then the governing equation is

$$q \cdot S_f = q \cdot S + V \cdot r \ , \tag{3.20}$$

where r is the rate of reaction per unit volume of the reactor.

Note: If r is expressed per unit mass then we multiply r by the enzyme concentration C_E, i.e., the mass of enzymes per unit volume of the reactor.

Dividing equation (3.20) by V gives us

$$\frac{q}{V} \cdot S_f = \frac{q}{V} \cdot S + r \ . \tag{3.21}$$

Here $D = q/V$ is the dilution rate measured in $(time)^{-1}$. It is the inverse of the residence time $\tau = V/q$. Therefore

$$D \cdot (S_f - S) = r \ . \tag{3.22}$$

Equation (3.22) can be easily solved: the nature of the solution(s) will depend on the rate of reaction function r. The right-hand side of equation (3.22) is called the **consumption function** $C(S) = r$, while the left-hand side $D(S_f - S)$ is called the **supply function** $S(S)$. One can solve this equation graphically for different types of kinetics, as shown in Figure 3.27.

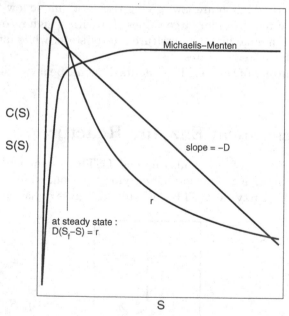

Consumption and removal
Figure 3.28

Figure 3.28 shows the consumption functions $C(S)$ for both nonmonotonic and Michaelis–Menten kinetics together with the removal function $R(S) = -D \cdot S$ in a line. The intersection(s) of $C(S)$ and $S(S)$ are the steady states. It is clear that for Michaelis–Menten[3] kinetics, i.e., for nonlinearity with saturation, see Figure 2.2, there is only one steady state for the whole range of D. This is the simplest case of a CSTR without bifurcation. However, for substrate-inhibited nonmonotonic kinetics as depicted by the nonmonotonic curve in Figure 3.28, more than one steady state may occur over a certain range of D values.

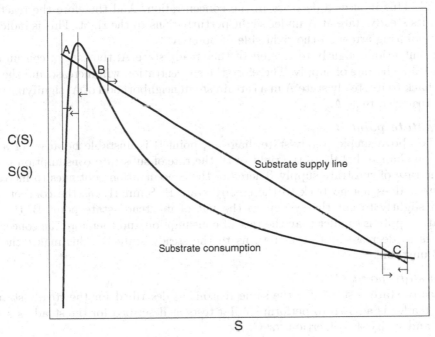

C(S)

S(S)

Substrate supply line

Substrate consumption

S

Local stability analysis of steady states
Figure 3.29

For nonmonotonic kinetics the stability details of the steady states A, B, and C are indicated by arrows in Figure 3.29. The stability behavior of the different steady states is explained below using chemico-physical reasoning. This applies to the earlier-mentioned adiabatic and nonadiabatic nonisothermal CSTRs as well.

Our definition of the stability of a steady state is as follows:

For a stable steady state:
If a small disturbance is made, the system will return to its initial steady state once the disturbance is removed or attenuated.

Unstable steady state:

---

[3]Leonor Michaelis, German chemist, 1875 – 1949
  Maude Leonora Menten, Canadian physician and biochemist, 1879 – 1960

If a small disturbance is made, the system will not return to its initial steady state when the disturbance is removed.

In Figure 3.29 we have indicated the stability behavior of the three steady states A, B, and C by horizontal arrows.

Steady-state point A

Using the simple steady-state diagram of Figure 3.29, point A is stable because slightly to the right of it, the rate of consumption of the substrate C is greater than the rate of supply S. This induces a decrease in the concentration. And therefore the reactor goes back to its steady state at A under slight perturbations to the right. This is indicated by the left-pointing arrow on the right side of point A.

For concentrations slightly to the left of the steady state A, the rate of consumption is smaller than the rate of supply. Therefore the concentration will increase and the reactor will go back to its steady state A in a certain small neighborhood of A, signifying stability of the steady state at A.

Steady state point B

From the above simple steady-state diagram, point B is unstable because for any concentration change slightly to the right of B, the rate of substrate consumption is smaller than the rate of substrate supply. Therefore the concentration continues to increase and the system does not go back to the steady state B. Similarly, as the concentration is lowered slightly to put the system to the left of its steady-state point B, the rate of substrate supply is smaller than the rate of consumption and therefore the concentration continues to decrease and never returns to the steady state B. This makes the steady state B unstable.

Steady state point C

This steady state is stable for the same reasons as described for the steady-state point A. The reader is advised to perform similar tests as described for the steady-state point A at C and verify this assertion for C.

One may be curious to know at which stable steady state a given CSTR is operating. This cannot be decided from the outside. The behavior of the reactor is determined by its previous history. This is a nonphysical and nonchemical initial condition for the reactor. The answer depends upon the dynamic behavior and the initial conditions of the system, and this will be discussed later.

3.4   Scalar Static Equations, *without Bifurcation*

As mentioned earlier, the approach of this book is to treat problems with bifurcations as the general case and problems without bifurcation as special cases.

3.4.1 Simple Examples of Reactions with No Possible Multiple Steady States

We should realize that certain chemical/biochemical problems can have no multiplicities of their steady states over their entire range of parameters. Consider, for example, a simple first-order reaction process $A \Rightarrow B$ with the rate equation

$$r = k \cdot C_A .$$

The simple design equation shall be

$$q \cdot C_A = q \cdot C_{A_f} - V \cdot k \cdot C_A ,$$

or in dimensionless form

$$C_{A_f} - C_A = \alpha \cdot C_A . \tag{3.23}$$

Thus the reactant consumption function $C(C_A)$ on the right-hand side of (3.23) is a straight line with slope α, or $C(C_A) = \alpha \cdot C_A$, and the reactant supply function $S(C_A)$ on the left-hand side of (3.23) is $S(C_A) = C_{A_f} - C_A$. We can easily solve (3.23) graphically to find the steady-state solution $C_{A_{ss}}$, since the steady state occurs when $C(C_{A_{ss}}) = S(C_{A_{ss}})$, or at the intersection of the two lines.

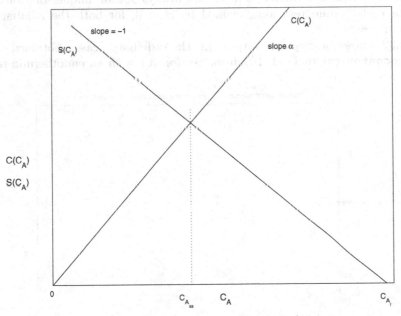

Simple consumption and removal
Figure 3.30

The change of $C_{A_{ss}}$, defined as the solution of (3.23) with respect to α, can be expressed by the simple function $C_A = C_{A_f}/(1 + \alpha)$. This equation is graphed in Figure 3.31.

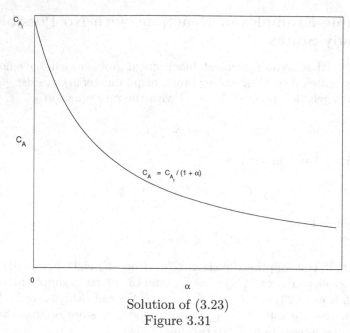

Solution of (3.23)
Figure 3.31

Note that for a nonisothermal CSTR we will always obtain unique solutions (no bifurcation) for endothermic reactions, defined by $\beta < 0$, for both the adiabatic and the nonadiabatic cases.

Figure 3.32 shows a typical output in the adiabatic case, obtained by calling `adiabNisocontourcurve (-.1,10)` from Section 3.1 with an endothermic reaction for $\beta = -0.1$.

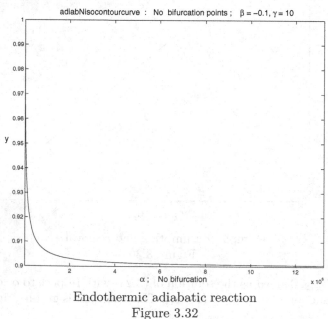

Endothermic adiabatic reaction
Figure 3.32

Similarly, the call of `runNadiabNisoalcurve(-.1,12,1,0,200,0.001,0,3.8*10^12)` from Section 3.3 gives this graph without any bifurcation for the endothermic nonadiabatic reaction when $\beta = -0.1$ is negative.

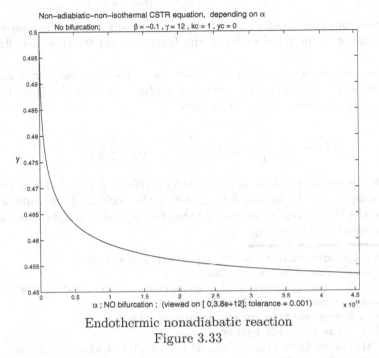

Endothermic nonadiabatic reaction
Figure 3.33

Note also that for biochemical systems, uniqueness of the solutions prevails over the whole range of systems with Michaelis–Menten kinetics.

3.4.2 Solving Some Static Transcendental and Algebraic Equations from the Chemical and Biological Engineering Fields

It is important to notice that chemical and biological engineers sometimes need to solve transcendental equations for a completely different purpose, such as when evaluating physical properties by solving a nonlinear (static) equation. In the following we investigate two such examples, one for determining friction factors and the other for finding specific volumes of ideal and nonideal gases.

The Colebrook Equation

The Colebrook[4] equation is used to calculate the friction of fluids in pipes, depending on

[4]Cyril Frank Colebrook, British civil engineer, 1910 – 1997

the diameter D of the pipe and on the Reynolds coefficient N_{Re} of the fluid as follows:

$$f = \left[-4 \cdot \log_{10} \left[\frac{\epsilon/D}{3.7} + \frac{1.256}{N_{Re} \cdot \sqrt{f}} \right] \right]^{-2} . \tag{3.24}$$

Here ϵ is the roughness coefficient, D is the inside diameter of the pipe, N_{Re} is the Reynolds number of the fluid, and f is the friction factor for turbulent flow inside the pipe.

This equation is transcendental in the unknown f. To solve for f with its two dimensionless parameters ϵ/D and N_{Re} given, we replace equation (3.24) by the equivalent equation in standard form $F(f) = 0$, namely

$$F(f) = f - \left[-4 \cdot \log_{10} \left[\frac{\epsilon/D}{3.7} + \frac{1.256}{N_{Re} \cdot \sqrt{f}} \right] \right]^{-2} = 0 . \tag{3.25}$$

This equation poses no problem at all for MATLAB's root-finder `fzero`, since F's zeros are always simple and the graph of F intersects the horizontal axis sufficiently steeply. Here is our MATLAB code `colebrookplotsolve.m`, which adapts itself automatically to the given inputs ϵ/D and N_{Re}.

```
function [xsol,iterations] = colebrookplotsolve(eps_D, NRe)
%          colebrookplotsolve(eps_D, NRe)
% Standard call : [xsol,iterations] = colebrookplotsolve(10^-4,10^5)
% Screen output : xsol = 0.00462915563470   iterations = 7
% Input : eps_D representing roughness/inside diameter;
%          NRe   the Reynolds number
% Adaptive algorithm that starts from start at .1 and finds an interval
% [bot top] with F(bot)*F(top) < 0.
% Then uses MATLAB's fzero to find the solution xsol.
% Blue curve represents F. Looking for its intersection with the red
% horizontal F = 0 line.

start = .1; A = colebrook(start,eps_D,NRe); iterations = -1; % adaptive start:
if A > 0, top = start; fb = 1; bot = start; % if F(start) > 0
    while fb == 1                           % searching for bot with F(bot) < 0
        bot = bot/2; c = colebrook(bot,eps_D,NRe); fb = sign(c);
        if c > 0, top = bot; end, end
elseif A < 0, bot = start; ft = -1; top = start; % if F(start) < 0
    while ft == -1                          % searching for top with F(top) > 0
        top = top*2; c = colebrook(top,eps_D,NRe); ft = sign(c);
        if c < 0, bot = top; end,  end
else, xsol = start; iterations = 0; end % if F(start) = 0 by sheer luck
if iterations < 0,                        % solve equation in interval [bot top]
    warning off,                          % turn off grandfather fzero warning
    [xsol,fval,exitflag,output] = fzero(@colebrook,[bot top],[],eps_D,NRe);
    iterations = output.iterations;       % solve and record # of iterations
    warning on, end
x = linspace(xsol*10^-5,xsol*2,200);    % x partition
```

```
F = x - 1./(-4 * log10(eps_D/3.7 + 1.256./(NRe*x.^0.5))).^2; % evaluate f(x)
plot(x,F,'b'), hold on, v = axis;           % plot F graph
ylabel('F  ','Rotation',0,'FontSize',14);
title(['Colebrook function plot  with  \epsilon/D = ',...
  num2str(eps_D,'%8.3g'),',',  N_{Re} = ',num2str(NRe,'%10.3g')],'FontSize',14),
plot([v(1) v(2)],[0 0],'r'),                % draw horizontal F = 0 line
xlabel(['f    (solution  f* = ',num2str(xsol,'%11.7g'),')'],'FontSize',14),
hold off

function F = colebrook(x,eps_D,NRe)        % Colebrook function
F = x - 1./(-4 * log10(eps_D/3.7 + 1.256./(NRe*x.^0.5))).^2;
```

The following numerical output is computed by calling `[xsol,iterations] =
colebrookplotsolve(10^-4,10^5)` and displayed on screen when `format long` is spec-
ified.

```
>> [xsol,iterations] = colebrookplotsolve(10^-4,10^5)
xsol =
   0.00462915563470
iterations =
   7
```

And the graphical output is shown in Figure 3.34.

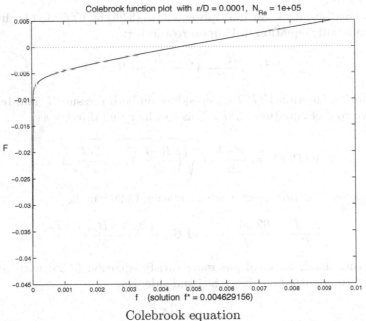

Colebrook equation

Figure 3.34

In the first ten lines of `colebrookplotsolve.m` the program searches for two values called
'top' and 'bot' with $F(top) \cdot F(bot) < 0$ in order to obtain an inclusion interval for the solution $bot < f^* < top$. The solution is then found (usually within ten or fewer iterations)

by MATLAB's `fzero` root finder, and the zero crossing of $F(f)$ is depicted graphically. MATLAB's `fzero` function is equipped to search for "top" and "bot" internally, but we rather determine "top" and "bot" adaptively ourselves.

Specific Volume of a Real Gas

The volume V of a real gas is related to the pressure P and temperature T according to the formula

$$P = \frac{R \cdot T}{V} + \frac{R \cdot T \cdot B}{V^2} . \tag{3.26}$$

Here B is calculated from the equation

$$\frac{B \cdot P_c}{R \cdot T_c} = B_0 + \omega \cdot B_1$$

for the critical pressure P_c and the critical temperature T_c of the gas. Note that

$$B_0 = 0.083 + \frac{0.422}{T_r^{1.6}}, \ B_1 = 0.139 - \frac{0.172}{T_r^{4.2}}, \ T_r = \frac{T}{T_c} ,$$

R is the gas constant $82.06 \ atm \cdot cm^3/(mol \cdot K)$, and ω is an empirical parameter, called the acentric factor.

In order to solve (3.26) for V, we multiply equation (3.26) by V^2/P to obtain the following normalized quadratic equation in V upon reordering:

$$V^2 - \frac{R \cdot T}{P} V - \frac{R \cdot T}{P} B = 0 . \tag{3.27}$$

Then the solution function $V(P,T)$, depending on both pressure P and temperature T, is the positive root of equation (3.27). This can be found directly as

$$V(P,T) = \frac{R \cdot T}{2P} + \sqrt{\left(\frac{R \cdot T}{2P}\right)^2 + \frac{R \cdot T}{P} B} . \tag{3.28}$$

Here the coefficient relations, given after equation (3.26), make

$$\frac{R \cdot T}{P} = \frac{82.06 \cdot T}{P} \ \text{ and } \mathcal{B} = \frac{(B_0 + \omega B_1) \, R \cdot Tc}{Pc} .$$

On the other hand, an ideal gas must satisfy equation (3.26) with its second, the quadratic V term, omitted on the right-hand side, i.e.,

$$P = \frac{R \cdot T}{V} . \tag{3.29}$$

This equation for the unknown volume V has the rather simple solution

$$V(P,T) = \frac{R \cdot T}{P} . \tag{3.30}$$

Note that the two volume formulas (3.28) and (3.30) differ only in the second term $(R \cdot T \cdot B)/P$ under the square root sign in (3.28).

For chemical and biological applications, one typically wants to compute the volume V of a real gas over a specified range of pressures and temperatures, such as for pressures and temperatures well below their critical points, but more often for pressures and temperatures above the critical points P_c and T_c of the gas. Notice that the condition of any "gas" above its critical point P_c and T_c is not really gaseous, nor is it a liquid. Such a "gas" is called a **supercritical fluid**.

Figures 3.35 – 3.37 show some plots, first of the volume V of an ideal gas under varying pressures $15 \leq P \leq 90$ and temperatures $215 \leq T \leq 1200$ using `idealgas3dplot.m` from our CD and calling `idealgas3dplot(15,90,215,1200);`.

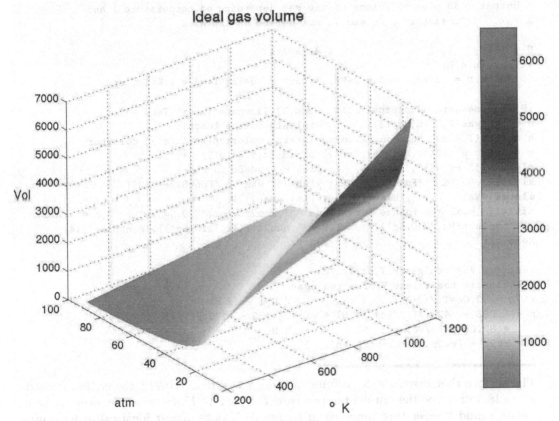

Volume of an ideal gas under pressure and temperature changes
Figure 3.35

As before, the "colors" of the computer-generated surface in Figure 3.35 denote the "height" above the T-P plane, i.e., the volume $V(P, T)$ of the gas for particular values of P and T. The volumes range from near $0 \ cm^3$ to well above $6000 \ cm^3$, and the height of a specific point on the volume surface can be read off the colorbar on the right edge of Figure 3.35.

We include the more complicated MATLAB m file `realgas3dplot.m`, which plots the volume as a surface, depending on the temperature and pressure for any real gas as shown in Figure 3.36.

```
function z = realgas3dplot(om,Pc,Tc,Pstart,Pend,Tstart,Tend)
%    z = realgas3dplot(om,Pc,Tc,Pstart,Pend,Tstart,Tend);
% Sample call   :  zr = realgas3dplot(.212,48.3,562.1,15,90,215,1200);
% Inputs: om (omega), critical pressure and temperature Pc, Tc;
%         optional: Pstart, Pend for pressure limits; [Pstart > 0 needed]
%                   Tstart and Tend for temperature limits.
%                   [Tstart > 215 needed]
% Output : 3D plot of volume of the gas depending on temperature T and
%           pressure P; Pc and Tc are marked on the axes.

N = 100;                        % defaults
if nargin < 7,
    Pstart = .2*Pc; Pend = 2*Pc; Tstart = .6*Tc; Tend = 1.8*Tc; end

[P T] = meshgrid(linspace(Pstart,Pend,N),linspace(Tstart,Tend,N));
z = realgas(P,T,om,Pc,Tc);      % evaluate volume function
h = surf(T,P,z); hold on; shading interp; colormap(hsv(128)); colorbar,
v = axis; plot3(Tc,Pc,v(5),'or'), plot3(Tc,v(3),v(5),'+r'),
plot3(v(1),Pc,v(5),'+r'),       % plot Pc and Tc in red +
xlabel(' ^o  K ','FontSize',12); ylabel(' atm  ','FontSize',12);
zlabel('Vol      ','Rotation',0,'FontSize',12);
title([' Real gas volume for \omega = ',num2str(om,'%10.5g'),' ,  P_c = ',...
    num2str(Pc,'%10.5g'),' ,   T_c = ',num2str(Tc,'%10.5g')],'FontSize',14);
hold off

function V = realgas(P,T,om,Pc,Tc)
% Evaluates the volume V of a real gas
rt2p = 82.06*T./(2*P);                  % R T / (2 P)
b = (.083 + .422./((T/Tc).^1.6) + om * (.139 - .172./((T/Tc).^4.2)) )...
    * 82.06 * Tc/Pc;            % B
V = rt2p + (rt2p.^2 + 2*rt2p.*b).^.5;
```

The surface that describes the volume of a real gas with $\omega = 0.212$, the critical pressure $P_c = 48.3\ atm$ and the critical temperature $T_c = 562.1°\ K$, for example, over the same pressure and temperature range as in Figure 3.35 looks almost identical in its central region to that of the ideal gas. And the difference between real gas and ideal gas volumes as a function of pressure and temperature seem to be rather small, except for very low temperatures, as seen by inspecting Figures 3.35 and 3.36 alone.

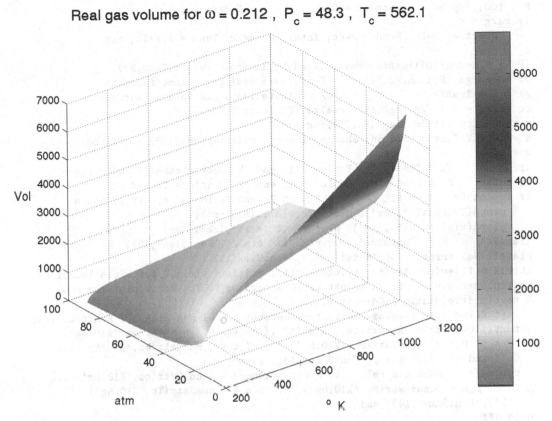

Volume of a real gas under pressure and temperature changes
Figure 3.36

However, the relative differences between the two surfaces in Figures 3.35 and 3.36 are rather large, in the neighborhood of 50%, for relatively high pressures combined with relatively low temperatures, and they are sizable, exceeding around 10% throughout. We can display this differing behavior of real and ideal gases with the help of idealrealgascompare.m.

```
function zc = idealrealgascompare(om,Pc,Tc,Pstart,Pend,Tstart,Tend)
%    zc = idealrealgascompare(om,Pc,Tc,Pstart,Pend,Tstart,Tend);
% Sample call    :  zc = idealrealgascompare(.212,48.3,562.1,15,90,215,1200);
% Inputs: om (omega), critical pressure and temperatures Pc, Tc;
%         optional: Pstart, Pend for pressure limits; [Pstart > 0 needed]
%                   Tstart and Tend for temperature limits.
%                   [Tstart > 215 needed]
% Output : 3D plot of relative volume ratio of an ideal and a real  gas with
%          given parameters om (omega), Pc, and Tc, depending on temperature
%          T and pressure P; Pc and Tc are marked on the axes and surface.
```

```
N = 100; Tcplot = 0; Pcplot = 0;                 % defaults
if nargin < 7,
     Pstart = .2*Pc; Pend = 2*Pc; Tstart = .6*Tc; Tend = 1.8*Tc; end

[P T] = meshgrid(linspace(Pstart,Pend,N),linspace(Tstart,Tend,N));
zr = realgas(P,T,om,Pc,Tc);      % evaluate real gas volume function
zi = idealgas(P,T);              % evaluate ideal gas volume function
zc = (zr - zi)./zr; rgcrit = realgas(Pc,Tc,om,Pc,Tc);
zcrit = (rgcrit-idealgas(Pc,Tc))/rgcrit;
h = surf(T,P,zc); hold on; shading interp; colormap(hsv(128)); colorbar,
v = axis;
if Tstart < Tc & Tend > Tc, Tcplot = 1; end  % check whether T and P range
if Pstart < Pc & Pend > Pc, Pcplot = 1; end  %    include Tc and Pc
if Tcplot*Pcplot == 1,           % plot Pc and Tc in red +,o if in range
     plot3(Tc,Pc,v(5),'or'), plot3(Tc,v(3),v(5),'+r'),
     plot3(v(1),Pc,v(5),'+r'), plot3(Tc,Pc,zcrit,'xk'), end
xlabel(' ^o  K ','FontSize',12); ylabel(' atm  ','FontSize',12);
zlabel('Rel error       ','Rotation',0,'FontSize',12);
ztitle = 1.4*v(6)-.4*v(5); xtitle = v(1); ytitle = v(4); % for title location
if Tcplot*Pcplot == 1,  % adjust title according to Tc, Pc in plotting range
  text(xtitle,ytitle,1.04*ztitle,...
     [' Relative gas volume deviations between a real gas with '],'FontSize',14);
  text(1.04*xtitle,.96*ytitle,.95*ztitle,['\omega = ',num2str(om,'%10.5g'),...
     '  ,  P_c = ',num2str(Pc,'%10.5g'),'  ,  T_c = ',num2str(Tc,'%10.5g'),...
     ' and an ideal gas'],'FontSize',14); else
  title(['   Real gas rel. vol. error  (\omega = ',num2str(om,'%10.5g'),...
     '  ,  P_c = ',num2str(Pc,'%10.5g'),'  ,  T_c = ',num2str(Tc,'%10.5g'),...
     ')'],'FontSize',14); end
hold off

function V = realgas(P,T,om,Pc,Tc)
% Evaluates the volume V of a real gas
rt2p = 82.06*T./(2*P);                   % R T / (2 P)
b = (.083 + .422./((T/Tc).^1.6) + om * (.139 - .172./((T/Tc).^4.2)) )...
  * 82.06 * Tc/Pc;                       % B
V = rt2p + (rt2p.^2 + 2*rt2p.*b).^.5;

function V = idealgas(P,T)
% Evaluates the volume V of an ideal gas
V = 82.06*T./P;                          % R T / P
```

Figure 3.37 is obtained by calling idealrealgascompare(.212,48.3,562.1,15,90,215, 1200);. Note that a real gas always exceeds an ideal gas in volume for a given pressure and temperature.

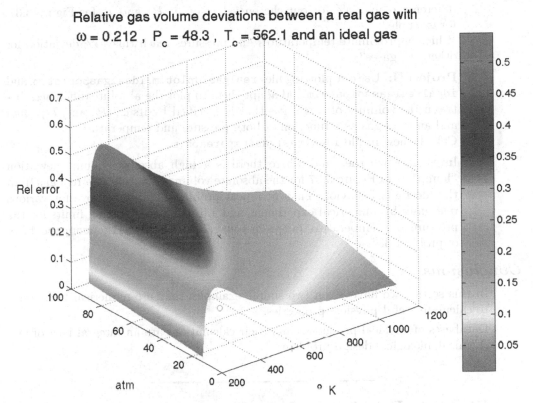

Relative gas volume deviations between a real gas with
$\omega = 0.212$, $P_c = 48.3$, $T_c = 562.1$ and an ideal gas

Relative volume differences of a real gas and an ideal gas
under pressure and temperature changes
Figure 3.37

In Figure 3.37, the relative gas volume deviations from the volume of an ideal gas are
expressed as percentages. A number such as 0.5 corresponds to a 50% increase, while 0.1
signifies a 10% increase in volume for the real gas.

Exercises for 3.4

1. **Project I:** Introduce a counter into our m file `colebrookplotsolve.m`, which
finds the number of function evaluations performed in the first 10 lines of
the code when trying to find an inclusion interval for the solution f^* of the
Colebrook equation.

   Which usually takes more effort, finding a root inclusion interval or solving
for f^* via `fzero` in MATLAB?

2. Investigate the high relative deviation "hump" in Figure 3.37 for various real
gases versus an ideal gas by adjusting the plotting limits for the pressure and
temperature ranges when calling `idealrealgascompare.m` in order to view
the hump and its location most clearly.

   For which approximate pressures and temperatures does the relative volume difference reach 75% for our chosen values of ω, P_c, and T_c for Figure 3.37, for example?

Which approximate temperatures and pressures reach this 75% deviation for other real gases?

3. **Project II:** Use our plotting files `realgas3dplot.m`, `idealgas3dplot.m`, and `idealrealgascompare.m` and adapt them to plot the absolute differences between the volumes of a real gas that is specified by its ω, P_c, and T_c values and an ideal gas as a function of both pressure and temperature.

   Call the new m file `idealrealgascompareabs.m`.

4. Investigate whether and where there is a high absolute volume deviation "hump" as in Figure 3.37 for the absolute volume deviation of a real gas from the ideal gas behavior. That is, repeat problems 2 and 3 above for various real gases by numerical experiment and adjusting the plotting limits for the pressure and temperature ranges in your calls of `idealrealgascompareabs.m` of problem 3.

Conclusions

In this section we have given two simple examples that use numerical techniques to calculate useful physical properties.

Databases of physical properties and their calculations are an integral part of any chemical/biological design problem.

Additional Problems for Chapter 3

1. Derive the steady-state mass balance equation for an isothermal CSTR in which a consecutive homogeneous reaction

$$ A \xrightarrow{k_1} B \xrightarrow{k_2} C $$

is taking place. Put the resulting equation in matrix form.

For $k_1 = 5.0\ sec^{-1}$ and $k_2 = 1.5\ sec^{-1}$, find the optimum residence time τ. (Hint: $\tau = V/q$, where V is the volume and q is the volumetric flow rate.)

2. A perfectly mixed adiabatic nonisothermal reactor carries out a simple first-order exothermic reaction in the liquid phase:

$$ A \longrightarrow B\ . $$

The product from this reactor is cooled from its output temperature to a temperature T_c and is then introduced into a separation unit in which the unreacted A is separated from the product B. The feed of the separation unit is split into two equal parts: top product and bottom product. Here the bottom product from the separation unit contains 95% of the unreacted A in the effluent of the reactor and

1% of B in the same stream. This bottom product (at the temperature T_c since the separation unit is iso-thermal) is recycled and remixed with the fresh feed to the reactor and the mixed stream is then heated to the reactor feed temperature T_f before being actually introduced into the reactor.

Write the steady-state mass and heat balance equations for this system, assuming constant physical properties and constant heat of reaction. (Note: Concentrate your modeling effort on the adiabatic nonisothermal reactor, and for the rest of the units, carry through a simple mass and heat balance in order to define the feed conditions for the reactor.)

Choose a set of physically feasible parameters for the system and write a MATLAB program to find the steady state.

3. Consider a system that initially consists of 1 mol of CO and 3 mol of H_2 at 1000 K. The system pressure is 25 atm. The following reactions are to be considered:

$$2\,CO + 2\,H_2 \quad \Leftrightarrow \quad CH_4 + CO_2 \tag{A}$$
$$CO + 3\,H_2 \quad \Leftrightarrow \quad CH_4 + H_2O \tag{B}$$
$$CO_2 + H_2 \quad \Leftrightarrow \quad H_2O + CO \tag{C}$$

When the equilibrium constants for the reactions (A) and (B) are expressed in terms of the partial pressure of the various species (in atm), the equilibrium constants for these reactions have the values $K_{P_A} = 0.046$ and $K_{P_B} = 0.034$. Determine the number of independent reactions, and then determine the equilibrium composition of the mixture, making use of a simple MATLAB program that you develop for this purpose.

4. The reaction of ethylene and chlorine in liquid ethylene dichloride solution is taking place in a CSTR. The stoichiometry of the reaction is $C_2H_4 + Cl_2 \rightarrow C_2H_4Cl_2$. Equimodular flow rates are used in the following experiment, which is carried out at $36°\ C$. The results of the experiment are tabulated in Table 3.1. Write a MATLAB program to carry out the following tasks:

   (a) Determine the overall order of the reaction and the reaction rate constant.

   (b) Determine the space time necessary for 65% conversion in a CSTR. Here "space time" is the residence time, i.e., the ratio between the volume of the reactor and the volumetric flow rate.

Experimental Data	
Space time	Effluent chlorine concentration
(s)	(mol/cm^3)
0	0.0116
300	0.0094
600	0.0081
900	0.0071
1200	0.0064
1500	0.0058
1800	0.00537

5. Some of the condensation reactions that take place when formaldehyde (F) is added to sodium paraphenolsulfonate (M) in an alkaline-aqueous solution have been studied. It was found that the reactions could be represented by the following eight equations:

$$
\begin{array}{lll}
(1) & F + M \rightarrow MA & k_1 = 0.15 \ \ l/(mol \cdot min) \\
(2) & F + MA \rightarrow MDA & k_2 = 0.49 \ \ l/(mol \cdot min) \\
(3) & MA + MDA \rightarrow DDA & k_3 = 0.14 \ \ l/(mol \cdot min) \\
(4) & M + MDA \rightarrow DA & k_4 = 0.14 \ \ l/(mol \cdot min) \\
(5) & MA + MA \rightarrow DA & k_5 = 0.04 \ \ l/(mol \cdot min) \\
(6) & MA + M \rightarrow D & k_6 = 0.056 \ \ l/(mol \cdot min) \\
(7) & F + D \rightarrow DA & k_7 = 0.50 \ \ l/(mol \cdot min) \\
(8) & F + DA \rightarrow DDA & k_8 = 0.50 \ \ l/(mol \cdot min)
\end{array}
$$

where M, MA, and MDA are monomers and D, DA, and DDA are dimers. The process continues to form trimers. The rate constants were evaluated using the assumption that the molecularity of each reaction is identical to its stoichiometry.

Derive a steady-state model for these reactions taking place in a single isothermal CSTR. Carefully define your terms and list your assumptions. Then write a MATLAB code to calculate the optimum residence time τ to give maximum yield of MA.

6. A Problem on Bifurcation

Derive the material and energy balance equations for a nonadiabatic CSTR where a single first-order reaction A $\rightarrow$ B takes place.

The reaction is exothermic. For $\beta = 1.2$ the exothermicity factor, and γ the dimensionless activities factor, find the range of α, the dimensionless preexponential factor, that gives multiplicity of steady states using a suitable MATLAB program with $K_c = 0$, the dimensionless heat transfer coefficient of the cooling jacket.

Choose a value of α in the multiplicity region and obtain the multiple steady-state dimensionless temperatures and concentrations.

For $y_c = 0.6$ find the minimum value of K_c, that makes the middle unstable saddle-type steady state unique and stable.

Develop a MATLAB program that gives you the values of α and K_c that result in multiple steady states.

Compute and plot the bifurcation diagram with α as the bifurcation parameter when $K_c = 0$. Discuss the physical meaning of the diagram.

Compute and plot the bifurcation diagram with K_c as the bifurcation parameter for $y_c = 0.6$ and the same value of α that you have chosen above to give multiplicity. Discuss the physical meaning of the diagram.

7. Another Problem on Bifurcation

Consider a consecutive reaction A $\rightarrow$ B $\rightarrow$ C in which both reactions are of first order and B is the desired product.

Derive the model equation for this reaction taking place in a liquid phase CSTR. Put the model equation into dimensionless form. Assume that the feed concentrations of B and C are zero. If $\beta_1 = 0.6$, $\beta_2 = 1.2$, $\gamma_1 = 18$, and $\gamma_2 = 27$ and the ratio between α_1 and α_2 is $\alpha_1/\alpha_2 = 1/1000$, find the range of α_1 (and α_2) that gives multiple steady states using a suitable MATLAB program.

Extend your program to find the optimum steady state that gives maximal yield of the desired product B. Is this steady state stable or unstable?

If it is unstable, develop and use a MATLAB program for a nonadiabatic CSTR and find the cooling jacket parameters K_c and y_c that will stabilize the unstable steady state.

8. **(Project A)**

Compare and try to combine the two solution methods of bisection and graphics to solve equation (3.3) in Section 3.1 for the adiabatic case.

(a) The graphics method generally has a larger residue $|f(x^*)| \approx 0$ at a solution x^* than the bisection method does if it successfully converges.

Experiment with increasing the number N of nodes used in the graphics method: What additional accuracy does one obtain by doubling N, for example, in adiabNisographsol.m? Learn to estimate the extra effort involved in doubling N by using the timing functions clock and etime of MATLAB. (Learn about these using >> help ... .) How effective is increasing N in your experiments?

(b) Design a hybrid program to solve equation (3.3) that uses the graphics solutions (for moderate N) from adiabNisographsol.m in Section 3.1 as seeds for a bisection algorithm modeled after solveadiabxy.m.

How close to the bifurcation limits does your bisection program succeed when the graphics solutions are used as starting points for fzero? What are the sizes of the residues in the computed solutions near the bifurcation points? Which of the proposed steady-state finders of part (a) or (b) do you prefer? Be careful and monitor your hybrid algorithm's effort via clock and etime.

9. **(Project B)**

In case the hybrid method of Project A above is successful and advisable for adiabatic CSTRs, repeat Project A for the nonadiabatic case and solve equation (3.12) (or any other equivalent form given in Section 3.2) using the MATLAB solvers NadiabNisoalgraphsol.m and solveNadiabxy.m of Section 3.2 in a modified, hybrid form. Is this hybrid method successful and efficient as well?

Start of the day on the Acropolis

Chapter 4

Initial Value Problems

Many chemical/biological engineering problems can be described by differential equations with known initial conditions, i.e., with known or given values of the state variables at the start of the process.

Different problems are modeled by two-point boundary value differential equations in which the values of the state variables are predetermined at both endpoints of the independent variable. These endpoints may involve a starting and ending time for a time-dependent process or for a space-dependent process, the boundary conditions may apply at the entrance and at the exit of a tubular reactor, or at the beginning and end of a counter-current process, or they may involve parameters of a distributed process with recycle, etc. Boundary value problems (**BVPs**) are treated in Chapter 5.

Initial value problems, abbreviated by the acronym **IVP**, can be solved quite easily, since for these problems all initial conditions are specified at only one interval endpoint for the variable. More precisely, for IVPs the value of the dependent variable(s) are given for one specific value of the independent variable such as the initial condition at one location or at one time. Simple numerical integration techniques generally suffice to solve IVPs. This is so nowadays even for *stiff differential equations*, since good stiff DE solvers are widely available in software form and in MATLAB.

Furthermore, IVPs have only one solution for the given initial values if there are sufficiently many initial conditions given, and therefore bifurcation plays no role once the initial conditions are specified. For dynamic systems, different initial conditions may, however, lead to different steady states; we refer to the fluidized bed reactor in Section 4.3, for example.

4.1   An Example of a Nonisothermal Distributed System

The problem of this section stems from a typical **distributed system**. In a distributed system the dependent state variables such as concentration and temperature vary in

135

terms of the independent variable such as length. This is the meaning of "distributed". Figure 4.1 shows a model of a tubular reactor at two incremental stages of the molar flow rate $n_i(l)$ for component i in terms of the variable length l.

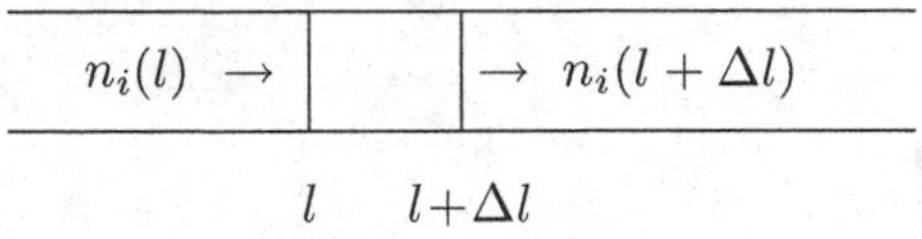

$n_i(l)\ \rightarrow$		$\rightarrow\ n_i(l + \Delta l)$

$$l \qquad l + \Delta l$$

Distributed system

Figure 4.1

For a distributed system the discrete mass balance design equation has the form

$$n_i(l + \Delta l) \;=\; n_i(l) + A_t \Delta l \sum_{j=1}^{N} \sigma_{ij} r_j \; , \qquad (4.1)$$

where n_i is the molar flow rate of component i, Δl is the length or thickness of the difference element, A_t is the total cross sectional area of the tube, N is the number of reactions, σ_{ij} is the stoichiometric number of component i in reaction j, and r_j is the generalized rate of reaction j.

If we form the difference quotient in (4.1) we obtain

$$\frac{n_i(l + \Delta l) - n_i}{\Delta l} \;=\; A_t \sum_{j=1}^{N} \sigma_{ij} r_j \; . \qquad (4.2)$$

If we take the limit of $\Delta l \longrightarrow 0$ in (4.2), we arrive at the equivalent differential equation

$$\frac{dn_i}{dl} \;=\; A_t \sum_{j=1}^{N} \sigma_{ij} r_j \qquad (4.3)$$

for $i = 1, 2, ..., M$, where M denotes the number of components.

For a single reaction, i.e., with $N = 1$, the above equation becomes

$$\frac{dn_i}{dl} \;=\; A_t \sigma_i r \; . \qquad (4.4)$$

To obtain the heat balance design equation, an enthalpy balance over the Δl element gives

$$\sum_{i=1}^{M} (n_i H_i)(l) + Q' \Delta l \;=\; \sum_{i=1}^{M} (n_i H_i)(l + \Delta l) \; , \qquad (4.5)$$

where

$M \quad = \quad$ total number of components involved (reactants plus products plus inerts), including components not in the feed but created during the reaction (which are the products)

$Q' \quad = \quad$ rate of heat removed or added per unit length of the reactor.

A simple rearrangement of (4.5) with $\Delta n_i H_i = (n_i H_i)(l + \Delta l) - (n_i H_i)(l)$ gives

$$\sum_{i=1}^{M} \frac{\Delta n_i H_i}{\Delta l} - Q' = 0 .$$

Taking the limit $\Delta l \longrightarrow 0$ gives rise to the corresponding differential equation

$$\sum_{i=1}^{M} \frac{d(n_i H_i)}{dl} - Q' = 0 .$$

The product rule of differentiation, i.e., $(f \cdot g)' = f \cdot g' + g \cdot f'$ for any two differentiable functions f and g, then leads to

$$\sum_{i=1}^{M} \left(n_i \frac{dH_i}{dl} + H_i \frac{dn_i}{dl} \right) - Q' = 0 . \qquad (4.6)$$

For a single-reaction we insert the differential equation (4.4) into (4.6) to obtain

$$\sum_{i=1}^{M} \left(n_i \frac{dH_i}{dl} + H_i A_t \sigma_i r \right) - Q' = 0 .$$

Rearrangement gives

$$\sum_{i=1}^{M} n_i \left(\frac{dH_i}{dl} \right) + \left(A_t r \sum_{i=1}^{M} H_i \sigma_i \right) - Q' = 0 .$$

Since

$$\sum_{i=1}^{M} \sigma_i H_i = (\Delta H) \equiv \text{ heat of reaction}$$

we get

$$\sum_{i=1}^{M} n_i \frac{dH_i}{dl} + A_t r (\Delta H) - Q' = 0 .$$

The above equation (for a single-reaction) can be written as

$$\sum_{i=1}^{M} n_i \frac{dH_i}{dl} = A_t r (-\Delta H) + Q' . \qquad (4.7)$$

For multiple reactions ($N > 1$) the corresponding equation is

$$\sum_{i=1}^{M} n_i \frac{dH_i}{dl} = \left(\sum_{j=1}^{N} A_t r_j (-\Delta H_j) \right) + Q' . \qquad (4.8)$$

To summarize, the design equations for the multiple-reactions case are (4.3) and (4.8), while for the single-reaction case the design equations are the equations (4.4) and (4.7).

4.1.1   Vapor-Phase Cracking of Acetone: An Example

Here we consider a nonisothermal, nonadiabatic tubular reactor as a distributed system. Our objective is to find its steady-state concentration and temperature profiles. Our specific example involves the vapor-phase cracking of acetone into ketone and methane, described by the endothermic reaction

$$CH_3COCH_3 \longrightarrow CH_2CO + CH_4 \ .$$

This takes place in a jacketed tubular reactor. Pure acetone enters the reactor at a temperature of $T_0 = 1030 \ K$ and a pressure of $P_0' = 160 \ kPa$. The temperature of the external cooling medium in the heat exchanger is constant at $T_J = 1200 \ K$. The other data is as follows:

Volumetric flow rate:	$q = q_f$ =	$0.003 \ m^3/s$
Volume of the reactor:	V_R =	$1.0 \ m^3$
Overall heat transfer coefficient between reactor and heat exchanger:	U' =	$110 \ W/(m^2 \ °K)$
Total heat transfer area between reactor and exchanger:	A =	$160 \ m^2/(m^3 \text{ of the reactor})$
Reaction rate constant:	k =	$3.56 \ e^{[34200 \cdot (1/1030 - 1/T)]} \ s^{-1}$
Heat of reaction:	ΔH_R =	$80700 + 6.7 \cdot (T - 298)$
		$-5.7 \cdot 10^{-3} \cdot (T^2 - 298^2)$
		$-1.27 \cdot 10^{-6} \cdot (T^3 - 298^3) \ J/mol$
Heat capacity of acetone :	C_{P_A} =	$26.65 + 0.182T - 45.82 \cdot 10^{-6}T^2 \ J/(mol \cdot K)$
Heat capacity of ketone :	C_{P_k} =	$20.05 + 0.095T - 31.01 \cdot 10^{-6}T^2 \ J/(mol \cdot K)$
Heat capacity of methane :	C_{P_M} =	$13.59 + 0.076T - 18.82 \cdot 10^{-6}T^2 \ J/(mol \cdot K)$

Our task is to determine the temperature and concentration (or conversion) profiles of the reacting gas along the length of the reactor. We assume constant pressure throughout the reactor.

4.1.2   Prelude to the Solution of the Problem

First we unify the units. The only parameters that need unification are U and P_0'. Since U' is given as $U' = 110 \ W/m^2$ it should be written as $U = 110 \ J/(m^2 \cdot sec \cdot K)$. With P_0' given as $160 \ kPa$, its units should be transformed to $P_0 = P_f = \frac{160}{101.3} = 1.58 \ atm$. The reaction is

$$A \longrightarrow B + C \ ,$$

where A is acetone, B is ketone, and C is methane.

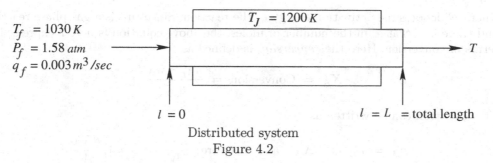

$$T_f = 1030\,K$$
$$P_f = 1.58\,atm$$
$$q_f = 0.003\,m^3/sec$$

Distributed system
Figure 4.2

For this system the material balance design equation in terms of the length l is

$$\frac{dn_A}{dl} = -A_t \cdot k_0 \cdot e^{-E/(RT)} \cdot C_A . \tag{4.9}$$

Since this equation uses the length l of the reactor as the independent variable, we have to find the diameter d_t of the tube in order to calculate the cross-sectional area A_t and the total required length L. This can be done as follows:

$$A = 160\,\frac{m^2}{m^3} = \frac{\pi \cdot d_t \cdot L}{(\pi \cdot d_t^2/4) \cdot L} = \frac{4}{d_t} .$$

Thus the diameter of the tube is $d_t = 4/A = 4/160 = 0.025\,m$ (or $2.5\,cm \approx 1\,inch$). To find the length we use

$$V_R = 1.0\,m^3 = \frac{\pi d_t^2}{4} \cdot L = \frac{\pi}{4} \cdot (0.025)^2 \cdot L ,$$

or

$$V_R = 1.0\,m^3 = 4.91 \cdot 10^{-4} \cdot L .$$

Hence $L = 1/4.91 \cdot 10^4\,m = 2036.36\,m$.

This very large length of more than $2\,km$, or about 1 mile and a quarter, can be divided into n connected tubes, each with a length of $2036.36/n\,m$. In this case each tube section has the volumetric flow rate $q(\text{per tube}) = 0.003/n\,m^3/(sec \cdot tube)$ and the number n can be 100 or 1000, for example, depending on the mechanical and locational factors. Alternatively the total length $L = 2036\,m$ can be achieved by a coiled reactor. The particular choice of apparatus is based on mechanical engineering factors.

It is also possible to circumvent these mechanical design decisions at the early process design stage by rewriting the differential equations (4.9) in terms of V, the volume of the reactor, instead of in terms of the length l. This is our chosen approach here. In it the independent variable V will vary from $V = 0$ to $V = V_R = V_{end} = 1.0\,m^3$.

4.1.3   Material Balance Design Equation in Terms of Volume

Analogous to (4.9), the material balance equation in terms of the volume V is

$$\frac{dn_A}{dV} = k_0 \cdot e^{-E/(R \cdot T)} \cdot C_A , \tag{4.10}$$

where E denotes the activation energy of the reaction. Since this is a gas phase reaction and there is a change in the number of moles, the above equation is better expressed in terms of conversion. Here the *conversion* is defined as

$$X_A \;=\; \text{Conversion} \;=\; \frac{n_{A_f} - n_A}{n_{A_f}} \,.$$

This equation can be written as

$$n_A \;=\; n_{A_f}(1 - X_A) \;\; \text{and therefore} \;\; \frac{dn_A}{dV} \;=\; -n_{A_f}\frac{dX_A}{dV} \tag{4.11}$$

and

$$C_A \;=\; \frac{n_A}{q} \;=\; \frac{n_{A_f}(1 - X_A)}{q} \,. \tag{4.12}$$

Here q is also a function of the total number of moles because

$$P \cdot q \;=\; n_t \cdot R \cdot T \;\; (\text{assuming } z \,=\, 1.0)\,, \tag{4.13}$$

where z is the compressibility factor. Since in an open system $P \,=\,$ constant, we have

$$q \;=\; n_t \cdot \frac{R \cdot T}{P}\,,$$

where n_t is the total number of moles at any conversion level, and n_t is given by

$$n_t \;=\; n_A + n_B + n_C \;=\; n_{A_f}(1 - X_A) + n_{A_f}X_A + n_{A_f}X_A \;=\; n_{A_f}(1 + X_A)\,.$$

Therefore

$$q \;=\; n_{A_f}(1 + X_A) \cdot \frac{R \cdot T}{P}\,. \tag{4.14}$$

Combining (4.12) and (4.14) we see that

$$C_A \;=\; \frac{n_{A_f}(1 - X_A)}{n_{A_f}(1 + X_A)\frac{R \cdot T}{P}} \;=\; \frac{(1 - X_A)P}{(1 + X_A) \cdot R \cdot T}\,. \tag{4.15}$$

Using the relations expressed in (4.11) and (4.14) in (4.10) gives us

$$n_{A_f}\frac{dX_A}{dV} \;=\; k_0 \cdot e^{-E/(RT)} \cdot \frac{n_{A_f}(1 - X_A)P}{n_{A_f}(1 + X_A)R \cdot T}\,.$$

Thus we obtain the differential equation for the conversion rate X_A in terms of V:

$$\frac{dX_A}{dV} \;=\; k_0 \cdot e^{-E/(RT)} \cdot \frac{(1 - X_A)P}{n_{A_f}(1 + X_A)R \cdot T} \;=\; F_1(X_A, T) \tag{4.16}$$

with the initial conditions $X_A \,=\, 0$ and $T \,=\, T_f \,=\, 1030\,K$ at $V = 0$. We may easily calculate n_{A_f} from

$$P \cdot q_f \;=\; n_{A_f}R \cdot T \qquad (P \,=\, \text{constant} \,=\, P_f \,=\, 1.58\,atm)\,.$$

The volumetric flow rate q_f in our setting is equal to $0.003 \ m^3/sec$ at the entrance. Thus

$$n_{A_f} = \frac{P \cdot q_f}{R \cdot T_f} \ . \tag{4.17}$$

Using equation (4.17) in (4.16) gives

$$\frac{dX_A}{dV} = k_0 \cdot e^{-E/(RT)} \cdot \frac{(1 - X_A)P}{\frac{P \cdot q_f}{R \cdot T_f}(1 + X_A)R \cdot T} = \frac{k_0 \cdot e^{-E/(RT)} \cdot (1 - X_A)}{q_f(1 + X_A)} \cdot \frac{T_f}{T}$$

with $q_f = 0.003 \ m^3/sec.$
From the given data for our problem we have

$$k_0 = 3.56 \cdot e^{\left(\frac{34200}{1030}\right)} = 9.37 \cdot 10^{14} \ sec^{-1} \text{ and } \frac{E}{R} = 34200.$$

Thus, the material balance design equation is

$$\frac{dX_A}{dV} = 9.37 \cdot 10^{14} \cdot e^{(-34200/T)} \cdot \frac{(1 - X_A)}{q_f(1 + X_A)} \cdot \frac{T_f}{T} \tag{4.18}$$

with the initial conditions $X_A = 0$ and $T = T_f = 1030 \ K$ at $V = 0$.

4.1.4   Heat Balance Design Equation in Terms of Volume

The heat balance design equation is

$$\sum_i n_i \frac{dH_i}{dV} = k_0 e^{-E/(RT)} \cdot C_A \cdot (-\Delta H) + Q'' = \text{R.H.S.} \ , \tag{4.19}$$

where we have abbreviated the expression $k_0 e^{-E/(RT)} \cdot C_A \cdot (-\Delta H) + Q''$ in (4.19) by the term R.H.S., or "right-hand side", and where Q'' denotes the heat transfer per unit volume, i.e., $Q'' = U \cdot A'' (T_J - T)$, $A'' = A = 160 \ m^2/m^3$, $U = 110 \ J/sec \cdot m^2 \cdot K$, and $T_J = T_c = 1200 \ K$. Since there is no phase change, equation (4.19) becomes

$$(n_A \cdot C_{P_A} + n_B \cdot C_{P_B} + n_C \cdot C_{P_C}) \cdot \frac{dT}{dV} = \text{R.H.S.} \tag{4.20}$$

for

$$
\begin{aligned}
C_{P_A} &= a_1 + b_1 T + c_1 T^2, & a_1 &= 26.65, & b_1 &= 0.182, & c_1 &= -45.82 \cdot 10^{-6}, \\
C_{P_B} &= a_2 + b_2 T + c_2 T^2, & a_2 &= 20.05, & b_2 &= 0.095, & c_2 &= -31.01 \cdot 10^{-6}, \\
C_{P_C} &= a_3 + b_3 T + c_3 T^2, & a_3 &= 13.59, & b_3 &= 0.076, & c_3 &= -18.82 \cdot 10^{-6}.
\end{aligned}
$$

We also know that $n_A = n_{A_f}(1 - X_A)$, $n_B = n_{A_f} X_A$, and $n_C = n_{A_f} X_A$. Thus equation (4.20) becomes

$$n_{A_f}(1 - X_A) \left[a_1 + b_1 T + c_1 T^2\right] \frac{dT}{dV} + n_{A_f} X_A \left[a_2 + b_2 T + c_2 T^2\right] \frac{dT}{dV} +$$

$$+ \ n_{A_f} X_A \left[a_3 + b_3 T + c_3 T^2\right] \frac{dT}{dV} = \text{R.H.S.} \ ,$$

or

$$n_{A_f} \left[(1 - X_A)(a_1 + b_1 T + c_1 T^2) + X_A(a_2 + b_2 T + c_2 T^2) + \right.$$
$$\left. + X_A(a_3 + b_3 T + c_3 T^2) \right] \frac{dT}{dV} =$$
$$= k_0 \cdot e^{-E/(RT)} \cdot \frac{n_{A_f}(1 - X_A)P}{n_{A_f}(1 + X_A)RT} \cdot (-\Delta H) + U \cdot A''(T_J - T) .$$

Hence

$$g(X_A, T) \frac{dT}{dV} = k_0 \cdot e^{-E/(RT)} \cdot \frac{(1 - X_A)P}{\frac{q_f \cdot P}{RT_f}(1 + X_A)RT} \cdot (-\Delta H) + \frac{U \cdot A''}{n_{A_f}}(T_J - T) .$$

And the heat balance design equation is

$$\frac{dT}{dV} = \frac{1}{g(X_A, T)} \left[k_0 \cdot e^{-E/(RT)} \cdot \frac{(1 - X_A)}{q_f \cdot (1 + X_A)} \cdot (-\Delta H) \frac{T_f}{T} + \frac{U \cdot A''}{\frac{P \cdot q_f}{R \cdot T_f}}(T_J - T) \right]$$
$$\tag{4.21}$$

with the initial conditions $X_A = 0$ and $T = T_f = 1030 \ K$ at $V = 0$.

4.1.5 Numerical Solution of the Resulting Initial Value Problem

We now have to solve the following system of two nonlinear coupled first-order ordinary differential equations for the given initial conditions:

$$\frac{dX_A}{dV} = 9.37 \cdot 10^{14} \cdot e^{\left(\frac{-34200}{T} \right)} \cdot \frac{(1 - X_A)}{q_f(1 + X_A)} \cdot \frac{T_f}{T} , \text{ and}$$

$$\frac{dT}{dV} = \frac{1}{g(X_A, T)} \left[\left[9.37 \cdot 10^{14} \cdot e^{\left(\frac{-34200}{T} \right)} \cdot \frac{(1 - X_A)}{q_f(1 + X_A)} \cdot \frac{T_f}{T} \cdot (-\Delta H) \right] + \frac{U \cdot A''}{\frac{P \cdot q_f}{RT_f}} \cdot (T_J - T) \right]$$
$$\tag{4.22}$$

with the initial conditions $X_A = 0$ and $T = T_f$ for $V = 0$, and with

$$g(X_A, T) = \left[(1 - X_A)(a_1 + 2b_1 T + 3c_1 T^2) \right] + \left[X_A(a_2 + b_2 T + c_2 T^2) \right] +$$
$$+ \left[X_A(a_3 + b_3 T + c_3 T^2) \right]$$

for
$$a_1 = 26.65, \ b_1 = 0.182, \ c_1 = -45.82 \cdot 10^{-6} ,$$
$$a_2 = 20.05, \ b_2 = 0.095, \ c_2 = -31.01 \cdot 10^{-6} ,$$
$$a_3 = 13.59, \ b_3 = 0.076, \ c_3 = -18.82 \cdot 10^{-6} .$$

Moreover,

$$(-\Delta H) = -80700 - 6.7(T - 298) + 5.7 \cdot 10^{-3}(T^2 - (298)^2) + 1.27 \cdot 10^{-6}(T^3 - (298)^3) .$$

We repeat the base set of parameters here:

$$q_f = 0.003 \; m^3/sec, \qquad V_R = V_{end} = 1.0 \; m^3, \quad U = 110 \; J/(sec \cdot m^2 \cdot K),$$
$$A'' = 160 \; m^2/m^3, \qquad\qquad T_J = 1200 \; K, \qquad\qquad T_f = 1030 \; K,$$
$$R = 82.06 \cdot 10^{-6} \; \tfrac{atm \cdot m^3}{mol \cdot K}, \qquad P = 1.58 \; atm \; .$$

To solve the initial value problem (4.22) numerically is a very simple procedure in MATLAB. MATLAB offers several integrators, all named ode.., for initial value problems of the form $dy(t)/dt = h(t,y)$, $y(t_0) = y_0$, such as ours. The alphanumeric designations ... that follow the root-name ode in MATLAB refer to the various methods of integration that are used. In our case, the simplest methods such as ode23 or ode45 which use explicit Runge-Kutta embedded integrator formulas (see Section 1.2) of orders two and three or four and five, respectively, may suffice. If the run times of the program exceed a few seconds, however, the integration problem is likely stiff, and we advise to use the stiff ODE solver ode15s of MATLAB instead. A formal definition of stiffness for DEs is given on p. 276 in Section 5.1; see also Section (H) of Appendix 1.

Here is our MATLAB program, specialized for acetone cracking, but easily adaptable for other chemico-physical processes.

```
function ...
[V, y] = fixedbedreact(qf,Adp,Tj,Tf,Vend,method,R,versus,S,halten,minTf,maxTf)
% fixedbedreact(qf,Adp,Tj,Tf,Vend,method,R,versus,S,halten,minTf,maxTf)
% Sample call :   [Note: Adapted for the specific data of acetone cracking.]
%      d = clock; fixedbedreact(.003,160,1200,1030,1,15,0,1); etime(clock,d)
% Inputs: physical constants qf, Adp, Tj, Tf;
%         Vend = end of volume interval, set to .1 if unspecified;
%         method = MATLAB ode function number; 15, 23, 45, or 113 are
%            supported. (Method 23 works fastest in general, method 15 refers to
%            the stiff ODE solver 15s in MATLAB which is used as default.)
%         If R = 0, we plot here; if R = 1, we are executing a run for multiple
%            input data and plot inside the respective ...run.m routines.
%         if versus = 1, we plot V in dependence of xA and T; else we plot
%            xA and T both in dependence of V (this generally gives more
%            meaningful graphics).
%         S is a color vector with possibly varying RGB values.
%         if halten = 1, we hold onto the subplots for multiple data graphs.
%         By default we set minTF = maxTF, unless in a call of Tfrun we want to
%            plot several curves for varying values of Tf whose graph are then
%            shifted a little to the right for a better view near V = 0.

if nargin == 4, Vend = .1; method = 15; R = 0; versus = 1; S = 'b';
   halten = 0; end
if nargin == 5, method = 15; R = 0; versus = 1; S = 'b'; halten = 0; end
if nargin == 6, R = 0; versus = 1; S = 'b'; halten = 0; end
if nargin == 7, versus = 1; S = 'b'; halten = 0; end   % setting default values
if nargin == 8, S = 'b'; halten = 0; end
if nargin == 9, halten = 0; end
if nargin <= 11, minTf = Tf; maxTf = Tf; end

Vspan = [0 Vend]; y0 = [0;Tf]; options = odeset('Vectorized','on'); % initialize
```

```
if method == 45, [V,y] = ode45(@fbr,Vspan,y0,options,qf,Adp,Tj,Tf); % integrate
elseif method == 23,  [V,y] = ode23(@fbr,Vspan,y0,options,qf,Adp,Tj,Tf);
elseif method == 113, [V,y] = ode113(@fbr,Vspan,y0,options,qf,Adp,Tj,Tf);
elseif method == 15, [V,y] = ode15s(@fbr,Vspan,y0,options,qf,Adp,Tj,Tf);
else disp('Error : Unsupported method number !'), return, end

if R == 0                            % no run if R = 0, in which case we plot:
subplot(2,1,1);                      % plot graph of solution to dxA/dV  IVP
if versus == 1, plot(y(:,1),V,'Color',S,'Linewidth',1); % switch graph axes
      if halten == 0, axis([0 1.1 -.1*Vend Vend]),  end
   else plot(V,y(:,1),'Color',S,'Linewidth',1);
      if minTf ~= maxTf, axis([-.1*Vend Vend 0 1.4]), end, end
if halten == 1, hold on, else hold off, end
if halten ~= 1,
   title(['          Tubular reactor (acetone): qf = ',num2str(qf,'%5.4g'),...
        ', A" = ', num2str(Adp,'%5.4g'),', Tj = ',num2str(Tj,'%5.4g'),...
        ', Tf = ',num2str(Tf,'%5.4g')],'FontSize',12), end
if versus == 1,
    xlabel('X_A','FontSize',12), ylabel('V     ','Rotation',0,'FontSize',12),
else xlabel('V','FontSize',12), ylabel('X_A     ','Rotation',0,'FontSize',12),
end
subplot(2,1,2);                      % plot graph of solution to dT/dV  IVP
if versus == 1, plot(y(:,2),V,'Color',S,'Linewidth',1); % switch graph axes
    if halten == 0, axis([.98*min(Tf,Tj) 1.01*max(Tf,Tj) -.1*Vend Vend]), end
   else plot(V,y(:,2),'Color',S,'Linewidth',1);
    if halten == 0, axis([0 Vend .98*min(Tf,Tj) 1.01*max(Tf,Tj)]), end,
      if minTf ~= maxTf, axis([-.1*Vend Vend minTf maxTf]), end, end
if halten == 1, hold on, else hold off, end
if versus == 1,
    xlabel('T','FontSize',12), ylabel('V     ','Rotation',0,'FontSize',12),
  else xlabel('V','FontSize',12), ylabel('T     ','Rotation',0,'FontSize',12),
end,
end              % end if run == 0

function dydt = fbr(V,y,qf,Adp,Tj,Tf)          % right hand side of ODE IVP
a1 = 26.65; a2 = 20.05; a3 = 13.59; b1 = .182; b2 = .095; b3 = .076;
c1 = -45.82*10^-6; c2 = -31.01*10^-6; c3 = -18.82*10^-6;
U = 110; largeten = 9.37*10^14; R = 82.06*10^-6; P = 1.58; fact = R*Tf/(P*qf);
dydt = [largeten*Tf*exp(-34200./y(2,:)).*(1-y(1,:))./(qf*(1+y(1,:)).*y(2,:));
   1./((1-y(1,:)).*(a1 + b1*y(2,:) + c1*y(2,:).^2) + ...
    y(1,:).*(a2 + a3 + (b2 + b3)*y(2,:) + (c2 + c3)*y(2,:).^2)).* ...
    (largeten*Tf*exp(-34200./y(2,:)).*(1-y(1,:))./(qf*(1+y(1,:)).*y(2,:)).* ...
    (-80700 - 6.7*(y(2,:)-298) + 5.7*10^-3*(y(2,:).^2 -298^2) + ...
    1.27*10^-6*(y(2,:).^3 -298^3))+ U*Adp*(Tj-y(2,:))*fact)];
```

This MATLAB code takes up to 12 inputs, of which the first four, namely the system parameters q_f, A'', T_J, and T_f, need to be user-specified. Depending on the specific problem, the remaining eight parameters may or may not be specified. In some of our

applications these parameters are internally set to their proper values in the multiple double graphing codes ...`run.m` that will be used later on for varying values of q_f, A'', T_J,
or T_f.

The lines that follow the initial % MATLAB comment lines in `fixedbedreact.m` set up
default values for the seven optional parameters. Then we prepare for the MATLAB
IVP solver `ode...` that solves our problem by using the function `dydt` to evaluate the
right-hand side of our IVP (4.22). Having solved (4.22) we plot two curves of the solution
to the two joint DEs.

The most demanding task of the above program involves importing and transferring
all the data constants correctly into the code and subfunction `dydt` and setting up the
right-hand side function `dydt` vectorially by replacing all multiplication, division, and
exponentiation signs in (4.22) by the corresponding dotted, i.e., by the vectorized `.*`, `./`,
and `.^` MATLAB operations for speedup wherever appropriate.

Figure 4.3 displays plots of the two-part solution to (4.22), graphed by plotting V
versus X_A and versus T from the initial value $V = 0$ until $V = V_{end} = 1 \ m^3$. The call of
`fixedbedreactor(.003,160,1200,1030,1,15,0,1)` creates Figure 4.3.

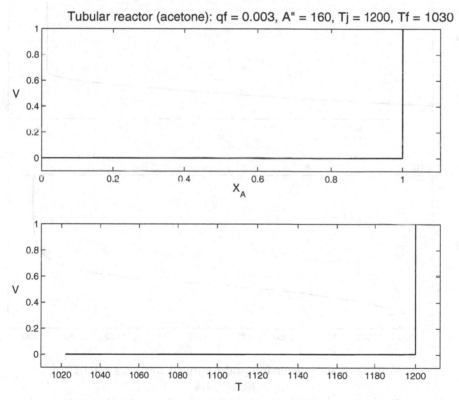

Volume versus conversion and temperature; $V_{end} = 1 \ m^3$
Figure 4.3

These graphs tell us immediately that for the given values of q_f and T_f the reaction goes to (near) full conversion $X_A \approx 1$ and to the maximal temperature $T = T_J = 1200\ K$ of the jacket at a very short distance from the start of the reactor. This means that with these conditions the reactor is extremely overdesigned. We also notice that the reactor temperature drops near the entrance of the reactor below the feed temperature of $T_f = 1030\ K$ to about $1022\ K$ according to the bottom graph in Figure 4.3. This means that in this small part at the entrance, the heat consumed by the endothermic reaction is larger than the heat transfer from the jacket.

It is advisable to graph and inspect the reaction more closely at the entrance of the reactor. After a few experiments with $V_{end} = 0.1$, 0.04, and 0.01, we have set the desired final volume to $V_{end} = 0.01\ m^3$ in Figure 4.4, which is drawn by the MATLAB command `fixedbedreact(.003,160,1200,1030,.01,15,0,1);` .

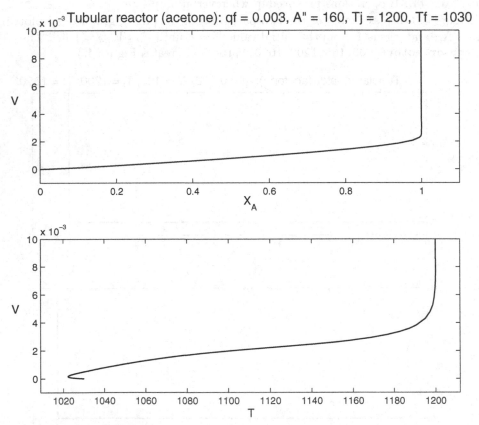

Volume versus conversion and temperature; $V_{end} = 0.01\ m^3$

Figure 4.4

The bottom plot of Figure 4.4 (with the V versus T graph) shows the initial drop in temperature at the entrance of the reactor more clearly than Figure 4.3.

Next we plot reversed graphs of the conversion X_A and the temperature T, both versus the volume V by specifying "versus = 2" in our `fixedbedreact.m` code with $V_{end} =$

0.006 and calling `fixedbedreact(.003,160,1200,1030,.006,15,0,2);`.

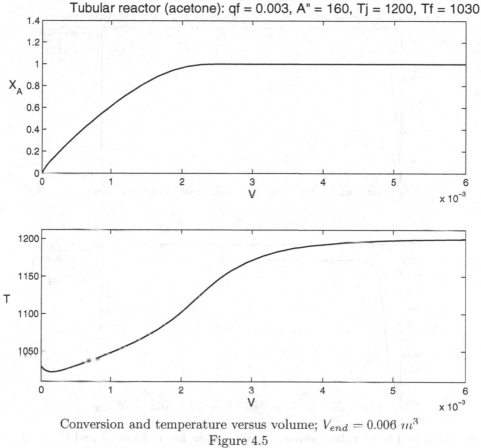

Conversion and temperature versus volume; $V_{end} = 0.006\ m^3$
Figure 4.5

Figure 4.5 again shows the temperature drop at the entrance. Since our feed volumetric flow rate is $q_f = 0.003\ m^3/sec$, which is very small, and the feed temperature $T_f = 1030\ K$ is quite high, we see from the plots of Figure 4.5 that the reaction reaches full conversion and maximal operating temperature when the reactor volume is approximately $0.006\ m^3$, or after the residence time $t = V/q_f$ reaches 2 seconds. In comparison, for the total volume $V_R = V_{end} = 1\ m^3$ the residence time is about $1/0.003\ sec$, or about 5 minutes and 40 seconds.

Next we want to investigate the effects of the feed rate q_f on the reactor. For $q_{f_1} = 0.0015$, $q_{f_2} = 0.003$, $q_{f_3} = 0.006$, and $q_{f_4} = 0.009$ we obtain the plots of Figure 4.6 using the program `qfrun.m`, which is printed below Figure 4.6. In the plots the solutions for varying values of q_f are color[1] coded as depicted in the **legend** on the right of each graph.

---

[1] References to color always refer to a computer generated color graph of the figure in question.

We advise our readers to replicate the graphs of Figures 4.6 through 4.11 on their computers for a better understanding.

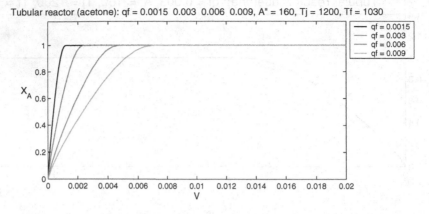

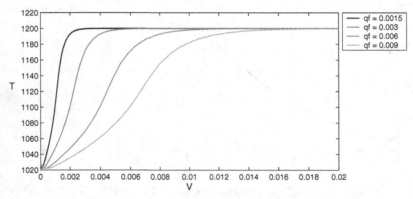

Conversion and temperature versus volume graphs for multiple q_f and $V_{end} = 0.02\ m^3$
Figure 4.6

Again we observe for all our chosen values of q_f that it takes less than 1 second of residence time for full conversion. Regarding the maximal temperature for $q_{f_4} = 0.009$, for example, full conversion is reached after a residence time of less than one second from the top graph in Figure 4.6, while the maximal temperature of $1200\ K$ is reached at the residence time of about $0.014/0.009 \approx 1.55\ sec$.

The actual MATLAB code `qfrun`.m for multiple q_f values is as follows:

```
function qfrun(qf,vend)
%       qfrun(qf,vend)
% Sample call : qfrun([.0015,.003,.006,.009],.02);
% Input : qf in vector form (increasing in value) with at most 6 entries,
%         vend = end of integration for the ODE
% Supporting m,file : fixedbedreact.m
% Output : plots two multiple line graphs of the fixed-bed-reactor xA and
%          T curves wrt V.

j = length(qf);          % maximally six qf values are supported
```

```
if j > 6, 'too many qf values', return, end,
Adp = 160; Tj = 1200; Tf = 1030; method = 15;  % standard settings (stiff DE)
for i = 1:j,            % create plot data for specified qf values
  [VV yy] = fixedbedreact(qf(i),Adp,Tj,Tf,vend,method,1,2,[],1);
  VV = VV'; yy = yy'; [k, lmax] = size(yy);
  if i > 1, % make size of new plot data compatible with previous plot data
      [k lm] = size(y1);
      if lmax > lm,        % pad old plot data in y1, y2, and V
          for m = lm+1:lmax
              y1(:,m) = NaN*ones(k,1); y2(:,m) = NaN*ones(k,1); end
          V = [V, NaN*ones(k,lmax-lm)]; end,
      if lmax < lm,        % plot new plot data in yy and VV
          for m = lmax+1:lm
              yy(:,m) = NaN*[1;1]; end;
          VV = [VV, NaN*ones(1,lm-lmax)]; end, end
  y1(i,:) = yy(1,:); y2(i,:) = yy(2,:); V(i,:) = VV; end % store new plot data
subplot(2,1,1);           % draw xA versus V plot
  plot(V',y1','Linewidth',1); v = axis; axis([v(1) v(2) 0 1.18]),
      title(['Tubular reactor (acetone): qf = ',num2str(qf,'%7.4g'),...
          ', A" = ',num2str(Adp,'%5.4g'),', Tj = ',num2str(Tj,'%5.4g'),...
          ', Tf = ',num2str(Tf,'%5.4g')],'FontSize',12),
  xlabel('V','FontSize',12), ylabel('X_A    ','Rotation',0,'FontSize',12),
  if j == 1       % include color coded legend (qf_1 = first qf value)
      legend(['qf = ',num2str(qf,'%7.4g')],0), end
  if j == 2
      legend({['qf = ',num2str(qf(1),'%7.4g')],...
              ['qf = ',num2str(qf(2),'%7.4g')]},0), end
  if j == 3
  legend({['qf = ',num2str(qf(1),'%7.4g')],['qf = ',num2str(qf(2),'%7.4g')],...
          ['qf = ',num2str(qf(3),'%7.4g')]},-1), end
  if j == 4
  legend({['qf = ',num2str(qf(1),'%7.4g')],['qf = ',num2str(qf(2),'%7.4g')],...
      ['qf = ',num2str(qf(3),'%7.4g')],['qf = ',num2str(qf(4),'%7.4g')]},-1), end
  if j == 5
  legend({['qf = ',num2str(qf(1),'%7.4g')],['qf = ',num2str(qf(2),'%7.4g')],...
      ['qf = ',num2str(qf(3),'%7.4g')],['qf = ',num2str(qf(4),'%7.4g')],...
      ['qf = ',num2str(qf(5),'%7.4g')]},-1), end
  if j == 6
legend({['qf = ',num2str(qf(1),'%7.4g')],['qf = ',num2str(qf(2),'%7.4g')],...
    ['qf = ',num2str(qf(3),'%7.4g')],['qf = ',num2str(qf(4),'%7.4g')],...
    ['qf = ',num2str(qf(5),'%7.4g')],['qf = ',num2str(qf(6),'%7.4g')]},-1), end
  hold off
subplot(2,1,2);              % draw T versus V plot
  plot(V',y2','Linewidth',1);
  xlabel('V','FontSize',12), ylabel('T    ','Rotation',0,'FontSize',12),
  if j == 1        % include color coded legend (qf_1 = first qf value)
      legend(['qf = ',num2str(qf,'%7.4g')],0), end
  if j == 2
      legend({['qf = ',num2str(qf(1),'%7.4g')],...
              ['qf = ',num2str(qf(2),'%7.4g')]},0), end
  if j == 3
  legend({['qf = ',num2str(qf(1),'%7.4g')],['qf = ',num2str(qf(2),'%7.4g')],...
          ['qf = ',num2str(qf(3),'%7.4g')]},-1), end
  if j == 4
  legend({['qf = ',num2str(qf(1),'%7.4g')],['qf = ',num2str(qf(2),'%7.4g')],...
      ['qf = ',num2str(qf(3),'%7.4g')],['qf = ',num2str(qf(4),'%7.4g')]},-1), end
```

```
if j == 5
legend({['qf = ',num2str(qf(1),'%7.4g')],['qf = ',num2str(qf(2),'%7.4g')],...
   ['qf = ',num2str(qf(3),'%7.4g')],['qf = ',num2str(qf(4),'%7.4g')],...
   ['qf = ',num2str(qf(5),'%7.4g')]},-1), end
if j == 6
legend({['qf = ',num2str(qf(1),'%7.4g')],['qf = ',num2str(qf(2),'%7.4g')],...
   ['qf = ',num2str(qf(3),'%7.4g')],['qf = ',num2str(qf(4),'%7.4g')],...
   ['qf = ',num2str(qf(5),'%7.4g')],['qf = ',num2str(qf(6),'%7.4g')]},-1), end
hold off
```

Note that for the ith value q_{f_i} of q_f this program stores the X_A and T graphing data in the ith rows of y1 and y2, respectively, and the corresponding V abscissas in the ith rows of V. To do so in the three matrices y1, y2, and V of our code, we have to fill in any data sets with shorter **support** than the maximal length support. Here by "support" we mean the set of data points. We do this by filling in short support data sets and setting every entry of the fill-in equal to NaN. These "not a number" NaN values are overlooked in plotting in MATLAB. Therefore we can plot the matrices Y_1^T and Y_2^T versus V^T directly without any ill effects of the variable lengths of support for the various inputs q_{f_i}. This unified matrix MATLAB plotting in turn allows us to add a coherent legend to the multiple plots. All other ..run.m programs in this section are designed similarly; for their codes we refer the user to the CD.

If we evaluate the first fractions of a second of residence time for the reactions with varying feeds q_f by calling qfrun([.0015,.003,.006,.009],.0003);, for example, we observe that in seemingly all instances of q_f, the minimal temperature that the reaction reaches is about 1022 K, as seen in Figure 4.7.

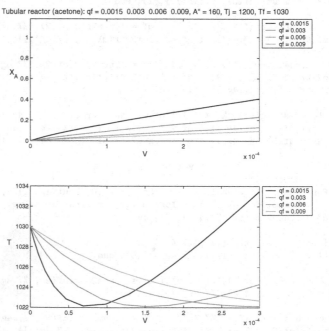

Conversion and temperature versus volume graphs for multiple q_f and
$$V_{end} = 0.0003 \ m^3$$
Figure 4.7

Next we study the effect of varying the heat transfer area A'' by setting $A_1'' = 300$, $A_2'' = 240$, $A_3'' = 160$, and $A_4'' = 80$ in the MATLAB call of Adprun([300,240, 160,80],.008). Again the curve for the first-mentioned value of A'' is denoted by A"$_1$, and so forth. Note that the solutions are increasing more slowly as A'' decreases, denoting a slower reaction for smaller areas A'' of heat transfer.

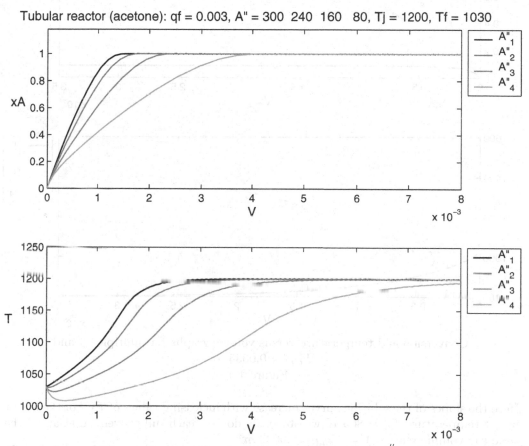

Conversion and temperature versus volume graphs for multiple A'' and $V_{end} = 0.008 \ m^3$
Figure 4.8

For $A'' = 300$ there is no apparent drop in temperature at the entrance part of the reactor, while for $A'' = 80$ the temperature drops the most, to around $1010 \ K$.

Next we show the effects observed by changing the jacket temperature T_J such as for $T_{J_1} = 800$, $T_{J_2} = 1200$, $T_{J_3} = 1400$, and $T_{J_4} = 2000 \ K$ while keeping all other parameters fixed as before. The calling sequence for this example is Tjrun([800,1200,1400,1600], 0.0035).

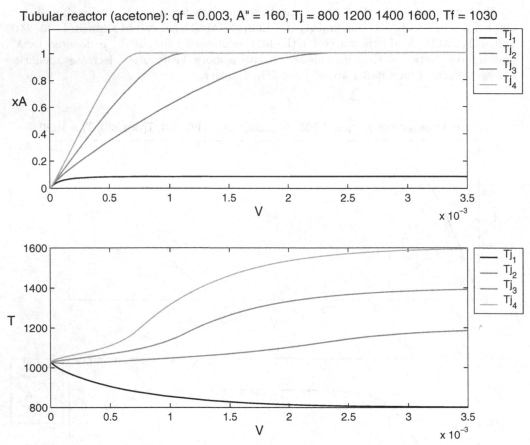

Conversion and temperature versus volume graphs for multiple T_J and
$$V_{end} = 0.0035 \ m^3$$
Figure 4.9

Here the slopes of the solution curves increase with increasing values of T_J. For the lowest jacket temperature $T_{J_4} = 800 \ K$ we obviously do not reach full conversion at all by the time the volume reaches $V = V_{end} = 0.0035 \ m^3$.

Note: *We advise our readers to recreate the multiple line graphs in Figures 4.6 to 4.11 in color on a computer screen and to inspect them individually. This will lead to a better visual understanding.*

The same advice holds for Sections 4.3, 5.1, 5.2, and 6.4.

Finally, we plot the effects of varying the feed temperature T_f for $T_{f_1} = 900$, $T_{f_2} = 1000$, $T_{f_3} = 1200$, $T_{f_4} = 1400$, and $T_{f_5} = 1600 \ K$ while keeping all other parameters fixed. The calling sequence is `Tfrun([900,1000,1200,1400,1600],.006)` for Figure 4.10.

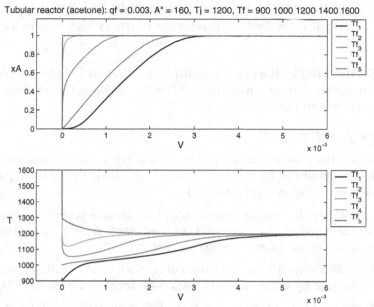

Conversion and temperature versus volume graphs for multiple T_f and $V_{end} = 0.006 \ m^3$

Figure 4.10

Figure 4.11 provides a closer look at the entrance of the reactor for various feed temperatures T_f by calling `Tfrun([900,1000,1200,1400,1600],.001)`.

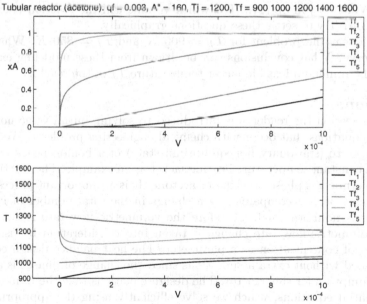

Conversion and temperature versus volume graphs for multiple T_f and $V_{end} = 0.001 \ m^3$

Figure 4.11

Note how instantaneously fast the switch from the initial condition $X_A = 0$ to full conversion $X_A = 1$ is when the feed temperature is relatively high such as $T_f = 1600\ K$ or $1400\ K$.

The respective MATLAB codes for multiple A'', T_J, and T_f value curves are all quite similar to the one given in qfrun.m above. Therefore we do not print these out here, but rather refer the user to the accompanying CD.

Exercises for 4.1

1. Study the code Tfrun.m on the CD and try to understand how and why we have shifted the left vertical axis $V = 0$ a little bit to the right in the plots such as in Figures 4.10 and 4.11.

2. Investigate the temperature drop at the entrance part of the reactor in Figure 4.5 more closely by running fixedbedreact.m for $V_{end} = 0.001\ m^3$ or less with the other parameters unchanged.

3. Redraw Figure 4.8 for low values of A'' such as $A'' = 40,\ 20$, and 10 until full conversion by using our MATLAB code Adprun.m from the CD.

4. Redraw Figure 4.9 for low values of T_f such as $T_f = 700$ and 500 until full conversion by using our MATLAB code Tfrun.m from the CD.

5. Use fixedbedract.m from the CD to plot the profiles for $q_f = 0.003$, $A'' = 160$, $T_J = 800$, $T_f = 1030$, and $V = 0.01,\ 0.1,\ 1.0,\ 10$, etc., thereby extending the two lowest graphs of Figure 4.9 until full conversion is reached if possible. What do you observe? For which volume will the reaction reach full conversion? Try to settle these questions graphically.
   Repeat the problem for $T_J = 900\ K$ and $T_J = 850\ K$. What do you observe? What conclusions can be drawn from these modeling experiments for the minimum feasible jacket temperature T_J of the reactor?

Conclusions

In this section the reader was introduced to solving initial value nonlinear differential equations that occur with chemical engineering problems. We have used the steady state (stationary nonequilibrium state) of a homogeneous tubular reactor with a constant temperature heating jacket as an example. The reaction used is the endothermic gas phase cracking of acetone. It is an important process in industry. This reaction is accompanied by a change in the total number of moles, and the reactor is nonisothermal. Therefore the volumetric flow rate is not constant and this volumetric flow rate change is taken into consideration by using conversion instead of concentration as a parameter. The heat balance design equations were developed without extra assumptions since "no change of phase" is a fact and not an assumption for this reactor. The design equations take the form of initial value differential equations, which we solve efficiently using the appropriate MATLAB IVP solvers. Our algorithms are used to investigate the effect of different operating and design parameters. It is clear that for the original operating parameters the reactor is overdesigned. The term "overdesigned" in this case means that the volume of the reactor is much more than necessary for the given parameters. As our graphs in Figures 4.3 to 4.8 show, a much smaller reactor of about 0.01 m^3 volume is sufficient.

4.2 Anaerobic Digester for Treating Waste Sludge

4.2.1 Process Description and Rate Equations

A continuous anaerobic digester that is used for treating municipal waste sludge is in principle like any other CSTR except that it involves multiphase biological reactions. In the digester the microorganisms grow biologically with the conversion. However, we treat such a system as a pseudohomogeneous system in this section. This means that we neglect all mass transfer resistances.

The feed material essentially consists of organisms with the concentration C_{X_f} measured in mol/m^3 and of substrate with the concentration C_{S_f} in mol/m^3. This is fed to the anaerobic digester at a daily volumetric feed flow rate of q m^3/day. The development of the model design equations involves the following assumptions:

1. The reactions of interest take place only in the liquid phase and the reactor's volume V in m^3 is constant.

2. The organism concentration C_X in the digester is uniform, as is the substrate concentration C_S. (The digester is perfectly mixed like a CSTR.)

We can obtain a model, i.e., the design equations for the digester process, by carrying out organism and substrate balances and using the following relationships:

1. The growth rate of the organism X is given by

$$r_X = \mu \cdot C_X \text{ in } g/(m^3 \cdot day) ,$$

where μ, the specific growth rate, is given by the Monod[2] function

$$\mu = \mu_0 \cdot \left(\frac{C_S}{K_S + C_S} \right) .$$

Here μ_0 is the maximum specific growth rate and K_S is the saturation constant.

2. The yield, i.e., the rate of growth of organisms with respect to the rate of consumption of substrate in the reactor, is given by

$$r_X = -Y_{XS} \cdot r_S ,$$

where Y_{XS} is the yield factor defined as the ratio of the microorganisms produced per consumed substrate.

---

[2] Jaques Monod, French biochemist, 1910 – 1976

4.2.2 Mathematical Modeling for a Continuous Anaerobic Digester

The process is shown schematically in Figure 4.12.

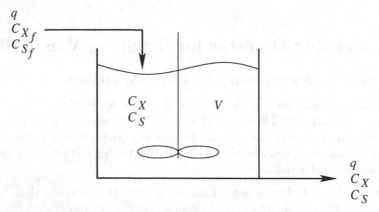

Schematic diagram of the anaerobic digestion process
Figure 4.12

The microbial reaction is given by $S \xrightarrow{X} P$. This is catalyzed by X, which also grows. The rate of production growth r_X of the microorganism X is given by

$$r_X = \mu \cdot C_X = \mu_0 \cdot \left(\frac{C_S}{K_s + C_S} \right) \cdot C_X \text{ in } g/(m^3 \cdot day) \ .$$

The rate of consumption r_s of substrate is given by

$$r_s = -\frac{1}{Y_{XS}} \cdot \mu_0 \cdot \left(\frac{C_S}{K_S + C_S} \right) \cdot C_X \text{ in } g/(m^3 \cdot day) \ ,$$

where the negative sign signifies that the substrate concentration decreases due to its consumption by the microorganisms. Note that r_X is positive because the microorganisms grow during the reaction.

An appropriate unsteady-state model is obtained from the microorganism and substrate material balances as follows:

For the substrate balance

$$V \cdot \frac{dC_S}{dt} = q \cdot C_{S_f} - q \cdot C_S - \frac{1}{Y_{XS}} \cdot \mu_0 \cdot \left(\frac{C_S}{K_S + C_S} \right) \cdot C_X \cdot V \ . \qquad (4.23)$$

And for the microorganism balance

$$V \cdot \frac{dC_X}{dt} = q \cdot C_{X_f} - q \cdot C_X + \mu_0 \cdot \left(\frac{C_S}{K_S + C_S} \right) \cdot C_X \cdot V \ . \qquad (4.24)$$

These two coupled differential equations are subject to the initial conditions $C_S(0) = C_{S_0}$ and $C_X(0) = C_{X_0}$ at $t = 0$.

Equations (4.23) and (4.24), together with their initial conditions, formulate the dynamic model for this microbial system. We note that the volume V of the digester is assumed to be constant.

A steady state for this system is reached when the variations in C_S and C_X have ceased, i.e., when their derivatives with respect to time t are equal to zero in equations (4.23) and (4.24). If we set $dC_S/dt = dC_X/dt = 0$ in (4.23) and (4.24), we obtain two coupled steady-state equations for the anaerobic digester, namely

$$q \cdot C_{S_f} - q \cdot C_{S_{ss}} - \frac{1}{Y_{XS}} \cdot \mu_0 \left(\frac{C_{S_{ss}}}{K_s + C_{S_{ss}}} \right) \cdot C_{X_{ss}} \cdot V = 0 \tag{4.25}$$

and

$$q \cdot C_{X_f} - q \cdot C_{X_{ss}} + \mu_0 \left(\frac{C_{S_{ss}}}{K_s + C_{S_{ss}}} \right) \cdot C_{X_{ss}} \cdot V = 0 . \tag{4.26}$$

Here $C_{S_{ss}}$ denotes the steady-state concentration of the substrate and $C_{X_{ss}}$ denotes the steady-state concentration of the microorganisms.

4.2.3 Solution of the Steady-State Equations

We shall try to express the volume V and the steady-state substrate conversion $X_{S_{ss}} := (C_{S_f} - C_{S_{ss}})/C_{S_f}$ in terms of the feed rates with the help of equations (4.25) and (4.26) for general and specific data.

From equation (4.26) we have

$$C_{X_{ss}} \cdot \left(q - \left(\frac{\mu_0 \cdot C_{S_{ss}}}{K_s + C_{S_{ss}}} \right) \cdot V \right) = q \cdot C_{X_f} ,$$

or

$$C_{X_{ss}} = \frac{q \cdot C_{X_f}}{\left(q - \left(\frac{\mu_0 \cdot C_{S_{ss}}}{K_s + C_{S_{ss}}} \right) \cdot V \right)} . \tag{4.27}$$

4.2.4 Steady-State Volume as a Function of the Feed Rate

Substituting (4.27) into (4.25) (multiplied by Y_{XS}) gives

$$q \cdot (C_{S_f} - C_{S_{ss}}) \cdot Y_{XS} = \mu_0 \left(\frac{C_{S_{ss}}}{K_s + C_{S_{ss}}} \right) \cdot V \cdot \frac{q \cdot C_{X_f}}{q - \left(\frac{\mu_0 \cdot C_{S_{ss}}}{K_s + C_{S_{ss}}} \right) \cdot V} . \tag{4.28}$$

Next we multiply (4.28) by its right-hand side denominator $q - (\mu_0 \cdot C_{S_{ss}} \cdot V)/(K_s + C_{S_{ss}})$. After multiplying out the resulting expression on the left-hand side we obtain

$$q^2 \cdot (C_{S_f} - C_{S_{ss}}) \cdot Y_{XS} - \frac{\mu_0 \cdot q \cdot (C_{S_f} - C_{S_{ss}}) \cdot Y_{XS} \cdot C_{S_{ss}} \cdot V}{K_s + C_{S_{ss}}} = \frac{\mu_0 \cdot q \cdot C_{S_{ss}} \cdot C_{X_f} \cdot V}{K_s + C_{S_{ss}}} . \tag{4.29}$$

Solving (4.29) for V gives us the following function in terms of the input parameter C_{X_f}:

$$V \;=\; V(C_{X_f}) \;=\; \frac{q \cdot (C_{S_f} - C_{S_{ss}}) \cdot Y_{XS} \cdot (K_s + C_{S_{ss}})}{\mu_0 \cdot C_{S_{ss}} \cdot (C_{X_f} + (C_{S_f} - C_{S_{ss}}) \cdot Y_{XS})} \;, \tag{4.30}$$

where C_{X_f} may range from 20 to 1000 g/m^3 and the other six parameters are specified in our example as given below. Note that C_{X_f} occurs only in the denominator of $V(C_{X_f})$. Thus $\lim_{C_{X_f} \to \infty} V(C_{X_f}) = 0$.

The physico-chemical parameters for our example are

$\mu_0 = 0.017 \; hr^{-1}$, $\;K_s = 32 \; g/m^3$, and $Y_{XS} = \frac{\text{grams of cells}}{\text{grams of substrate}} = 0.1$,

while the operating parameters are

$q = 7000 \; m^3/hr$, $\;C_{S_f} = 1400 \; g/m^3$, and C_{X_f} ranges between 20 and 1000 $\;g/m^3$.

If we consider a conversion (removal of substrate) rate of 99%, then there is 1% of the substrate left at the output, or $C_{S_{ss}} = 14 \; g/m^3$ since the digester receives $C_{S_f} = 1400 \; g/m^3$ of substrate at its input. Notice that the variable in (4.30) is $C_{S_{ss}}$ and that all parameters in (4.30) are now specified.

To reformulate the problem from a biological engineering point of view, we are given all the input and physical parameters, as well as the required conversion of the substrate S as $X_S = (1400 - 14)/1400 = 0.99$, or 99 % conversion. And we want to find how the size of the bioreactor changes with a change of the feed concentration of the microorganisms.

With our particular data we therefore study the equation

$$V(C_{X_f}) \;=\; \frac{4.46292 \cdot 10^7}{32.987 + 0.238 \cdot C_{X_f}} \quad (\text{in } m^3). \tag{4.31}$$

First we plot the graph of $V(C_{X_f})$ for C_{X_f} ranging from 20 g/m^3 to 1000 g/m^3 via our MATLAB file VversusCxf.m. Note that in VversusCxf.m the parameters are entered in l, i.e, "liters", while the output V is given in terms of cubic meters m^3 for convenience by dividing the expression for V in (4.30) or (4.31) by 1000 inside the program.

```
function [V, Cxf] = VversusCxf(Cxfstart,Cxfend,mu0,Ks,Yxs,q,Csf,Csss)
%   [V, Cx] = VversusCxf(Cxfstart,Cxfend,mu0,Ks,Yxs,q,Csf,Csss)
% Sample call :
%      VversusCxf(20,1000,.017,32,.1,7000,1400,14);
% Inputs : Cxfstart, Cxfend are the limits for Cxf values;
%               if they are identical, we just evaluate V(Cxf).
%          mu0, Ks, Yxs, q, Csf, and Csss are systems
%               parameters (in m^3)
% Output: V = V(Cxf) (in m^3) plotting data and plot.

top = q*(Csf-Csss)*Yxs*(Ks+Csss); bot1 = mu0*Csss*(Csf-Csss)*Yxs;
bot2 = mu0*Csss;             % preassembling constants used in (4.29), (4.30)
              % if the Cxf values are identical, we evaluate V(Cxf) only
if Cxfstart == Cxfend, Cxf = Cxfstart; V = top/(1000*(bot1+bot2*Cxf));
```

```
else             % if the Cxf values differ, we plot the V(Cxf) curve
   Cxf=linspace(Cxfstart,Cxfend,100); % creating Cx nodes
   V = top./(bot1+bot2.*Cxf);  %     and corresponding V values (in m^3)
   plot(Cxf,V,'LineWidth',1),   xlabel('C_{xf}','Fontsize',12),
   ylabel('V (m^3)             ','Fontsize',12,'Rotation',0),
   title(['              Anaerobic digester: \mu_0 = ',num2str(mu0,'%7.4g'),...
      ', K_s = ',num2str(Ks,'%5.4g'),',', Y_{xs} = ',num2str(Yxs,'%5.4g'),...
      ', q = ',num2str(q,'%5.4g'),',', C_{sf} = ',num2str(Csf,'%5.4g'),...
      ', C_{sss} = ',num2str(Csss,'%5.4g')],'FontSize',12), end
```

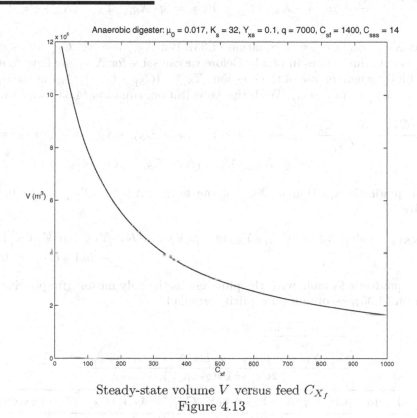

Steady-state volume V versus feed C_{X_f}
Figure 4.13

Figure 4.13 shows that the volume of the bioreactor necessary for achieving 99% sub-strate conversion decreases nonlinearly as the feed concentration of the microorganisms increases.

4.2.5   Steady-State Conversion in Terms of the Feed Concentration of the Microorganisms

Recall the expression $X_{S_{ss}} = (C_{S_f} - C_{S_{ss}})/C_{S_f}$ for the steady-state conversion. Here we want to express $X_{S_{ss}}$ as a function of the concentration of microorganisms C_{X_f} for a constant volume V of the digester. This requires us to modify (4.30) as follows.

We multiply equation (4.30) by the denominator of the right-hand side and divide the resulting equation by C_{S_f} throughout to obtain

$$V \cdot \frac{\mu_0 \cdot C_{S_{ss}} \cdot (C_{X_f} + (C_{S_f} - C_{S_{ss}}) \cdot Y_{XS})}{C_{S_f}} = \frac{q \cdot (C_{S_f} - C_{S_{ss}}) \cdot Y_{XS} \cdot (K_s + C_{S_{ss}})}{C_{S_f}} .$$

(4.32)

With $X_{S_{ss}} = (C_{S_f} - C_{S_{ss}})/C_{S_f}$, equation (4.32) becomes

$$\mu_0 \cdot V \frac{C_{S_{ss}} \cdot C_{X_f}}{C_{S_f}} + \mu_0 \cdot V \cdot X_{S_{ss}} \cdot C_{S_{ss}} \cdot Y_{XS} = q \cdot X_{S_{ss}} \cdot Y_{XS} \cdot (K_s + C_{S_{ss}}) . \quad (4.33)$$

Note that $X_{S_{ss}}$ occurs twice in equation (4.33). But $X_{S_{ss}}$ involves $C_{S_{ss}}$ in its definition and $C_{S_{ss}}$ occurs three times in (4.33). Before we can solve for $X_{S_{ss}}$, we have to eliminate $C_{S_{ss}}$ in (4.33) by making use of the relation $X_{S_{ss}} = (C_{S_f} - C_{S_{ss}})/C_{S_f}$ in the equivalent form $C_{S_{ss}} = C_{S_f} - X_{S_{ss}} \cdot C_{S_f}$. With this substitution, equation (4.33) becomes

$$\mu_0 \cdot V \frac{(C_{S_f} - X_{S_{ss}} \cdot C_{S_f}) \cdot C_{X_f}}{C_{S_f}} + \mu_0 \cdot V \cdot X_{S_{ss}} \cdot (C_{S_f} - X_{S_{ss}} \cdot C_{S_f}) \cdot Y_{XS} =$$

$$= q \cdot X_{S_{ss}} \cdot Y_{XS} \cdot (K_s + C_{S_f} - X_{S_{ss}} \cdot C_{S_f}) . \quad (4.34)$$

This is a quadratic equation in $X_{S_{ss}}$ of the form $a X_{S_{ss}}^2 + b X_{S_{ss}} + c = 0$, or more specifically

$$[C_{S_f} \cdot Y_{XS}(q - \mu_0 V)] X_{S_{ss}}^2 - (C_{S_f} \cdot Y_{XS}(q - \mu_0 V) + q \cdot K_s \cdot Y_{XS} + \mu_0 V \cdot C_{X_f}) X_{S_{ss}} +$$

$$+ \mu_0 V \cdot C_{X_f} = 0 . \quad (4.35)$$

Using the quadratic formula with the plus sign as the only meaningful positive solution for equation (4.35), we obtain an explicit formula for

$$X_{S_{ss}}(C_{X_f}) = \frac{-b + \sqrt{b^2 - 4ac}}{2a} = \quad (4.36)$$

$$= \frac{C_{S_f} \cdot Y_{XS}(q - \mu_0 V) + q \cdot K_s \cdot Y_{XS} + \mu_0 V \cdot C_{X_f}}{2 C_{S_f} \cdot Y_{XS}(q - \mu_0 V)} +$$

$$+ \frac{\sqrt{\left(C_{S_f} \cdot Y_{XS}(q - \mu_0 V) + q \cdot K_s \cdot Y_{XS} + \mu_0 V \cdot C_{X_f}\right)^2 - 4\mu_0 V \cdot C_{X_f} \cdot C_{S_f} \cdot Y_{XS}(q - \mu_0 V)}}{2 C_{S_f} \cdot Y_{XS}(q - \mu_0 V)} .$$

Here the variable C_{X_f} occurs only in the numerator, hence $\lim_{C_{X_f} \to \infty} X_{S_{ss}}(C_{X_f}) = \infty$.

We can apply this formula to an existing digester with a volume of $V = 7500\ m^3$ and with $q = 318\ m^3/hr$. All other parameters in this example are the same as before, namely $\mu_0 = 0.017\ hr^{-1}$, $K_s = 32\ g/m^3$, $Y_{XS} = 0.1$, and $C_{S_f} = 1400\ g/m^3$. We want to plot the steady-state conversion $X_{S_{ss}} = (C_{S_f} - C_{S_{ss}})/C_{S_f}$ for different values of C_{X_f} from 20 to 2000 g/m^3. Here is the MATLAB code XsssversusCxf.m for our problem.

```
function [Xsss, Cxf] = XsssversusCxf(Cxfstart,Cxfend,V,mu0,Ks,Yxs,q,Csf)
% [V, Cx] = XsssversusCxf(Cxfstart,Cxfend,V,mu0,Ks,Yxs,q,Csf)
% Sample call :
%     XsssversusCxf(20,2000,7500,.017,32,.1,318,1400);
% Inputs : Cxfstart, Cxfend are the limits for Cxf values;
%              if they are identical, we just evaluate Xsss(Cxf).
%          mu0, V, Ks, Yxs, q, and Csf are systems
%              parameters (in m^3)
% Output : Xsss = Xsss(Cxf) plotting data and plot (if Cxfstart /= Cxfend).

a = Csf*Yxs*(q-mu0*V);  % preassembling constant a used in (4.36)
if Cxfstart == Cxfend,  % for a single Cxf value we evaluate Xsss(Cxf) only
    Cxf = Cxfstart; minusb = a + q*Ks*Yxs + mu0*V*Cxf; c = mu0*V*Cxf;
    Xsss = (minusb + (minusb^2- 4*a*c)^0.5)/2*a;
else                    % for an interval of Cxf values we plot the V(Cxf) curve
  Cxf=linspace(Cxfstart,Cxfend,100);                 % creating Cx nodes
  minusb = a + q*Ks*Yxs + mu0*V*Cxf; c = mu0*V*Cxf; % polynomial coefficients
  Xsss = (minusb + (minusb.^2- 4*a*c).^0.5)/2*a;    % compute Xsss values
  plot(Cxf,Xsss,'LineWidth',1),   xlabel('C_{xf}','Fontsize',12),
  ylabel('X_{sss}          ','Fontsize',12,'Rotation',0),
  titlo([{' Anaerobic digester:'},{['V = ',num2str(V,'%5.4g'),...
      ', \mu_0 = ',num2str(mu0,'%7.4g'),', K_s = ',num2str(Ks,'%5.4g'),...
      ', Y_{xs} = ',num2str(Yxs,'%5.4g'),', q = ',num2str(q,'%5.4g'),...
      ', C_{sf} = ',num2str(Csf,'%5.4g')]}],'FontSize',12), end
```

This code generates a plot shown in Figure 4.14, of the steady-state conversion $X_{S_{ss}}$ for the above data.

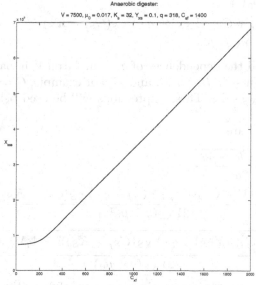

Steady-state conversion $X_{S_{ss}}$ versus feed C_{X_f}
Figure 4.14

Equation (4.36) makes the steady-state conversion $X_{S_{ss}}$ weakly nonlinear for varying feed rates C_{X_f}. This is so since C_{X_f} appears both in the leading linear term "$-b/2a$" and under the square root sign of $X_{S_{ss}}(C_{X_f})$ in equation (4.36).

Our last task in this subsection is to plot the steady-state substrate and organism concentrations $C_{S_{ss}}$ and $C_{X_{ss}}$ in terms of the volume V of the digester.

To do so, we take a different approach now and solve equation (4.25) for $C_{X_{ss}}$ first to obtain

$$C_{X_{ss}} = \frac{q(C_{S_f} - C_{S_{ss}})Y_{XS}(K_S + C_{S_{ss}})}{\mu_0 \cdot C_{S_{ss}} \cdot V} . \tag{4.37}$$

Next we plug this expression for $C_{X_{ss}}$ into equation (4.26) and obtain

$$q \cdot C_{X_f} - \frac{q^2(C_{S_f} - C_{S_{ss}})Y_{XS}(K_S + C_{S_{ss}})}{\mu_0 \cdot C_{S_{ss}} \cdot V} + q(C_{S_f} - C_{S_{ss}})Y_{XS} = 0 . \tag{4.38}$$

Equation (4.38) is a quadratic polynomial in $C_{S_{ss}}$ that depends on the variable V and the system parameters q, C_{X_f}, C_{S_f}, Y_{XS}, K_S, and μ_0. To see this, we first divide equation (4.38) by its common factor q and multiply by the denominator $\mu_0 \cdot C_{S_{ss}} \cdot V$. Then we rewrite it in the standard form $a \cdot C_{S_{ss}}^2 + b \cdot C_{S_{ss}} + c = 0$ of a quadratic polynomial as

$$
\begin{aligned}
Y_{XS}(q - \mu_0 V) \cdot C_{S_{ss}}^2 & \\
+ \big[\mu_0 \cdot V(C_{X_f} + C_{S_f} \cdot Y_{XS}) - q \cdot Y_{XS}(C_{S_f} - K_S)\big] \cdot C_{S_{ss}} & \\
- q \cdot C_{S_f} \cdot Y_{XS} \cdot K_S & = 0 .
\end{aligned}
\tag{4.39}
$$

This polynomial $a \cdot C_{S_{ss}}^2 + b \cdot C_{S_{ss}} + c = 0$ has two solutions $(-b \pm \sqrt{b^2 - 4ac})/(2a)$ with

$$
\begin{aligned}
a &= Y_{XS}(q - \mu_0 V) && (= A \cdot V + B), \\
b &= \mu_0 \cdot V(C_{X_f} + C_{S_f} \cdot Y_{XS}) - q \cdot Y_{XS}(C_{S_f} - K_S) && (= C \cdot V + D), \\
c &= -q \cdot C_{S_f} \cdot Y_{XS} \cdot K_S && (= E) .
\end{aligned}
$$

Here we have expressed the dependences of a, b, and c on V in parenthetical expressions for appropriate constants A, B, C, D, and E. For example, $C = \mu_0 \cdot (C_{X_f} + C_{S_f} \cdot Y_{XS})$ and $E = -q \cdot C_{S_f} \cdot Y_{XS} \cdot K_S$. These expressions will be used below to find the limit of $C_{S_{ss}}$ as $V \to \infty$.

The solutions to (4.39) are

$$
\begin{aligned}
C_{S_{ss}1,2}(V) &= \frac{-b \pm \sqrt{b^2 - 4ac}}{2a} = \tag{4.40} \\
&= \frac{-\mu_0 \cdot V(C_{X_f} + C_{S_f} \cdot Y_{XS}) - q \cdot Y_{XS}(C_{S_f} - K_S)}{2Y_{XS}(q - \mu_0 V)} \pm
\end{aligned}
$$

$$
\pm \frac{\sqrt{(\mu_0 \cdot V(C_{X_f} + C_{S_f} \cdot Y_{XS}) + q \cdot Y_{XS}(C_{S_f} - K_S))^2 + 4q Y_{XS}^2 C_{S_f} K_S(q - \mu_0 V)}}{2Y_{XS}(q - \mu_0 V)},
$$

of which only the one with the $+$ sign gives a meaningful positive steady-state substrate concentration $C_{S_{ss}}$.

The value of $C_{S_{ss}}$ depends on the variable V and the six system parameters q, C_{X_f}, C_{S_f}, Y_{XS}, K_S, and μ_0.

To find $\lim_{V \to \infty} C_{S_{ss}}(V)$ we write $C_{S_{ss}} = (-b \pm \sqrt{b^2 - 4ac})/(2a)$ in terms of the simplified expressions $a = A \cdot V + B$, $b = C \cdot V + D$, and $c = E$ as

$$C_{S_{ss}}(V) = \frac{-(C \cdot V + D) + \sqrt{(C \cdot V + D)^2 - 4E(A \cdot V + B)}}{2(A \cdot V + B)} . \tag{4.41}$$

Multiplying this simplified expression for $C_{S_{ss}}(V)$ by the conjugate expression

$$\frac{-(C \cdot V + D) - \sqrt{(C \cdot V + D)^2 - 4E(A \cdot V + B)}}{-(C \cdot V + D) - \sqrt{(C \cdot V + D)^2 - 4E(A \cdot V + B)}} ,$$

i.e., by 1, leads to the equivalent representation of $\lim_{V \to \infty} C_{S_{ss}}(V)$ as

$$\lim_{V \to \infty} C_{S_{ss}}(V) = \lim_{V \to \infty} \frac{4E(A \cdot V + B)}{2(A \cdot V + B) \cdot (-2)(C \cdot V + D)} = \lim_{V \to \infty} \frac{-E}{C \cdot V + D} = 0. \tag{4.42}$$

We will use this expansion of $C_{S_{ss}}$ in terms of V subsequently when studying the limit behavior of $C_{X_{ss}}$ as $V \to \infty$.

Equation (4.37) expresses $C_{X_{ss}}$ as a function of V that depends on the system parameters q, C_{X_f}, C_{S_f}, Y_{XS}, K_S, and μ_0, as well as the expression $C_{S_{ss}}(V)$ of (4.40), namely

$$C_{X_{ss}}(V) = \frac{q(C_{S_f} - C_{S_{ss}}(V))Y_{XS}(K_S + C_{S_{ss}}(V))}{\mu_0 \cdot C_{S_{ss}}(V) \cdot V} . \tag{4.37}$$

When contemplating the asymptotic behavior of $C_{X_{ss}}(V)$ as $V \to \infty$ in formula (4.37), we note that the numerator converges to $qC_{S_f} \cdot Y_{XS} \cdot K_S$ since $C_{S_{ss}}$ converges to zero as $V \to \infty$.

What about the term $C_{S_{ss}}(V) \cdot V$ in the denominator of (4.37)?

According to (4.42), $C_{S_{ss}}(V)$ behaves as $-E/(C \cdot V + D)$ does for large V. Thus

$$\lim_{V \to \infty} (C_{X_{ss}}(V) \cdot V) = \lim_{V \to \infty} \frac{-E \cdot V}{C \cdot V + D} = \frac{-E}{C} = \frac{qC_{S_f} \cdot Y_{XS} \cdot K_S}{\mu_0(C_{X_f} + C_{S_f} \cdot Y_{XS})} .$$

Hence

$$\lim_{V \to \infty} C_{X_{ss}}(V) = \frac{(qC_{S_f} \cdot Y_{XS} \cdot K_S) \cdot (\mu_0(C_{X_f} + C_{S_f} \cdot Y_{XS}))}{\mu_0 \cdot q \cdot C_{S_f} \cdot Y_{XS} \cdot K_S} = C_{X_f} + C_{S_f} \cdot Y_{XS} .$$

Figure 4.15 shows two plots of $C_{S_{ss}}(V)$ and $C_{X_{ss}}(V)$ for $1000 \leq V \leq 350000 \, m^3$, and our previous data set $q = 1000 \, m^3/hr$, $\mu_0 = 0.017 \, hr^{-1}$, $K_s = 32 \, g/m^3$, $Y_{XS} = 0.1$, $C_{S_f} = 1400 \, g/m^3$, and $C_{X_f} = 800 \, g/m^3$. Note that for this data,

$$\lim_{V \to \infty} C_{X_{ss}}(V) = C_{X_f} + C_{S_f} \cdot Y_{XS} = 800 + 1400 \cdot 0.1 = 940 \, g/m^3 .$$

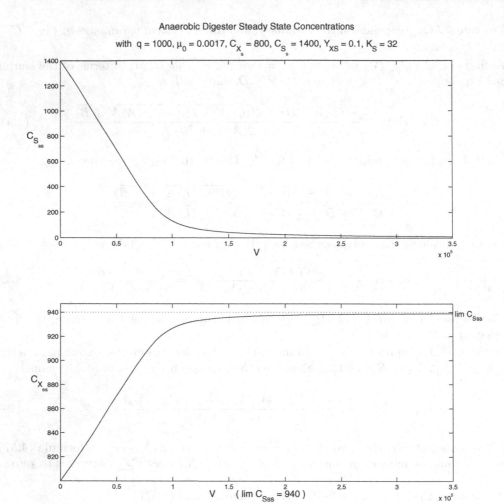

Steady-state concentration of organisms X and of substrate S in terms of volume V
Figure 4.15

This pair of plots was drawn by the MATLAB file `steadystCsCx.m`.

```
function limCSss = steadystCsCx(Vstart,Vend,q,m0,Cx,Cs,YXS,KS)
%   steadystCsCx(Vstart,Vend,q,m0,Cx,Cs,YXS,KS)
% sample call : steadystCsCx(10,350000,1000,.0017,800,1400,.1,32)
% Input : limits Vstart and Vend for the volume;
%         system parameters q, m0, Cx, Cf, YXS, and KS
% Output : two plots of CSss(V) and CXss(V)

n = 100; V = linspace(Vstart,Vend,n);        % initialize
a = YXS*(q-m0*V);                            % quadratic polynomial coeff.
b = m0*V*(Cx+Cs*YXS)-q*YXS*(Cs-KS);
c = -q*Cs*YXS*KS;
```

```
CS = (-b + (b.^2 - 4*a*c).^0.5)./(2*a);          % CSss data
CX = ((q*(Cs-CS)*YXS).*(KS+CS))./(m0*CS.*V);
limCSss = Cx + Cs*YXS;                % finding optimal CXss plot region limits :
ymaxCX = max(max(CX),limCSss); yminCX = min(min(CX), limCSss);
if ymaxCX == limCSss, yCXtop = ymaxCX + 0.05*(ymaxCX-yminCX);
else, yCXtop = ymaxCX; end
if yminCX == limCSss, yCXbot = yminCX - 0.05*(ymaxCX-yminCX);
else, yCXbot = yminCX; end
subplot(2,1,1), plot(V,CS), xlabel('V','Fontsize', 14),
ylabel('C_{S_{ss}}    ','Fontsize', 14,'Rotation',0),
title([{'Anaerobic Digester Steady State Concentrations'},{' '},...
       {['with  q = ', num2str(q,'%5.4g'),', \mu_0 = ',num2str(m0,'%5.4g'),...
         ', C_{X_s} = ',num2str(Cx,'%5.4g'),', C_{S_s} = ',...
       num2str(Cs,'%5.4g'),', Y_{XS} = ',num2str(YXS,'%5.4g'),...
       ', K_{S} = ',num2str(KS,'%5.4g')]}],'FontSize',12)

subplot(2,1,2), plot(V,CX), v = axis; axis([v(1),v(2),yCXbot,yCXtop]);
    xlabel(['V          ( lim C_{Sss} = ',...
           num2str(limCSss,'%5.4g'),' )'],'Fontsize', 14), hold on,
plot([v(1), v(2)],[limCSss, limCSss],':r');    % asymptotic line
text(v(2),0.97*(limCSss-v(3))+v(3),' lim C_{Sss}','Fontsize',12,'Color','r'),
ylabel('C_{X_{ss}}    ','Fontsize', 14,'Rotation',0),  hold off
```

4.2.6 The Unsteady-State Behavior of the Digester and the Solution of the Initial Value Differential Equations

Having studied the steady-state behavior of the digester, we now turn to its unsteady state or dynamic behavior.

If we divide the two differential equations (4.23) and (4.24) by V we obtain the following simpler version:

$$\frac{dC_S}{dt} = \frac{1}{V}\left[q \cdot C_{S_f} - q \cdot C_S - \frac{1}{Y_{XS}} \cdot \mu_0 \cdot \left(\frac{C_S}{K_s + C_S}\right) \cdot C_X \cdot V\right]$$

and   (4.43)

$$\frac{dC_X}{dt} = \frac{1}{V}\left[q \cdot C_{X_f} - q \cdot C_X + \mu_0 \cdot \left(\frac{C_S}{K_s + C_S}\right) \cdot C_X \cdot V\right] .$$

This gives rise to an IVP if we specify $C_S(0) = C_{S_0}$ and $C_X(0) = C_{X_0}$ at $t = 0$.

Here is a MATLAB code that solves (4.43) for any given initial values $C_S(0)$, $C_X(0)$ and for any system parameters q, C_{S_f}, Y_{XS}, μ_0, K_s, V, and C_{X_f} by plotting the time graphs of C_S and C_X from $t = 0$ until $t = t_{end}$.

```
function anaerdigest(tend,Cs0,Cx0,q,Csf,Yxs,mu0,Ks,V,Cxf,method)
%     anaerdigest(tend,Cs0,Cx0,q,Csf,Yxs,mu0,Ks,V,Cxf,method)
% Sample call : anaerdigest(240,1000,400,150,1400,0.1,0.0017,32,4000,800,45);
% Input : tend = final time (hours)
%         Cs0, Cx0 = intimal values for Cs and Cx
%         q, Csf, Yxs, mu0, Ks, V, Cxf = parameters for the system,
%              all scaled to be in m^3.
% Output: Plot of the Cs and Cx curves from time = 0 to tend in hours.

if nargin == 10, method = 23; end                % default method
                                                 % initialize :
tspan = [0 tend]; y0 = [Cs0;Cx0]; options = odeset('Vectorized','on');
                                                 % integrate :
if method == 45,
    [t,y] = ode45(@ADDE,tspan,y0,options,q,Csf,Yxs,mu0,Ks,V,Cxf);
elseif method == 23,
    [t,y] =  ode23(@ADDE,tspan,y0,options,q,Csf,Yxs,mu0,Ks,V,Cxf);
elseif method == 113,
    [t,y] = ode113(@ADDE,tspan,y0,options,q,Csf,Yxs,mu0,Ks,V,Cxf);
else disp('Error : Unsupported method number !'), return, end

subplot(2,1,1);                        % plot graph of solution to dCs/dt  IVP
plot(t,y(:,1),'Linewidth',1)
xlabel(['t hours      (with C_S(0) = ',num2str(Cs0,'%5.4g'),')'],...
     'FontSize',12), ylabel('C_S    ','Rotation',0,'FontSize',12),
title(['Anaerobic digester: q = ',num2str(q,'%5.4g'),', C_{S_f} = ',...
        num2str(Csf,'%5.4g'),', Y_{xs} = ',num2str(Yxs,'%5.4g'),...
       ', \mu_0 = ',num2str(mu0,'%5.4g'),', K_s = ',num2str(Ks,'%5.4g'),...
       ', V = ',num2str(V,'%5.4g'),', C_{X_f} = ',num2str(Cxf,'%5.4g')],...
     'FontSize',12),
subplot(2,1,2);                        % plot graph of solution to dCx/dt  IVP
plot(t,y(:,2),'Linewidth',1)
xlabel(['t hours      (with C_X(0) = ',num2str(Cx0,'%5.4g'),')'],...
     'FontSize',12), ylabel('C_X    ','Rotation',0,'FontSize',12),

function dydt = ADDE(t,y,q,Csf,Yxs,mu0,Ks,V,Cxf)    % r.h.s of the DE
  inter = y(1,:)./(Ks + y(1,:)) .* y(2,:);
  dydt = 1/V * [q*(Csf - y(1,:)) - mu0/Yxs *V .* inter;
                q*(Cxf - y(2,:)) + mu0*V .* inter];
```

We close with two plots drawn by `anaerdigest.m`. The first one uses the parameters $t_{end} = 240$ *hours* (10 *days*), $C_S(0) = 1000$ g/m^3, $C_X(0) = 400$ g/m^3, $q = 150$ m^3/hr, $C_{S_f} = 1400$ g/m^3, $Y_{XS} = 0.1$, $\mu_0 = 0.017$ hr^{-1}, $K_s = 32$ g/m^3, $V = 4000$ m^3, and $C_{X_f} = 800$ g/m^3. The call of `anaerdigest(240,1000,400,150,1400,0.1,0.0017,32,4000,800,15);` produces Figure 4.16 for this data.

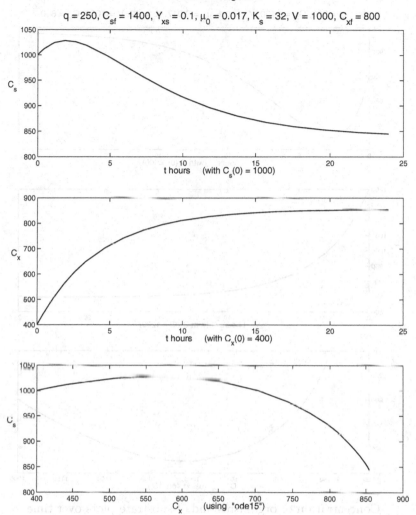

Anaerobic digester:

q = 250, C$_{sf}$ = 1400, Y$_{xs}$ = 0.1, μ$_0$ = 0.017, K$_s$ = 32, V = 1000, C$_{xf}$ = 800

Concentration of organisms and of substrate plots over time
and phase plot of C_{S_f} versus C_{X_f}
Figure 4.16

Figure 4.17 shows a second plot for a much higher growth rate μ_0 of the organisms
with the parameters $t_{end} = 24\ hours$, $C_S(0) = 900\ g/m^3$, $C_X(0) = 1200\ g/m^3$, $q =$
$300\ m^3/hr$, $C_{S_f} = 1400\ g/m^3$, $Y_{XS} = 0.1$, $\mu_0 = 0.017\ hr^{-1}$, $K_s = 32\ g/m^3$, $V =$
$1000\ m^3$, and $C_{X_f} = 800\ g/m^3$. For this set of parameters, the digester reaches its steady
state in about 1 day, or ten times faster than in the example of Figure 4.16. Figure 4.17 is
drawn by calling **anaerdigest(24,900,1200,300,1400,0.1,0.017,32,1000,800,15);**.

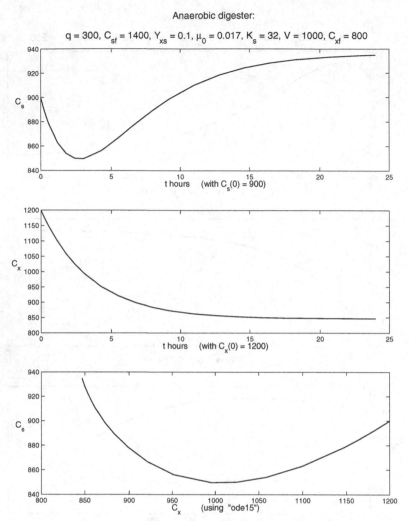

Concentration of organisms and of substrate plots over time
and phase plot of C_{S_f} versus C_{X_f}
Figure 4.17

Exercises for 4.2

1. (a) Study our `VversusCxf.m` code and learn how to use it to find the numerical value of the steady-state volume V for the data used for (4.31) when
   $C_{X_f} = 20 \ g/m^3$ and $1000 \ g/m^3$.

   (b) For the feed rate of $C_{X_f} = 1000 \ g/m^3$ in part (a), a waste-treatment plant
   is to be build in the form of 22 identical digesters. If the digesters are to
   have cylindrical form with a radius of $10 \ m$, how tall must each digester
   be?

2. Use `VversusCxf.m` for the data $\mu_0 = 0.198\ hr^{-1}$, $K_s = 250\ g/m^3$, $q = 7000\ m^3/hr$, $C_{S_f} = 1400\ g/m^3$, $Y_{XS} = 0.1$, and $C_{S_{ss}} = 14\ g/m^3$ when C_{X_f} ranges from $20\ g/m^3$ to $2000\ g/m^3$.

3. Find the steady-state volume V when $C_{X_f} = 20\ g/m^3$ and $2000\ g/m^3$ for the previous problem.

4. For exercises 2 and 3 above, show how during startup the system approaches its steady state.

Conclusions

In this section we have presented an important biological engineering example, namely waste-water treatment (WWT). The enzyme digester is treated like a CSTR, while the catalyzing microorganisms are growing with the reaction. A steady-state analysis and computer programs using either algebraic equations or IVPs have been introduced, discussed, and applied to standard problems.

4.3   Bubbling Fluidized Bed Catalytic Reactors (Heterogeneous Two-Phase System)

Gas-solid fluidized beds are very important in a number of petrochemical and petroleum refining processes. One of the most important one in the petroleum refining industry is the fluid catalytic cracking (FCC) process for the production of high-octane gasoline through the cracking of gas oil. In the petrochemical industry fluidized bed reactors are used in the UNIPOL® process for producing polyethelyne and polypropylene. These two industrial processes are presented and analyzed in some detail in Chapter 7 of this book.

4.3.1   Mathematical Modeling and Simulation of Fluidized Beds

When a gas stream is passed through a bed of fine powder in a tube (as shown in Figure 4.18), we observe the following:

1. In a certain low range of the flow rate, the system will behave like a fixed bed, i.e., the gas stream will flow through a fixed porous medium.

2. The minimum fluidization condition refers to the situation in which ΔP, the pressure drop across the bed, is equal to the bed weight.
   A slight expansion occurs in this case (see Figure 4.18). The bed will behave very similarly to a rough liquid formed by the gas and the fine powder, which together form a pseudophase. For example, this pseudophase can be transferred between two containers to make the levels equal in the two containers in a manner similar to liquids.

3. A freely bubbling fluidized bed has a higher flow rate of the gas than the minimum fluidization limit. It will reach the case of three phases: solid, gas in contact with solid, and gas in bubbles inside the tube.

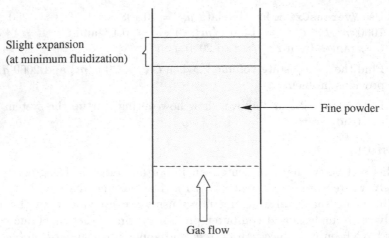

Slight expansion
(at minimum fluidization)

Fine powder

Gas flow

Fine powder inside a tube starting to fluidize due to gas flow rate
Figure 4.18

In this context we adhere to the two-phase theory of fluidization, which states that "almost all the gas in excess of that necessary for minimum fluidization will appear as gas bubbles" (Figure 4.19).

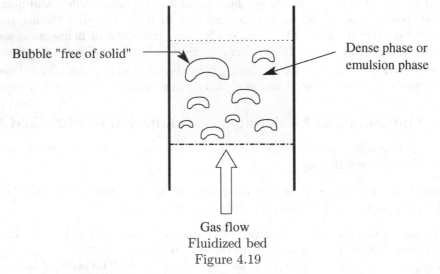

Bubble "free of solid"

Dense phase or
emulsion phase

Gas flow
Fluidized bed
Figure 4.19

There are advantages and disadvantages of fluidized beds for catalytic reactions.

Advantages of fluidized beds (freely bubbling)

1. Perfect mixing of solids occurs due to the presence of bubbles (isothermality).

2. Good heat transfer characteristics (high heat transfer coefficient).

3. The resistance of the pellets to heat and mass transfer to the surrounding is very small, making the rate of reaction closer to the intrinsic rate.

4. All the advantages of minimum fluidization conditions.

Disadvantages of fluidized beds

1. Bypassing of bubbles through the bed.

2. Bubble explosion on the surface causes high entrainment.

3. Difficult mechanical design; for example, some industrial FCC units have to handle 7000 tons of solids.

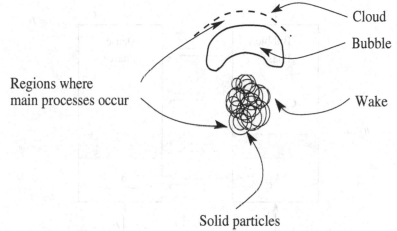

Diagram of the main regions around a bubble
Figure 4.20

Furthermore, we should note the following:

1. Bypassing of bubbles is compensated partlially by diffusion between dense and bubble phases (i.e., by delayed addition of reactants).

2. There is a solid exchange between the wake, the cloud, and the dense phase. This is accounted for in three-phase models.

3. Although the dense phase is perfectly mixed, the bubble phase is almost in plug flow.

4. It is possible to break the bubbles by using baffles or redistributors. Stirrers are not recommended because of vortex formation.

Mathematical formulation (steady state)

Using the two-phase model, a fluidized catalytic bed reactor can be divided into two regions, one for the dense phase, i.e., the emulsion phase, and another for the bubble phase, with associated mass and heat transfer between the two regions and phases. For illustration purposes we consider a simple reaction

$$A \rightarrow D .$$

We refer to Figure 4.21, where C_A is constant for all heights, i.e., A is in perfectly mixed condition independent of the height h, and $C_{AB} = f(h)$, since the bubble phase is in plug flow and therefore the concentration C_{AB} depends on the height h. Here C_A denotes the concentration of component A in the dense phase, and C_{AB} is the concentration of component A in the bubble phase.

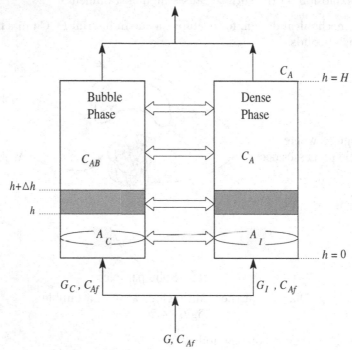

Schematic diagram of a two phase fluidized bed
Figure 4.21

Steady-state modeling of the dense phase

This is a heterogeneous system formed of two phases. For general principles of modeling heterogeneous systems, we refer to Chapter 6.

The molar balance on the dense phase gives us

$n_{if}+$ transfer of component i from the bubble phase to the dense phase $= n_i+V{\cdot}\sigma_i{\cdot}r$.

For component A with a constant volumetric flow rate G_I, we can write the above equation as

$$G_I \cdot C_{Af} + \int_0^H K_g a(C_{AB} - C_A)A_c \, dh \; = \; G_I C_A + A_I H \rho_b k C_A \, , \qquad (4.44)$$

where ρ_b is the bulk density of the solid at minimum fluidization conditions,
$a \; = \;$ (external surface area of bubbles)/(volume of bubble phase), and K_g is the mass

transfer coefficient between the bubble phase and the dense phase in $mol/(cm^2 \cdot sec)$.

Steady-state modeling of the bubble phase

Assuming that there is a negligible rate of reaction for an element in plug flow mode in the bubble phase, the molar balance can be expressed as

$$G_C C_{AB} = G_C(C_{AB} + \Delta C_{AB}) + K_g a(C_{AB} - C_A)A_c \Delta h \,,$$

see Figure 4.22. After a few manipulations and by taking the limit of $\Delta C_{AB}/\Delta h$ as $\Delta h \to 0$, we obtain the DE

$$G_C \frac{dC_{AB}}{dh} = - \underbrace{K_g a \, A_c}_{Q_E}(C_{AB} - C_A) \tag{4.45}$$

with the initial condition $C_{AB} = C_{Af}$ at $h = 0$.

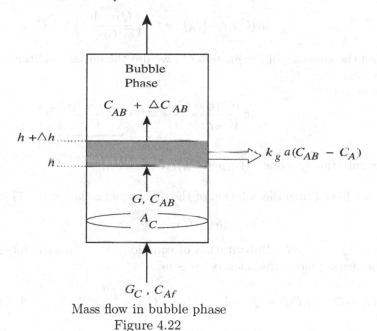

Mass flow in bubble phase
Figure 4.22

Thus, the molar balance gives us the following two equations:
For the dense phase: an integral equation according to (4.44),

$$G_I \cdot (C_{Af} - C_A) + Q_E \cdot A_C \int_0^H (C_{AB} - C_A)dh = A_I \cdot H \cdot \rho_b \cdot k \cdot C_A \,. \tag{4.46}$$

For the bubble phase: a differential equation according to (4.45),

$$G_C \frac{dC_{AB}}{dh} = -Q_E \cdot A_c \cdot (C_{AB} - C_A) \tag{4.47}$$

with the initial condition $C_{AB} = C_{Af}$ at $h = 0$. These two coupled integrodifferential equations (IDE) can be manipulated analytically as we shall see.

4.3.2   Analytical Manipulation of the Joint Integrodifferential Equations

Analytical solution of the differential equation (4.47)

By separation of variables we obtain

$$\frac{dC_{AB}}{(C_{AB} - \underbrace{C_A}_{\text{constant}})} = -\left(\frac{Q_E \cdot A_C}{G_C}\right) dh \ .$$

By integration we find that

$$\ln(C_{AB} - C_A) = -\left(\frac{Q_E \cdot A_C}{G_C}\right) h + C_1 \ .$$

To find the constant of integration C_1, we use the initial condition at $h = 0$ and obtain $C_1 = \ln(C_{Af} - C_A)$.

Thus we get

$$\ln \frac{(C_{AB} - C_A)}{(C_{Af} - C_A)} = -\underbrace{\left(\frac{Q_E \cdot A_C}{G_C}\right)}_{\bar{\alpha}} h \ .$$

Exponentiation gives us $(C_{AB} - C_A)/(C_{Af} - C_A) = e^{-\bar{\alpha}h}$.

Thus we have found the solution of the differential equation (4.47) to be

$$C_{AB}(h) - C_A = (C_{Af} - C_A) \cdot e^{-\bar{\alpha}h} \tag{4.48}$$

for $\bar{\alpha} = Q_E \cdot A_C/G_C$. Substitution of equation (4.48) into the integral equation (4.46) for the dense phase subsequently gives us

$$G_I \cdot (C_{Af} - C_A) + Q_E \cdot A_C \cdot (C_{Af} - C_A) \cdot \int_0^H e^{-\bar{\alpha}h} dh = A_I \cdot H \cdot \rho_b \cdot k \cdot C_A \ .$$

Since $\int_0^H e^{-\bar{\alpha}h} dh = (1 - e^{-\bar{\alpha}H})/\alpha$ and $\bar{\alpha} = Q_E \cdot A_C/G_C$ we get

$$G_I \cdot (C_{Af} - C_A) + G_C \cdot (C_{Af} - C_A) \cdot \left(1 - e^{-\frac{Q_E A_C}{G_C} H}\right) = A_I \cdot H \cdot \rho_b \cdot k \cdot C_A \ .$$

In more simplified and reorganized form this becomes

$$\left(G_I + G_C \cdot \left(1 - e^{-\frac{Q_E A_C}{G_C} H}\right)\right) \cdot (C_{Af} - C_A) = A_I \cdot H \cdot \rho_b \cdot k \cdot C_A \ . \tag{4.49}$$

Note the following:

1. The analogy between the above equations and the CSTR model can be easily realized. From this analogy we can define "the modified flow rate" with the following physical significance:

    (a) At very high values of Q_E the pseudo flow rate

$$G_I + G_C \cdot \left(1 - e^{-\frac{Q_E A_C H}{G_C}} \right)$$

    converges to $G = G_I + G_C$; thus equation (4.49) approaches that of a CSTR.

    (b) If $Q_E \approx 0$, then $G \approx G_I$; thus we have complete segregation.

2. When $G = G_I + G_C$, the output concentration C_{Aout} of A at $h = H$ is calculated from the relation

$$G_I \cdot C_A + G_C \cdot C_{AB} \, |_{h=H} \; = \; G \cdot C_{Aout} \, .$$

Heat balance design equations for fluidized beds

For a reaction $A \to D$ with the reaction constant $k = k_0 \cdot e^{-\frac{E}{RT}}$ we have the situation of Figure 4.23.

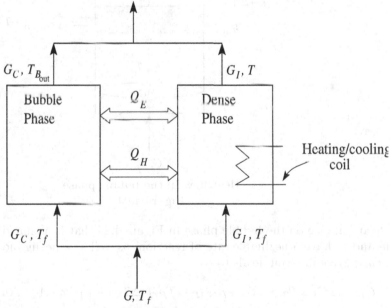

Heat transfer in the fluidized bed
Figure 4.23

The heat balance equation applied to the dense phase gives us

$$G_I \cdot \rho \cdot C_P \cdot (T - T_f) \; = \; A_I H \rho_b k C_A (-\Delta H) + \left(\int_0^H h_B a A_C (T_B - T) dh \right) - U A_J (T - T_J) \, .$$

$$(4.50)$$

Note that we consider that the heat supply/removal coil affects only the dense phase balance. Here

T_B = bubble phase temperature (variable),
T_J = jacket temperature,
T = dense phase temperature (constant),
U = heat transfer coefficient between the jacket and dense phase,
A_J = jacket area available for heat transfer, and
h_B = heat transfer coefficient between the bubble and the dense phase per unit volume of the bubble phase (in $(K \cdot J)/(sec \cdot cm^3)$).

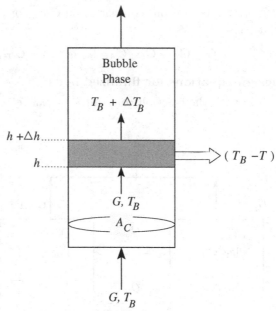

Heat flow in the bubble phase
Figure 4.24

The heat balance on the bubble phase in Figure 4.24 that is assumed to be in plug flow mode and to have a negligible rate of reaction, as well as a negligible heat transfer with the heating/cooling coil, leads to

$$G_C \cdot \rho \cdot C_P \cdot T_B \; = \; G_C \cdot \rho \cdot C_P \cdot (T_B + \Delta T_B) - h_B \cdot a \cdot A_C \cdot \Delta h \cdot (T - T_B) \,.$$

After rearranging and taking the limit of $\Delta T_B / \Delta h$ as $\Delta h \to 0$ we obtain the corresponding DE

$$G_C \cdot \rho \cdot C_P \cdot \frac{dT_B}{dh} \; = \; h_B \cdot a \cdot A_C \cdot (T - T_B) \,. \tag{4.51}$$

Analytical solution of (4.51)

By separating the variables in equation (4.51) we get

$$\frac{dT_B}{(T - T_B)} = \beta \cdot dh \,,$$

where $\beta = (h_B \cdot a \cdot A_C)/(G_C \cdot \rho \cdot C_P)$. Integration gives us $-\ln(T - T_B) = \beta \cdot h + C_2$. The integration constant C_2 can be calculated from the initial condition $T_B = T_f$ at $h = 0$ as $C_2 = -\ln(T - T_f)$.
Substituting this value back in, we get

$$\ln\left(\frac{T - T_B}{T - T_f}\right) = -\beta \cdot h \,.$$

And exponentiation finally gives us

$$\frac{T - T_B}{T - T_f} = e^{-\beta \cdot h} \,.$$

We use the above relation in the heat balance equation for the dense phase (4.50) in order to calculate the integral heat transfer term:

$$G_I \cdot \rho \cdot C_P(T - T_f) =$$

$$= A_I \cdot H \cdot \rho_b \cdot k_0 \cdot e^{-\frac{E}{R \cdot T}} \cdot C_A(-\Delta H) - h_B \cdot a \cdot A_C(T - T_f) \int_0^H e^{-\beta \cdot h} dh + U \cdot A_J(T - T_J) \,.$$

Note that

$$\int_0^H e^{-\beta \cdot h} dh = \frac{1}{\beta}(1 - e^{-\beta \cdot H}) = \frac{G_C \cdot \rho \cdot C_P}{h_B \cdot a \cdot A_C}(1 - e^{-\beta \cdot H}) \,.$$

Therefore we can write

$$\left[G_I \rho C_P + G_C \rho C_P (1 - e^{-\beta \cdot H})\right](T - T_f) = A_I H \rho_b k_0 e^{-\frac{E}{R \cdot T}} C_A(-\Delta H) + U A_J(T - T_J) \,.$$

For a fluidized bed without chemical reaction, the heat transfer equation thus will be

$$\rho C_P \left[G_I + G_C \left(1 - e^{-\beta \cdot H}\right)\right](T - T_f) = U \cdot A_J(T - T_J) \,. \tag{4.52}$$

We shall make use of these model equations at the end of Section 4.4 for our numerical computations.

4.3.3 Bifurcation and Dynamic Behavior of Fluidized Bed Catalytic Reactors

4.3.4 Dynamic Models and Chemisorption Mechanisms

From the mass balance equation (4.49) we can obtain the unsteady-state equations for a dynamic model by using an accumulation term as follows.

$$\left(G_I + G_C \cdot \left(1 - e^{-\frac{Q_B A_C}{G_C} H}\right)\right)(C_{Af} - C_A) = A_I \cdot H \cdot \rho_b \cdot k \cdot C_A + \text{Accumulation} \,. \tag{4.53}$$

The number of moles $\bar{n}_A$ of component A inside the reactor is

$$\bar{n}_A = A_I \cdot H \cdot \varepsilon \cdot C_A + A_I \cdot H \cdot \rho_b \cdot C_{AS} , \qquad (4.54)$$

where

$\bar{n}_A$	=	number of moles of component in the fluidized bed reactor,
ρ_b	=	bulk density of the catalyst = $\frac{g \text{ catalyst}}{cm^3 \text{ bed}}$ = $\rho_S(1-\varepsilon)$,
ρ_S	=	solid density of the catalyst = $\frac{g \text{ catalyst}}{cm^3 \text{ bed}}$.

Differentiation of $\bar{n}_A$ in equation (4.54) with respect to time t gives us

$$\frac{d\bar{n}_A}{dt} = A_I \cdot H \left(\varepsilon \cdot \frac{dC_A}{dt} + \rho_b \cdot \frac{dC_{AS}}{dt} \right) = \text{Accumulation} . \qquad (4.55)$$

Substituting (4.55) into (4.53) yields

$$\left(G_I + G_C \cdot \left(1 - e^{-\frac{Q_E A_C}{G_C}H} \right) \right) (C_{Af} - C_A) = A_I \cdot H \cdot \rho_b \cdot k \cdot C_A + A_I \cdot H \left(\varepsilon \cdot \frac{dC_A}{dt} + \rho_b \cdot \frac{dC_{AS}}{dt} \right) .$$
$$(4.56)$$

Thus we have obtained a differential equation in which two variables, C_A and C_{AS}, are functions of time. To solve for these coupled functions, either we have to find a simple relation between C_A and C_{AS}, or we need to develop an additional differential equation for C_{AS}.

We will follow the first path and build on the following physically justified details regarding the relation between C_A and C_{AS}. We consider the chemisorption step of the gas-solid catalytic reaction $A \to D$ as follows:

$$A + S \;\leftrightarrow\; AS \qquad \text{at equilibrium} ,$$
$$AS \;\to\; D + X .$$

Thus we can write

$$K_A = \frac{C_{AS}}{C_A \cdot C_S'} ,$$

where $C_S' = \bar{C}_m - C_{AS}$ and $\bar{C}_m$ is the total concentration of active sites, and we get

$$K_A \cdot C_A(\bar{C}_m - C_{AS}) = C_{AS} .$$

Thus

$$C_{AS} = \frac{K_A \cdot \bar{C}_m \cdot C_A}{1 + K_A \cdot C_A} .$$

This is a simple Langmuir[3] isotherm.

Considering low concentrations of C_A, i.e., a linear relation between C_A and C_{AS}, we obtain

$$C_{AS} = (K_A \cdot \bar{C}_m)C_A . \qquad (4.57)$$

---

[3]Irving Langmuir, US chemist and engineer, 1881-1957

By differentiating equation (4.57) with respect to time t (and assuming constant K_A and $\bar{C}_m$) we get

$$\frac{dC_{AS}}{dt} = (K_A \cdot \bar{C}_m)\frac{dC_A}{dt} \ . \tag{4.58}$$

Therefore, the final unsteady-state mass balance equation, or the dynamic model derived earlier in (4.56), takes the form

$$\left(G_I + G_C \cdot \left(1 - e^{-\frac{Q_E A_C}{G_C}H}\right)\right)(C_{Af} - C_A) = A_I \cdot H \cdot \rho_b \cdot k \cdot C_A + A_I \cdot H\left(\varepsilon + K_A \cdot \bar{C}_m \cdot \rho_b\right)\frac{dC_A}{dt} \ . \tag{4.59}$$

Note: It is always difficult to find usable values of K_A and $\bar{C}_m$, except for very common processes or reactions.

A nonlinear isotherm will complicate the capacitance term as follows:

$$\frac{dC_{AS}}{dt} = K_A \cdot \bar{C}_m\left((1 + K_A \cdot C_A)^{-1}\frac{dC_A}{dt} + C_A(-1)(1 + K_A \cdot C_A)^{-2}K_A\frac{dC_A}{dt}\right) \ .$$

In this case we have

$$\frac{dC_{AS}}{dt} = K_A \cdot \bar{C}_m\left(\frac{1}{1 + K_A \cdot C_A} - \frac{K_A \cdot C_A}{(1 + K_A \cdot C_A)^2}\right)\frac{dC_A}{dt} \ . \tag{4.60}$$

For a nonisothermal system the unsteady-heat balance equation (dynamic model) is given by

$$\bar{\gamma} \cdot (T - T_f) = A_I \cdot H \cdot k_0 e^{-\frac{E}{R \cdot T}}\rho_b \cdot C_A(-\Delta H) - U \cdot A_J(T - T_J) - \frac{d\bar{Q}}{dt} \ , \tag{4.61}$$

where $\bar{Q}$ is the heat content of the fluidized bed reactor,

$$\begin{aligned}\bar{\gamma} &= G_I \cdot \rho \cdot C_P + \left(G_C \cdot \rho \cdot C_P(1 - e^{-\beta \cdot H})\right) \ , \\ \bar{Q} &= A_I \cdot H \cdot \varepsilon \cdot \rho \cdot C_P \cdot T_g + A_I \cdot H \cdot \rho_b \cdot C_{PS} \cdot T_S \ , \end{aligned} \tag{4.62}$$

and ρ and C_P denote the gas phase density and the specific heat, respectively. Assuming negligible heat transfer resistances between the gas and solid, we have

$$T_g = T_S = T.$$

By differentiating equation (4.62) with respect to t we obtain

$$\frac{d\bar{Q}}{dt} = A_I \cdot H\left(\underbrace{\varepsilon \cdot \rho \cdot C_P}_{\text{negligible}} + \rho_b \cdot C_{PS}\right)\frac{dT}{dt} \ .$$

Therefore

$$\frac{d\bar{Q}}{dt} = A_I \cdot H \cdot \rho_b \cdot C_{PS}\frac{dT}{dt} \ . \tag{4.63}$$

The expression for $d\bar{Q}/dt$ in equation (4.63) can be substituted into equation (4.61) to obtain our final unsteady-state heat balance relation, namely

$$\gamma \cdot (T - T_f) = A_I \cdot H \cdot k_0 e^{-\frac{E}{R \cdot T}} \rho_b \cdot C_A(-\Delta H) - U \cdot A_J(T - T_J) - A_I \cdot H \cdot \rho_b \cdot C_{PS} \frac{dT}{dt} . \quad (4.64)$$

With these expressions, the unsteady-state equations become:
For the mass balance equation:

$$A_I \cdot H \cdot (\epsilon + K_A \bar{C}_m \rho_b) \frac{dC_A}{dt} = \gamma_M(C_{Af} - C_A) - A_I \cdot H \cdot \rho_b \cdot k \cdot C_A , \quad (4.65)$$

where $\gamma_M = G_I + G_C(1 - e^{-\bar{\alpha}H})$ for $\bar{\alpha} = Q_E \cdot A_C/G_C$.
And *for the heat balance equation:*

$$\frac{A_I \cdot H \cdot \rho_S \cdot C_{Ps}}{\rho \cdot C_P} \frac{dT}{dt} = \gamma_H(T_f - Y) + \frac{A_I \cdot H \cdot \rho_b \cdot k \cdot C_A(-\Delta H)}{\rho \cdot C_P} + \frac{U \cdot A}{\rho \cdot C_P}(T_J - T) \quad (4.66)$$

with $\gamma_H = G_I + G_C(1 - e^{-\beta H}) = \bar{\gamma}/(\rho \cdot C_P)$ and $k = k_0 \cdot e^{-E/RT}$.

In normalized time $\tau = t \cdot \rho \cdot C_P/(A_I \cdot H \cdot \rho_b \cdot C_{Ps})$, equations (4.65) and (4.66) take the form

$$\frac{dT}{d\tau} = \gamma_H(T_f - T) + \frac{A_I \cdot H \cdot \rho_b \cdot k_0}{\rho \cdot C_P} \cdot e^{-E/RT} \cdot C_A \cdot (-\Delta H) + \frac{U \cdot A}{\rho \cdot C_P}(T_J - T) \quad (4.67)$$

and

$$\frac{1}{Le_A} \frac{dC_A}{d\tau} = \gamma_M(C_{Af} - C_A) - A_I \cdot H \cdot \rho_b \cdot k_0 \cdot e^{-E/RT} \cdot C_A \quad (4.68)$$

with the Lewis[4] number

$$Le_A = \frac{\rho_b \cdot C_{PS}}{(\varepsilon + K_A \cdot \bar{C}_m \cdot \rho_b)\rho \cdot C_P} .$$

The dimensionless Lewis number measures the ratio of the heat capacitance term and the mass capacitance term.

With the normalized parameters

$$y = \frac{T}{T_{ref}}, \quad y_f = \frac{T_f}{T_{ref}}, \quad y_J = \frac{T_J}{T_{ref}}, \quad \beta = \frac{(-\Delta H) \cdot C_{Aref}}{\rho \cdot C_P \cdot T_{ref}}, \quad \gamma = \frac{E}{R \cdot T_{ref}},$$

$$x_A = \frac{C_A}{C_{Aref}}, \quad x_{Af} = \frac{C_{Af}}{C_{Aref}}, \quad \bar{K} = \frac{U \cdot A}{\rho \cdot C_p}, \quad \text{and} \quad \alpha = A_I \cdot H \cdot \rho_b \cdot k_0,$$

we finally rewrite the DEs (4.67) and (4.68) as

$$\frac{dy}{d\tau} = \gamma_H(y_f - y) + \alpha \cdot e^{-\gamma/y} \cdot x_A \cdot \beta + \bar{K}(y_J - y) \quad (4.69)$$

and

$$\frac{1}{Le_A} \frac{dx_A}{d\tau} = \gamma_M(1 - x_A) - \alpha \cdot e^{-\gamma/y} \cdot x_A . \quad (4.70)$$

---

[4]Warren K. Lewis, US chemical engineer, 1882-1975

4.3.5 Fluidized Bed Catalytic Reactor with Consecutive Reactions

The procedure used in the previous section to develop a mathematical model for a catalytic bubbling fluidized bed reactor with a simple reaction $A \rightarrow B$ can be extended to develop a model for the practically important consecutive reaction $A \xrightarrow{k_1} B \xrightarrow{k_2} C$, where B is the desired product.

The maximum productivity of the desired product B usually occurs at the middle unstable saddle-type steady state. In order to stabilize the unstable steady state, a simple proportional-feedback-controlled system can be used, and we shall analyze such a controller now. A simple feedback-controlled bubbling fluidized bed is shown in Figure 4.25.

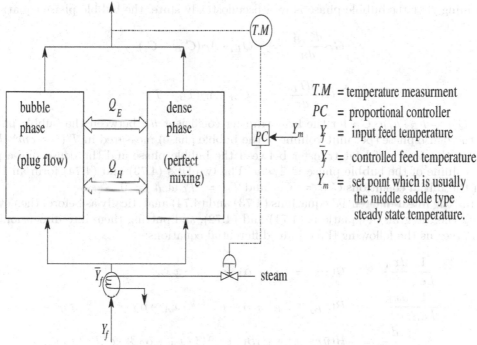

$T.M$	= temperature measurment
PC	= proportional controller
Y_f	= input feed temperature
$\bar{Y}_f$	= controlled feed temperature
Y_m	= set point which is usually the middle saddle type steady state temperature.

Simulation model for two-phase fluidized bed reactor with single input single output proportional (SISOP) feedback control

Figure 4.25

Using the usual assumptions, we obtain the following dimensionless unsteady-state material and energy balance equations for the dense (emulsion) phase of the bubbling fluidized bed:

$$\hat{\Phi}_i \cdot A_I \cdot H \cdot \frac{dC_i}{dt} = G_I(C_{if} - C_i) + Q_{Ei} \cdot A_C \int_0^H (C_{iB} - C_i)dh - \rho_s(1 - \epsilon)A_I \cdot H \cdot \sum_{j=1}^{2} \sigma_{ij} r_j \,,$$

$$(4.71)$$

where $\sum_{j=1}^{2} \sigma_{ij} r_j$ is the net rate of reaction of component i. Notice further that $\rho_b = \rho_s(1 - \epsilon)$.

Likewise we obtain the unsteady-state heat balance equation

$$\hat{\Phi}_H \cdot A_I \cdot H \cdot \frac{dT}{dt} = \tag{4.72}$$

$$= G_I \cdot \rho \cdot C_P (\bar{T}_f - T) + \rho \cdot C_P \cdot Q_H \cdot A_C \int_0^H (T_B - T) dh + A_I \cdot H \cdot \rho_b \sum_{j=1}^{2} r_j (-\Delta H_j),$$

where $\hat{\Phi}_i = \epsilon + K_i \bar{C}_m \rho_b$ is the specific mass capacity of component i per unit volume of the dense phase, $\hat{\Phi}_H = \rho_S C_{PS}$ is the specific heat capacity per unit volume of the dense phase, and r_1 denotes the rate of reaction for $A \to B$ and r_2 that for $B \to C$.

Assuming that the bubble phase is in a pseudosteady state, the bubble phase equations are

$$G_C \frac{dC_{iB}}{dn} = -Q_{Ei} \cdot A_C (C_{iB} - C_i) \tag{4.73}$$

and

$$G_C \frac{dT_B}{dn} = -Q_H \cdot A_C (T_B - T) \tag{4.74}$$

with $Q_H = h_B \cdot a / \rho \cdot C_P$ for the heat transfer coefficient h_B between the bubble phase and the dense phase (per unit volume of the bubble phase) measured in $J/(sec \cdot cm^3 \cdot K)$, and for the area a of heat transfer between the bubble phase and the dense phase per unit volume of the bubble phase in $1/cm$. The two DEs (4.73) and (4.74) form an IVP with the initial conditions $C_{iB} = C_{if}$ and $T_B = \bar{T}_f$ at $h = 0$.

Solving the linear differential equations (4.73) and (4.74) analytically as before, then evaluating the integrals in equations (4.71) and (4.72), and putting them into dimensionless form gives us the following three joint differential equations:

$$\frac{1}{Le_A} \frac{dx_A}{d\tau} = \bar{B}(x_{Af} - x_A) - \alpha_1 \cdot e^{-\gamma_1/y} \cdot x_A , \tag{4.75}$$

$$\frac{1}{Le_B} \frac{dx_B}{d\tau} = \bar{B}(x_{Bf} - x_B) + \alpha_1 \cdot e^{-\gamma_1/y} \cdot x_A - \alpha_2 \cdot e^{-\gamma_2/y} \cdot x_B , \tag{4.76}$$

$$\frac{dy}{d\tau} = \bar{B}(\bar{y}_f - y) + \alpha_1 \beta_1 \cdot e^{-\gamma_1/y} \cdot x_A + \alpha_2 \beta_2 \cdot e^{-\gamma_2/y} \cdot x_B , \tag{4.77}$$

where τ is the dimensionless time $\tau = t \cdot \rho \cdot C_P / (A_I \cdot H \cdot \rho_S \cdot C_{PS})$ and Le_A and Le_B are the Lewis numbers

$$Le_A = \frac{\rho_S \cdot C_{PS}}{(\epsilon + K_A \cdot \bar{C}_m \cdot \rho_b) \rho \cdot C_P} \quad \text{and} \quad Le_B = \frac{\rho_S \cdot C_{PS}}{(\epsilon + K_B \cdot \bar{C}_m \cdot \rho_b) \rho \cdot C_P} .$$

Moreover, $\bar{B} = G_I + G_C(1 - e^{-\alpha' H})$ for $\alpha' = Q_{Ei} \cdot A_C / G_C$ and $Q_{Ei} = Q_{EA} = Q_{EB} = Q_H$, and $G_C = G_B [1.0 + \epsilon/(\alpha_f - 1.0)]$, $G_B = A(U - U_{mf})$, $G_I = A \cdot U - G_C$, $A_C = \epsilon \cdot G_B / [(\alpha_f - 1.0) U_{mf}]$, and $A_I = A - A_C$.

For this case we use the following parameter values:

$A = 0.3\ m^2,\ \ U = 0.1\ m/sec,\ \ U_{mf} = 0.00875\ m/sec\ ,\ \ \alpha_f = 19.5,\ \ \epsilon = 0.4,$

$Q_{Ei} = Q_{EA} = Q_{EB} = Q_H = 0.4\ sec^{-1},\ \ H = 1.0\ m\ .$

And for the dimensionless group we take

$$\alpha_1 = 10^8\ sec^{-1},\ \ \alpha_2 = 10^{11}\ sec^{-1},\ \ \beta_1 = 0.4,\ \ \beta_2 = 0.6,$$

$$\gamma_1 = 18.0,\ \ \gamma_2 = 27.0,\ \ Le_A = 1.0,\ \ Le_B = 0.454545,$$

$$x_{Af} = 1.0,\ \ x_{Bf} = 0.0,\ \ x_{Cf} = 0.0,$$

with the dimensionless preliminary feed temperature $y_f = 0.47396$ and the preliminary setpoint $y_m = 0.86446$.

[In the next section we shall learn why and how the feed temperature and setpoint values were found for optimal performance of the system under feedback control.]

The central PID controlled $\bar{y}_f$ is given as

$$\bar{y}_f = y_f + K(y_m - y) + K_I \int_0^t (y_m - y)d\tau + K_D \frac{dy}{d\tau}\ .$$

Here the acronym **PID** refers to proportional, integral, and derivative. We will concentrate on proportional control, i.e., use $\bar{y}_f$ of the form $\bar{y}_f = y_f + K(y_m - y)$ with $K \geq 0$.

The steady-state equations corresponding to (4.75), (4.76), and (4.77) are

$$\bar{B}(x_{Af} - x_A) = \alpha_1 \cdot e^{-\gamma_1/y} \cdot x_A\ , \tag{4.78}$$
$$\bar{B}(x_{Bf} - x_B) = -\alpha_1 \cdot e^{-\gamma_1/y} \cdot x_A + \alpha_2 \cdot e^{-\gamma_2/y} \cdot x_B\ , \tag{4.79}$$

and

$$\bar{B}(\bar{y}_f - y) + \alpha_1\beta_1 \cdot e^{-\gamma_1/y} \cdot x_A + \alpha_2\beta_2 \cdot e^{-\gamma_2/y} \cdot x_B = 0\ , \tag{4.80}$$

where $\bar{y}_f = y_f + K(y_m - y)$ for $K \geq 0$. Here the parameters α_1, α_2, β_1, β_2, γ_1, γ_2, x_{Af}, x_{Bf}, y_f, and y_m are all given above, and the proportional gain K of the controller will be used as the bifurcation parameter.

How do we calculate the only other remaining parameter $\bar{B}$?

Note that

$$\bar{B} = G_I + G_C(1 - e^{-\alpha' \cdot H})\ ,$$

where each parameter on the right-hand side depends on our supplied data and is given by the formulas following equation (4.77). In particular, the number $\bar{B}$ is readily computed for our data as $\bar{B} = 0.019372\ m^3/sec\ .$

The steady-state equations (4.78), (4.79), and (4.80) can be further reduced to one equation in y and to two additional simple relations. These relations (4.81) for x_A in terms of y and (4.82) for x_B in terms of y help us calculate x_A and x_B as soon as y is determined from the single equation (4.84) in y and K.

From (4.78) we have

$$\bar{B}x_{Af} = x_A\left(\bar{B} + \alpha_1 e^{-\gamma_1/y}\right) \quad \text{or} \quad x_A = \frac{\bar{B}x_{Af}}{\bar{B} + \alpha_1 e^{-\gamma_1/y}} \, . \tag{4.81}$$

Substituting x_A from (4.81) into (4.79) with x_{Bf} set to zero gives us

$$\bar{B}x_B = \alpha_1 e^{-\gamma_1/y} \cdot \frac{\bar{B}x_{Af}}{\bar{B} + \alpha_1 e^{-\gamma_1/y}} - \alpha_2 e^{-\gamma_2/y}x_B \, ,$$

and by rearrangement we get

$$x_B\left(\bar{B} + \alpha_2 e^{-\gamma_2/y}\right) = \frac{\alpha_1 e^{-\gamma_1/y}\bar{B}x_{Af}}{\bar{B} + \alpha_1 e^{-\gamma_1/y}} \, .$$

Thus

$$x_B = \frac{\alpha_1 e^{-\gamma_1/y}\bar{B}x_{Af}}{\left(\bar{B} + \alpha_1 e^{-\gamma_1/y}\right) \cdot \left(\bar{B} + \alpha_2 e^{-\gamma_2/y}\right)} \, . \tag{4.82}$$

Substituting the expressions for x_A and x_B from equations (4.81) and (4.82) into (4.80) gives us the following equation for the heat removal on the left and for the heat generation on the right-hand side:

$$\bar{B}(y-\bar{y}_f) = \frac{\alpha_1\beta_1 e^{-\gamma_1/y}\bar{B}x_{Af}}{\bar{B} + \alpha_1 e^{-\gamma_1/y}} + \frac{\alpha_1\alpha_2\beta_2 e^{-\gamma_1/y}e^{-\gamma_2/y}\bar{B}x_{Af}}{\left(\bar{B} + \alpha_1 e^{-\gamma_1/y}\right) \cdot \left(\bar{B} + \alpha_2 e^{-\gamma_2/y}\right)} \quad (= \tilde{G}(y)) \, . \tag{4.83}$$

Using our controller of the form $\bar{y}_f = y_f + K(y_m - y)$ for $K \geq 0$, the left-hand-side heat removal function in (4.83) becomes

$$\bar{B}(y - (y_f + K(y_m - y))) = \bar{B}((1 + K)y - (y_f + Ky_m)) = \tilde{G}(y) \, , \tag{4.84}$$

the heat generation function $\tilde{G}(y)$ on the right-hand side of (4.83).

We can normalize equation (4.84) by dividing by its common factor $\bar{B}$ to obtain the simpler form

$$(1 + K)y - (y_f + Ky_m) = \tilde{G}(y)/\bar{B} = G(y) \, . \tag{4.85}$$

4.3.6 Numerical Treatment of the Steady-State and Dynamical Cases of the Bubbling Fluidized Bed Catalytic Reactor with Consecutive Reactions

First we look for the solutions of equation (4.84). To do so we divide equation (4.84) by its common factor $\bar{B}$ and bring it into standard form $F(y, K) = 0$ as follows:

$$F(y,K) = (1+K)y - (y_f + Ky_m) - \frac{\alpha_1\beta_1 e^{-\gamma_1/y}x_{Af}}{\bar{B} + \alpha_1 e^{-\gamma_1/y}} - \frac{\alpha_1\alpha_2\beta_2 e^{-\gamma_1/y}e^{-\gamma_2/y}x_{Af}}{\left(\bar{B} + \alpha_1 e^{-\gamma_1/y}\right) \cdot \left(\bar{B} + \alpha_2 e^{-\gamma_2/y}\right)} = 0 \, .$$
$$\tag{4.86}$$

To solve equation (4.86) for y and $K \geq 0$ we use the graphical level set method that we have introduced in Chapter 3 for the adiabatic and nonadiabatic CSTRs and draw the surface $z = F(y, K)$, as well as the y versus K curve of solutions to equation (4.86) in order to exhibit and study the bifurcation behavior of the underlying system.

First we list the code of our graphing controlhetss.m function for $F(y, K)$, which plots the surface $z = F(y, K)$v and the set of solutions of (4.86) in the white zero contour curve $F(y, K) = 0$ on the surface, as well as as another copy of contour on the ground plane below the surface in Figure 4.26. This program is written similarly as adiabNisosurf.m in Chapter 3.

```
function controlhetss(ystart,yend,Kstart,Kend,a1,a2,b1,b2,ga1,ga2,xaf,N,yf,ym)
%    controlhetss(ystart,yend,Kstart,Kend,a1,a2,b1,b2,ga1,ga2,xaf,N,yf,ym)
% Sample call:
%    (full parameter call with yf and ym values chosen near the automatically
%                    computed ones) :
% controlhetss(.3,1.5,0,6,10^8,10^11.5,0.4,0.6,18,27,1.0,100,0.488,0.84)
%    (reduced (i.e., normal) short parameter list call) :
%        controlhetss(0.3,1.5,0,6,10^8,10^11.5,0.4,0.6,18,27,1.0);
%     [Both plots almost agree.]
% Plots the heterogeneous control equation surface and the zero contour curve
% Input : limiting values for y and K,
%         seven system parameters  a1, a2, b1, b2, ga1, ga2, xaf,
%         as well as the (optional) size N of the y and K partitions,
%         and yf anf ym (again optional).
%         If the three parameters N, yf, and ym are not set, contolhetss will
%         use the default N = 200 and compute the parameters yf and ym in
%         hetcontbifrange.m automatically.
% Output: 3D surface plot of the control equation surface with xero contours.
%         [For best view, rotate figure until the level zero bifurcation
%         curves are in good view.]

hold off,                          % extra constants to compute Bbar
A = 0.3; U = 0.1; Umf = 0.00875; alf = 19.5; eps = 0.4; QE = 0.4; H = 1.0;
if nargin == 11,                   % default setting
    N = 200;
    [yf,ym] = hetcontbifrange(yend,a1,a2,b1,b2,ga1,ga2,xaf,0,0,400); end,
if nargin == 12,
    [yf,ym] = hetcontbifrange(yend,a1,a2,b1,b2,ga1,ga2,xaf,0,0,400); end,


                             % make grids for y and K :
[y,K] = meshgrid([ystart:(yend-ystart)/N:yend],...
    [Kstart:(Kend-Kstart)/N:Kend]);
Bbar = Bfun(A,U,Umf,alf,eps,QE,H);          % evaluate Bbar from extra constants
z = Ffun(y,K,a1,a2,b1,b2,ga1,ga2,xaf,yf,ym,Bbar);  % z = het. control function
h = surf(K,y,z); hold on;
shading interp; colormap(hsv(128)); colorbar,% plot surface
a = get(gca,'zlim'); zpos = a(1);        % Always put 0 contour below the plot
[cc,hh] = contour3(K,y,z,[0 0],'-k'); % draw zero contour data on surface
```

```
[ccc,hhh] = contour3(K,y,z,[0 0],'-b'); % put 0 contour data at the bottom
for i = 1:length(hhh)                    % size zpos to match the data
    zzz = get(hhh(i),'Zdata');
    set(hhh(i),'Zdata',zpos*ones(size(zzz))); end

xlabel('K','FontSize',12),              % annotate top 3D plot
ylabel('y','Rotation',0,'FontSize',12),
zlabel('controlhetss(Y,K,a1,a2,b1,b2,ga1,ga2,xaf,yf,ym,N)','FontSize',12),
title([{'Heterogeneous Solid State Control Problem'},{' '},...
        {['            \alpha_1 = ',num2str(a1,'%10.5g'),', \alpha_2 = ',...
            num2str(a2,'%10.5g'),'; \beta_1 = ',num2str(b1,'%10.5g'),...
            ', \beta_2 = ',num2str(b2,'%10.5g'),'; \gamma_1 = ',...
            num2str(ga1,'%10.5g'),', \gamma_2 = ',num2str(ga2,'%10.5g'),...
        '; x_{Af} = ',num2str(xaf,'%10.5g'),', y_f = ',num2str(yf,'%10.5g'),...
        ', y_m = ',num2str(ym,'%10.5g'),', and Bbar = ',...
    num2str(Bbar,'%10.5g')]}],'FontSize',12),  hold off

function f = Ffun(y,K,a1,a2,b1,b2,ga1,ga2,xaf,yf,ym,Bbar) % heterog. c. funct
f = (1+K).*y - (yf+K*ym) - (a1*b1*exp(-ga1./y)*xaf)./(Bbar+a1*exp(-ga1./y)) ...
    - (a1*a2*b2*exp((-ga1-ga2)./y)*xaf)./...
                ((Bbar + a1*exp(-ga1./y)).*(Bbar + a2*exp(-ga2./y)));

function g = Bfun(A,U,Umf,alf,eps,QE,H)                  % evaluate Bbar
GB = A*(U-Umf); AC = eps*GB/((alf-1)*Umf); GC = GB*(1+eps/(alf-1));
alprime = QE*AC/GC; g = A*U - GC*exp(-alprime*H);
```

Note that this section and our programs herein deal only with the case that the feed rates for B and C are zero, i.e., $x_{Bf} = x_{Cf} = 0$ as specified on p. 183.

Figure 4.26 shows the (rotated) surface and contour zero plot obtained by calling `controlhetss(0.3,1.5,0,6,10^8,10^11,0.4,0.6,18,27,1,200,.47396,.86446)` for the data that was specified on p. 183. The displayed MATLAB figure at first hides parts of the bifurcation curves. Therefore, before printing or saving a 3D MATLAB figure, it is best to use the rotation button of the toolbar in MATLAB's Figure window, by clicking on it and then dragging the mouse over the figure until a good view is obtained.

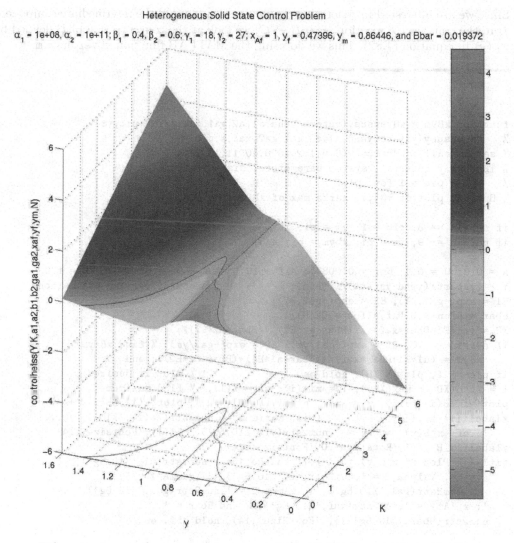

Bifurcation curve for y values in terms of the feedback controller gain K

Figure 4.26

Figure 4.26 indicates that there are three steady states for the system with associated values of $y \approx 0.4...$, $0.8...$, and $1.4...$ when no feedback ($K = 0$) is applied. In the y-K ground plane of Figure 4.26 the reader will notice a "horizontal" line at $y = y_m = 0.86446$ for all values of K. The contour curve in Figure 4.26 reminds us of Figure 3.21 in Section 3.2 and the pitchfork bifurcation for a nonadiabatic, nonisothermal CSTR with one stable steady state at the set point. As in Figure 3.21 for the CSTR, here we observe a fixed steady state of our system for all values of K. This steady state lies at the set point $y = y_m = 0.86446$ of the system. The reason for this and its meaning will become apparent when we study equation (4.88) on p. 192 in Figure 4.29.

Since we are interested in generating the maximal amount of the intermediate component B in the consecutive reaction $A \to B \to C$, we look at the concentration of B defined by $x_B(y)$ in equation (4.82). This we do using the MATLAB function xBversusy.m.

```
function xBym = xBversusy(ystart,yend,a1,a2,ga1,ga2,xaf,N,pic,ym)
%    xBversusy(ystart,yend,a1,a2,ga1,ga2,xaf,N,pic,ym)
% sample call : xBversusy(0.6,1.2,10^8,10^11,18,27,1,400)
% Input : y interval, system parameters, size of partition N;
%         pic = 1 for  plot
% Output: plot of xB(y), marks max of xB on the plot

if nargin == 8, pic = 1; ym = pi; xBym = pi; end
if nargin == 9, ym = pi; xbym = pi; end

A = 0.3; U = 0.1; Umf = 0.00875; alf = 19.5; eps = 0.4; QE = 0.4; H = 1.0;
Y = [ystart:(yend-ystart)/N:yend];              % preparations and evaluations
E1 = exp(-ga1./Y); E2 = exp(-ga2./Y);
Bbar = Bfun(A,U,Umf,alf,eps,QE,H);
XB = (a1*E1*Bbar*xaf)./((Bbar+a1*E1).*(Bbar+a2*E2));
if nargin == 10, E1 = exp(-ga1/ym); E2 = exp(-ga2/ym); % find xB(ym)
    xBym = (a1*E1*Bbar*xaf)/((Bbar+a1*E1)*(Bbar+a2*E2)); end
if pic == 1, plot(Y,XB); hold on,              % plot if desired
i = find(XB == max(XB)); i = min(i); yf = Y(i);  % find maximum
plot(yf,XB(i),'+r'), v = axis; plot(yf,v(3),'+k'), plot(v(1),XB(i),'+k')
xlabel([{' '},{['y             (with maximum  ',num2str(XB(i),'%10.5g'),...
    ' of x_B(y) at  y = ',num2str(yf,'%10.5g'),' )']}],'Fontsize',14)
ylabel('x_B    ','Rotation',0,'Fontsize',14)
title([{'Plot of x_B(y) for' },{['\alpha_1 = ',num2str(a1,'%10.5g'),...
        ', \alpha_2 = ',num2str(a2,'%10.5g'),'; \gamma_1 = ',...
    num2str(ga1,'%10.5g'),', \gamma_2 = ',num2str(ga2,'%10.5g'),...
    '; x_{Af} = ',num2str(xaf,'%10.5g'),', and Bbar = ',...
num2str(Bbar,'%10.5g')]}],'FontSize',14), hold off, end

function g = Bfun(A,U,Umf,alf,eps,QE,H)            % evaluate Bbar
GB = A*(U-Umf); AC = eps*GB/((alf-1)*Umf); GC = GB*(1+eps/(alf-1));
alprime = QE*AC/GC; g = A*U - GC*exp(-alprime*H);
```

The call of xBversusy(0.001,1.6,10^8,10^11,18,27,1,400) produces the graph of x_B in Figure 4.27 for the very same parameter data from p. 183 that was used before to obtain Figure 4.26.

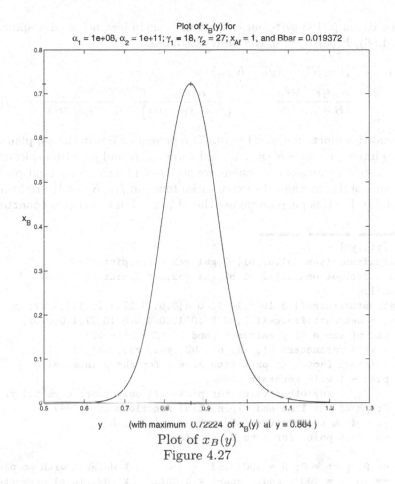

Plot of $x_B(y)$ for
$\alpha_1 = 1e{+}08$, $\alpha_2 = 1e{+}11$; $\gamma_1 = 18$, $\gamma_2 = 27$; $x_{Af} = 1$, and Bbar = 0.019372

(with maximum 0.72224 of $x_B(y)$ at $y = 0.864$)

Plot of $x_B(y)$

Figure 4.27

Figure 4.27 describes the concentration x_B of B in terms of y according to formula
(4.82) on p. 184. This concentration is very low (almost zero) at the two stable steady
states associated with $y \approx 0.4...$ and $1.4....$ But it has a very desirable value of 72% when
$y = 0.86....$ Thus we should try to operate the system at this middle steady state with the
dimensionless temperature y at around $0.86....$ Unfortunately, this middle steady state
is a saddle-type steady state and therefore it is unstable. The system will always try to
move to one of the stable steady states with, unfortunately, a negligibly low rendition of
B, when no or low feedback $K \leq 3.8$ is applied. But as Figure 4.26 indicates, if we apply
a sufficient amount of feedback $K > 4$ in this example, then the resulting system has a
unique steady-state solution with $y \approx 0.86...$ and the system yields a high concentration
of B. Moreover, once the steady state is made unique by feedback $K > 4$, this steady
state is necessarily globally stable and the system will move toward it for all (reasonable)
starting values of x_A, x_B, and y. We shall depict the dynamic behavior of the associated
IVP consisting of the differential equations (4.75), (4.76), and (4.77) later on, in Figures
4.34 to 4.37.
But first we investigate how to find the "magic" values of the feed temperature y_f and

the setpoint y_m on p. 183 systematically. This is again best achieved graphically. Equation (4.86), $F(y,K) = 0$, can be rewritten as

$$g(y,K) = (1+K)y - (y_f + Ky_m) = \tag{4.87}$$

$$= \frac{\alpha_1\beta_1 e^{-\gamma_1/y} x_{Af}}{\bar{B} + \alpha_1 e^{-\gamma_1/y}} + \frac{\alpha_1\alpha_2\beta_2 e^{-\gamma_1/y} e^{-\gamma_2/y} x_{Af}}{\left(\bar{B} + \alpha_1 e^{-\gamma_1/y}\right) \cdot \left(\bar{B} + \alpha_2 e^{-\gamma_2/y}\right)} = f(y,K) \ .$$

The left-hand-side function $g(y,K)$ in (4.87) represents a line in the y-g plane with slope $1+K$ and g intercept $-y_f - K \cdot y_m$. To find values of y_f and y_m with associated multiple (or a unique) steady states of the system, we need to find instances of multiple (or unique) crossing points of this line and the exponential function $f(y,K)$ on the right-hand side of equation (4.87). For this purpose we use the MATLAB m function hetcontbifrange.m.

```
function [yf, ym] = ...
    hetcontbifrange(yend,a1,a2,b1,b2,ga1,ga2,xaf,K,pict,N)
%   hetcontbifrange(yend,a1,a2,b1,b2,ga1,ga2,xaf,K,pict,N)
% sample call:
%   b = hetcontbifrange(1.3,10^8,10^11,0.4,0.6,18,27,1.0,.1,1,400); or
%   [yf, ym] = hetcontbifrange(1.3,10^8,10^11,0.4,0.6,18,27,1.0,0,0);
% Input : end of range of y values ( yend >= yf + b1 + b2);
%         system parameters a1, a2, b1, b2, ga1, ga2, xaf,
%         feedback factor K; partition size N for the y interval
%         pict = 1 will generate plots;
%                  any other value for pict will only compute ym and yf
% Output: Graph of the line and exponential functions in (4.84);
%         yf = feed temperature for max rendition of B
%         ym = set point for A to B reaction

if nargin == 9, pict = 0; N = 400; end          % default with no plot
if nargin == 10, N = 400;  end, ystart = 0.001; % additional parameters :
A = 0.3; U = 0.1; Umf = 0.00875; alf = 19.5; eps = 0.4; QE = 0.4; H = 1.0;
Bbar = Bfun(A,U,Umf,alf,eps,QE,H);        % evaluate Bbar from extra constants
Y = [ystart:(yend-ystart)/N:yend];        % create y partition
E1 = exp(-ga1./Y); E2 = exp(-ga2./Y);     % evaluate exponential functions
XB = (a1*E1*Bbar*xaf)./((Bbar+a1*E1).*(Bbar+a2*E2)); % evaluate xB(y) curve
m = min(find(XB == max(XB))); ym = Y(m);  % find ym from xB curve
Gym = ffun(ym,a1,a2,b1,b2,ga1,ga2,xaf,Bbar);  % find yf from f(ym,K) curve
yf = ym - Gym; G = yline(Y,K,yf,ym);      %    and get line data
F = ffun(Y,a1,a2,b1,b2,ga1,ga2,xaf,Bbar);     % get f(Y,K) data

if pict == 1,                        % plot only if desired
  subplot(2,1,1)                     % plot line and exp curve on top graph
  plot(Y,F), hold on, plot(Y,G,'r'), plot(ym,yline(ym,K,yf,ym),'or');
  axis([ystart,yend,-.1*min(max(F),max(G)),1.1*min(max(F),max(G))]), v = axis;
  plot(ym,v(3),'ok')                 % mark steady state at ym
  xlabel('y','FontSize',12),         % label and title with data
  ylabel('f(y,K) in blue, g(y,K) in red','FontSize',12)
```

```
  title([{'Graphic zeros of equations (4.84), (4.85) for :'},{' '},...
      {['\alpha_1 = ',num2str(a1,'%10.5g'),', \alpha_2 = ',...
      num2str(a2,'%10.5g'),'; \beta_1 = ',num2str(b1,'%10.5g'),...
      ', \beta_2 = ',num2str(b2,'%10.5g'),'; \gamma_1 = ',...
      num2str(ga1,'%10.5g'),', \gamma_2 = ',num2str(ga2,'%10.5g'),...
    '; x_{Af} = ',num2str(xaf,'%10.5g'),', K = ',num2str(K,'%10.5g'),...
    ', Bbar = ',num2str(Bbar,'%10.5g')]}],'FontSize',12),  hold off % top plot

  subplot(2,1,2)                                    % plot xB(y) curve
  plot(Y,XB,'b'); w = axis; axis([v(1),v(2),w(3),w(4)]); hold on
  plot(ym,XB(m),'+r'), plot(ym,w(3),'+k')           % mark maximum
  xlabel([{' '},{['y      ( maximum ',num2str(XB(m),'%10.5g'),...
      ' of x_B(y)    at  y_m = y_{opt} = ',num2str(ym,'%10.5g'),' )']}],...
      'FontSize',12),
  title(['Computed values :   y_m = ',num2str(ym,'%10.5g'),...
      ' ,   y_f = ',num2str(yf,'%10.5g')],'FontSize',12),
  ylabel('x_B   ','Rotation',0,'FontSize',12), hold off, % bottom plot
end

function g = yline(y,K,yf,ym)                       % evaluate line graph
  g = (1+K)*y - (yf+K*ym);

function f = ffun(y,a1,a2,b1,b2,ga1,ga2,xaf,Bbar) % evaluate exp graph
  f = (a1*b1*exp(-ga1./y)*xaf)./(Bbar+a1*exp(-ga1./y)) + ...
        (a1*a2*b2*exp((-ga1-ga2)./y)*xaf)./...
                ((Bbar + a1*exp(-ga1./y)).*(Bbar + a2*exp(-ga2./y)));

function g = Bfun(A,U,Umf,alf,eps,QE,H)             % evaluate Bbar
GB = A*(U-Umf); AC = eps*GB/((alf-1)*Umf); GC = GB*(1+eps/(alf-1));
```

Note that the code line m = min(find(XB == max(XB))); ym = Y(m); in the beginning
of the above code is rather crude in determining the maximum of x_B. This is so since
the maximum value of x_B and the position y_m in our code clearly depend on the chosen
y partition, namely on its endpoints y_{start} and y_{end}, as well as on the step size, i.e., the
chosen N. But this approach suffices to obtain 2 or 3 valid digits for y_f and y_m, which
is good enough for all practical purposes here.

We mention our lax maximum search so that users will not be worried about slightly
varying y_f and y_m outputs of controlhetss.m, hetcontbifrange.m, and xBversusy.m
when the inputs y_{start}, y_{end}, or N are varied while the system parameters are kept fixed.
Compare with Exercise 1 at the end of this section.

A call of [yf,ym] = hetcontbifrange(1.6,10^8,10^11,0.4,.6,18,27,1.0,0,1,400)
gives us the image (in Figure 4.28) of the line and exponential curve on the two sides of
equation (4.87) for $K = 0$ and $0 < y \le 1.6$. The plot indicates three steady states for
$K = 0$ at $y \approx 0.4...$, $0.86...$, and $1.4...$, i.e., at the intersection points of the line and the
curve in the top plot. The bottom plot repeats the plot of $x_B(y)$ of Figure 4.27.

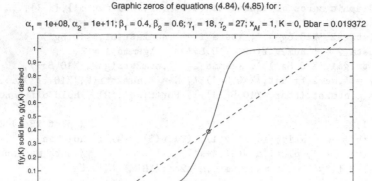

Graphic zeros of equations (4.84), (4.85) for :

$\alpha_1 = 1e+08$, $\alpha_2 = 1e+11$; $\beta_1 = 0.4$, $\beta_2 = 0.6$; $\gamma_1 = 18$, $\gamma_2 = 27$; $x_{Af} = 1$, $K = 0$, Bbar = 0.019372

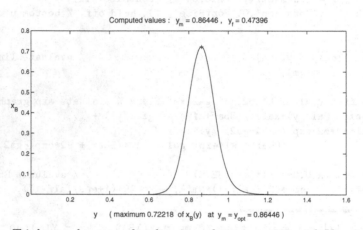

Computed values : $y_m = 0.86446$, $y_f = 0.47396$

y   (maximum 0.72218 of $x_B(y)$ at $y_m = y_{opt} = 0.86446$)

Triple steady states for the given data on p. 183 and $K = 0$
Figure 4.28

The graph of x_B versus y in Figure 4.28 is based on the relation (4.82). Note that the steady states with $y \approx 0.4...$ and $1.4...$ correspond to rather low values of x_B, while the middle steady state at $y \approx 0.8...$ gives us the maximal rendition of around 72% of component B when $K = 0$.

What happens if we change K? In equation (4.87) we were given the equation

$$g(y, K) \ = \ (1 + K)y - (y_f + K y_m) \tag{4.88}$$

of the line depicted in the top graph of Figure 4.28. Let us evaluate g at y_m for any K:

$$g(y_m, K) \ = \ (1 + K)y_m - (y_f + K y_m) \ = \ y_m - y_f \ = \ g(y_m, 0) \, .$$

Thus for any feedback $K \geq 0$, the line (4.88) passes through the point

$$(y_m, g(y_m, K)) \ = \ (y_m, y_m - y_f) \, ,$$

which we have marked by a circle in the topmost plot of Figure 4.28. If we increase K, the line (4.88) increases its slope $1 + K$ while still passing through the point associated with the middle (unstable) steady state for $K = 0$. Therefore for large enough K, when the line becomes steep enough, there is only one intersection with the exponential curve $f(y, K)$ of equation (4.87) in Figure 4.28. Therefore for K sufficiently large, the feedback system has a unique steady state. Due to uniqueness, this steady state is stable, and moreover, at this steady state the system yields the optimal amount of component B. Thus the location y_m of the maximal concentration x_B of B on the graph of $x_B(y)$ gives the optimal output level $y_m = y_{opt}$ for B that can be achieved globally. This optimal output level can be maintained stably only via an appropriate feedback control, since the associated steady state of the uncontrolled system is unstable. Recall the "horizontal line" at $y_m = y_{opt}$ on the ground plane of Figure 4.26.

We illustrate the behavior of the line $g(y, K)$ on the left-hand side of (4.87) and the right-hand-side exponential function $f(y, K)$ of equation (4.87) in Figure 4.29, which includes five versions of the topmost graph in Figure 4.28 for varying feedback values $K = 0, 2, 4, 6$, and 10 in varying colors.

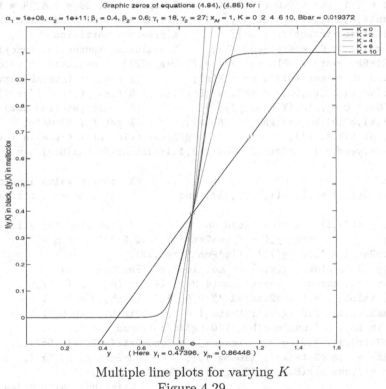

Multiple line plots for varying K
Figure 4.29

Notice that when K clearly exceeds 4, a visual inspection of Figure 4.29 shows that the line intersects the exponential black curve from the right-hand side of (4.87) only once, signifying a unique and statically stable steady state of the controlled system.

Figure 4.29 was obtained via the MATLAB code hetcontbifmultiK.m which we have
derived from hetcontbifrange.m. It features an auxiliary plot of the exponential curve
to find the proper axes limits for the plot first. Note also the elaborate sequence of "legend" commands that we use.

```
function hetcontbifmultiK(yend,a1,a2,b1,b2,ga1,ga2,xaf,K,N)
%    hetcontbifmultiK(yend,a1,a2,b1,b2,ga1,ga2,xaf,K,N)
% sample call:
%   hetcontbifmultiK(1.6,10^8,10^11,0.4,0.6,18,27,1,[0,2,4,6,10],400)
% Input : end of range of y values ( yend >= yf + b1 + b2; ystart = 0.001);
%         system parameters a1, a2, b1, b2, ga1, ga2, xaf,
%         feedback factor K (a vector of length M <= 6 only);
%         partition size N for the y interval
% Output: Graph of multiple lines and exponential function in (4.84);

if nargin == 9, N = 400; end, ystart = 0.001;  % defaults
M = length(K); if M > 6, 'too many K values', return, end,
                                    % additional parameters :
A = 0.3; U = 0.1; Umf = 0.00875; alf = 19.5; eps = 0.4; QE = 0.4; H = 1.0;
Bbar = Bfun(A,U,Umf,alf,eps,QE,H);       % evaluate Bbar from extra constants
Y = [ystart:(yend-ystart)/N:yend];       % create y partition
E1 = exp(-ga1./Y); E2 = exp(-ga2./Y);    % evaluate exponential functions
XB = (a1*E1*Bbar*xaf)./((Bbar+a1*E1).*(Bbar+a2*E2)); % evaluate xB(y) curve
m = min(find(XB == max(XB))); ym = Y(m);  % find ym from xB curve
Gym = ffun(ym,a1,a2,b1,b2,ga1,ga2,xaf,Bbar);  % find yf from f(ym,K) curve
yf = ym - Gym; G = yline(Y,K(1),yf,ym);   %   and get line data
F = ffun(Y,a1,a2,b1,b2,ga1,ga2,xaf,Bbar);  % get f(Y,K) data
figure(1), plot(Y,F,'k'),              % first trial plot for axis info only
axis([ystart,yend,-.1*min(max(F),max(G)),1.1*min(max(F),max(G))]), v = axis;

for i = 1:M                              % other K value line plots
  k = K(i); G(i,:) = yline(Y,k,yf,ym); end  % get line data for each K(i)

figure(2), plot(Y,G), axis(v), hold on   % plot line for all k = K(i)
xlabel(['y        ( Here  y_f = ',num2str(yf,'%10.5g'),',  y_m = ',...
        num2str(ym,'%10.5g'),' )'],'FontSize',12),
ylabel('f(y,K) in black, g(y,K) in multicolor','FontSize',12)
title([{'Graphic zeros of equations (4.84), (4.85) for :'},{' '},...
        {['\alpha_1 = ',num2str(a1,'%10.5g'),',  \alpha_2 = ',...
        num2str(a2,'%10.5g'),'; \beta_1 = ',num2str(b1,'%10.5g'),...
        ', \beta_2 = ',num2str(b2,'%10.5g'),'; \gamma_1 = ',...
        num2str(ga1,'%10.5g'),',  \gamma_2 = ',num2str(ga2,'%10.5g'),...
      '; x_{Af} = ',num2str(xaf,'%10.5g'),',  K = ',num2str(K,'%3.2g'),...
      ', Bbar = ',num2str(Bbar,'%10.5g')]}],'FontSize',12),
if M == 1                               % include color coded legend
      legend(['K = ',num2str(K,'%7.4g')],0), end
if M == 2
  legend({['K = ',num2str(K(1),'%7.4g')],['K = ',...
            num2str(K(2),'%7.4g')]},-1), end
```

```
if M == 3
  legend({['K = ',num2str(K(1),'%7.4g')],['K = ',...
         num2str(K(2),'%7.4g')],['K = ',num2str(K(3),'%7.4g')]},-1), end
  if M == 4
  legend({['K = ',num2str(K(1),'%7.4g')],['K = ',...
         num2str(K(2),'%7.4g')],['K = ',num2str(K(3),'%7.4g')],...
         ['K = ',num2str(K(4),'%7.4g')]},-1), end
  if M == 5
  legend({['K = ',num2str(K(1),'%7.4g')],['K = ',...
         num2str(K(2),'%7.4g')],['K = ',num2str(K(3),'%7.4g')],...
         ['K = ',num2str(K(4),'%7.4g')],['K = ',...
         num2str(K(5),'%7.4g')]},-1), end
  if M == 6
  legend({['K = ',num2str(K(1),'%7.4g')],['K = ',...
         num2str(K(2),'%7.4g')],['K = ',num2str(K(3),'%7.4g')],...
         ['K = ',num2str(K(4),'%7.4g')],['K = ',...
         num2str(K(5),'%7.4g')],['K = ',num2str(K(6),'%7.4g')]},-1), end
plot(Y,F,'k'),                                   % plot exponential curve
plot(ym,v(3),'ok'); plot(ym,yline(ym,K,yf,ym),'or'); % mark pivot for lines
hold off, close(1)

function g = yline(y,K,yf,ym)                            % evaluate line graph
  g = (1+K)*y - (yf+K*ym);

function f = ffun(y,a1,a2,b1,b2,ga1,ga2,xaf,Bbar) % evaluate exp graph
  f = (a1*b1*exp(-ga1./y)*xaf)./(Bbar+a1*exp(-ga1./y)) + ...
      (a1*a2*b2*exp((-ga1-ga2)./y)*xaf)./...
         ((Bbar + a1*exp(-ga1./y)).*(Bbar + a2*exp(-ga2./y)));

function g = Bfun(A,U,Umf,alf,eps,QE,H)                  % evaluate Bbar
GB = A*(U-Umf); AC = eps*GB/((alf-1)*Umf); GC = GB*(1+eps/(alf-1));
alprime = QE*AC/GC; g = A*U - GC*exp(-alprime*H);
```

Let us repeat the same problem as before, but for the following altered α and γ parameters, namely for

$$\alpha_1 = 10^{10}, \quad \alpha_2 = 10^{7.5}, \quad \gamma_1 = 8, \quad \text{and} \quad \gamma_2 = 12.$$

All other parameters are left unchanged from p. 183. For this we use a simplified call of our surface and zero-contour plotting function controlhetss, which omits specifying the last three inputs N, y_f, and y_m. Instead, the shortened call sequence uses the default setting of $N = 200$ points for the partition of the y and K intervals, and controlhetss computes the optimal values for y_f and y_m by calling hetcontibifrange internally. The call of controlhetss(0.01,1.1,0,2,10^10,10^7.5,0.4,0.6,8,12,1) renders a surface and level curve plot with fivefold bifurcation for all small controller gains $0 \leq K \leq 1.7$, one of which occurs at the set point $y = y_m = 0.41038$. In Figure 4.30, we again display a manually rotated 3D picture that give a better view of the bifurcation curves, refer to the earlier comments on using the MATLAB rotate button on p. 186.

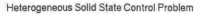

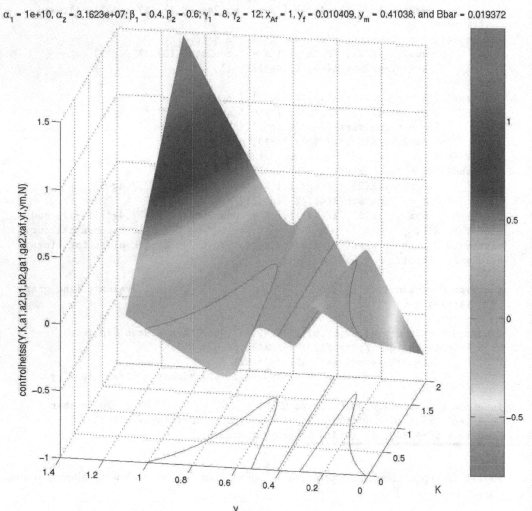

A fivefold bifurcation curve for y values in terms of the controller gain K
Figure 4.30

The line and exponential curve graphs resulting from a call of `hetcontbifrange(1.1,` `10^10,10^7.5,0.4,.6,8,12,1,0,1,400)` look as follows for the modified data of p. 195.

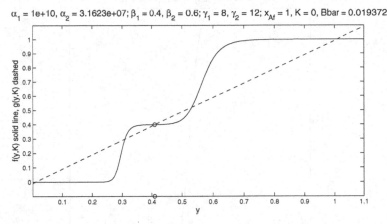

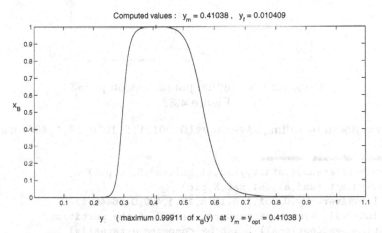

Five steady states for the modified data and $K = 0$
Figure 4.31

Here the central steady state at $y \approx 0.4...$ and the two extremal steady states with $y \approx 0.0...$ and $1.0...$ are stable, while the two remaining steady states at $y \approx 0.3...$ and $0.5...$ are of saddle type and therefore unstable. Of these, the central one gives the optimal concentration for component B. The unstable ones that are adjacent to the middle one give moderate concentrations for B, while the extremal steady states with $y \approx 0.0...$ and $1.0...$ produce hardly any amount of B according to the bottom plot of $x_B(y)$ in Figure 4.31. We shall return to this example in Figure 4.38 with further comments on the robustness of the optimal central steady state and on the beneficial role of feedback in chemical/biological systems.

If we also draw out the curve of $x_A(y)$ as given in equation (4.81), we can find approximations of the coordinates x_A, x_B, and y for each steady state. Figure 4.32 gives the plot of $x_A(y)$ for our original data on p. 183.

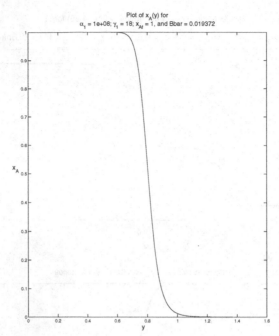

$x_A(y)$ for the original parameters on p. 183
Figure 4.32

Figure 4.32 was drawn by calling xAversusy(0.001,1.6,10^8,18,1,400) using the following code:

```
function xAym = xAversusy(ystart,yend,a1,ga1,xaf,N,ym,pic)
%    xAversusy(ystart,yend,a1,ga1,xaf,N,pic)
% sample call : xAversusy(0.001,1.6,10^8,18,1,400,0.86446,1)
% Input : y interval, system parameters, size N of the partition,
%         optimal ym (optional) [must be computed externally]
%         pic = 1 : draw plot; otherwise compute xAym only
% Output: plot of xA(y)
%         xAym = value of xA(ym) if ym has been specified,
%                else yAm = pi = 3.14...

if nargin == 6, ym = pi; pic = 1; end  % dummy input if ym is not specified
A = 0.3; U = 0.1; Umf = 0.00875; alf = 19.5; eps = 0.4; QE = 0.4; H = 1.0;
Y = linspace(ystart,yend,N); E1 = exp(-ga1./Y);  % set up and evaluate
Bbar = Bfun(A,U,Umf,alf,eps,QE,H);
XA = (Bbar*xaf)./(Bbar+a1*E1);
if pic == 1, plot(Y,XA); hold on, end       % plot if desired
xAym = (Bbar*xaf)/(Bbar+a1*exp(-ga1/ym)); % in case ym is known, find xA(ym)
if nargin == 6, xAym = pi; end             % dummy output
if pic == 1,
  if xAym == pi,  xlabel([{'y      '}],'Fontsize',14),
  else  xlabel(['y       (y_m = ',num2str(ym,'%10.5g'),')'],'Fontsize',14), end
```

```
ylabel('x_A      ','Rotation',0,'Fontsize',14)
title([{'Plot of x_A(y) for' },{['\alpha_1 = ',num2str(a1,'%10.5g'),...
      '; \gamma_1 = ',num2str(ga1,'%10.5g'),'; x_{Af} = ',...
      num2str(xaf,'%10.5g'),', and Bbar = ',num2str(Bbar,'%10.5g')]}],...
      'FontSize',14), hold off, end

function g = Bfun(A,U,Umf,alf,eps,QE,H)                % evaluate Bbar
GB = A*(U-Umf); AC = eps*GB/((alf-1)*Umf); GC = GB*(1+eps/(alf-1));
alprime = QE*AC/GC; g = A*U - GC*exp(-alprime*H);
```

The call of xAversusy(0.1,0.5,10^10,8,1,400,0.41038,1) for the modified data on
p. 195 and with the optimal value of $y_m = 0.41038$ taken from the fivefold bifurcation of
Figure 4.31 gives us the plot of x_A versus y shown in Figure 4.33.

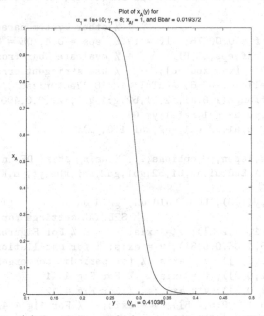

$x_A(y)$ for the modified parameters on p. 195

Figure 4.33

Now we return to the earlier case with three steady states for the parameter data
from p. 183. Looking at Figures 4.28 and 4.32 jointly allows us to approximate the three
steady-state coordinates $(x_A,\ x_B,\ y)$ of this specific system in the following table, or-
dered by increasing values of the third coordinate y:

x_A	1	0.17619...	0
x_B	0	0.72225...	0
y	0.47...	0.86446...	1.47...

Knowing how to obtain this data from the graphs will help us in understanding the
dynamical behavior of the system given by the differential equations (4.75), (4.76), (4.77).

After the steady-state analysis, we now use the differential equations for the system (4.75), (4.76), (4.77) to draw 3D and 2D phase plots of the underlying dynamical system behavior for the original parameters given on p. 183.

Our IVP solving and solution-plotting MATLAB code is as follows:

```
function fluidbed(tend,xao,xbo,yo,LeA,LeB,a1,a2,b1,b2,ga1,ga2,xaf,xbf,K)
%       fluidbed(tend,xao,xbo,yo,LeA,LeB,a1,a2,b1,b2,ga1,ga2,xaf,xbf,K)
% sample call:
%       fluidbed(10,1,1,0.1,1,45/99,10^8,10^11,.4,.6,18,27,1,0,0)
% Input: tend = end of time interval for the integration
%        xao, yao, yo = initial values for xA, xB, y
%        LeA, LeB, a1, a2, b1, b2, ga1, ga2, xaf, xbf system parameters
%        K = feedback gain (K >= 0)
% Output: 3D plot of solution to the DEs in (4.74), (4.75), and (4.76)


                              % other system parameters :
A = 0.3; U = 0.1; Umf = 0.00875; alf = 19.5; eps = 0.4; QE = 0.4; H = 1.0;
Bbar = Bfun(A,U,Umf,alf,eps,QE,H);      % evaluate Bbar from extra constants
Tspan = [0 tend]; y0 = [xao;xbo;yo];    % use stringent error bounds :
options = odeset('RelTol',10^-6,'AbsTol',10^-8,'Vectorized','on');
[yf,ym] = hetcontbifrange(1.6,a1,a2,b1,b2,ga1,ga2,xaf,0,0,400); % find ym
xAym = xAversusy(yo,yo,a1,ga1,xaf,1,ym,0);                % xA(ym)
xBym = xBversusy(yo,yo,a1,a2,ga1,ga2,xaf,1,0,ym);         % xB(ym)

[V,y] = ode15s(@frhs,Tspan,y0,options,... % using stiff DE integrator ode15s
              LeA,LeB,a1,a2,b1,b2,ga1,ga2,xaf,xbf,yf,ym,K,Bbar);

plot3(y(:,1),y(:,2),y(:,3),'b'), hold on, grid on,
                       % SPECIAL settings for AXES: ADJUST !!
%axis([-.05,.6,-.05,1,.4,2.5]); v = axis;        % For Figures 4.34 - 4.37
%axis([-.04,0.2,-.05,1.05,0.5,3]), v = axis; % for oscillations Fig 4.46
%axis([.08,0.2,0,1,0.5,1]); v = axis; % for periodic convergence
%axis([-.05,1,-.05,1,0,2]); v = axis;   % For Fig 4.41
%axis([-.05,1,-.05,1,0,3]); v = axis;   % For Fig 4.42
%axis([-.05,0.2,-.05,1.2,0.5,2.5]); v = axis;    % For Fig 4.47, 4.48
%axis([-.05,0.2,-.05,1.2,0.5,2.5]); v = axis;    % For Fig 4.48
%axis([-.05,1,-.05,1,0,1.2]); v = axis;  % For high Lewis numbers
plot3(y(1,1),y(1,2),y(1,3),'or'),       % start mark
plot3(y(end,1),y(end,2),y(end,3),'xr') % end mark of solution
plot3(xAym,xBym,ym,'*k')                % attractor mark
xlabel('x_A','Fontsize',12), ylabel('x_B','Fontsize',12),
zlabel('y','Rotation',0,'Fontsize',12)
title([{['Dynamic solutions to the IVP (4.74), (4.75), and (4.76) with ',...
  'T_{end} = ',num2str(tend,'%10.5g'),', \alpha_1 = ',num2str(a1,'%10.5g'),...
  ', \alpha_2 = ',num2str(a2,'%10.5g'),';']},{' '},{[' \beta_1 = ',...
  num2str(b1,'%10.5g'),', \beta_2 = ',num2str(b2,'%10.5g'),'; \gamma_1 = ',...
     num2str(ga1,'%10.5g'),', \gamma_2 = ',num2str(ga2,'%10.5g'),...
  '; x_{Af} = ',num2str(xaf,'%10.5g'),', K = ',num2str(K,'%10.5g'),...
  '; Le_A = ',num2str(LeA,'%10.5g'),', Le_B = ',num2str(LeB,'%10.5g'),...
```

```
'   for optimal y_f  and y_m']}],'Fontsize',12)

function dydt = frhs(V,y,LeA,LeB,a1,a2,b1,b2,ga1,ga2,xaf,xbf,yf,ym,K,Bbar)
F1 = a1*exp(-ga1./y(3,:)).*y(1,:); F2 = a2*exp(-ga2./y(3,:)).*y(2,:);
dydt = [LeA*(Bbar*(xaf-y(1,:))-F1);
        LeB*(Bbar*(xbf-y(2,:))+F1-F2);
        Bbar*(yf+K*(ym-y(3,:))-y(3,:)) + b1*F1 + b2*F2];

function g = Bfun(A,U,Umf,alf,eps,QE,H)                     % evaluate Bbar
GB = A*(U-Umf); AC = eps*GB/((alf-1)*Umf); GC = GB*(1+eps/(alf-1));
alprime = QE*AC/GC; g = A*U - GC*exp(-alprime*H);
```

This code is typically called by a MATLAB command such as fluidbed(4000,.4,.9,1,1, 45/99,10^8,10^11,.4,.6,18,27,1,0,10), which uses a large time limit of 4000 dimensionless time units so that the individual solution curves of the IVPs have time to run to the global steady state at (0.17619, 0.72225, 0.86446). We do this with the original parameter data of p. 183, a controller gain of $K = 10$, and several different initial values. Since the value of K is greater than 4, the middle steady state with maximal x_B yield is both unique and (statically) stable.

Note further that inside fluidbed.m we use the stiff integrator ode15s instead of one of our previous favorites ode23 or ode45, since the latter two integrators take an inordinate amount of time to solve our IVPs.

Remark on Experimental IVP Stiffness and IVP Solver Validation:

MATLAB allows the user to choose the IVP integrator, such as ode23 or ode45 etc, and to select a stiff or nonstiff integrator, each as warranted by the specific problem. Moreover, each of the MATLAB's ODE solvers ode... allows us to specify certain "options", as done in fluidbed.m in the fourth MATLAB command line

  options = odeset('RelTol',10^-6,'AbsTol',10^-8,'Vectorized','on');

for example.

How do we choose a MATLAB IVP solver, and how do we decide whether to use a stiff integrator such as ode15s or a standard nonstiff one?

Recall that the stiff solver ode15s was the default integrator in fixedbedreact.m of section 4.1.5, while ode23 was sufficient for section 4.2.6. If we compare the shape of the solutions to the DEs in section 4.1.5 and Figures 4.3 to 4.11 with those of section 4.2.6 and Figures 4.16 and 4.17, we notice a generally sharp change of directions in the solution curve, or a step function look in many of the plots of section 4.1.5, while the solution curves of section 4.2.6 are all smoother and more gentle in their variations. This in essence separates the stiff and the nonstiff differential equations: stiff DEs are visually distinguished by very sharp variations and they require stiff ODE solvers for finding accurate solutions quickly; nonstiff DEs generally have smooth, flowing solutions and can be solved by nonstiff solvers most efficiently. In MATLAB, the user can easily decide from the graph of an ODE solution and from the time it takes to compute it, whether a given DE is stiff or not. Users should always experiment with a few IVP solvers of either type before deciding on which one to use for a given class of problems.

Once the stiffness of an ODE has been decided upon by numerical experiment, the final task involves finding a solution stably. This can be achieved by varying the tolerances 'RelTol' and 'AbsTol' in the "options" line of code and observing the solution closely: If the tolerances are chosen too large, the solution may not be correct. This shows itself in variations of the solution graphs for varying specified tolerances. When decreasing the tolerances for a stable ODE problem in steps of powers of 10, the solution graphs will eventually become stationary or identical. At that point the true solution has been found. In our specific MATLAB "options" line mentioned earlier, the two tolerances are set to 10^{-6} and 10^{-8}, respectively. It would be a good exercise for our readers to loosen these tolerances to 10^{-2} or to tighten them to 10^{-12} etc. in our codes fluidbed.m and fluidbedprofiles.m and to observe all resulting plots for the rest of this subsection, as well as doing the same experiment with neurocycle.m in subsection 4.4.6 to make sure that chaos really occurs for the neurocycle model.

Specifically for the following examples, we run fluidbed.m for the eight corners of the cube

$$C = [0.1, 0.4] \times [0.5, 0.9] \times [0.5, 0.9] \in \mathbb{R}^3$$

as initial values $(x_A(0), x_B(0), y(0))$ in Figures 4.34 to 4.37. Note that the unique steady state (0.17619, 0.72225, 0.86446) for the system of IVPs (4.75), (4.76), and (4.77) with controller gain $K = 10$ lies inside our chosen cube C.

First we draw the 3D plot of the $(x_A(t), x_B(t), y(t))$ profiles, followed by the corresponding x_A-x_B, x_A-y, and x_B-y 2D phase plots. Note in all our graphs in Figures 4.34 to 4.37 that the controlled system has only one steady state, marked by an asterisk in the plots, to which all our plotted solution profiles converge.

Here the coordinates of the eight corners of the cube C serve as the initial values x_{A0}, x_{B0}, and y_0. These are marked by small circles in each instance. The eight trajectories proceed from the corners of the cube C to the only steady state of the system, marked by a gray $*$. Their final positions at the end of the considered time interval are marked by eight x symbols. After $T = 600$ time units the trajectories in our example have all converged to the unique steady state of Figure 4.34.

A detailed reading of the fluidbed.m program reveals that a hold on MATLAB command is put onto the plot that is never released. This helps us draw multiple trajectories onto one 3D plot such as in Figure 4.34. Note that Figure 4.34 depicts the actual MATLAB output after a suitable rotation as explained before on p. 186. If the user wants to have a fresh plotting start, a call of fluidbed(....   ...) (and many other plotting routines) should be preceded by MATLAB's hold off command to clear the current figure window.

Dynamic solutions to the IVP (4.74), (4.75), and (4.76) with T_{end} = 600, α_1 = 1e+08, α_2 = 1e+11;

β_1 = 0.4, β_2 = 0.6; γ_1 = 18, γ_2 = 27; x_{Af} = 1, K = 10; Le_A = 1, Le_B = 0.45455 for optimal y_f and y_m

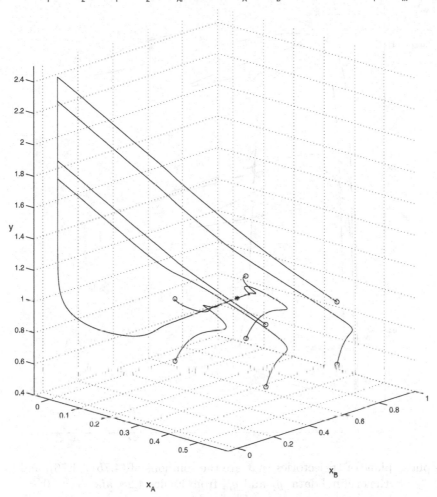

Set of trajectories that are the solutions of (4.75), (4.76), and (4.77) for the original
data on p. 183, y_f and y_m from Figure 4.28, and $K = 10$ (after a manual rotation)
Figure 4.34

We note in Figure 4.34 that, depending on the initial values, some of the IVP solutions
converge directly to the steady state of the system, such as for the four initial values with
$x_A(0) = 0.1$, while the other four sets of initial conditions with $x_A(0) = 0.4$ first go to
relatively high values of y with $x_A = x_B = 0$, before converging to the steady state once
y has dropped to less than 1.
Note further in Figures 4.34 - 4.37 that apparently there are exactly two trajectories in
(opposite) directions for solutions to follow when approaching the system's steady state
∗ with the controller gain set to $K = 10$.

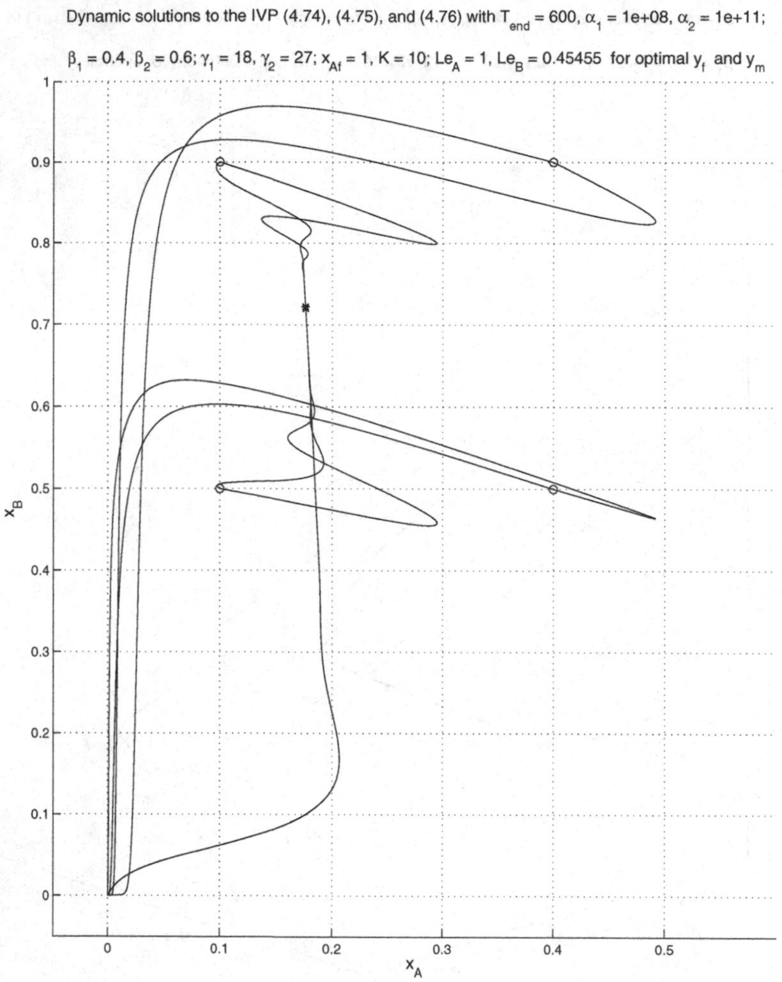
Dynamic solutions to the IVP (4.74), (4.75), and (4.76) with T_end = 600, α_1 = 1e+08, α_2 = 1e+11;
β_1 = 0.4, β_2 = 0.6; γ_1 = 18, γ_2 = 27; x_Af = 1, K = 10; Le_A = 1, Le_B = 0.45455 for optimal y_f and y_m
x_B
x_A
0
0.1
0.2
0.3
0.4
0.5
0.6
0.7
0.8
0.9
1

x_A-x_B phase plane of trajectories that are the solutions of (4.75), (4.76), and (4.77) for the original data, y_f and y_m from Figure 4.28, and $K = 10$

Figure 4.35

We emphasize that in each of our phase plots, Figures 4.35 through 4.37, we look at the cube C face on from one cardinal direction and use C's eight corners as our starting values. This is achieved simply by rotating the MATLAB plot of Figure 4.34 so that one coordinate plane becomes parallel to the screen, giving us a 2D coordinate planar plot, see p. 186 for details on rotating MATLAB figures via the mouse. In particular, each of the four circles in a 2D phase plot depicts two different initial values.
[We note that it is mathematically impossible for two different trajectories to emerge from one set of initial values of an IVP.]
Therefore, when looking at the x_A-x_B 2D phase plot of Figure 4.35 for example, we must

be aware that the upper rightmost o mark depicts the two initial values

$$(x_A(0), \ x_B(0), \ y(0)) \ = \ (0.4, \ 0.9, \ 0.5) \quad \text{and} \quad (0.4, \ 0.9, \ 0.9)$$

for the differing initial $y(0)$ values of 0.5 and 0.9. Looking back at the 3D plot in Figure 4.34, we observe that the trajectory starting from the initial value (0.4, 0.9, 0.5) with $y(0) = 0.5$ reaches higher x_B values first, making its trajectory in the x_A-x_B phase plot of Figure 4.35 go up and to the left from the upper rightmost o mark, while the trajectory that starts at (0.4, 0.9, 0.9) in Figure 4.35 with the higher initial temperature $y(0) = 0.9$ goes down and to the right first from the same o mark in the x_A-x_B phase plot.

Similar observations will help the reader interpret the phase plots that follow in Figures 4.36 and 4.37.

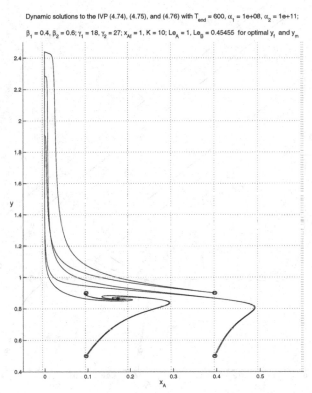

x_A-y phase plane of trajectories that are the solutions of (4.75), (4.76), and (4.77) for
the original data, y_f and y_m from Figure 4.28, and $K = 10$
Figure 4.36

Tracing the trajectories of this 2D phase plot as before, we see that the initial values of

$$(x_A(0), \ x_B(0), \ y(0)) \ = \ (0.4, \ 0.5, \ 0.9) \quad \text{and} \quad (0.4, \ 0.9, \ 0.9)$$

are depicted by the right and topmost circle o in Figure 4.36. Comparing with the 3D trajectories emanating from these initial values in Figure 4.34, we can distinguish the two topmost branching curves of Figure 4.36: according to Figure 4.34, the initial value (0.4, 0.9, 0.9) reaches the highest heat $y > 2.4$. Thus the topmost curve in Figure 4.36 starts at (0.4, 0.9, 0.9), and the trajectory from the initial value (0.4, 0.5, 0.9) that is depicted by the same top right o mark in Figure 4.36 reaches a maximal heat of $y \approx 1.9$, as can be seen by tracing the lower branch trajectory emanating from $(x_A(0),\ x_B(0)) = (0.4,\ 0.9)$ until $x_A = x_B = 0$ in Figure 4.36.

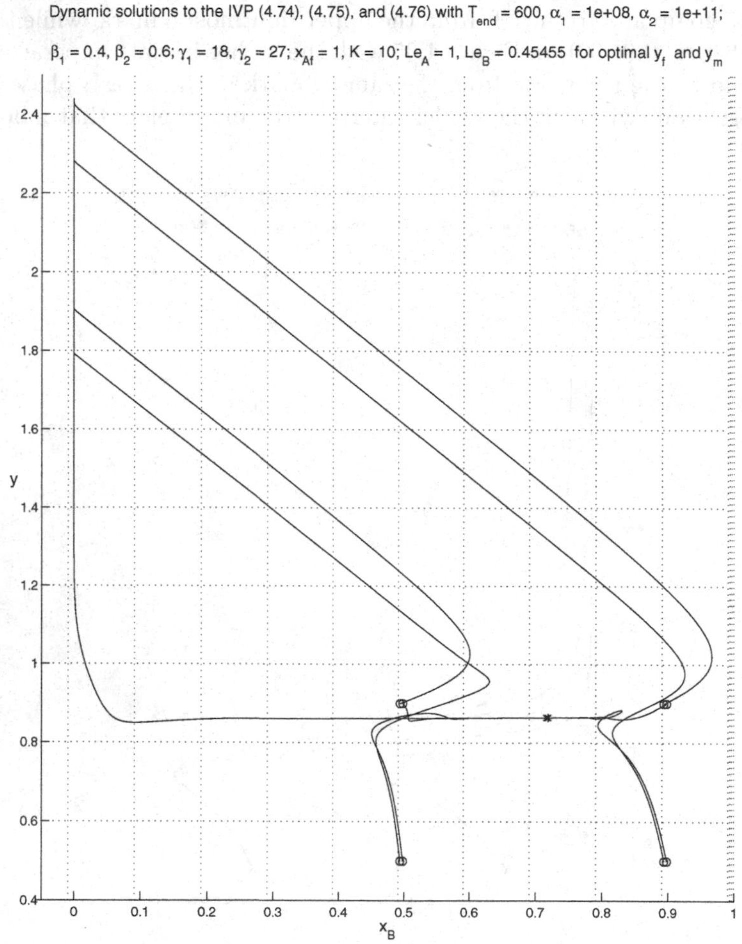
Dynamic solutions to the IVP (4.74), (4.75), and (4.76) with T_end = 600, α_1 = 1e+08, α_2 = 1e+11;
β_1 = 0.4, β_2 = 0.6; γ_1 = 18, γ_2 = 27; x_Af = 1, K = 10; Le_A = 1, Le_B = 0.45455 for optimal y_f and y_m
y
x_B
2.4
2.2
2
1.8
1.6
1.4
1.2
1
0.8
0.6
0.4
0
0.1
0.2
0.3
0.4
0.5
0.6
0.7
0.8
0.9
1

x_B-y phase plane of trajectories that are the solutions of (4.75), (4.76), and (4.77) for the original data, y_f and y_m from Figure 4.28, and $K = 10$

Figure 4.37

The last three phase plots were all created from the same 3D graphics MATLAB window of Figure 4.34 by clicking on the "Rotate" MATLAB figure icon, then dragging the mouse and thereby rotating the 3D image in the window until the base plane becomes one of the three 2D phase planes.

Our readers should experiment with `fluidbed.m` and their own data.

To complete the picture, here are the three profiles of $x_A(t)$, $x_B(t)$, and $y(t)$ for $0 \le t \le 600$ for the example depicted in Figures 4.34 to 4.37 for the three initial values

$$(x_A(0), x_B(0), y(0)) = (0.1, 0.5, 0.9), \ (0.1, 0.9, 0.5), \ \text{and} \ (0.4, 0.9, 0.9) \,.$$

We use feedback ($K = 10$). The plots are best drawn and viewed in color on a monitor by calling `fluidbedprofiles(600,[.1,.1,.4],[.5,.9,.9],[.9,.5,0.9],1,45/99,10^8, 10^11,.4,.6,18,27,1,0,10,1)`.

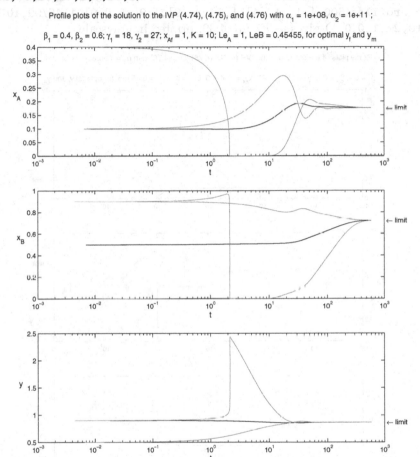

x_A, x_B, and y profiles of the trajectories that solve (4.75), (4.76), and (4.77) for the original data, y_f and y_m from Figure 4.28, and $K = 10$

Figure 4.38

Note that the subplots of Figure 4.38 use a logarithmic scale for the horizontal time axis by invoking the MATLAB plot command `semilogx` in `fluidbedprofiles.m`. This allows us to see the system development near the start very well while shortening the long asymptotic time behavior scale optically. In a **logarithmic MATLAB plot** such as ours, the "powers of ten" axis points $10^{-1}, 10^0, 10^1, 10^2$, etc. are written out and carry a large tick mark. Moreover, eight multiplicative intermediate values are marked in a logarithmic, nonlinear order in between, such as in the sequence ..., 0.8, 0.9, $10^0 =$ 1, 2, 3, 4, 5, 6, 7, 8, 9, $10^1 = 10$, 20, 30, ..., 80, 90, $10^2 = 100$, 200, 300, ... with small tick marks on the t axes in Figure 4.38. This notation differs markedly from a **linear plot** with nine linearly and equally spaced tick marks at 1, 1.1, 1.2, 1.3, 1.4, ..., 1.8, 1.9, 2, for example between the written out numbers 1 and 2 on the axis.

If we do not use feedback, i.e., for $K = 0$, the output of the similar call `fluidbedprofiles(600,[.1,.1,.4],[.5,.9,.9],[.9,.5,0.9],1,45/99,10^8,10^11,` `.4,.6,18,27,1,0,0,1)` gives us Figure 4.39, printed in grayscale.

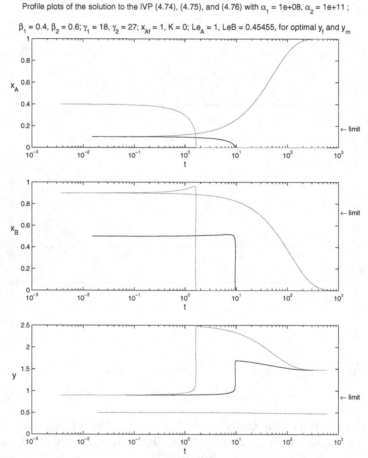

Profile plots of the solution to the IVP (4.74), (4.75), and (4.76) with $\alpha_1 = 1e+08$, $\alpha_2 = 1e+11$;

$\beta_1 = 0.4$, $\beta_2 = 0.6$; $\gamma_1 = 18$, $\gamma_2 = 27$; $x_{Af} = 1$, $K = 0$; $Le_A = 1$, $LeB = 0.45455$, for optimal y_f and y_m

x_A, x_B, and y profiles of the trajectories that solve (4.75), (4.76), and (4.77) for the original data, y_f and y_m from Figure 4.28, and $K = 0$

Figure 4.39

In each subplot of Figures 4.38 and 4.39 we indicate the location of the middle steady state of the system by an arrow on the right margin. As noted earlier on p. 199, this steady state is unstable when $K = 0$, and the three profiles bear this out: x_A converges to either 1 or 0, x_B converges to zero, and y converges either to 0.47... or 1.47..., just as was computed on p. 199 for the steady-state coordinates.

The code of `fluidbedprofiles.m` starts below:

```
function ...
 fluidbedprofiles(tend,xao,xbo,yo,LeA,LeB,a1,a2,b1,b2,ga1,ga2,xaf,xbf,K,log)
% fluidbedprofiles(tend,xao,xbo,yo,LeA,LeB,a1,a2,b1,b2,ga1,ga2,xaf,xbf,K,log)
% sample call:
%    fluidbedprofiles(500,[1,0,0.5],[1,0.8,0.2],[0.1,0.4,0.9],...
%                     1,45/99,10^8,10^11,.4,.6,18,27,1,0,0,1)
% Input: tend = end of time interval for the integration
%        xao, yao, yo = initial values for xA, xB, y
%                       (in equal sized row vector form)
%        LeA, LeB, a1, a2, b1, b2, ga1, ga2, xaf, xbf = system parameters
%        K = feedback gain (K >= 0)
%        log : if log = 1, we plot time in a log10 scale; else linear scale
% Output: profile plots xA, xB, and y versus t of solution to the DEs in
%         (4.74), (4.75), and (4.76)

                                    % other system parameters :
A = 0.3; U = 0.1; Umf = 0.00875; alf = 19.5; eps = 0.4; QE = 0.4; H = 1.0;
Bbar = Bfun(A,U,Umf,alf,eps,QE,H);       % evaluate Bbar from extra constants
ma = length(xao); mb = length(xbo); my = length(yo);
if ma ~= mb | ma ~= my,               % check that vector input matches
    'Length of x_Ao, x_Bo, and  yo do not match! Abandon', return, end
m = ma; TT = NaN*zeros(4000,m); XA = TT; XB = TT; Y = TT;  % preparations

for i = 1:m,                            % cycle through the given data
  Tspan = [0 tend]; y0 = [xao(i);xbo(i);yo(i)]; % use stringent error bounds :
  options = odeset('RelTol',10^-6,'AbsTol',10^-8,'Vectorized','on');
[yf,ym] = hetcontbifrange(1.6,a1,a2,b1,b2,ga1,ga2,xaf,0,0,400); % find ym, yf
  xAym = xAversusy(yo(i),yo(i),a1,ga1,xaf,1,ym,0);         % xA(ym)
  xBym = xBversusy(yo(i),yo(i),a1,a2,ga1,ga2,xaf,1,0,ym); % xB(ym)
  [V,y] = ode15s(@frhs,Tspan,y0,options,... % using stiff DE integrator ode15s
                 LeA,LeB,a1,a2,b1,b2,ga1,ga2,xaf,xbf,yf,ym,K,Bbar);
  TT(1:length(V),i) = V; XA(1:length(y(:,1)),i) = y(:,1);
  XB(1:length(y(:,2)),i) = y(:,2); Y(1:length(y(:,3)),i) = y(:,3); end

subplot(3,1,1)                % xA plot, with or without log10 time scale
if log == 1, semilogx(TT,XA), else, plot(TT,XA), end, hold on, v = axis;
text(v(2),xAym,' \leftarrow limit'),    % marking optimal steady state
xlabel('t','Fontsize',12), ylabel('x_A    ','Fontsize',12,'Rotation',0),
title([{['Profile plots of the solution to the IVP (4.74), (4.75),'...
    ' and (4.76) with \alpha_1 = ',num2str(a1,'%10.5g'),', \alpha_2 = ',...
    num2str(a2,'%10.5g'),' ;']},...
    {' '},{['\beta_1 = ',num2str(b1,'%10.5g'),...
        ', \beta_2 = ',num2str(b2,'%10.5g'),'; \gamma_1 = ',...
        num2str(ga1,'%10.5g'),', \gamma_2 = ',num2str(ga2,'%10.5g'),...
```

```
';  x_{Af} = ',num2str(xaf,'%10.5g'),',  K = ',num2str(K,'%10.5g'),...
';  Le_A = ',num2str(LeA,'%10.5g'),',  LeB = ',num2str(LeB,'%10.5g'),...
', for optimal y_f and y_m']}],'Fontsize',12)

subplot(3,1,2)                  % xB plot, with or without log10 time scale
if log == 1, semilogx(TT,XB), else, plot(TT,XB), end, hold on, v = axis;
text(v(2),xBym,' \leftarrow limit'),   % marking optimal steady state
xlabel('t','Fontsize',12), ylabel('x_B  ','Fontsize',12,'Rotation',0),

subplot(3,1,3)                  % y plot, with or without log10 time scale
if log == 1, semilogx(TT,Y), else plot(TT,Y), end, hold on, v = axis;
text(v(2),ym,' \leftarrow limit'),   % marking optimal steady state
xlabel('t','Fontsize',12), ylabel('y  ','Fontsize',12,'Rotation',0),
                                % free the plot windows :
subplot(3,1,1), hold off, subplot(3,1,2), hold off, subplot(3,1,3), hold off,
                                % right hand side of IVP
function dydt = frhs(V,y,LeA,LeB,a1,a2,b1,b2,ga1,ga2,xaf,xbf,yf,ym,K,Bbar)
F1 = a1*exp(-ga1./y(3,:)).*y(1,:); F2 = a2*exp(-ga2./y(3,:)).*y(2,:);
dydt = [LeA*(Bbar*(xaf-y(1,:))-F1);
        LeB*(Bbar*(xbf-y(2,:))+F1-F2);
        Bbar*(yf+K*(ym-y(3,:))-y(3,:)) + b1*F1 + b2*F2];

function g = Bfun(A,U,Umf,alf,eps,QE,H)                      % evaluate Bbar
GB = A*(U-Umf); AC = eps*GB/((alf-1)*Umf); GC = GB*(1+eps/(alf-1));
alprime = QE*AC/GC; g = A*U - GC*exp(-alprime*H);
```

This program offers our users the option to plot the profiles in linear or in $\log_{10}$ time. The $\log_{10}$ time plots spread out the earlier reactions more clearly, while a linear time scale gives every second the same horizontal axis space.

For the novice, our 3D plots in Figures 4.34 to 4.37 can be quite confusing to interpret correctly. The corresponding individual 2D profiles that were plotted in Figures 4.38 and 4.39 are useful for interpreting the 3D plots.

For relatively large Lewis numbers and with appropriate feedback, we observe that convergence to the system's unique stable steady state is quite swift and straightforward from any initial value. To illustrate, we now plot the 3D trajectories that emanate from the eight corners of the (x_A, x_B, y) unit cube

$$C_u = [0,1] \times [0,1] \times [0,1] \in \mathbb{R}^3$$

for the same system data as was used in Figures 4.34 to 4.39, but with the relatively large Lewis numbers

$$Le_A = Le_B = 10$$

instead.

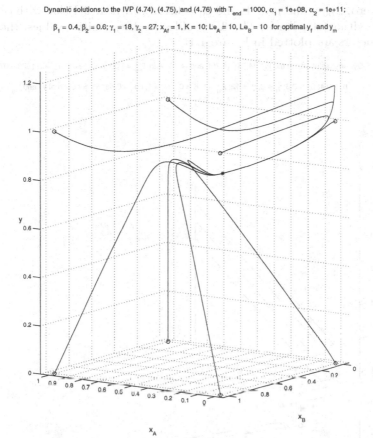

Dynamic solutions to the IVP (4.74), (4.75), and (4.76) with $T_{end} = 1000$, $\alpha_1 = 1e{+}08$, $\alpha_2 = 1e{+}11$; $\beta_1 = 0.4$, $\beta_2 = 0.6$; $\gamma_1 = 18$, $\gamma_2 = 27$; $x_{Af} = 1$, $K = 10$; $Le_A = 10$, $Le_B = 10$ for optimal y_f and y_m

Set of trajectories that solve (4.75), (4.76), and (4.77) for the original data, y_f and y_m
from Figure 4.28, high Lewis numbers, and $K = 0$

Figure 4.40

The eight trajectories of Figure 4.40 are obtained from eight separate calls of `fluidbed`
with `hold on` set, namely from each of the eight corners of the unit cube as starting
points and after a customary manual figure rotation. All trajectories in Figure 4.40 converge to the unique steady state of the system marked by $*$ from the eight corners o of
the unit cube C_u. Again there are exactly two final approaches to $*$ for the trajectories,
while only three trajectories with the initial high relative heat of $y(0) = 1$ exceed this
value for a short period.

Let us now revisit the modified example with the α_i and γ_i parameter data as described on p. 195. We recall that for this system there are five steady states prior to
feedback. Three of these are stable at the approximate $(x_A,\ x_B,\ y)$ locations of $(1, 0,
0)$, $(0, 1, 0.41...)$, and $(0, 0, 1)$. Of these three, the middle stable one yields the maximal
amount of component B and therefore it is the most desirable steady state. Figure 4.41
shows a set of 3D profile plots with initial values from a neighborhood of the middle
stable steady state $(x_A, x_B, y) = (0, 1, 0.41038)$ (taken from Figure 4.31) with $x_{A0} = 0.1$

or 0.2 near zero, $x_{B0} = 0.8$ near 1, and $y_0 = 0.1, ..., 0.7$ near 0.41038 with controller gain $K = 0$, i.e., without feedback. The 3D profiles for these initial values, the system and initial parameters are plotted in Figure 4.41.

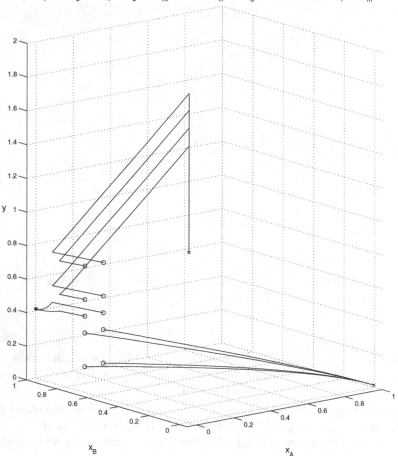

Dynamic solutions to the IVP (4.74), (4.75), and (4.76) with $T_{end} = 600$, $\alpha_1 = 1e+10$, $\alpha_2 = 3.1623e+07$;

$\beta_1 = 0.4$, $\beta_2 = 0.6$; $\gamma_1 = 8$, $\gamma_2 = 12$; $x_{Af} = 1$, $K = 0$; $Le_A = 1$, $Le_B = 0.45455$ for optimal y_f and y_m

Set of trajectories that solve (4.75), (4.76), and (4.77) for the modified data on p. 195, y_f and y_m from Figure 4.31, and $K = 0$
Figure 4.41

The ten trajectories from our ten initial values start at the o marks and end after 600 time units at the x marks. Each of the ten trajectories converges to one of the three steady states of the system, marked by x in Figure 4.41. From Figure 4.41 it is painfully obvious that the region of attraction for the middle stable steady state, marked by * in the plot, is very small since in our plot only the two trajectories that start with $y(0) = 0.4$ converge to the middle steady state with $y = 0.41038$ (near the left edge of Figure 4.41). For all initial values outside a very small attracting region for *, the depicted trajectories converge to one of the two other stable steady states at $(x_A, x_B, y) = (1, 0, 0)$ or $(0, 0, 1)$, both of which give no useful rendition of component B. Small perturbations of the uncontrolled system will cause the system to slip away from the maximal-yield middle steady state quickly, even if the system is operating at this stable steady state. Thus it is very difficult for the uncontrolled system $(K = 0)$ to produce much of the component B reliably over a long time.

Running a system at a high yielding stable steady state is not a cure-all! It is much more advisable to use feedback to create a controlled system with a unique high-yield steady state. We exemplify this in a plot of the 3D solution trajectories of the IVP (4.75), (4.76), and (4.77) for the modified data from p. 195 and initial values (x_A, x_B, y) from the whole cube $CC = [0.1, 0.9] \times [0.1, 0.9] \times [0.1, 0.9]$ with feedback set to $K = 10$.

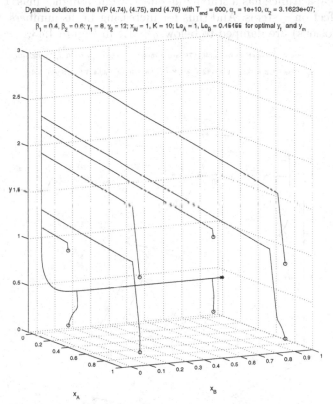

Set of trajectories that solve (4.75), (4.76), and (4.77) for the modified data on p. 195, y_f and y_m from Figure 4.31, and $K = 10$

Figure 4.42

In Figure 4.42, every trajectory that starts at one of the eight corners of the cube CC of (x_A, x_B, y) initial values ends up at the optimal stable middle steady state marked by $*$ when sufficient feedback $K = 10$ is used. Most trajectories in this example go through a relatively high temperature spike and $x_A = x_b = 0$ phase before reaching the unique steady state under feedback, except for the two trajectories that start at $(0.1, 0.1, 0.1)$ and $(0.1, 0.9, 0.1)$. These two trajectories reach the steady state directly without a temperature explosion. The cause of these sudden increases in temperature on the other trajectories is the simultaneous occurrence of high temperatures and high concentrations of the reactants. This causes a very high rate of reaction accompanied by a very high rate of heat production. Whether such explosions occur for given high α_i and β_i values is linked to (a) the initial conditions, as is evident from Figure 4.42, and (b) the relation between the Lewis numbers. The latter link is shown in Figure 4.43, where we have increased both Lewis numbers Le_A and Le_B from Figure 4.42 by a factor of 10 to become $Le_A = 10$ and $Le_B = 4.54545$. From the definition of Lewis numbers as heat capacitance divided by mass capacitance, increased Lewis numbers signify a larger heat capacitance or a lower mass capacitance, or a combination thereof. In general, an increased heat capacitance has a stabilizing effect, while a higher mass capacitance has a destabilizing effect on systems. And thus for increasing Lewis numbers, the temperature runaways (explosions) are damped to below $y = 1.1$ in Figure 4.43.

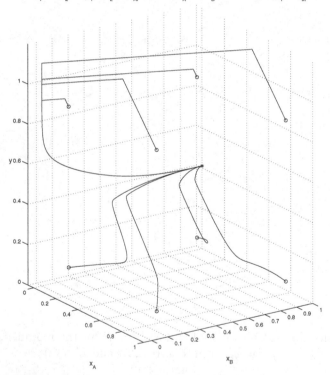

Dynamic solutions to the IVP (4.74), (4.75), and (4.76) with $T_{end} = 600$, $\alpha_1 = 1e{+}10$, $\alpha_2 = 3.1623e{+}07$;

$\beta_1 = 0.4$, $\beta_2 = 0.6$; $\gamma_1 = 8$, $\gamma_2 = 12$; $x_{Af} = 1$, $K = 10$; $Le_A = 10$, $Le_B = 4.5455$ for optimal y_f and y_m

Set of trajectories that solve (4.75), (4.76), and (4.77) for the modified data on p. 195, y_f and y_m from Figure 4.31, and $K = 10$

Figure 4.43

Temperature explosions are also more likely for high α_i values, which signify high rates of reactions, as well as for high β_i values, indicating high exothermic heat accompanying the reactions. As explained before, the effect of the Lewis numbers on explosions is more complex. In order to make this clear, we repeat that Lewis numbers express the ratio of heat capacitance and mass capacitance. Therefore, small Lewis numbers correspond to large mass capacitances or to low heat capacitances. Small Lewis numbers tend to make instability and temperature explosions more likely because high mass capacitances of the reactants correspond to the possible release of the chemoabsorbed mass, causing an increase in the rate of reaction and subsequent heat production. Moreover, for small Lewis numbers the low heat capacitance does not allow for an efficient dissipation of the generated heat.

Therefore we conclude not only that *feedback control* is *useful to stabilize an optimal unstable steady state* such as depicted in Figures 4.34 to 4.37 for the original set of parameter data, but *feedback control can* also *help ensure the robustness of an otherwise stable optimal steady state over a larger region of parameters and system perturbations.* Proper feedback control is also helpful in damping temperature explosions.

All locations of the steady states for the two-phase reaction model are determined by the eight system parameters α_1, α_2, β_1, β_2, γ_1, γ_2, x_{Af}, y_f, as well as the additional seven parameters on p. 183 that are needed to find $\bar{B}$. None of the steady states depends on or is influenced by the Lewis numbers Le_A and Le_B of the apparatus, since the Lewis numbers occur only on the left-hand side of the system of DEs (4.75), (4.76), and (4.77). If the Lewis numbers are not chosen carefully for the apparatus, then the corresponding chemical/biological system may be subject to oscillations even when we use feedback control to limit the number of steady states to one optimal stable steady state as we have done since Figure 4.34.

The call of `fluidbed(1200,0.1,0.1,0.5,.12,18,10^8,10^11,.4,.6,18,27,1,0,10)` produces Figure 4.34. The reader will immediately notice that we use our original 8 plus 7 parameter data of p. 183. Feedback gain is set to $K = 10$, so that according to Figure 4.34, the (x_A, x_B, y) point $(0.17619, 0.7225, 0.86446)$ is the optimal and unique steady state of this system. However, a careful inspection of the results in Figure 4.44 shows that the trajectory from the initial value $(x_A(0), x_B(0), y(0)) = (0.1, 0.1, 0.5)$, marked by o in the plot, proceeds in 1200 time units to the x mark near the unique steady state of the system that we have indicated by $*$. Figure 4.44 shows that the system converges to the optimal B conversion steady state $*$, but it does so only in a slow oscillatory fashion, circling the steady-state location $(0.17619, 0.7225, 0.86446)$ eight times in ever tighter loops during 1200 time units.

The corresponding profiles for x_A, x_B, and y are depicted in Figure 4.45 using `fluidprofiles.m` for the data and calling fluidbedprofiles(1200,[1,0,0.5],[1,0.8,0.2],[0.1, 0.4,0.9],.12,18,10^8,10^11,.4,.6,18,27,1,0,10,1).

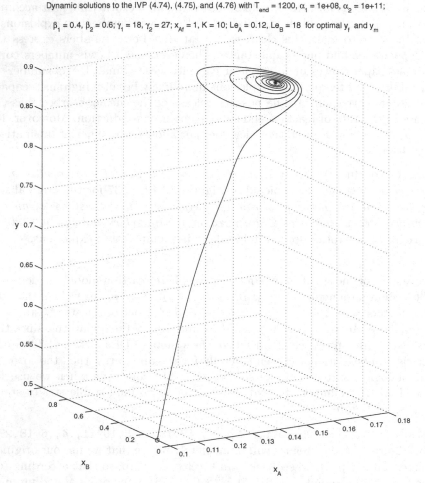

Dynamic solutions to the IVP (4.74), (4.75), and (4.76) with T_{end} = 1200, α_1 = 1e+08, α_2 = 1e+11;

β_1 = 0.4, β_2 = 0.6; γ_1 = 18, γ_2 = 27; x_{Af} = 1, K = 10; Le_A = 0.12, Le_B = 18 for optimal y_f and y_m

Trajectory for the IVP (4.75), (4.76), and (4.77) for the original data on p. 183 with
$Le_A = 0.12$, $Le_B = 18$, $K = 10$, and the initial value $(x_{A0},\ x_{B0},\ y_0) = (0.1, 0.1, 0.5)$
over 1,200 time units

Figure 4.44

Notice in Figure 4.45 that only the set of profiles that emanate from the initial value $(x_{A0},\ x_{B0},\ y_0) = (0.0, 0.8, 0.9)$ reaches the steady state without an explosion. Our other chosen initial value profiles result in explosions with temperature spikes of y around 5, 9, and 3 near 1, 20, and 100 time units, respectively. For all three initial value profiles note the oscillations with shrinking amplitudes around the steady-state values of x_A, x_B, and y after about 200 time units until convergence in Figure 4.45. On the other hand, the temperature profiles (except around the time of the explosions) settle down to near their steady-state value very smoothly after about 100 time units. This corroborates our 3D image in Figure 4.44 perfectly, in which the trajectory is depicted as oscillating around the steady-state coordinates on a near-level temperature surface.

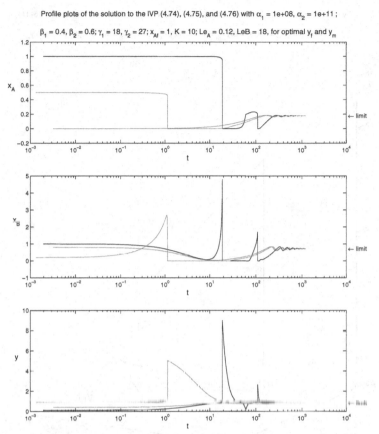

Profile plots of the solution to the IVP (4.74), (4.75), and (4.76) with α_1 = 1e+08, α_2 = 1e+11 ;

β_1 = 0.4, β_2 = 0.6; γ_1 = 18, γ_2 = 27; x_{Af} = 1, K = 10; Le_A = 0.12, LeB = 18, for optimal y_f and y_m

Solution profiles for the IVP (4.75), (4.76), and (4.77) for the original data on p. 183 with $Le_A = 0.12$, $Le_B = 18$, $K = 10$, and three initial values while running for 1,200 time units

Figure 4.45

Figures 4.44 and 4.45, best viewed in color, show a benign complication of the problem caused by the Lewis numbers. If, however, we reduce the Lewis number Le_A further to 0.07, the system trajectories indicate periodic explosions of the underlying system throughout all time, and the trajectories do not converge to the steady state at all, even with what we thought to be proper feedback. The trajectory that these curves settle at is called a **periodic attractor** of the system in contradistinction to the earlier encountered **point attractor** of Figures 4.43 or 4.44, for example. A point attractor, or more accurately a fixed-point attractor, is a more commonly encountered steady state in chemical and biological engineering systems. It could be called a stationary nonequilibrium state to distinguish it from the stationary equilibrium states associated with closed or isolated batch processes.

Figure 4.46 shows another (rotated) 3D picture of a periodic attractor, this one for a system with $Le_A = 0.07$ and for which all other parameters are as before.

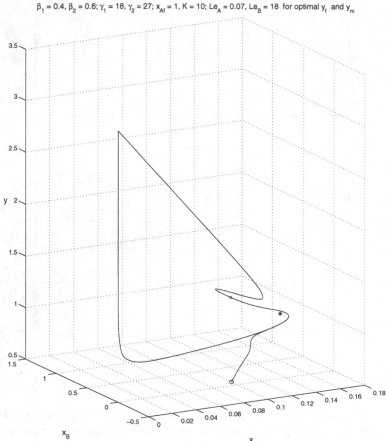

Trajectory for the IVP (4.75), (4.76), and (4.77) for the original data on p. 183 with
$Le_A = 0.07$, $Le_B = 18$, $K = 10$, and the initial value $(x_{A0},\ x_{B0},\ y_0) = (0.1, 0.1, 0.5)$
over 2,000 time units

Figure 4.46

The 3D trajectory in Figure 4.46 starts from the circle o at the initial value $(x_A(0), x_B(0),$
$y(0)) = (0.1, 0.1, 0.5)$ and begins similarly to the one in Figure 4.44. But it never converges to the steady state marked by $*$ in the plot. It rather goes through an intricate
identical periodic counterclockwise loop many times. This loop, also called a **limit cycle**,
is a **periodic attractor** for the system. At $\tau = 2000$ it has reached the spot marked
by x near the steady state $*$. Since the plot in Figure 4.46 gives us no information on
the number of periods within our time frame of 2,000 time units, we now plot three
profiles by themselves in a linear time scale for the initial values $(x_A(0), x_B(0), y(0)) =$
(0.1,0.1,0.5), (0.4,0.3,0.8), and (0.6,0.6,0.2) with the standard system data using the command `fluidbedprofiles(2000,[.1,.4,.6],[.1,.3,.6],[.5,.8,.2],.07,18,10^8,`
`10^11,.4,.6,18,27,1,0,10,0)`.

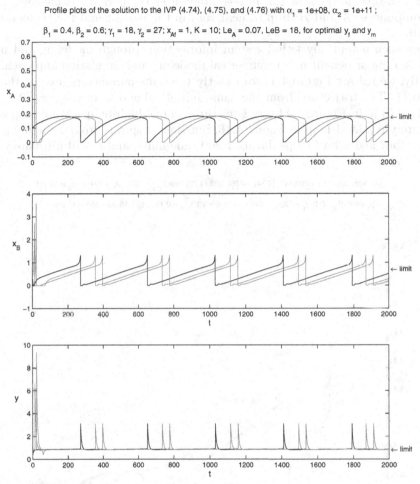

Solution profiles for the IVP (4.75), (4.76), and (4.77) for the original data on p. 183
with $Le_A = 0.07$, $Le_B = 18$, $K = 10$, and three initial values while running for 2,000
time units
Figure 4.47

In Figure 4.47, best viewed in color, we observe a parallel time-shifted behavior of the
three profiles that start from our three different initial values after the first 50 time
units. This seems to indicate that for infinite time, the solutions that start at any initial
value will eventually travel with the same period along the same unique loop that is
depicted in three dimensions in Figure 4.46, but time-shifted one from the other like
trains on the same track. More specifically, the profiles plot of Figure 4.47 indicates
that in 2,000 time units, this periodic attractor loop is traversed about five and a half
times, or approximately once every 380 time units. The system reaches an explosive state
eventually (after an early possible higher-level explosion, depending on the choice of the
initial value) every 380 time units. These periodic explosions occur with temperature

$y \approx 3$, $x_A \approx 0.18$, and $x_B \approx 1.3$. Immediately after the periodic explosions, the levels of the components A and B drop to near zero in the reactor and the temperature y to about 0.9.

Obtaining such potentially lethal system information through modeling and numerical analysis is a major benefit of our numerical modeling and simulation approach.

Finally, we redraw Figure 4.44 for exactly the same parameters, except that we use $Le_A = 0.11$. The trajectory from the same initial value $o = (x_A(0),\ x_B(0),\ y(0)) = (0.1,\ 0.1,\ 0.5)$ as in Figure 4.44 now goes through one high-temperature loop similar to the infinitely repeated loop in Figure 4.46, but then it spirals around the unique steady state $*$ in four and a half loops during 1,200 time units, and it will ultimately settle at the steady state $*$.

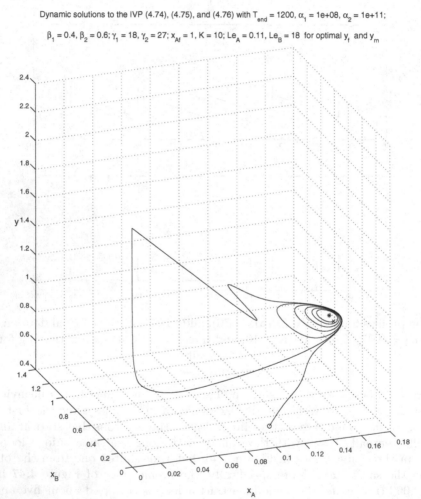

Dynamic solutions to the IVP (4.74), (4.75), and (4.76) with $T_{end} = 1200$, $\alpha_1 = 1e+08$, $\alpha_2 = 1e+11$; $\beta_1 = 0.4$, $\beta_2 = 0.6$; $\gamma_1 = 18$, $\gamma_2 = 27$; $x_{Af} = 1$, $K = 10$; $Le_A = 0.11$, $Le_B = 18$ for optimal y_f and y_m

Trajectory for the IVP (4.75), (4.76), and (4.77) for the original data on p. 183 with $Le_A = 0.11$, $Le_B = 18$, $K = 10$, and the initial value $(x_{A0},\ x_{B0},\ y_0) = (0.1, 0.1, 0.5)$ over 1,200 time units

Figure 4.48

We encourage our readers to draw out their own version of Figure 4.48 with our data in MATLAB via our program `fluidbed.m` while using the suggested `axis` settings for this figure inside the program.

Having this figure on their computer screen, the reader can use the special "magnifying glass" feature ⊕ and the rotation tool of MATLAB's graphics window to look at the behavior of the solution curve in more and better detail. We suggest that one magnify the regions where the single "wayward" high-temperature solution branch splits off at first from the later convergent oval solution loops and also where it joins them. This will exhibit several differing trajectories where Figure 4.48 appears to draw only one. Mathematically this must be so, since each location on the solution trajectory acts as an initial condition for continuing the solution and since IVPs always have unique solutions. Therefore the solution must differ on each of the oval loops, if only by a tiny amount.

Finally, in order to make sure that the high temperature "wayward" curve in Figure 4.48 is not a computational figment, the reader should verify its existence even when the error tolerances `RelTol` and `AbsTol` in `fluidbed`'s integrator `ode15s` are tightened to 10^{-12} or 10^{-14}.

Exercises for 4.3

1. Refresh your calculus skills and your MATLAB graphics skills:

    (a) Find the first derivative of $x_A(y)$ from equation (4.81) with respect to y and show that x_A is monotonically decreasing.

    (b) Find the first derivative of $x_B(y)$ from equation (4.82) with respect to y. Set up a numerical scheme to find y_m with $dx B(y_m)/dy = 0$ or use the symbolic toolbox of MATLAB (if available) to solve the equation $dx B(y)/dy = 0$ symbolically for the optimal setpoint y_m and general input parameters α_i, β_i, ..., etc.

    (c) Revise the simple maximum finding line
    ```
    m = min(find(XB == max(XB))); ym = Y(m);  % find ym from xB curve
    ```
    in `hetcotbifrange.m`, so that `ym` is always computed to an accuracy of 5 to 6 digits, rather than the 2 or 3 that we achieve in the current version of `hetcotbifrange.m`.

    (d) How would you compare the amount of effort (programming and computational) of parts (b) and (c) with the improved results they achieve?

2. **Project:**
    Create a MATLAB m file that draws the zero-contour curves $F(y, K) = 0$ of Figures 4.26 or 4.30 directly.
    (Hint: Model your code on the contour-curve-plotting codes `adiabNisocolor contour.m` and `adiabNisocontourcurve.m` of Section 3.1.)

3. Draw a 3D plot like Figure 4.34 or 4.42 for the original data, except with $Le_A = 0.07$, using `fluidbed.m` and verify that the 3D loop in Figure 4.46 is attracting from all initial values $(x_A(0), x_B(0), y(0))$.

4. Redraw Figure 4.47 for a few decreasing values of $Le_A < 0.07$ and $Le_B = 18$ and for a few increasing values of $Le_B > 18$ and $Le_A = 0.07$ to see how the

values of the Lewis numbers affect the period length of the periodic attractor, as well as the maximal temperature levels of the periodic explosions. What can you conclude?

5. Create the solution profiles of Figure 4.48 using `fluidbedprofiles.m` and interpret the behavior of the solution for the initial value problem trajectory of Figure 4.48.

6. (a) Verify our uniqueness assertions on p. 221 about the differing branches of the IVP solution in Figure 4.48 for your own plot of this figure.

   (b) What can one say in this respect about the points on the limit cycle depicted in Figure 4.46? Are they the same from one pass to the next? Why or why not?

7. Modify `fluidbed.m` to use several different MATLAB integrators `ode...` in turn and compare the results.

Conclusions

In this section we have developed steady-state and dynamic models for a heteregeneous system. Specifically, we have chosen the bubbling fluidized bed catalytic reactor, which has many industrial applications. We have built the model for a consecutive reaction $A \rightarrow B \rightarrow C$ with the component B being the desired component.

The static bifurcation behavior and its practical implications have been investigated. We have also formulated the unsteady-state dynamic model and we have used it to study the dynamic behavior of the system by solving the associated IVP numerically. Both the controlled and the uncontrolled cases have been investigated. Two particular reactions have been studied, one with three steady states, and one with five steady states.

4.4   A Biomedical Example: The Neurocycle Enzyme System

Chemical engineering principles and models play an important role in describing and analyzing biological (biochemical and/or biomedical) systems, where biological phenomena interact with the purely chemical and physical ones of standard chemical engineering. We have earlier encountered biological examples in the form of the enzyme reactor in Section 3.3 and the anaerobic digester in Section 2 of this chapter, and we will study biological fermentors in Chapter 7.

During the last few decades, many chemical engineering departments have changed their names to departments of chemical and biological engineering. What distinguishes biological engineering from biochemical engineering is that biological engineering includes both biochemical and biomedical engineering. In this section we develop a model and numerical solutions of a problem of biomedical engineering.

4.4.1 Fundamentals

The chemical synapse is a highly specialized structure that has evolved for exquisitely controlled voltage-dependent secretions. The chemical messengers, stored in vesicles, are released from the presynaptic cell following the arrival of an action potential that triggers the vesicular release into the presynaptic terminal. Once released from the vesicles, the transmitter diffuses across a narrow synaptic cleft, then binds to specific receptors in the postsynaptic cell, and finally initiates an action potential event in the nerve-muscle cell membrane by triggering muscle contractions.

Acetylcholine plays a recognized role in the nerve excitation scenario that we have described above very briefly. It is found in cholinergic synapses that provide stimulatory transmissions in the nervous system. Its complete neurocycle implies a coupled two-enzymes/two-compartments model with two strongly coupled events as follows:
The activation event: Acetylcholine is synthesized from choline and acetyl coenzyme A (Acetyl-CoA) by the enzyme choline acetyltransferase ($ChAT$) and is immediately stored in small vesicular compartments closely attached to the cytoplasmic side of the presynaptic membranes.
The degradation event: Once acetylcholine has completed its activation duty, the synaptic cleft degradation begins to remove the remaining acetylcholine. This occurs through the destruction of acetylcholine by hydrolysis that uses the acetylcholinesterase enzyme ($AchE$) to form choline and acetic acid.

Diseases such as Alzheimer's and Parkinson's are the result of an imbalance of the cholinergic system considered above, with devastating consequences to human health. The above simplified sequence of events suggests that we might obtain some insight into such a cycle by using a simple diffusion-reaction model that simulates the nonlinear interaction between these events. Membrane models that are targeted to physiological problems can be a promising heuristic way for tackling complex biomedical systems such as the one considered above.

4.4.2 The Simplified Diffusion-Reaction Two Enzymes/Two Compartments Model

In the following we attempt to describe the acetylcholinesterase/choline acetyltransferase enzyme system inside the neural synaptic cleft in a simple fashion; see Figure 4.49. The complete neurocycle of the acetylcholine as a neurotransmitter is simulated in our model as a simple two-enzymes/two-compartments model. Each compartment is described as a constant-flow, constant-volume, isothermal, continuous stirred tank reactor (CSTR). The two compartments (I) and (II) are separated by a nonselective permeable membrane as shown in Figure 4.50.

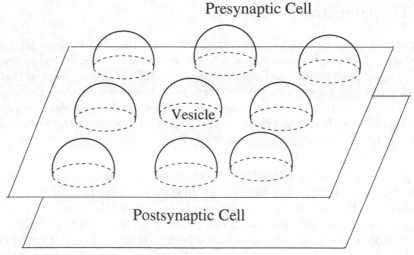

The synaptic cleft
Figure 4.49

Assuming that all the events are homogeneous in all vesicles, and using the proper dimensionless state variables and parameters, we consider the behavior for a single synaptic vesicle as described by this simple two-compartment model, where (I) and (II) denote the two compartments.

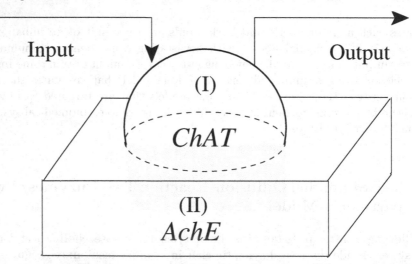

The two enzymes/two compartments model
Figure 4.50

A schematic presentation is given in Figure 4.51 for the simplified diffusion-reaction two-enzymes/two-compartments model. From an enzyme kinetics point of view, we consider the most general case, in which both enzymes have nonmonotonic dependence on the substrate and hydrogen-ion concentrations.

We assume that acetylcholine is synthesized in compartment (I) by $ChAT$ due to the activation reaction R_1 in which the stimulatory neurotransmitter acetylcholine is synthesized:

$$R_1 : \quad \text{choline} + \text{acetyl} - CoA \overset{ChAT}{\longrightarrow} \text{acetylcholine} + CoA . \tag{4.89}$$

Acetylcholine is destroyed (hydrolyzed) in compartment (II) by $AchE$ in the degradation reaction R_2 where the stimulatory neurotransmitter acetylcholine is degraded:

$$R_2 : \quad \text{acetylcholine} + \text{water} \overset{AchE}{\longrightarrow} \text{choline} + \text{acetate} + H^+ . \tag{4.90}$$

Both reactions are considered to be substrate-inhibited and hydrogen-ion rate-dependent. This leads to a nonmonotonic dependence of the reaction rates on both the substrates and pH.

The following reaction rate expressions can be derived.

The acetylcholine synthesis reaction rate:

$$R_1 = \left\{ \frac{V_{M1} \cdot [S_2]_1}{[S_2]_1 + \dfrac{K_{S2}}{[H^+]_1}\left(Kh_2 + [H^+]_1 + \dfrac{[H^+]_1^2}{Khh_2}\right) + \dfrac{[S_2]_1^2}{Ki_2}} \right\} \cdot$$

$$\cdot \left\{ \frac{[S_3]_1}{[S_3]_1 + \dfrac{K_{S3}}{[H^+]_1}\left(Kh_3 + [H^+]_1 + \dfrac{[H^+]_1^2}{Khh_3}\right) + \dfrac{[S_3]_1^2}{Ki_3}} \right\} . \tag{4.91}$$

The acetylcholine hydrolysis reaction rate:

$$R_2 = \frac{V_{M2} \cdot [S_1]_2}{[S_1]_2 + \dfrac{K_{S1}}{[H^+]_2}\left(Kh_1 + [H^+]_2 + \dfrac{[H^+]_2^2}{Khh_1}\right) + \dfrac{[S_1]_2^2}{Ki_1}} . \tag{4.92}$$

4.4.3 Dynamic Model Development

The schematic diffusion-reaction structure shown in Figure 4.51 utilizes our suggested two enzymes/two-compartments model. In this structure, both the feed and the output are to/from compartment (I), while the input and output of compartment (II) both occur though membrane diffusion.

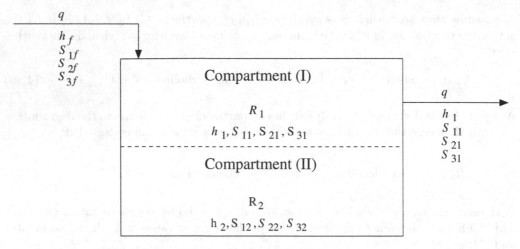

Simplified diffusion-reaction two enzymes/two compartments model
Figure 4.51

Next we formulate the dynamic model differential equations for the different components in the two compartments:

(A) The hydrogen-ion dynamic mole balances in the two compartments are given by

$$V_j \frac{d[H^+]_j}{dt} = a_{1j} \cdot q \cdot \left([H^+]_f - [H^+]_1\right) - a_{2j} \cdot \alpha'_{H^+} \cdot A_M \left([H^+]_1 - [H^+]_2\right) +$$
$$+ V_j \left(a_{4j} \cdot R_2 \cdot \overline{AchE} - Rw_j\right) \qquad (4.93)$$

for the indices $j = 1, 2$, which denote the compartments (I) and (II), respectively. Here $a_{11} = 1$, $a_{12} = 0$, $a_{21} = 1$, $a_{22} = -1$, $a_{41} = 0, a_{42} = 1$; $[H^+]_f$ is the concentration of hydrogen ions in the feed, and Rw_j is the rate of water formation in compartment j. The parameter α'_{H^+} is the membrane permeability for hydrogen ions and A_M is the active membrane area.

(B) The dynamic mole balances for the hydroxyl ions in the two compartments are given by

$$V_j \frac{d[OH^-]_j}{dt} = a_{1j} \cdot q \cdot \left([OH^-]_f - [OH^-]_1\right)$$
$$- a_{2j} \cdot \alpha'_{OH^-} \cdot A_M \left([OH^-]_1 - [OH^-]_2\right) - V_j \cdot Rw_j \quad (4.94)$$

for $j = 1, 2$ as before. Here $[OH^-]_f$ is the concentration of hydroxyl ions in the feed and α'_{OH^-} is the membrane permeability for hydroxyl ions.

(C) The acetylcholine dynamic mole balances in the two compartments are given by

$$V_j \frac{d[S_1]_j}{dt} = a_{1j} \cdot q \cdot \left([S_1]_f - [S_1]_1\right) - a_{2j} \cdot \alpha'_{S_1} \cdot A_M \left([S_1]_1 - [S_1]_2\right) +$$
$$+ V_j \left(a_{3j} \cdot R_1 \cdot \overline{ChAT} - a_{4j} \cdot R_2 \cdot \overline{AchE}\right) \qquad (4.95)$$

for $j = 1, 2$ with $a_{31} = 1$, $a_{32} = 0$. Here $[S_1]_f$ is the concentration of acetylcholine in the feed and α'_{S_1} is the membrane permeability for acetylcholine.

(D) The choline dynamic mole balances in the two compartments are given by

$$V_j \frac{d[S_2]_j}{dt} = a_{1j} \cdot q \cdot ([S_2]_f - [S_2]_1) - a_{2j} \cdot \alpha'_{S_2} \cdot A_M \left([S_2]_1 - [S_2]_2\right) +$$
$$+ V_j \left(-a_{3j} \cdot R_1 \cdot \overline{ChAT} + a_{4j} \cdot R_2 \cdot \overline{AchE}\right) \tag{4.96}$$

for $j = 1, 2$, where $[S_2]_f$ is the concentration of choline in the feed and α'_{S_2} is the membrane permeability for choline.

(E) Finally, the acetate dynamic mole balances in the two compartments are given by

$$V_j \frac{d[S_3]_j}{dt} = a_{1j} \cdot q \cdot ([S_3]_f - [S_3]_1) - a_{2j} \cdot \alpha'_{S_3} \cdot A_M \left([S_3]_1 - [S_3]_2\right) +$$
$$+ V_j \left(-a_{3j} \cdot R_1 \cdot \overline{ChAT} + a_{4j} \cdot R_2 \cdot \overline{AchE}\right) \tag{4.97}$$

for $j = 1, 2$, where $[S_3]_f$ is the concentration of acetate in the feed and α'_{S_3} is the membrane permeability for acetate.

The pseudosteady-state assumption for the hydroxyl ions gives us

$$\frac{d[OH^-]}{dt} = 0 . \tag{4.98}$$

Assuming that the hydrogen and hydroxyl ions are at equilibrium yields the equation

$$K_w = [H^+] \cdot [OH^-] \tag{4.99}$$

for K_w, the equilibrium constant of water reversible dissociation.
Subtracting equation (4.94) from equation (4.93) and substituting equations (4.98) and (4.99) into equation (4.93) results in the equation

$$V_j \frac{d[H^+]_j}{dt} = a_{1j} \cdot q \cdot \left\{ ([H^+]_f - [H^+]_1) - K_w \left(\frac{1}{[H^+]_f} - \frac{1}{[H^+]_1} \right) \right\}$$
$$- a_{2j} \cdot \alpha'_{H^+} \cdot A_M \left([H^+]_1 - [H^+]_2\right) +$$
$$+ a_{2j} \cdot A_M \left\{ \alpha'_{OH^-} \cdot K_w \left(\frac{1}{[H^+]_1} - \frac{1}{[H^+]_2} \right) \right\} +$$
$$+ V_j \cdot a_{4j} \cdot R_2 \cdot \overline{AchE} \tag{4.100}$$

for $j = 1, 2$. The two equations (4.100) for $j = 1, 2$ replace the four equations (4.93) and (4.94) for $j = 1, 2$ under the above assumptions.

Before proceeding further, we list all notations used in the above equations (4.91) to (4.100).

Notations:

The subscripts $j = 1, 2$ denote the compartments (I) and (II), respectively. The subscript f denotes the feed conditions.

A_M	active area of membrane between compartments; in m^2
$AchE$	acetylcholinesterase
$\overline{AchE}$	concentration of acetylcholinesterase in compartment (II); in kg/m^3
$B_1 = V_1 V_{M1} \overline{ChAT}/q$	parameter for the acetyltransferase enzyme activity; $kmol/m^3$
$B_2 = V_2 V_{M2} \overline{AchE}/q$	parameter for the acetylcholinesterase enzyme activity; in $kmol/m^3$
$ChAT$	choline acetyltransferase
$\overline{ChAT}$	concentration of choline acetyltransferase in compartment (I); in kg/m^3
CoA	coenzyme A
$\overline{CoA}$	concentration of coenzyme A in compartment (I); in kg/m^3
E_N	enzyme N
$[H^+]$	hydrogen ion concentration; in $kmol/m^3$
$K_{S_1}, \ Ki_1, \ Kh_1, \ Khh_1$	kinetic constants for the choline acetyltransferase catalyzed reaction; in $kmol/m^3$
$K_{S_2}, \ Ki_2, \ Kh_2, \ Khh_2$	kinetic constants for the coenzyme A catalyzed reaction; in $kmol/m^3$
$K_{S_3}, \ Ki_3, \ Kh_3, \ Khh_3$	kinetic constants for the acetylcholinesterase catalyzed reaction; in $kmol/m^3$
K_w	equilibrium constant of water; in $kmol^2/m^6$
$[OH^-]$	hydroxyl ions concentration; in $kmol/m^3$
P_N	reaction product N, the product of catalyzing S_N by E_N
q	volumetric flow rate into compartment (I); in m^3/sec
R_j	rate of reaction in compartment j; in $kmol/(m^2 \cdot sec)$
Rw_j	rate of water formation in compartment j; $kmol/(m^3 \cdot sec)$
$[S_1]$	acetylcholine concentration; in $kmol/m^3$
$[S_2]$	choline concentration; in $kmol/m^3$
$[S_3]$	acetate concentration; in $kmol/m^3$
S_N	substrate N, catalyzed by the enzyme N
t	time; in sec
V_j	volume of compartment j; in m^3

Dimensionless parameters:

$h_j = [H^+]_j/Kh_1$	dimensionless hydrogen ion concentration
$s_{1j} = [S_1]_j/K_{S_1}$	dimensionless acetylcholine concentration
$s_{2j} = [S_2]_j/[S_2]_{ref}$	dimensionless choline concentration
$s_{3j} = [S_3]_j/[S_3]_{ref}$	dimensionless acetate concentration

$V_R = V_1/V_2$ ratio of the volume of the two compartments

$T = q \cdot t/V_1$ dimensionless time

Dimensionless kinetic parameters for the acetylcholinesterase catalyzed reaction:

$$\gamma = \frac{Kw}{Ki_1}, \quad \delta = \frac{Kh_1}{Khh_1}, \quad \alpha = \frac{K_{S_1}}{Ki_1}.$$

Dimensionless kinetic parameters for the choline acetyltransferase catalyzed reaction:

$\theta_1, \; ..., \; \theta_8.$

Greek letters:

$\alpha_{H+} = \alpha'_{H+} \cdot A_M/q$ dimensionless membrane permeability for hydrogen ions

$\alpha_{OH} = \alpha'_{OH-} \cdot A_M/q$ dimensionless membrane permeability for hydroxyl ions

$\alpha_{S_1} = \alpha'_{S_1} \cdot A_M/q$ dimensionless membrane permeability for acetylcholine

$\alpha_{S_2} = \alpha'_{S_2} \cdot A_M/q$ dimensionless membrane permeability for choline

$\alpha_{S_3} = \alpha'_{S_3} \cdot A_M/q$ dimensionless membrane permeability for acetate

α'_{H+} membrane permeability for hydrogen ions; in m/sec

α'_{OH-} membrane permeability for hydroxyl ions; in m/sec

α'_{S_1} membrane permeability for acetylcholine ions; in m/sec

α'_{S_2} membrane permeability for choline ions; in m/sec

α'_{S_3} membrane permeability for acetate ions; in m/sec

4.4.4   Normalized Form of the Model Equations

With the notations of Section 4.4.3, the equations (4.91), (4.92), (4.95) to (4.97), and (4.100) can be expressed in normalized form. The differential equations (4.95) to (4.97) and (4.100) can be written in matrix form as

$$\frac{d\Psi_j}{dt} = F_j(\Psi) \tag{4.101}$$

for $j = 1, 2$ with the known initial conditions $\Psi(0)$ at $t = 0$.

In the matrix representation (4.101) we use the following settings:

$$\Psi_j(t) = \Psi_j = \begin{pmatrix} h_j \\ s_{1j} \\ s_{2j} \\ s_{3j} \end{pmatrix}, \quad \Psi = \begin{pmatrix} \Psi_1 \\ \Psi_2 \end{pmatrix}, \text{ and } F_j(\Psi) = \begin{pmatrix} F_{1j}(\Psi) \\ F_{2j}(\Psi) \\ F_{3j}(\Psi) \\ F_{4j}(\Psi) \end{pmatrix}, \tag{4.102}$$

both for $j = 1$ and $j = 2$, and the feed condition

$$\Psi_{feed} = \begin{pmatrix} \Psi_{1feed} \\ \Psi_{2feed} \end{pmatrix} = \begin{pmatrix} h_f \\ s_{1f} \\ s_{2f} \\ s_{3f} \\ 0 \\ 0 \\ 0 \\ 0 \end{pmatrix}.$$

Note that the entries of Ψ_{feed} have already been incorporated into the equations **(A)** to **(E)** on p. 226 to 227. The eight component functions F_{kj} for $k = 1,...,4$ and $j = 1, 2$ are the dynamic mole balance equations in normalized form. They are given in equations (4.103) to (4.106) as follows.

Normalized equations (A') and (B'), replacing the earlier equations **(A)** and **(B)**

The DEs (4.93) and (4.94) have earlier been combined into equation (4.100) for the hydrogen and hydroxyl ions balances. This latter DE has the normalized right-hand side

$$F_{1j}(\Psi) = a_{1j}\left\{(h_f - h_1) - \gamma\left(\frac{1}{h_f} - \frac{1}{h_1}\right)\right\}$$
$$- b_j \cdot a_{2j}\left\{\alpha_{H^+} \cdot (h_1 - h_2) - \gamma \cdot \alpha_{OH^-} \cdot \left(\frac{1}{h_1} - \frac{1}{h_2}\right)\right\} +$$
$$+ b_j \cdot a_{4j} \cdot B_2 \cdot r_2/Kh_1 \qquad (4.103)$$

for $j = 1, 2$. Here the B_i are defined in Section 4.4.3 on p. 228, while $b_1 = 1$ and $b_2 = V_R$. Here and in the following, the a_{ij} are as given on p. 226.

Normalized equation (C'), instead of **(C)**

The DE (4.95) for acetylcholine can be expressed with the normalized right-hand side

$$F_{2j}(\Psi) = a_{1j}(s_{1f} - s_{11}) - b_j \cdot a_{2j} \cdot \alpha_{S_1}(s_{11} - s_{12}) +$$
$$+ b_j (a_{3j} \cdot B_1 \cdot r_1/Ks_1 - a_{4j} \cdot B_2 \cdot r_2/Ks_1) \qquad (4.104)$$

for $j = 1, 2$ and b_j and a_{ij} as above.

Normalized equation (D'), instead of **(D)**

The DE (4.96) for choline can be expressed with the normalized right-hand side

$$F_{3j}(\Psi) = a_{1j}(s_{2f} - s_{21}) - b_j \cdot a_{2j} \cdot \alpha_{S_2}(s_{21} - s_{22}) +$$
$$+ b_j (-a_{3j} \cdot B_1 \cdot r_1/[S_2]_{ref} + a_{4j} \cdot B_2 \cdot r_2/[S_2]_{ref}) \qquad (4.105)$$

for $j = 1, 2$ and b_j and a_{ij} as before.

Normalized equation (E'), instead of **(E)**

The DE (4.97) for acetate can be expressed with the normalized right-hand side

$$F_{4j}(\Psi) = a_{1j}(s_{3f} - s_{31}) - b_j \cdot a_{2j} \cdot \alpha_{S_3}(s_{31} - s_{32}) +$$
$$+ b_j (-a_{3j} \cdot B_1 \cdot r_1/[S_3]_{ref} + a_{4j} \cdot B_2 \cdot r_2/[S_3]_{ref}) \qquad (4.106)$$

for $j = 1, 2$ and b_j and a_{ij} as before.

Normalized reaction rates (4.89) **and** (4.90)

$$r_1 = \left\{\frac{s_{21}}{s_{21} + \theta_1/h_1 + \theta_2 + \theta_3 h_1 + \theta_4 s_{21}^2}\right\} \cdot \left\{\frac{s_{31}}{s_{31} + \theta_5/h_1 + \theta_6 + \theta_7 h_1 + \theta_8 s_{31}^2}\right\} \qquad (4.107)$$

and

$$r_2 = \frac{s_{12}}{s_{12} + 1/h_2 + 1 + \delta h_2 + \alpha s_{12}^2} . \tag{4.108}$$

The relatively simple two-enzymes/two-compartments model is thus represented in (4.101) via the above set of eight coupled ordinary nonlinear differential equations (4.103) to (4.106). This system of IVPs has the eight state variables $h_j(t)$, $s_{1j}(t)$, $s_{2j}(t)$, $s_{3j}(t)$ for $j = 1, 2$ that depend on the time t. The normalized reaction rates $r_j(t)$ are given in equations (4.107) and (4.108). The system has 26 parameters that describe the dynamics for all compounds considered in the two compartments. A specific list of validated experimental parameter values follows in Section 4.4.5.

4.4.5   Identification of Parameter Values

Because of the lack of good experimental data in human brain chemistry, our presentation is limited to the use of carefully chosen normalized experimental parameters in order to reproduce the basic static and dynamic characteristics of this coupled enzymes system.

Most of the data is taken from earlier experimental work. The concentrations in the feed and reference values are taken from mouse and rat brain data for the sake of illustration. The membrane permeability parameters for s_2 and s_3 are assumed equal to the value for s_1. The normalized parameter B_1 is taken equal to B_2, which was found earlier experimentally. The kinetic parameters θ_m for $m = 1, .., 8$ of reaction (1) are chosen by using a known dissociation constant and by keeping the experimentally found proportion for reaction (2) on substrate-inhibition and hydrogen-ion effects.

Table of 26 experimentally obtained numerical parameter values:

V_r	1.2		Kh_1	$1.0066 \cdot 10^{-6}$ $kmol/m^3$
B_1	$5.033 \cdot 10^{-5}$ $kmol/m^3$		Ks_1	$5.033 \cdot 10^{-7}$ $kmol/m^3$
B_2	$5.033 \cdot 10^{-5}$ $kmol/m^3$		$[S_2]_{ref}$	0.0001 $kmol/m^3$
α_{H^+}	2.25		$[S_3]_{ref}$	0.000001 $kmol/m^3$
α_{OH^-}	0.5		θ_1	32000
α_{S_1}	1		θ_2	4
α_{S_2}	1		θ_3	0.125
α_{S_3}	1		θ_4	0.125
s_{1f}	2.4		θ_5	84500
s_{2f}	1.15		θ_6	65
s_{3f}	3.9		θ_7	76.72
α	0.5		θ_8	0.00769
γ	0.01			
δ	0.1			

4.4.6   Numerical Considerations

Our numerical task is to solve the IVP given by the eight DEs (4.103) to (4.106) both for $j = 1$ and $j = 2$ with the given 26 system parameters and the two rate equations (4.107) and (4.108) for any physically feasible set of initial values $\Psi(0)$ and feed parameters Ψ_{feed} numerically.

A numerical solution of the problem may consist of a plot of all eight profiles $h_1(t)$, $s_{11}(t)$, $s_{21}(t)$, $s_{31}(t)$, $h_2(t)$, $s_{12}(t)$, $s_{22}(t)$, and $s_{33}(t)$ for a certain time interval $0 \leq t \leq T_{end}$. Or it may involve phase plots, such as that of the acetylcholine concentration in compartment (I) versus that in compartment (II). Our aim in the computations that follow is to show the variations in the quality of the solutions for differing values of h_f, the hydrogen ion concentration of the feed to compartment (I). Another dependency of the solutions is explored in the exercises.

Throughout this section we work with the initial values vector

$$
\Psi(0) \;=\; \begin{pmatrix} 0.09594 \\ 1.27 \\ 1.155 \\ 4.405 \\ 0.7 \\ 0.2 \\ 1.16 \\ 4.8 \end{pmatrix},
$$

which uses physiologically feasible starting values for the chemistry of the human brain.

Here is the MATLAB program `neurocycle.m`, which upon the user's specification and using MATLAB's (un)commenting feature, which places the symbol % at the start of comment lines, either plots all eight profiles, or one profile and one phase plot, or only one phase plot for relatively high values of the time parameter t, when the system has reached its periodic limit cycle.

The commenting or uncommenting of MATLAB code line blocks can best be achieved from the MATLAB text editor window for an m file. Simply highlight a block of code lines via a mouse drag in the MATLAB text editor window, then click on the "Text" entry of the editor's toolbar and click "Comment" or "Uncomment" as appropriate. This action makes % commenting marks appear at or disappear from the front of each code line of the highlighted block.

```
function [t, y] = neurocycle(hf,Tend,y0,S,standard)
%   [t, y] = neurocycle(hf,Tend,y0,S)
% Sample call : [t,y] = neurocycle(0.004554,20);
% Input : hf = hydrogen ion concentration of input
%         Tend = end of time interval for integration
%         y0 = initial value column vector,
%         with entries for h_1, s_11, s_21, s_31, h_2, s_12, s_22, s_32
%         S = color choice for graphs, such as 'r' for red, 'g' for green etc
```

```
%            standard : if "standard" is set to 1, we use the standard initial
%                       values from the book (section 4.4);
%                       if "standard" is set different from 1, we use special
%                       initial values and annotate the graphs differently.
% Output: Three different plots are obtainable by (un)commenting :
%         First plotting block active:
%            Graphs of all 8 profiles for 0 <= t <= Tend in an 8 window plot
%         Second plotting block active:
%            Phase plot of acetylcholine concentrations in (I) versus (II),
%            after limit cycle has been reached (for t >= 0.4 * Tend)
%         Third plotting block active:
%            s_12 profile and phase plot for 0 <= t <= Tend
%         Use the MATLAB text editor ("Text" help button) to (un)comment the
%            three plotting blocks so that ONLY ONE is active.

if nargin == 2, y0 = [0.09594; 1.27; 1.155; 4.405; 0.7; 0.2; 1.16; 4.8];
   S = 'b'; standard = 1; end                    % setting defaults
if nargin == 3, S = 'b'; standard = 1; end
if nargin == 4, standard = 1; end


tspan = [0 Tend];           % De IVP solver
options = odeset('RelTol',10^-6,'AbsTol',10^-7,'Vectorized','on');
[t,y] = ode15s(@FjPsi,tspan,y0,options,hf); % using  DE integrator ode...

% clf, subplot(4,2,1)                % uncomment this block for 8 profiles
% plot(t,y(:,1),S), v = axis;
% xlabel('t','FontSize',11); ylabel('h 1     ','FontSize',12,'Rotation',0);
% title('Hydrogen ion concentration in (I)','FontSize',12);
% if standard == 1,
%     text(v(1)+0*(v(2)-v(1)),v(4)+0.26*(v(4)-v(3)),...
%     ['Neurocycle enzyme system with h_f = ',num2str(hf,'%10.7g'),...
%        ' and the standard initial value \Psi(0)'],'FontSize',14);
% else,
%     text(v(1)+0*(v(2)-v(1)),v(4)+0.32*(v(4)-v(3)),...
%     ['Neurocycle enzyme system with h_f = ',num2str(hf,'%10.7g'),...
%        ' and initial value'],'FontSize',14);
% text(v(1)+0.1*(v(2)-v(1)),v(4)+0.21*(v(4)-v(3)),...
%     ['\Psi(0) = ',num2str(y0,'%7.5g')],'FontSize',14); end
% subplot(4,2,3)
% plot(t,y(:,2),S)
% xlabel('t','FontSize',11); ylabel('s_{11}     ','FontSize',12,'Rotation',0);
% title('Acetylcholine concentration in (I)','FontSize',12);
% subplot(4,2,5)
% plot(t,y(:,3),S)
% xlabel('t','FontSize',11); ylabel('s_{21}     ','FontSize',12,'Rotation',0);
% title('Choline concentration in (I)','FontSize',12);
% subplot(4,2,7)
% plot(t,y(:,4),S)
```

```
% xlabel('t','FontSize',11); ylabel('s_{31}      ','FontSize',12,'Rotation',0);
% title('Acetate concentration in (I)','FontSize',12);
%
% subplot(4,2,2)
% plot(t,y(:,5),S)
% xlabel('t','FontSize',11); ylabel('h_2      ','FontSize',12,'Rotation',0);
% title('Hydrogen ion concentration in (II)','FontSize',12);
% subplot(4,2,4)
% plot(t,y(:,6),S)
% xlabel('t','FontSize',11); ylabel('s_{12}      ','FontSize',12,'Rotation',0);
% title('Acetylcholine concentration in (II)','FontSize',12);
% subplot(4,2,6)
% plot(t,y(:,7),S)
% xlabel('t','FontSize',11); ylabel('s_{22}      ','FontSize',12,'Rotation',0);
% title('Choline concentration in (II)','FontSize',12);
% subplot(4,2,8)
% plot(t,y(:,8),S)
% xlabel('t','FontSize',11); ylabel('s_{32}      ','FontSize',12,'Rotation',0);
% title('Acetate concentration in (II)','FontSize',12);

% clf, subplot(2,1,1),  % uncomment this block for s_12 profile and phase plot
% plot(t,y(:,6),S), v = axis;
% xlabel('t','FontSize',11); ylabel('s_{12}      ','FontSize',12,'Rotation',0);
% if standard == 1,
%    text(v(1)+0*(v(2)-v(1)),v(4)+0.11*(v(4)-v(3)),...
%    ['Neurocycle enzyme system with h_f = ',num2str(hf,'%10.7g'),...
%       ' and the standard initial value \Psi(0)'],'FontSize',14);
% else,   text(v(1)+0.05*(v(2)-v(1)),v(4)+0.135*(v(4)-v(3)),...
%          ['Neurocycle enzyme system with h_f = ',num2str(hf,'%10.7g'),...
%          ' and initial value'],'Fontsize',14)
% text(v(1)+0.1*(v(2)-v(1)),v(4)+0.082*(v(4)-v(3)),...
%          ['\Psi(0) = ',num2str(y0,'%7.5g')],'FontSize',14); end
% title('Acetylcholine concentration in (II)','FontSize',12);
% subplot(2,1,2)
% plot(y(:,2),y(:,6),S)
% xlabel([{' '},{['s_{11}     ( for  0 \leq t \leq ',...
%          num2str(Tend,'%7.5g'),' )']}],'FontSize',12)
% ylabel('s_{12}     ','FontSize',12,'Rotation',0);
% title('Phase plot of acetylcholine concentrations in (I) versus in (II)',...
%      'FontSize',12)

                  % uncomment this block for terminal limit cycle plot
clf, k = length(t); Tstart = floor(0.4*k);
plot(y(Tstart:end,2),y(Tstart:end,6)), v  = axis;
if standard == 1,
   text(v(1)+0*(v(2)-v(1)),v(4)+0.046*(v(4)-v(3)),...
   ['Neurocycle enzyme system with h_f = ',num2str(hf,'%10.7g'),...
       ' and the standard initial value \Psi(0)'],'FontSize',14);
```

```
else,
  text(v(1)+0.08*(v(2)-v(1)),v(4)+0.056*(v(4)-v(3)),...
        ['Neurocycle enzyme system with h_f = ',num2str(hf,'%10.7g'),...
        ' and initial value'],'Fontsize',14)
  text(v(1)+0.1*(v(2)-v(1)),v(4)+0.036*(v(4)-v(3)),...
        ['\Psi(0) =  ',num2str(y0','%7.5g')],'FontSize',14); end
xlabel([{' '},{['s_{11}     ( for ',num2str(t(Tstart),'%7.5g'),...
        ' \leq t \leq ',num2str(Tend,'%7.5g'),' )']}],'FontSize',12);
ylabel('s_{12}   ','FontSize',12,'Rotation',0);
title('Phase plot of acetylcholine concentrations in (I) versus in (II)',...
  'FontSize',12);

function dydx = FjPsi(Psi,y,hf)        % right hand side of DE function
thet1=32000; thet2=4; thet3=0.125; thet4=0.125; thet5=84500; thet6=65;
thet7=76.72;  thet8=0.00709,       al=0.5; ga=0.01; del=0.1;    VR=1.2;
a11=1; a12=0; a21=1; a22=-1; a31=1; a32=0; a41=0; a42=1;    b1=1; b2=VR;
alHp=2.25; alHm=0.5; alS1=1; alS2=1; alS3=1;  B1=5.033*10^-5; B2=5.033*10^-5;
Kh1=1.0066*10^-6; Ks1=5.033*10^-7;  S2ref=0.0001; S3ref=0.000001;
s1f=2.4; s2f=1.15; s3f=3.9;              % list with parameter values
r1 = (y(3,:).*y(4,:))./((y(3,:) + thet1./y(1,:) + thet2 + thet3*y(1,:) + ...
        thet4*y(3,:).^2) .* (y(4,:) + thet5./y(1,:) + thet6 + thet7*y(1,:) + ...
        thet8*y(4,:).^2));        % rate equations
r2 = y(6,:)./(y(6,:) + 1./y(5,:) + 1 + del*y(5,:) + al*y(6,:).^2);
dydx = [a11*((hf - y(1,:)) - ga*(1/hf - 1./y(1,:))) - ...   % 8 DEs
          b1*a21*(alHp*(y(1,:)-y(5,:)) - ga*alHm*(1./y(1,:)-1./y(5,:))) + ...
          b1*a41*B2*r2/Kh1;
        a11*(s1f-y(2,:))  b1*a21*alS1*(y(2,:)-y(6,:)) + b1*(a31*B1*r1/Ks1 ...
          - a41*B2*r2/Ks1);
       a11*(s2f-y(3,:)) - b1*a21*alS2*(y(3,:)-y(7,:)) + b1*(-a31*B1*r1/S2ref ...
          + a41*B2*r2/S2ref);
       a11*(s3f-y(4,:)) - b1*a21*alS3*(y(4,:)-y(8,:)) + b1*(-a31*B1*r1/S3ref ...
          + a41*B2*r2/S3ref);
        a12*((hf - y(1,:)) - ga*(1/hf - 1./y(1,:))) - ...
          b2*a22*(alHp*(y(1,:)-y(5,:)) - ga*alHm*(1./y(1,:)-1./y(5,:))) + ...
          b2*a42*B2*r2/Kh1;
        a12*(s1f-y(2,:)) - b2*a22*alS1*(y(2,:)-y(6,:)) + b2*(a32*B1*r1/Ks1 ...
          - a42*B2*r2/Ks1);
       a12*(s2f-y(3,:)) - b2*a22*alS2*(y(3,:)-y(7,:)) + b2*(-a32*B1*r1/S2ref ...
          + a42*B2*r2/S2ref);
       a12*(s3f-y(4,:)) - b2*a22*alS3*(y(4,:)-y(8,:)) + b2*(-a32*B1*r1/S3ref ...
          + a42*B2*r2/S3ref)];
```

Next we display 13 single, double, or multiple plots drawn by our MATLAB program neurocycle.m of (a) the acetylcholine concentration profile in compartment (II) above the phase plot of the acetylcholine concentration in compartment (I) versus that in compartment (II), or (b) the limit cycle plot, or (c) the plot of all 8 profiles. We include interpretative comments on the solution's behavior in each case.

In the plots we vary the hydrogen-ion feed concentration h_f from 0.0044 to 0.0065, or by roughly 50%.

Note that all s_{11}-s_{12} phase plots in this section start at $(s_{11}(0),\ s_{12}(0)) = (1.27,\ 0.2)$ as specified earlier in the initial condition vector $\Psi(0)$.

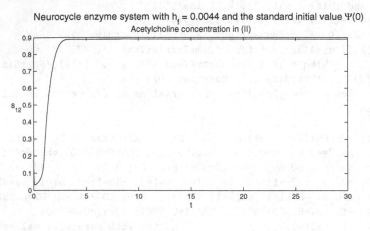

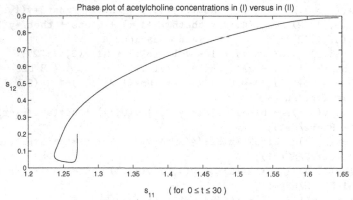

Profile of $s_{12}(t)$ and phase plot of s_{12} versus s_{11};
Convergence to a fixed steady state, no periodicity
Figure 4.52

For all values $h_f \leq 0.0044$, the plots have the same shape as displayed in Figure 4.52. Note that the phase plot starts out from the end of the lower left hook and proceeds toward the upper right in the bottom image as t increases.

When we increase h_f, small but increasing oscillations of the concentration profiles occur, as the following plots for $h_f = 0.004544$ indicate. Note how the spiral at the temporal end of the phase plot in Figure 4.53, bottom, increases in diameter with time, as does the oscillation amplitude of the acetylcholine concentration in compartment II in the top graph of Figure 4.53, bottom plot. For an enlarged display of the spiral in the bottom plot of Figure 4.53, we have used the magnifying glass tool $\overset{\oplus}{\diagup}$ of MATLAB's figure window toolbar before saving and printing.

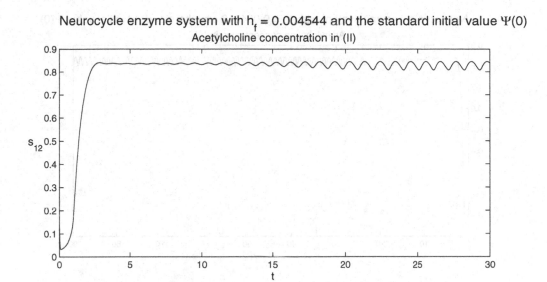

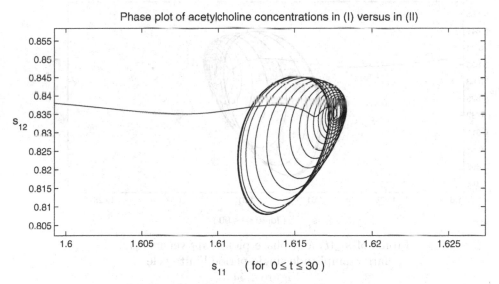

Profile of $s_{12}(t)$ and phase plot of s_{12} versus s_{11};
Increasing oscillatory behavior with a single period
Figure 4.53

For $h_f = 0.004546$, the solution $s_{12}(t)$ becomes more irregular at first until it settles into one stable limit cycle; see Figure 4.54.

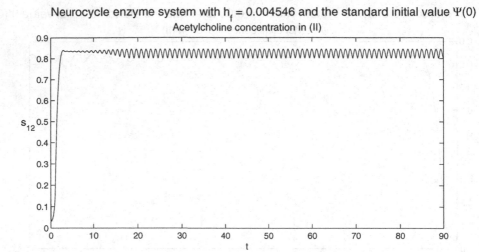

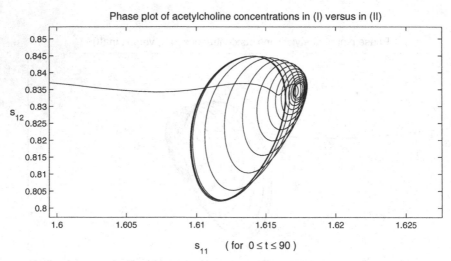

Profile of $s_{12}(t)$ and phase plot of s_{12} versus s_{11};
larger-amplitude single-period limit cycle
Figure 4.54

Note that in the previous two phase plots of Figures 4.53 and 4.54 we have zoomed in on the oscillatory part in MATLAB's figure window using the ⊕ tool. We shall do so repeatedly in the future for added clarity.

Our next-higher value of $h_f = 0.004552$ exhibits a double-loop limit cycle, or a period-two periodic attractor in the phase plot of Figure 4.55.

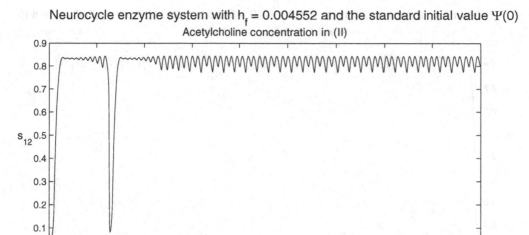

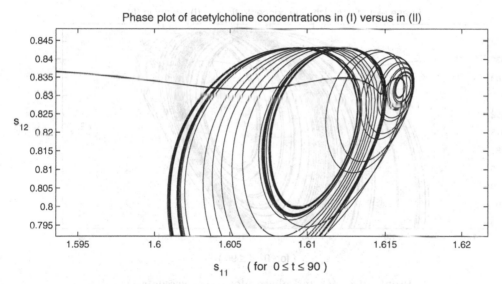

Profile of $s_{12}(t)$ and phase plot of s_{12} versus s_{11};
Larger-amplitude single period with a one double-loop limit cycle
Figure 4.55

If we increase h_f further to 0.0045525, we observe chaotic behavior of the $s_{12}(t)$ concentration in Figure 4.56. This also manifests itself in the phase plot, which consists of random loops until our plotting ends at 90 time units.

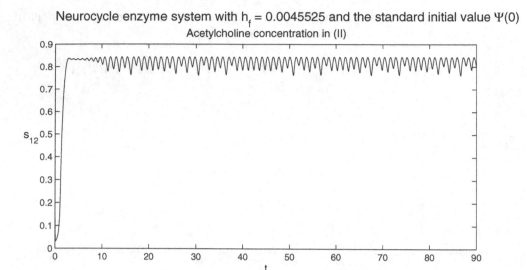

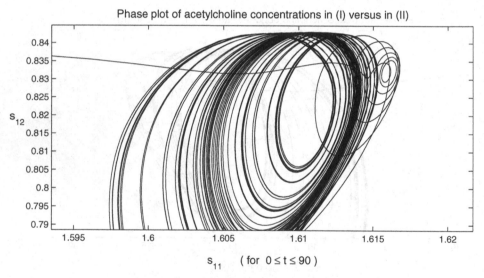

Profile of $s_{12}(t)$ and phase plot of s_{12} versus s_{11};
chaotic behavior
Figure 4.56

The profile and phase plots show oscillatory behavior of Figure 4.56, but no pattern or periods can be seen therein. This is called a **strange attractor** in modern nonlinear dynamics theory. A strange attractor can be chaotic or nonchaotic (high-dimensional torus). Differentiating between chaotic and nonchaotic strange attractors is beyond the scope of this undergraduate book.

A slight increase to $h_f = 0.0045526$ ends the chaotic behavior after around 20 time units, and an orderly 9-periodic behavior (period-nine periodic attractor) sets in for the solution from then on, as seen in Figure 4.57.

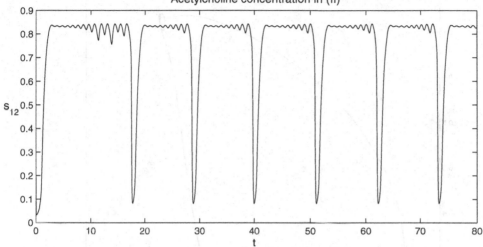

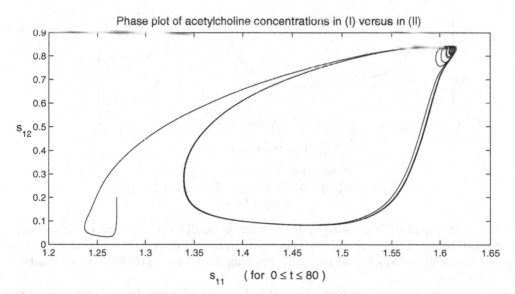

Profile of $s_{12}(t)$ and phase plot of s_{12} versus s_{11};
after a short period of chaos, periodic behavior begins
Figure 4.57

In order to display the asymptotic limit cycle in the phase plot of Figure 4.57 more clearly, we separate this limit cycle by using the third plotting block inside `neurocycle.m` to obtain Figure 4.58 for the same data.

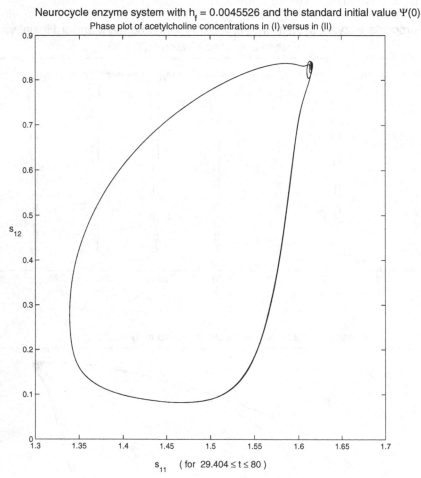

Neurocycle enzyme system with $h_f = 0.0045526$ and the standard initial value $\Psi(0)$

Phase plot of acetylcholine concentrations in (I) versus in (II)

s_{11} (for $29.404 \leq t \leq 80$)

Phase plot of s_{12} versus s_{11};
limit cycle detail for large t of Figure 4.57
Figure 4.58

Notice that this limit cycle goes through a number of small loops, each one corresponding to one small bump of the profile of $s_{12}(t)$ in the top plot of Figure 4.57, followed by a wide swing in the profile of $s_{12}(t)$ and the corresponding large limit cycle loop approximately once every 11 seconds.

Let us count the number of periods of $s_{12}(t)$ in Figure 4.57, once the periodic oscillatory steady-state loop has been reached. For $t > 20$ the system has stabilized in its varying behavior: there are 9 separate periods in the top graph of Figure 4.57 in each complete limit cycle of length about 11 time units. This can be verified by counting the number of local maxima between subsequent large amplitude drops in the top profile curve of

Figure 4.57. Note that there are also exactly 9 corresponding maxima in the enlarged phase plot in Figure 4.59. Figure 4.59 shows four and a half complete periodic loops for $29.404 \le t \le 80$ according to the top profile graph of Figure 4.57.

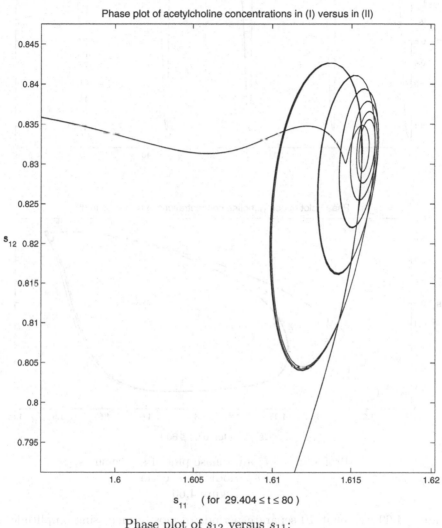

Phase plot of acetylcholine concentrations in (I) versus in (II)

Phase plot of s_{12} versus s_{11};
enlarged detail of Figure 4.58
Figure 4.59

When $h_f = 0.0045539$, the initial chaotic behavior has stopped and the system's limit cycle is reached on the second pass through a large-amplitude drop in the acetylcholine concentration in compartment (II), as shown in Figure 4.60.

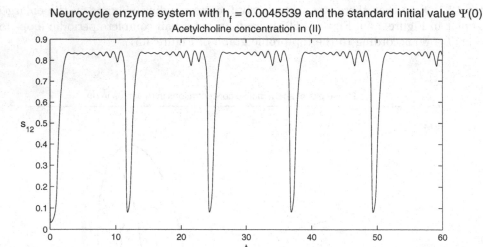

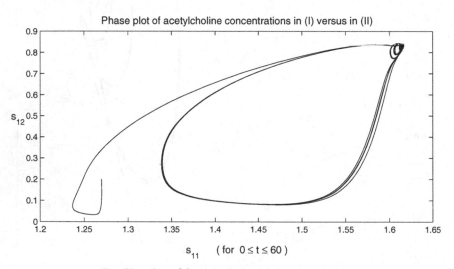

Profile of $s_{12}(t)$ and phase plot of s_{12} versus s_{11};
10-periodic limit cycle
Figure 4.60

In Figure 4.60 we count 10 separate periods of mostly increasing amplitudes until the cycle repeats.

For $h_f = 0.004556$ the number of periods drops to 8 in each cycle, while for $h_f = 0.004558$ there are only 7 periods in the limit cycle. For $h_f = 0.00457$ there are 4 separate periods, for $h_f = 0.00458$ there are only 3 in Figure 4.61. For $h_f = 0.0046$ the number of separate periods decreases to 2, and for $h_f \geq 0.00465$, there is only a single period until h_f becomes significantly larger.

These assertions can be verified by our readers using the MATLAB code `neurocycle.m`.

As an illustration we now plot the graph for $h_f = 0.00458$ that generates a 3-periodic limit cycle (period-three periodic attractor).

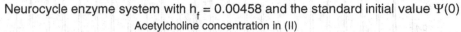

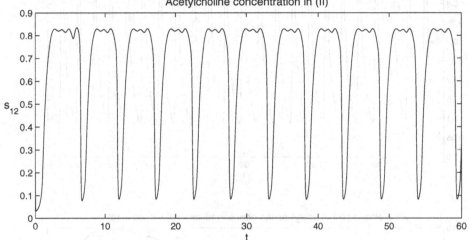

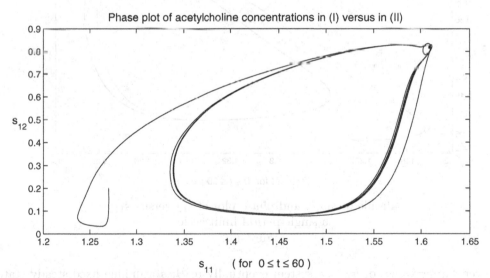

Profile of $s_{12}(t)$ and phase plot of s_{12} versus s_{11};
3-period limit cycle
Figure 4.61

For much larger h_f values such as $h_f = 0.006325$, the limit cycle has exactly one maximum, and this is reached after the first complete cycle, as shown in Figure 4.62.

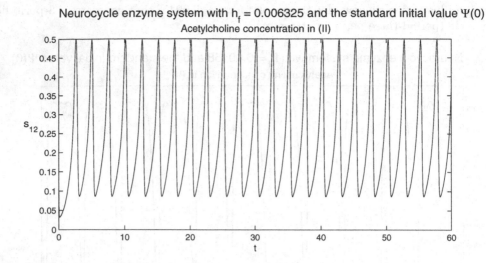

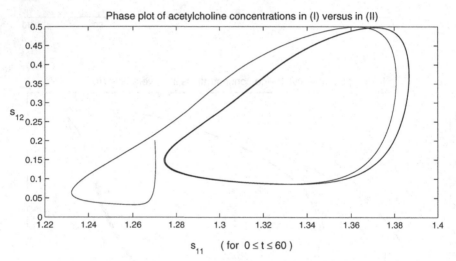

Profile of $s_{12}(t)$ and phase plot of s_{12} versus s_{11};
a single-period limit cycle
Figure 4.62

For even larger values of h_f, the system eventually reaches a unique fixed steady state that is stationary and involves no limit cycle at all, just as we have seen to be the case for small values of h_f in Figure 4.52. For example, for $h_f = 0.0065$, the phase plot starts at $s_{11} = 1.27$ and $s_{12} = 0.2$ in the bottom plot of Figure 4.63 and moves in two spiral loops toward the asymptotic steady state with $s_{11} \approx 1.285$ and $s_{12} \approx 0.17$, as depicted in Figure 4.63.

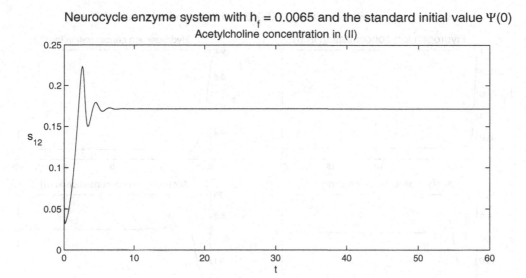

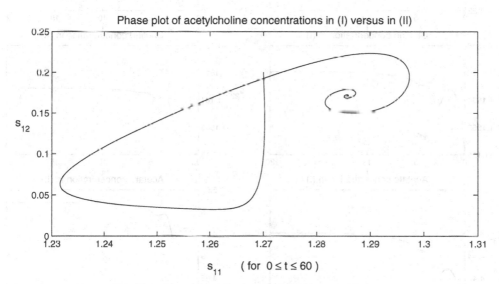

Profile of $s_{12}(t)$ and phase plot of s_{12} versus s_{11};
a unique steady state for large h_f values
Figure 4.63

The following unified eight profile plot verifies our assertion that the system reaches a unique fixed steady state when $h_f = 0.0065$ and when the initial value $\Psi(0)$ is as before. This figure is drawn by uncommenting the first large plotting block in **neurocycle.m** to make it active, while commenting out the other two plotting blocks with % signs and thereby making them inactive.

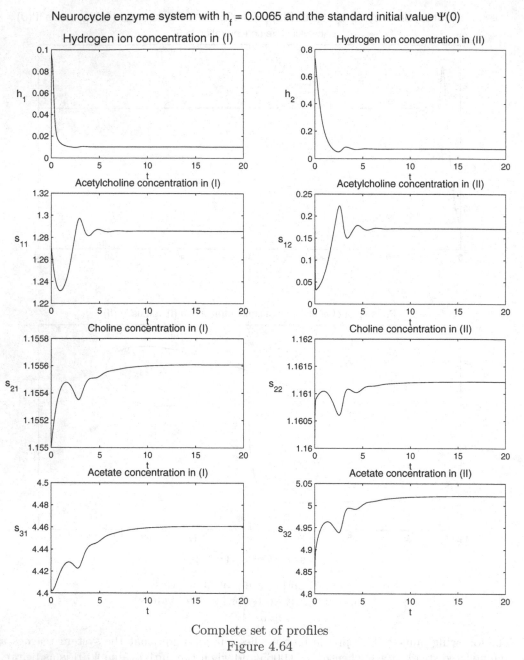

Neurocycle enzyme system with h_f = 0.0065 and the standard initial value $\Psi(0)$

Complete set of profiles
Figure 4.64

The previous 13 figures show several transition stages in the behavior of the solution to the given IVP from having one fixed asymptotic steady-state solution for low values of h_f; through small oscillation for all times, to limit cycles, irregularity, and chaos; back to repeated oscillations with ever-decreasing numbers of periods; and then back to one fixed asymptotic steady-state solution again as we vary one parameter, h_f, from 0.0044 to 0.0065.

Exercises for 4.4

1. Recall the remark on stiffness and IVP solver validation of p. 201 and vary the tolerances inside neurocycle.m to verify that the acetylcholine concentration of the neurocycle enzyme system with $h_f = 0.0045525$ behaves chaotically as depicted in Figure 4.56.

2. Verify the assertions about the diminishing periods of the limit cycles (periodic attractors) in the range $0.004556 \leq h_f \leq 0.0046$ on p. 244 using neurocycle.m.

3. For $h_f = 0.0045539$ (see Figure 4.60) investigate whether the periodic limit cycle (periodic attractor) is attracting from various initial values $\Psi(0)$ by running the limit cycle phase plot of neurocycle.m from varying initial values of $\Psi(0)$ in varying colors S. (Hint: To achieve various colors, refer to qfrun.m in Section 4.1 and the creation of differently colored graphs in one plot there.)

4. Investigate how the single-period solution limit cycle (period-one periodic attractor) for $h_f = 0.006325$ in Figure 4.62 transforms to an asymptotically constant solution for $h_f = 0.007$ similar to Figure 4.63. Does this transition involve oscillations? Describe and explain your findings.

5. **Project**

   Modify the program neurocycle.m to become neurocycleB2.m with the first MATLAB code line function neurocycleB2(B2,Tend,y0,S).

   In the program neurocycleB2.m, use $h_f = 0.0055$ as a fixed parameter and vary $B_2 = V_2 \cdot V_{M2} \cdot \overline{AchE}/q$ between 0 and $20 \cdot 10^{-5}$ $kmol/m^3$. Observe the behavior of the solution profiles and phase plots and interpret your results.

   The parameter B_2 is chosen for this project in order to gain some insight into possible consequences of varying the capability of the acetylcholinesterase to hydrolyze the neurotransmitter. Imbalances in this capability give rise to devastating diseases such as Alzheimer's and Parkinson's. The enzyme activity is included in the grouped parameter B_2, which includes the maximum reaction velocity in reaction 2. The parameter B_2 itself includes the enzyme activity together with three constants for the enzyme system, namely the concentration of acetylcholinesterase in compartment (II), the volume V_2 of compartment (II), and the flow rate q.

Conclusions

We have developed, solved, and analyzed an eight-dimensional model for a coupled acetylcholinesterase/choline acetyltransferase enzyme system. The complex dynamic characteristics, both stable and unstable, and the chaotic behavior of this IVP system have been investigated with some reference to acetylcholine neural transmission.

The variation of the hydrogen-ion feed concentration h_f as bifurcation parameter has a strong effect on the state variables at low concentrations, in contrast to its weak effect at high concentrations. At low concentrations it is found that a complex dynamic behavior with period doubling, period adding, and period subtracting dominate the dynamics of the system. Checking for a possible correspondence to physiological values for the pH values, we can say that compartment (II) (where the pH level is between 6.73 and 7.97 for the studied variations of h_f) has a pH value near to the expected value for the human brain. This represents a complex biological example that sheds some light on the relation between enzyme activities in the brain and Alzheimer's and Parkinson's diseases.

Exercises for Chapter 4

1. A vertical cylindrical tank is filled with well water at $65°$ F. The tank is insulated at the top and bottom, but is exposed at its vertical sides to cold night air at $10°F$. The tank's diameter is 2 ft and its height 3 ft. The overall heat transfer coefficient is 20 $Btu/(h \, °F \, ft^2)$. Neglect the metal wall of the tank and assume that the water in the tank is perfectly mixed.

   (a) Write out a differential equation that models the temperature of the water in the tank over time.

   (b) Solve the DE of part (a) analytically.

   (c) Solve the DE of part (a) numerically via MATLAB, and compare the results. (Hint: look up the Differential Equations Examples MATLAB browser to learn about the MATLAB DE solvers `ode....m`. Simply type `odeexamples('ode')` at the MATLAB prompt to start this browser.)

   (d) Write a MATLAB program to determine how long it will take for the water in the tank to freeze completely. The heat of fusion of water is 144 Btu/lb$_m$.

2. A cylindrical tank is fed with water at a flow rate of 2.3 $m^3/hour$ and is equipped with an output control valve at the bottom. The steady-state height in the tank is 2 m. What is the valve coefficient (and its units) under these conditions? The tank diameter is 3 m and the total height of the tank is 5 m.
   If the feed to the tank increases from 2.3 $m^3/hour$ to 3.4 $m^3/hour$ and the valve opening remains the same, i.e., the valve coefficient remains the same, calculate and plot the change of height with time. Find the final height using the dynamic model and the steady-state model and make sure that they both give the same result.

3. If in the above problem it is required to keep the height at the same value of 2 m when the flow rate changes from 2.3 to 3.4 $m^3/hour$, we can use feedback control to achieve this.
   The feedback control loop consists in measuring the height, comparing it with the set point, i.e., the height for the input flow rate of 2.3 $m^3/hour$, and using the

difference in height as a proportional drive to change the valve opening in order to compensate for this change.

Calculate the behavior using this proportional feedback control with your choice of several different suitable values of the proportional gain K_p. Calculate and plot the change of height with time when this controller is used and find the final height for each K_p. Moreover, calculate the offset for each value of K_p.

Choose the best value of K_p and compare the behavior with the behavior for the open loop system without controls, i.e., for $K_p = 0.0$. Show that it is possible to remove the offset by using a proportional plus integral (PI) controller and find the best value of K_I that can be used with the best value for K_p as obtained above. Plot the change of height with time and compare it with the results of the open-loop system in the case without control and also when the system has only proportional control.

4. An industrial system consists of a nonisothermal CSTR (with a cooling jacket) and a tubular adiabatic reactor in series. The reaction is a first-order irreversible reaction:

$$A \longrightarrow B .$$

The plant manager requires the following:

(a) Formulate a rigorous dynamic model for the system.

(b) Suggest a solution algorithm to find the steady state(s) of the system.

(c) Suggest a solution algorithm to investigate the dynamic behavior of the system.

(d) Discuss the main steady-state and the dynamic characteristics of the system and their practical implications.

5. Formulate the unsteady-state equations (dynamic model) for a nonadiabatic CSTR where an irreversible first-order exothermic reaction

$$A \longrightarrow B$$

takes place. Put the dynamic equations in dimensionless form with the feed temperature as reference temperature and the feed concentration as reference concentration.

For the thermicity factor $\beta = 1.2$, the dimensionless activation energy $\gamma = 18$, and with the constant dimensionless cooling jacket temperature $y_c = 0.85$, compute the following:

(a) The set of values of the dimensionless preexponential factor α and of the dimensionless heat transfer coefficients K_c that give multiple steady states. Formulate a numerical algorithm in MATLAB that finds the dimensionless temperature and concentrations at each of the three steady states for given α and K_c values.

(b) If the middle steady state obtained when $K_c = 0.0$ is the desirable steady state, show how to stabilize it.

(c) Construct phase planes for a case with multiple steady states and a case with a unique steady state.

(d) In the multiplicity region, choose one of the cases and linearize the equations in a neighborhood of each of the three steady states. Compute the eigenvalues (characteristic roots) of the linearized model using the built-in MATLAB function eig. From the eigenvalue information determine the stability characteristics of each of the three steady states.

(e) For the stabilized unstable middle steady state in part (b) find the relation between the stability characteristics and the values of K_c.

The general assumptions for this model are:

(1) The stirrer and wall heat capacities are negligible and the volumetric flow rate is constant.

(2) There is no change of phase.

(3) The average specific heat and density of the mixture is constant.

[Hints: Start your investigation of part (a) with the adiabatic case of $K_c = 0.0$. You can use dimensionless units in your dynamics and stability part of the investigation by using the dimensionless time $t' = t/\tau$, where t is the real time, $\tau = V/q$ is the residence time, V is the active volume of the reactor, and q is the constant volumetric flow rate.]

Water–air boundary ripples

Chapter 5

Boundary Value Problems, *with and without Bifurcation*

A distributed model is usually described by differential equations. Such a model differs from a lumped model that is generally described by transcendental equations. In chemical and biological engineering distributed systems often arise with tubular equipment. When a one-dimensional model is used for a distributed system there are two types of models:

1. If mixing, diffusion, and conduction are neglected, then the system is described by the so called **plug flow model**, expressed in terms of initial value ODEs, i.e., by initial value problems, or IVPs.

2. If the model accounts for the effects of axial dispersion, then the system is described by an **axial dispersion model** in terms of two-point boundary ODEs, i.e., by boundary value problems, or BVPs.

5.1 The Axial Dispersion Model

For plug flow, only the flow and the processes other than mixing, diffusion, and conduction are considered. These have been studied in Chapter 4. In a plug flow tubular reactor model we consider only the convective one-dimensional flow and the chemical reaction as shown in Figure 5.1, where n_i is the convective molar flow rate for the constant volumetric flow rate q_i of component i. These two rates are connected by the equation $n_i = q \cdot C_i$ for the concentration C_i.

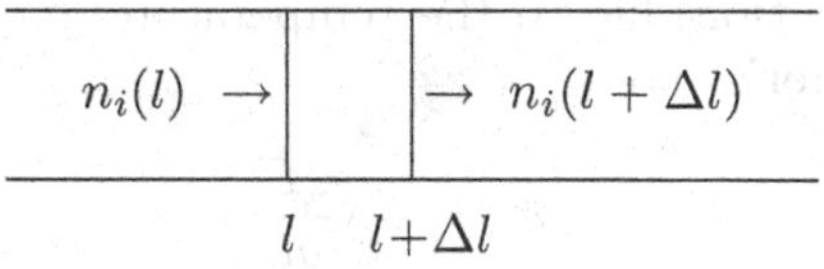

$$n_i(l) \quad \rightarrow \qquad \rightarrow \; n_i(l + \Delta l)$$

$$l \quad l + \Delta l$$

Convective flow

Figure 5.1

If axial dispersion is also considered in the model, then the diffusive flow due to differences in concentration (in the isothermal case) is also taken into account as shown in Figure 5.2.

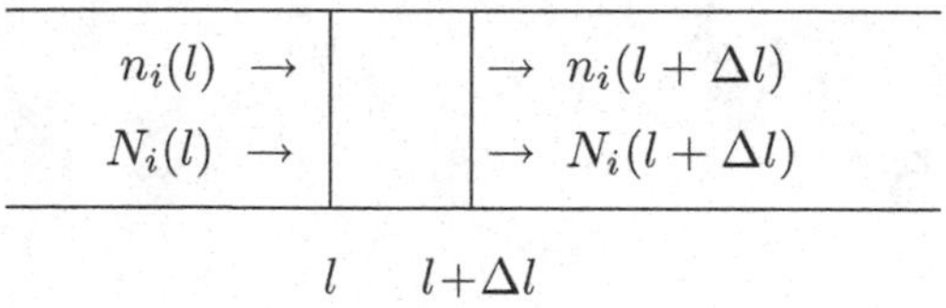

$$
\begin{array}{c|c}
n_i(l) \ \rightarrow & \rightarrow \ n_i(l + \Delta l) \\
N_i(l) \ \rightarrow & \rightarrow \ N_i(l + \Delta l)
\end{array}
$$

$$l \qquad l + \Delta l$$

Convective and diffusive flows
Figure 5.2

Here N_i is the diffusion flux. It is most simply expressed for constant diffusion coefficients by Fick's[1] law

$$N_i \ = \ -D_i \cdot \frac{dC_i}{dl} \ . \tag{5.1}$$

In the nonisothermal case, the plug flow model accounts only for the convective heat flow as shown in Figure 5.3.

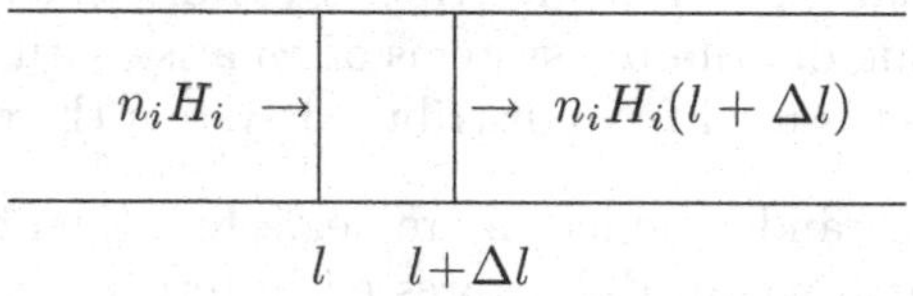

$$
\begin{array}{c|c}
n_i H_i \ \rightarrow & \rightarrow \ n_i H_i(l + \Delta l)
\end{array}
$$

$$l \qquad l + \Delta l$$

Convective heat flow
Figure 5.3

Here H_i is the enthalpy (energy per mole) for component i.
When the axial conductivity of heat. i.e., the heat dispersion, is also considered, the situation becomes as depicted in Figure 5.4.

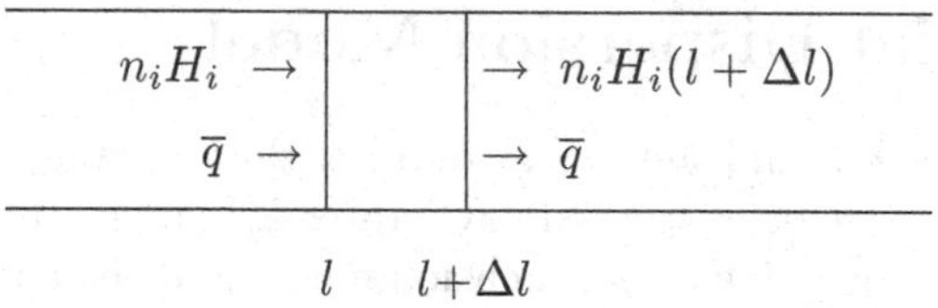

$$
\begin{array}{c|c}
n_i H_i \ \rightarrow & \rightarrow \ n_i H_i(l + \Delta l) \\
\bar{q} \ \rightarrow & \rightarrow \ \bar{q}
\end{array}
$$

$$l \qquad l + \Delta l$$

Convective and heat dispersion flows
Figure 5.4

Here $\bar{q}$ is the heat conduction due to the temperature gradient. The simplest way to express this is using Fourier's[2] law

$$\bar{q} \ = \ -\lambda \frac{dT}{dl} \ , \tag{5.2}$$

---

[1] Adolf Eugen Fick, German physiologist, 1829 – 1901
[2] Jean Baptiste Joseph Fourier, French mathematician, 1768 – 1830

where λ is the thermal conductivity coefficient.

Needless to say, the assumption of plug flow is not always appropriate. In plug flow we assume that the convective flow, i. e., the flow at velocity $q/A_t = v$ that is caused by a compressor or pump, is dominating any other transport mode. In practice this is not always so and dispersion of mass and heat, driven by concentration and temperature gradients are sometimes significant enough to need to be included in the model. We will discuss such a model in detail, not only because of its importance, but also because the techniques used to handle the ensuing boundary value differential equations are similar to those used for other diffusion-reaction problems such as catalyst pellets, as well as for counter-current processes.

5.1.1 Formulation of the Axial Dispersion Model

The simplest description of axial dispersion of mass for constant diffusivities is given by Fick's law (5.1)

$$N_i = -D_i \cdot \frac{dC_i}{dl} \, ,$$

where N_i is the mass flux of component i, measured in $moles/(cm^2 \cdot sec)$, D_i is the diffusion coefficient of component i in cm^2/sec, C_i is the concentration of component i in $moles/cm^3$, and l is the length co-ordinate in cm .

The axial dispersion of heat (axial heat conduction) is described by Fourier's law (5.2)

$$\overline{q} = -\lambda \frac{dT}{dl} \, ,$$

where $\overline{q}$ is the heat flux in $J/cm^2 \, sec$, λ is the thermal conductivity in $J/(cm \cdot sec \cdot K)$, and T is the temperature in K.

We introduce the axial dispersion of mass and heat for a single reaction in a tubular reactor that operates at a steady state with the generalized rate of reaction r' per unit volume.

In Figure 5.2 the mass balance for the convective flows n_i and the diffusion flows N_i have been depicted.

By A_t we denote the cross sectional area of the reactor tube. Then the steady-state mass balance with axial dispersion gives us the equation

$$n_i(l + \Delta l) + A_t \cdot N_i(l + \Delta l) = n_i(l) + A_t \cdot N_i(l) + \sigma_i \cdot A_t \cdot \Delta l \cdot r' \, . \qquad (5.3)$$

After dividing this equation by Δl, rearranging to obtain difference quotients for n_i and N_i, and then taking $\lim_{\Delta l \to 0}$, we get

$$\frac{dn_i}{dl} + A_t \frac{dN_i}{dl} = \sigma_i \cdot A_t \cdot r' \, . \qquad (5.4)$$

This differential equation can be put into dimensionless form by introducing one simple assumption such as a constant volumetric flow rate q, i. e.,

$$\frac{dn_i}{dl} = q \frac{dC_i}{dl} \, .$$

This assumption means that $n_i = q \cdot C_i$ for a constant value of q. For all liquid-phase systems the volumetric flow rate q is constant and not merely assumed so, and likewise for gas-phase systems with a constant number of total moles running at constant temperature and pressure. For all other systems the constancy of the volumetric flow rate q is an extra assumption.

For a constant volumetric flow rate q we can rewrite (5.4) as

$$q\frac{dC_i}{dl} + A_t\frac{dN_i}{dl} = \sigma_i \cdot A_t \cdot r' . \tag{5.5}$$

If we further assume that the diffusion coefficient D_i is constant and then differentiate Fick's law (5.1) with respect to l we obtain

$$\frac{dN_i}{dl} = -D_i\frac{d^2C_i}{dl^2} . \tag{5.6}$$

And the mass balance design equation thus becomes

$$q\frac{dC_i}{dl} - A_t \cdot D_i\frac{d^2C_i}{dl^2} = \sigma_i \cdot A_t \cdot r' .$$

Dividing by A_t gives us

$$v\frac{dC_i}{dl} - D_i\frac{d^2C_i}{dl^2} = \sigma_i \cdot r' ,$$

where $v = q/A_t$ is the velocity of the flow. Next we define a dimensionless length, namely $\omega = l/L$, where L denotes the total length of the tubular reactor. This gives us

$$\frac{v}{L}\frac{dC_i}{d\omega} - \frac{D_i}{L^2}\frac{d^2C_i}{dl^2} = \sigma_i \cdot r' .$$

By rearranging, we obtain

$$\frac{1}{Pe_M}\frac{d^2C_i}{dl^2} - \frac{dC_i}{d\omega} + \sigma_i \cdot \frac{L}{v} \cdot r' = 0 , \tag{5.7}$$

where $Pe_M = L \cdot v/D_i$ is the Peclet[3] number for mass.

Applying the same procedure for the heat balance with axial dispersion (axial conduction) of heat, we get

$$\sum_i n_i(l) \cdot H_i(l) + A_t \cdot \overline{q}(l) + Q' \cdot \Delta l = \sum_i n_i(l+\Delta l) \cdot H_i(l+\Delta l)) + A_t \cdot \overline{q}(l+\Delta l) , \tag{5.8}$$

where $\overline{q}$ is the heat flux and Q' is the external heat added per unit length of the tubular reactor.

---

[3] Jean Claude Péclet, French scientist, 1793 − 1857

By rearranging, forming difference quotients, and taking the limit of $\Delta l \to 0$ as before, we transform (5.8) to

$$\sum_i \frac{d(n_i H_i)}{dl} + A_t \frac{\overline{q}}{dl} = Q' .$$

And the product rule of differentiation, applied to $d(n_i H_i)/dl$, yields

$$\sum_i n_i \frac{dH_i}{dl} + \sum_i H_i \frac{dn_i}{dl} + A_t \frac{d\overline{q}}{dl} = Q' . \tag{5.9}$$

Equation (5.4) can be rewritten as

$$\frac{dn_i}{dl} = \sigma_i \cdot A_t \cdot r' - A_t \frac{dN_i}{dl} .$$

By substituting this expression for dn_i/dl into the heat balance design equation (5.9) we obtain

$$\sum_i n_i \frac{dH_i}{dl} + \sum_i H_i \left(\sigma_i \cdot A_t \cdot r' - A_t \frac{dN_i}{dl} \right) + A_t \frac{d\overline{q}}{dl} = Q' ,$$

which can be rearranged to become

$$\sum_i n_i \frac{dH_i}{dl} + A_t \cdot r' \sum_i \sigma_i H_i - A_t \sum_i H_i \frac{dN_i}{dl} - A_t \cdot \lambda \frac{d^2 T}{dl^2} = Q' .$$

Here we have used Fourier's law (5.2) $\overline{q} = -\lambda \cdot dT/dl$ for constant λ, differentiated once more as $d\overline{q}/dl = -\lambda \cdot d^2 T/dl^2$.

With the reasonable approximations for a one-phase system that have been used earlier for the plug flow system, we can approximate the above differential equation by

$$n_t \cdot C'_{P_{mix}} \frac{dT}{dl} + A_t \cdot r' \cdot (\Delta H) - A_t \sum_i H_i \frac{dN_i}{dl} - A_t \cdot \lambda \frac{d^2 T}{dl^2} = Q' ,$$

where $C'_{P_{mix}}$ denotes the average molar specific heat of the mixture and n_t is the average total molar flow rate. Our main assumptions in this approximation are the four assumptions that we have used before, namely: no change in phase (only sensible heat is involved), average the concentrations C_p for each component i, then average for the whole mixture, and finally average the total number of moles of the system.

The last equation can be rewritten in the form

$$A_t \lambda \frac{d^2 T}{dl^2} - q\rho C_{P_{mix}} \frac{dT}{dl} + A_t r'(-\Delta H) + A_t \sum_i H_i \frac{dN_i}{dl} = Q' , \tag{5.10}$$

where $C_{P_{mix}}$ is the average mass specific heat for the mixture, q is the average volumetric flow rate, and ρ is the average density of the mixture.

A problematic term :

In equation (5.10) there is the term

$$A_t \sum_i H_i \frac{dN_i}{dl} \tag{5.11}$$

which is usually neglected in the literature, but is most problematic. This term represents the enthalpy carried with the axial dispersion of mass. We can make several approximations for this term that help to handle it without blindly neglecting it.

1. *Most simple approximation :*
    We may consider that the term $A_t \sum_i (H_i \cdot dN_i/dl)$, that accounts for the heat transferred with the axial dispersion of mass is accounted for through a small empirical correction in λ by replacing λ by λ_e in (5.10).

2. *Less simple approximation :*
    The term $A_t \sum_i (H_i \cdot dN_i/dl)$ can be expressed as $-A_t \sum_i (H_i D_i \cdot d^2C_i/dl^2)$ according to (5.1). The latter approximately equals

    $$-A_t \cdot D_{av} \sum_i H_i \frac{d^2C_i}{dl^2} \tag{5.12}$$

    for the average value D_{av} of the diffusion coefficient D_i.
    From the mass balance equation for an assumed plug flow we have

    $$q \frac{dC_i}{dl} = \sigma_i \cdot A_t \cdot r'$$

    which we rewrite as

    $$\frac{dC_i}{dl} = \left(\frac{A_t}{q} \right) \sigma_i \cdot r' . \tag{5.13}$$

    Using equation (5.13) in equation (5.12) gives us

    $$-A_t \cdot D_{av} \sum_i H_i \left(\frac{A_t}{q} \right) \sigma_i \frac{dr'}{dl}$$

    as an approximation for the expression (5.11). This in turn reduces to

    $$-A_t \cdot D_{av} \left(\frac{A_t}{q} \right) \frac{dr'}{dl} \sum_i \sigma_i H_i .$$

    Finally, approximating $\sum_i \sigma_i H_i$ by (ΔH) makes the problematic expression (5.11) approximately equal to

    $$-\frac{A_t^2 \cdot D_{av}}{q} \frac{dr'}{dl} (\Delta H) \left(\approx A_t \sum_i H_i \frac{dN_i}{dl} \right) . \tag{5.14}$$

Using the expression (5.14) inside the heat balance equation (5.10) now gives us

$$A_t \cdot \lambda \frac{d^2 T}{dl^2} - q \cdot \rho \cdot C_{Pmix} \frac{dT}{dl} + A_t \cdot r'(-\Delta H) + \frac{A_t^2 \cdot D_{av}}{q} \frac{dr'}{dl} = -Q' \,.$$

This we rearrange to become

$$A_t \cdot \lambda \frac{d^2 T}{dl^2} - q \cdot \rho \cdot C_{Pmix} \frac{dT}{dl} + A_t(-\Delta H) \left(r' + \frac{A_t}{q} D_{av} \frac{dr'}{dl} \right) = -Q' \,. \tag{5.15}$$

Dividing equation (5.15) by A_t and noticing that $A_t/q = 1/v$ leads to

$$\lambda \frac{d^2 T}{dl^2} - v \cdot \rho \cdot C_{Pmix} \frac{dT}{dl} + (-\Delta H) \left(r' + \frac{D_{av}}{v} \frac{dr'}{dl} \right) = -Q' \,.$$

For most systems we have $r' \gg (D_{av}/v) \cdot (dr'/dl)$. If this is so, then we can rewrite the last equation in the simplified form

$$\lambda \frac{d^2 T}{dl^2} - v \cdot \rho \cdot C_{Pmix} \frac{dT}{dl} + r' \cdot (-\Delta H) = -Q' \tag{5.16}$$

by omitting the $(D_{av}/v) \cdot (dr'/dl)$ term. Note that for cases where r' is not much larger than $(D_{av}/v) \cdot (dr'/dl)$, this term cannot be neglected and thus it must be included in the equation.

Now we will use the simple approach of using an effective λ_e term instead. By setting $\omega = l/L$, equation (5.16), now in dimensionless form, becomes

$$\frac{\lambda_e}{L^2} \frac{d^2 T}{d\left(\frac{l}{L}\right)^2} - \frac{v \cdot \rho \cdot C_{Pmix}}{L} \frac{dT}{d\left(\frac{l}{L}\right)} + r'(-\Delta H) = -Q' \,. \tag{5.17}$$

Multiplying both sides by $L/(v \cdot \rho \cdot C_{Pmix})$ gives us

$$\frac{\lambda_e}{L^2} \frac{L}{v \cdot \rho \cdot C_{Pmix}} \frac{d^2 T}{d\omega^2} - \frac{dT}{d\omega} + r' \cdot (-\Delta H) \frac{L}{v \cdot \rho \cdot C_{Pmix}} = -Q' \frac{L}{v \cdot \rho \cdot C_{Pmix}} \,.$$

Since $\lambda_e/(L \cdot v \cdot \rho \cdot C_{Pmix}) = 1/P_{eH}$ for the dimensionless Peclet number P_{eH} for heat transfer, we get

$$\frac{1}{P_{eH}} \frac{d^2 T}{d\omega^2} - \frac{dT}{d\omega} + \frac{r' \cdot (-\Delta H) \cdot L}{v \cdot \rho \cdot C_{Pmix}} = -\overline{Q'} \,, \tag{5.18}$$

where $\overline{Q'} = Q' \cdot L/(v \cdot \rho \cdot C_{Pmix})$.

Let us consider that the rate of reaction r' is that of a simple first-order irreversible reaction, namely $r' = k_0 \cdot e^{-\frac{E}{RT}} \cdot C_A$. We define dimensionless temperature and concentration by introducing $y = T/T_f$ for the dimensionless temperature and $x_A = C_A/C_{A_f}$ for the dimensionless concentration. With these settings, equation (5.18) becomes

$$\frac{1}{P_{eH}} \frac{d^2 y}{d\omega^2} - \frac{dy}{d\omega} + \frac{(-\Delta H) \cdot C_{A_f}}{\rho \cdot C_{Pmix} \cdot T_f} D_a \cdot e^{-\gamma/y} \cdot x_A = -\frac{\overline{Q'}}{T_f} \,,$$

where γ is the dimensionless activation energy $\gamma = E/(R \cdot T_f)$ and D_a is the Damköhler[4] number $D_a = k_0 \cdot L/v$. With the thermicity factor $\beta = \left[(-\Delta H) \cdot C_{A_f} \right] / (\rho \cdot C_{Pmix} \cdot T_f)$ and $\hat{Q} = \overline{Q}'/T_f$, the last differential equation reads as follows:

$$\frac{1}{P_{eH}} \frac{d^2 y}{d\omega^2} - \frac{dy}{d\omega} + \beta \cdot D_a \cdot e^{-\gamma/y} \cdot x_A = -\hat{Q} . \tag{5.19}$$

Similarly for the mass Peclet number P_{eM}, the mass balance design equation for this case is

$$\frac{1}{P_{eM}} \frac{d^2 x_A}{d\omega^2} - \frac{dx_A}{d\omega} - D_a \cdot e^{-\gamma/y} \cdot x_A = 0 . \tag{5.20}$$

Notice that both mass and heat balance design equations (5.19) and (5.20) are second-order two-point boundary value differential equations. Therefore each one requires two boundary conditions. These boundary conditions can be derived as shown in the example that follows.

We will develop the boundary conditions for the mass balance design equation only. For the nonisothermal case, the boundary conditions for the heat balance design equation can be found similarly and are left as an exercise for the reader.

5.1.2   Example of an Axial Dispersion Model. Linear and Nonlinear Two-point Boundary Value Problems (BVPs)

In this example we develop the isothermal model and its boundary conditions for a case in which the differential equation is linear and can be solved analytically. Then we work on the nonlinear model.

The Linear Case

For a steady state, isothermal, homogeneous tubular reactor consider a simple reaction

$$A \quad \rightarrow \quad B$$

taking place in the tubular reactor where the axial dispersion is not negligible and the Peclet number P_e equals 15.0. We shall try to answer the following design questions.

1. If the reaction is first-order, what is the value of the Damköhler number D_a in order to achieve a conversion of 0.75 at the exit of the reactor?

2. If the reaction is second-order and the numerical value of the Damköhler number D_a is the same as in part 1, find the exit conversion using the solution of the nonlinear two point boundary value differential equation.

To answer these design questions we first formulate the model equations for first- and second-order reactions.

---

[4]Gerhard Damköhler, German chemist, 1908 – 1944

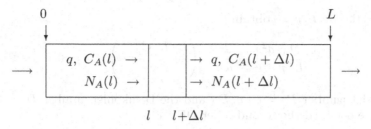

Axial dispersion mass flow
Figure 5.5

For an nth order reaction, the mass balance design equation for the isothermal case with constant volumetric flow rate q is

$$q \cdot C_A + A_t \cdot N_A(l) = q \; C_A(l + \Delta l) + A_t \cdot N_A(l + \Delta l) + A_t \cdot \Delta l \cdot k \cdot C_A^n \; ,$$

where n is the order of the reaction of our problem and $n = 1$ or 2. Canceling similar terms from both sides of the equation gives us

$$0 = q \cdot \Delta C_A + A_t \cdot \Delta N_A + A_t \cdot \Delta l \cdot k \cdot C_A^n \; .$$

Dividing by Δl and taking $\lim_{\Delta l \to 0}$ makes

$$0 = q \frac{dC_A}{dl} + A_t \frac{dN_A}{dl} + A_t \cdot k \cdot C_A^n \; . \tag{5.21}$$

By Fick's law for diffusion for the component A with its diffusion coefficient D_A we have

$$N_A = D_A \frac{dC_A}{dl} \; ,$$

and upon differentiation

$$\frac{dN_A}{dl} = -D_A \frac{d^2 C_A}{dl^2} \; . \tag{5.22}$$

Combining (5.21) and (5.22) we get

$$0 = q \frac{dC_A}{dl} - A_t \cdot D_A \frac{d^2 C_A}{dl^2} + A_t \cdot k \cdot C_A^n \; . \tag{5.23}$$

We divide both sides by A_t and notice that the flow velocity v is equal to q/A_t. Thus we obtain

$$0 = v \frac{dC_A}{dl} - D_A \frac{d^2 C_A}{dl^2} + kC_A^n \; ,$$

or

$$D_A \frac{d^2 C_A}{dl^2} - v \frac{dC_A}{dl} - kC_A^n = 0 \; .$$

With the following dimensionless variables: $\omega = l/L$ for a dimensionless length with L denoting the total length of the reactor and $x = C_A/C_{A_f}$, a dimensionless concentration with C_{A_f} denoting the feed concentration, we get

$$\frac{D_A}{L^2} \frac{d^2 \left(C_A/C_{A_f} \right)}{d(l/L)^2} - \frac{v}{l} \frac{d \left(C_A/C_{A_f} \right)}{d(l/L)} - k \cdot C_{A_f}^{n-1} \frac{C_A^n}{C_{A_f}^n} = 0 \; .$$

Multiplying this by L/v we obtain

$$\frac{D_A}{L \cdot v}\frac{d^2x}{d\omega^2} - \frac{dx}{d\omega} - \left(\frac{k \cdot L \cdot C_{A_f}^{n-1}}{v}\right)x^n = 0 \; .$$

For the Peclet number $P_e = L \cdot v/D_A$ and the Damköhler number $D_a = k \cdot L \cdot C_{A_f}^{n-1}/v$ we thus have obtained the second-order DE

$$\frac{1}{P_e}\frac{d^2x}{d\omega^2} - \frac{dx}{d\omega} - D_a \cdot x^n = 0 \; , \tag{5.24}$$

where for a first-order reaction (with $n = 1$) we use $D_a = k \cdot L/v$ and for a second-order reaction (with $n = 2$) we have $D_a = k \cdot L \cdot C_{A_f}/v$.
The boundary conditions for (5.24) are as follows at the exit :

$$\text{for} \;\; \omega = 1 \; : \qquad \frac{dx}{d\omega} = 0 \; , \tag{5.25}$$

and at the entrance :

$$\text{for} \;\; \omega = 0 \; : \qquad \frac{1}{P_e}\frac{dx}{d\omega} = x - 1 \; . \tag{5.26}$$

For example, the boundary condition (5.26) at the entrance can be obtained as shown in Figure 5.6.

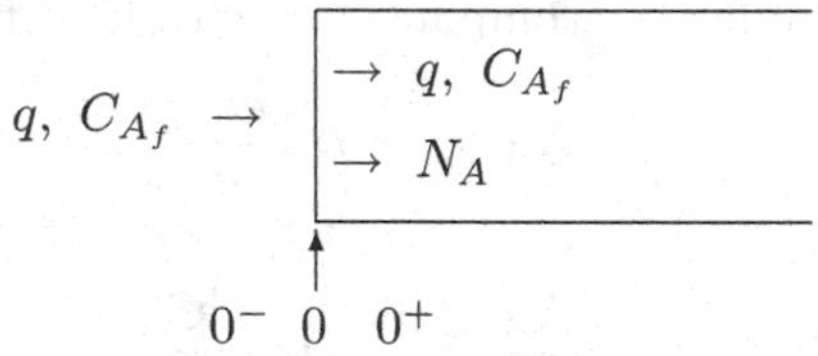

$$\begin{array}{c} q, \; C_{A_f} \; \rightarrow \end{array} \quad \begin{array}{|l} \rightarrow \; q, \; C_{A_f} \\ \rightarrow \; N_A \end{array}$$

$$0^- \;\; 0 \;\; 0^+$$
Mass flow at the entrance
Figure 5.6

The mass balance at the entrance gives us for $l = 0$

$$q \cdot C_{A_f} = q \cdot C_A + A_t \cdot N_A \; .$$

Dividing by A_t and using Fick's law once more, we get

$$v \cdot C_{A_f} = v \cdot C_A - D_A\frac{dC_A}{dl}$$

with the diffusion coefficient D_A of component A, or

$$1 = x - \frac{D_A}{l \cdot v}\frac{dC_A}{dl} \; ,$$

and ultimately

$$\frac{1}{P_e}\frac{dx}{d\omega} = x - 1 \; ,$$

as stated in (5.26). Thus the complete model equations for the nth order reaction are

$$\frac{1}{P_e}\frac{d^2x}{d\omega^2} - \frac{dx}{d\omega} - D_a \cdot x^n = 0 \tag{5.24}$$

for $n = 1, 2$, together with the earlier boundary conditions (5.25) and (5.26).

The present problem is a design problem, since the required conversion rate of 0.75 is given and we want to find the value of D_a that achieves this. But for educational purposes, before we present the solution of this design problem, we first present the simulation problem where D_a is given and the conversion is unknown.

Analytic Solution of the Linear Case

Here we assume that $D_a - 1.4$ and that the reaction is first-order. The solution of the simulation problem is as follows.

Solution of the second-order differential equation (5.24) for the first-order reaction

For a first-order reaction ($n = 1$), equation (5.24) becomes

$$\frac{d^2x}{d\omega^2} - (P_e) \cdot \frac{dx}{d\omega} - (P_e \cdot D_a)x = 0 \ . \tag{5.27}$$

This is a linear second order ordinary differential equation that can be solved explicitly by using the characteristic equation

$$\lambda^2 - P_e\lambda - P_e \cdot D_a = 0 \ . \tag{5.28}$$

The roots of this second-degree polynomial are

$$\lambda_{1,2} = \frac{P_e \pm \sqrt{(P_e)^2 + 4P_e \cdot D_a}}{2} \ . \tag{5.29}$$

Thus for $P_e = 15$ and $D_a = 1.4$ we have

$$\lambda_{1,2} = \frac{15 \pm 17.58}{2} = 16.29 \ \text{or} \ -1.29$$

by using a calculator, or via MATLAB's polynomial `root` command for the vector `[1,-15,-15*1.4]` of the polynomial coefficients for λ^2, $\lambda^1 = \lambda$, and the constant term $\lambda^0 = 1$ in (5.28).

```
>> roots([1 -15 -15*1.4])
ans =
  16.28919791562348
  -1.28919791562347
```

From the theory of linear differential equations, the solution of the second-order linear differential equation (5.27) has the general form

$$x(\omega) = C_1e^{\lambda_1\omega} + C_2e^{\lambda_2\omega} = C_1e^{16.29\omega} + C_2e^{-1.29\omega} \ . \tag{5.30}$$

The constants C_1 and C_2 in (5.30) can be calculated from our boundary conditions. The derivative of x in (5.30) is

$$\frac{dx}{d\omega} = C_1\lambda_1 e^{\lambda_1\omega} + C_2\lambda_2 e^{\lambda_2\omega} .$$

The boundary condition (5.25) at the end ($\omega = 1$) yields the equation

$$C_1 \cdot 16.29 \cdot e^{16.29} = -C_2(-1.29)e^{-1.29} ,$$

or $C_2 = (5.4479 \cdot 10^8) \cdot C_1$. From the boundary condition (5.26) at the entrance ($\omega = 0$) we deduce

$$\frac{1}{P_e}\frac{dx(0)}{d\omega} = \frac{1}{P_e}(C_1\lambda_1 + C_2\lambda_2) = (C_1 + C_2) - 1 = x(0) - 1 .$$

For our data the inner equation above reads as

$$\frac{1}{15}(16.29 \cdot C_1 + (-1.29) \cdot 5.448 \cdot 10^8 \cdot C_1) = C_1 + 5.448 \cdot 10^8 \cdot C_1 - 1 ,$$

making $C_1 = 16.9 \cdot 10^{-10}$. Since $C_2 = (5.448 \cdot 10^8) \cdot C_1$, we obtain $C_2 = 0.9207$. Thus the solution of the boundary value problem is

$$x(\omega) = 16.9 \cdot 10^{-10} e^{16.29\omega} + 0.9207 e^{-1.29\omega} .$$

At the exit when $\omega = 1$ the dimensionless concentration of component A is

$$x(1) = 16.9 \cdot 10^{-10} e^{16.29} + 0.9207 e^{-1.29} = 0.27351 .$$

For $P_e = 15.0$ and $D_a = 1.4$, $x = C_A/C_{A_f}$, and the conversion $x_A = (C_{A_f} - C_A)/C_{A_f} = 1 - x$, the actual value of the conversion at the exit is $1 - 0.27351 = 0.72649$ or about 73% , i.e., we have reached close to 75% conversion at the reactor exit.

Now we study the design problem where D_a is unknown, the conversion x_A is to be equal to 0.75, and therefore the dimensionless concentration at the exit is $x(1) = 0.25$ ($= 1 - x_A$). We follow the same procedure above until we reach the expression (5.29) for λ_1 and λ_2. This expression now contains an unknown D_a and the variable P_e.

$$\lambda_{1,2} = \lambda_{1,2}(D_a, P_e) = \frac{P_e \pm \sqrt{P_e^2 + 4 \cdot P_e \cdot D_a}}{2} .$$

With $x(\omega) = C_1 e^{\lambda_1\omega} + C_2 e^{\lambda_2\omega}$ and $dx/d\omega = C_1 \cdot \lambda_1 \cdot e^{\lambda_1\omega} + C_2 \cdot \lambda_2 \cdot e^{\lambda_2\omega}$, the boundary condition at the exit ($\omega = 1$) becomes $dx(1)/d\omega = C_1 \cdot \lambda_1 \cdot e^{\lambda_1} + C_2 \cdot \lambda_2 \cdot e^{\lambda_2} = 0$. Therefore

$$C_2 = -\frac{C_1 \cdot \lambda_1 \cdot e^{\lambda_1}}{\lambda_2 \cdot e^{\lambda_2}} . \tag{5.31}$$

The boundary condition at the start ($\omega = 0$) is $(1/P_e) \cdot (dx(0)/d\omega) = x(0) - 1$, and thus for our specific case

$$\frac{1}{P_e}(C_1 \cdot \lambda_1 + C_2 \cdot \lambda_2) = (C_1 + C_2) - 1 . \tag{5.32}$$

By combining (5.31) and (5.32) we obtain

$$\frac{1}{P_e}\left(C_1 \cdot \lambda_1 - \frac{C_1 \cdot \lambda_1 \cdot e^{\lambda_1}}{\lambda_2 \cdot e^{\lambda_2}}\lambda_2\right) = \left(C_1 - \frac{C_1 \cdot \lambda_1 \cdot e^{\lambda_1}}{\lambda_2 \cdot e^{\lambda_2}}\right) - 1 \, .$$

Solving for C_1 we finally get

$$C_1 = \frac{-1}{\left(\frac{\lambda_1}{P_e} - 1\right) + \left(\frac{\lambda_1}{\lambda_2}e^{\lambda_1 - \lambda_2}\right)\left(1 - \frac{\lambda_2}{P_e}\right)} \, . \tag{5.33}$$

Note that $\lambda_1 - \lambda_2 = \left(P_e/2 + \sqrt{(P_e/2)^2 + P_e \cdot D_a}\right) - \left(P_e/2 - \sqrt{(P_e/2)^2 + P_e \cdot D_a}\right) = \sqrt{P_e^2 + 4 \cdot P_e \cdot D_a}$. Thus C_1 can be expressed in dependence on the Damköhler number D_a and the Peclet number P_e as follows.

$$C_1 = F_1(D_a, P_e) = \frac{-P_e}{\lambda_1 - P_e + \frac{\lambda_1}{\lambda_2}e^{\lambda_1 - \lambda_2} \cdot (P_e - \lambda_2)} \, . \tag{5.34}$$

Using (5.31), namely $C_2 = -C_1\frac{\lambda_1}{\lambda_2}e^{\lambda_1 - \lambda_2}$, we furthermore have the functional relation

$$C_2 = F_2(D_a, P_e) = -F_1(D_a)\frac{P_e + \sqrt{P_e^2 + 4 \cdot P_e \cdot D_a}}{P_e - \sqrt{P_e^2 + 4 \cdot P_e \cdot D_a}}\, e^{\sqrt{P_e^2 + 4 \cdot P_e \cdot D_a}}$$

that expresses C_2 in terms of D_a and P_e.

Next we consider the dimensionless concentration equation $x(\omega) = C_1 e^{\lambda_1 \omega} + C_2 e^{\lambda_2 \omega}$ in light of the boundary conditions. At the exit $\omega = 1$ we have

$$\begin{aligned}
x(1) &= C_1 e^{\lambda_1} + C_2 e^{\lambda_2} \\
&= F_1(D_a, P_e)e^{P_e/2 + \sqrt{(P_e/2)^2 + P_e \cdot D_a}} + F_2(D_a, P_e)e^{P_e/2 - \sqrt{(P_e/2)^2 + P_e \cdot D_a}} \\
&= F_1(D_a, P_e) \cdot f_1(D_a, P_e) + F_2(D_a, P_e) \cdot f_2(D_a, P_e)
\end{aligned}$$

for $f_{1,2}(D_a, P_e) = e^{P_e/2 \pm \sqrt{(P_e/2)^2 + P_e \cdot D_a}}$ and F_i as before for $i = 1, 2$.

As we desire to find D_a for the given Peclet number $P_e = 15$ and the desired conversion $x_A(1) = 1 - x(1) = 0.75$ at the end, we need to solve

$$F(D_a, P_e) = F_1(D_a, P_e) \cdot f_1(D_a, P_e) + F_2(D_a, P_e) \cdot f_2(D_a, P_e) - x(1) = 0 \tag{5.35}$$

for D_a using the desired exit concentration $x(1) = 1 - x_A(1) = 0.25$. Equation (5.35) is a transcendental equation in D_a for the given values of P_e and $x(1)$ (or $x_A(1)$). It can be solved easily in MATLAB using its `fzero.m` root finder. Here is a program that solves (5.35) for all values of P_e and all desired exit conversions $x_A(1) = 1 - x(1)$ (or all desired exit concentrations $x(1) = 1 - x_A(1)$). The program starts out with 12 lines of comments, followed by default settings and the call of `fzero` that solves (5.35) for the given input data. The next block of lines contains the plotting preparations and commands, if plotting is desired. The final subfunction `DaPe` evaluates the function F in (5.35) for use inside the call of `fzero` in the main program.

```
function [Da, err] = conversionDa(Pe,percent,start,plotting)
%      [Da, err] = conversionDa(Pe,percent,plot)
% Sample call : [Da, err] = conversionDa(15,.75,1)
% Input: Pe : Peclet number; in the range 1 to 25;
%         percent : percentage of conversion xA(1) desired at the end;
%                   range .01 to 0.99 (or 1 to 99 %)
%         start:    starting estimate of Da. Default Da = 2;
%                   try start = 1, 20, 100, or 1000 if NaNs appear.
%         plotting: set to 1 if a plot is desired for F(Da).
% Output: Da : the desired value for the Damkoehler number;
%          err: the deviation from zero of F(Da), the error.
%          Plot of the F(Da) curve near the root, if desired (plotting = 1)

if nargin == 2, start = 2; plotting = 0; end
if nargin == 3, plotting = 0; end

warning off; ltol = 10^-14;  % Default settings; call fzero from Da = start:
Da = fzero(@DaPe,start,optimset('dis',[],'tolx',ltol),Pe,percent);
err = DaPe(Da,Pe,percent);

if plotting == 1              % if plot is desired
   x = linspace(Da/3,3*Da,100); hperc = 100*percent;
   plot(x,DaPe(x,Pe,percent),'LineWidth',1.5), hold on, v = axis;
   plot([v(1) v(2)],[0 0],'-r','LineWidth',1.5), plot(Da,v(3),'r+'),
   title(['Finding D_a with F(D_a) = 0    (for P_e = ',num2str(Pe,'%5.3g'),...
        ' and ',num2str(hperc,'%8.6g'),' %   conversion)'],'FontSize',12)
   xlabel(['D_a     (solution at D_a = ',num2str(Da,'%9.6g'),' )'],...
            'FontSize',12);
   ylabel('F(D_a)     ','FontSize',12,'Rotation',0); hold off, end

function F = DaPe(x,Pe,percent)
Da = x; int = (Pe^2 + 4*Pe*Da).^0.5;      % prepare data and constants
F1 = -Pe./((Pe+int)/2 - Pe +...           % form the four ingredients
     ((Pe + int)./(Pe - int)).*exp(int).*(Pe+int)/2);
f1 = exp(Pe/2 + int/2);
f2 = exp(Pe/2 - int/2);
F2 = -F1.*(Pe + int)./(Pe - int).*exp(int);
F = F1.*f1 + F2.*f2 - (1 - percent);      % form F
```

Specifically, for $P_e = 15.0$ and a desired exit conversion rate $x_A(1)$ of 75% we compute the corresponding Dahmköhler number as $D_a = 1.506$ after setting start = 1 in the above code conversionDa.m and using percent $= x_A(1) = 1 - x = 0.75$.

We can validate our formulas and code against the earlier computed conversion rate of 72.64% at the end of the tubular reactor for the Damköhler number $D_a = 1.4$ and $P_e = 15.0$ by running >> conversionDa(15,0.7264) with the inputs $P_e = 15$ and the conversion rate percent $= 0.7264$ ($= x_A$) in MATLAB. This call computes the Damköhler number $D_a = 1.4007$ correctly to within 0.05%.

The MATLAB function multiconversionDa.m is designed to compute and display the value of the Damköhler number D_a that solve the equation (5.35) for various inputs of the desired concentration $x_A(1)$. Here is its code.

```
function [Da, err] = multiconversionDa(Pe,Percent,start)
%      [Da, err] = multiconversionDa(Pe,percent,start)
% Sample call : [Da, err] = multiconversionDa(15,0.01:0.02:.99);
% Input: Pe : Peclet number; in the range 1 to 25;
%        Percent : vector of percentage of conversion xA(1) desired at the end;
%                  range .01  to 0.99 ( 1% to 99% )
%        start:   starting estimate of Da. Default Da = 2;
%                  try start = 1, 20, 100, or 1000 if NaNs appear.
% Output: Da : the desired vector of values for the Damkoehler number;
%         err: the deviation vector from zero of F(Da), the error.
%         Plot of Da versus xA(1) curve

if nargin == 2, start = 2; end

m = length(Percent); Da = zeros(m,1); err = zeros(m,1); hperc = 100*Percent;
warning off; ltol = 10^-14;  % Default settings; call fzero from Da = start:
for i = 1:m              % Find Da for all given percentage rates
  Da(i) = fzero(@DaPe,start,optimset('dis',[],'tolx',ltol),Pe,Percent(i));
  err(i) = DaPe(Da(i),Pe,Percent(i));   end

plot(Percent,Da)           % plot Da versus xA
  title(['Finding D_a with F(D_a) = 0    for P_e = ',...
    num2str(Pe,'%5.3g'),'   and   ',num2str(min(Percent),'%8.4g'),...
    ' \leq x_A(1) \leq ',num2str(max(Percent),'%8.4g'),'  conversion'],...
    'FontSize',12), xlabel('x_A(1)','FontSize',12);
    ylabel('D_a    ','FontSize',12,'Rotation',0); hold off,

function F = DaPe(x,Pe,percent)
Da = x; int = (Pe^2 + 4*Pe*Da).^0.5;     % prepare data and constants
F1 = -Pe./((Pe+int)/2 - Pe +...          % form the four ingredients
  ((Pe + int)./(Pe - int)).*exp(int).*(Pe+int)/2);
f1 = exp(Pe/2 + int/2);
f2 = exp(Pe/2 - int/2);
F2 = -F1.*(Pe + int)./(Pe - int).*exp(int);
F = F1.*f1 + F2.*f2 - (1 - percent);     % form F
```

Figure 5.7 shows the plot obtained by calling `multiconversionDa(15,0.01:0.02:.99);` for example.

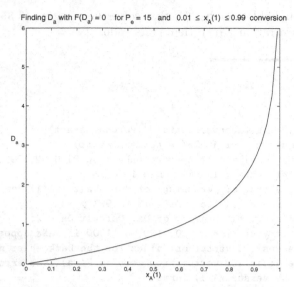

Damköhler solutions D_a of (5.35) for varying output conversions $x_A(1)$ and $P_e = 15$

Figure 5.7

Next we study equation (5.35) for a fixed value of D_a. Here we try to solve the equation for P_e as $x_A(1)$ varies. The call of `multiconversionPe(1.5,0.5:0.002:0.9);` creates the following plot of P_e as a function of the output conversion $x_A(1)$.

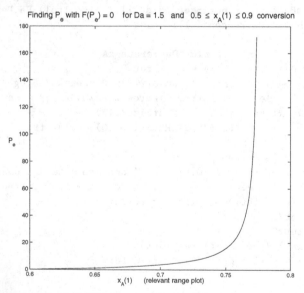

Peclet number solutions P_e of (5.35) for varying output conversions $x_A(1)$ and $D_a = 1.5$

Figure 5.8

The code of `multiconversionPe.m` is much more involved than `multiconversionDa.m` and is printed below.

```
function [Pe, err] = multiconversionPe(Da,Percent)
%     [Pe, err] = multiconversionPe(Da,percent)
% Sample call : [Pe, err] = multiconversionPe(1.5,0.5:0.002:0.9);
% Input: Da : Damkoehler number; in the range of 0.1 to 10;
%          Percent : vector of percentage of conversion xA(1) desired at the end;
%                    range .01  to 0.99 ( 1% to 99% )
%          start:    starting estimate of Pe. Default Pe = 12;
%                    try start = 1, 20, 100, or 1000 if NaNs appear.
% Output: Pe : the desired vector of values for the Peclet number;
%          err: the deviation vector from zero of F(Pe), the error.
%          Plot of Pe versus xA(1) curve
%             If plot is discontinuous, refine your Percent vector to cover the
%             relevant range

m = length(Percent); Pe = zeros(m,1); err = zeros(m,1); start = 1;
warning off; ltol = 10^-10;  % Default settings; call fzero from Da = start:
for i = 1:m                   % Find Da for all given percentage rates
  if Da >= 1.3 & Da <= 1.56  % set starting values for fzero
     plim = 0.44*Da; PLim = plim*1.05; end
  if Da > 1.56, plim = 0.37*Da; PLim = plim*1.05; end
  if Da > 2, plim = Da*0.3; PLim = plim*1.05; end
  if Da < 1.3 & Da > 0.8, plim = Da/2; PLim = plim*1.05; end
  if Da <= 0.8, plim = Da*.6; PLim = 1.05*plim; end
  if Da < 0.6, plim = Da*.7; PLim = plim*1.05; end
  if Da < 0.41, plim = Da*.73; PLim = plim*1.05; end
  if Percent(i) <= PLim/1.05 & Percent(i) >= plim/1.05, start = 1; end
  if Percent(i) <= PLim & Percent(i) >= plim, start = 2; end
  if Percent(i) <= PLim*1.05 & Percent(i) >= PLim, start = 5; end
  if Percent(i) <= PLim*1.05^2 & Percent(i) >= PLim*1.05, start = 14; end
  if Percent(i) <= PLim*1.05^3 & Percent(i) >= PLim*1.05^2, start = 30; end
  if i > 1 & Pe(i-1) > 0, start = Pe(i-1); end % use last Pe value as start
  try                       % circumnavigate failures
    Pe(i) = fzero(@DaPe,start,optimset('Display','off','TolX',ltol),...
                  Da,Percent(i));  err(i) = DaPe(Da,Pe(i),Percent(i));
    Pe(i) = max(Pe(i),-1);
    if imag(err(i)) ~= 0,  Pe(i) = -1; err(i) = -1; end
  catch                     % set failure data equal to unplotted NaNs
    Pe(i) = -1; err(i) = -1;
  end
  if imag(Pe(i)) ~= 0, Pe(i) = -1; end, end

j = find(Pe ~= -1); Peplot = Pe(j);
Percentplot = Percent(j);          % select relevant data
plot(Percentplot,Peplot)           % plot Da versus xA in relevant range
 title(['Finding P_e with F(P_e) = 0    for Da = ',...
   num2str(Da,'%6.4g'),'   and   ',num2str(min(Percent),'%8.4g'),...
```

```
'   \leq  x_A(1)  \leq ',num2str(max(Percent),'%8.4g'),'  conversion'],...
'FontSize',12), xlabel('x_A(1)        (relevant range plot)','FontSize',12);
ylabel('P_e      ','FontSize',12,'Rotation',0); hold off,
```

```
function F = DaPe(x,Da,percent)
Pe = x; int = (Pe.^2 + 4*Pe*Da).^0.5;        % prepare data and constants
F1 = -Pe./((Pe+int)/2 - Pe +...              % form the four ingredients
    ((Pe + int)./(Pe - int)).*exp(int).*(Pe+int)/2);
f1 = exp(Pe/2 + int/2);
f2 = exp(Pe/2 - int/2);
F2 = -F1.*(Pe + int)./(Pe - int).*exp(int);
F = F1.*f1 + F2.*f2 - (1 - percent);         % form F
```

multiconversionPe.m gives valuable output for all D_a with $0.15 \leq D_a \leq 15$. Its first 16 lines of MATLAB code adjust the starting value start that is used initially in the MATLAB built-in root finder fzero for the function DaPe, given at the end of the code. Our function DaPe represents the equation (5.35) in standard form. When solving (5.35) via fzero over a broad interval of $x_A(1)$ values, the algorithm usually encounters many failures for $x_A(1)$ values that are out of range. This data is cleaned up before we can plot $P_e(x_A(1))$ in its relevant, i.e., data yielding range. Once fzero has successfully solved equation (5.35), we use the previously computed value for P_e as the starting guess for the next call of fzero. This method of "continuation" works quite well.

For another approach to solving equation (5.35) for D_a or P_e, see Problem 1 of the Exercises for this section.

The Nonlinear Case

Solution of the second-order two-point boundary value problem (5.24) for a second-order reaction

For a second-order reaction ($n = 2$), equation (5.24) becomes

$$\frac{d^2x}{d\omega^2} - (P_e) \cdot \frac{dx}{d\omega} - (P_e \cdot D_a)x^2 = 0 \tag{5.36}$$

with $D_a = k \cdot L \cdot C_{A_f}/v$. This is a nonlinear second-order ordinary differential equation due to the appearance of x^2. Nonlinear DEs can generally not be solved explicitly. Equation (5.36) comes with the two boundary conditions

$$\frac{dx}{d\omega} = 0 \quad \text{at} \quad \omega = 1$$

and

$$\frac{1}{P_e}\frac{dx}{d\omega} = x - 1 \quad \text{at} \quad \omega = 0 .$$

As explained in Section 1.2.2, the numerical solution of second- or higher-order DEs is generally approached numerically from an equivalent but enlarged first-order system of

DEs. Therefore our first step is to put the second-order differential equation (5.36) into the form of two first-order ones. For this we set $x_1 = x$ and $x_2 = dx/d\omega$. Then

$$\frac{d^2 x}{d\omega^2} = \frac{d^2 x_1}{d\omega^2} = \frac{dx_2}{d\omega}$$

and the original second-order differential equation (5.36) can be rewritten as a system of two first-order differential equations, namely

$$\frac{dx_1}{d\omega} = x_2$$

and   (5.37)

$$\frac{dx_2}{d\omega} = P_e \cdot x_2 + P_e \cdot D_a \cdot x_1^2$$

with the boundary conditions $x_2(1) = 0$ at $\omega = 1$ and $x_2(0) = P_e \cdot (x_1(0) - 1)$ at $\omega = 0$.

Let us investigate how MATLAB handles boundary value problems such as (5.24) with the boundary conditions (5.25) and (5.26) for first- and second-order reactions. The MATLAB program linquadbvp.m has been designed for this purpose.

```
function linquadbvp(Pe,Da)
%   linquadbvp(Pe,Da)
% sample call : linquadbvp(15,1.4)
% Input : Pe = Peclet number; Da = Damkoehler number
% Output: Plots of the solutions (theoretical [if known] and numerical)
%         to the 2 point BVP (5.24) for first and second order right
%         hand sides, and their derivatives.

if Pe == 15 & Da == 1.4,        % If theoretical solution is known :
  om = linspace(0,1,100);       % plot theoretical solution and its derivative
  xtheor = 16.9*10^-10*exp(16.29*om) + 0.9207*exp(-1.29*om);
  xprimetheor = 16.29*16.9*10^-10*exp(16.29*om) - 1.29*0.9207*exp(-1.29*om);
  plot(om,xtheor,'+r'), hold on
  plot(om,xprimetheor,'g'), end

sol = bvpinit(linspace(0,1,24),@initial);  % initialize boundary value solver
options = bvpset('NMax',1000,'Vectorized','on'); % solve with first order RHS
solone = bvp4cfsinghouseqr(@frhslin,@frandbedlin,sol,options,Pe,Da);
xint = linspace(0,1); Sxintone = deval(solone,xint);
plot(xint,Sxintone(1,:),'b'), hold on         % plot numerical solution
                            % solve with second order RHS and plot :
soltwo = bvp4cfsinghouseqr(@frhsquad,@frandbedquad,sol,options,Pe,Da);
xint = linspace(0,1); Sxinttwo = deval(soltwo,xint);
plot(xint,Sxinttwo(1,:),'k'),
if Pe ~= 15 | Da ~= 1.4, plot(xint,Sxintone(2,:),'g'), end
plot(xint,Sxinttwo(2,:),':k'),
if Pe == 15 & Da == 1.4,
  legend('theor. first order DE sol.','theor. first order DE slope',...
```

```
      'num. first order DE sol.','num. second order DE sol.',...
      'num. second order DE slope',0), else
   legend('num. first order DE sol.','num. second order DE sol.',...
      'num. first order DE slope', 'num. second order DE slope',0), end
title(['Solutions to the BVP (5.24) for first and second order reactions',...
   '      (for P_e = ',num2str(Pe,'%6.4g'),'  and  D_a = ',...
      num2str(Da,'%6.4g'),' )'],'Fontsize',12), xlabel('\omega','Fontsize',12),
ylabel([{' '},{' '},{' '},{' '},{'x(\omega)    '},{' '},{' '},{' '},...
         {' '},{' '},{' '},{'xprime(\omega)    '}],'Fontsize',12,'Rotation',0),
hold off


function dydx = frhslin(x,y,Pe,Da)        % Right hand side of first order DE
dydx = [ y(2,:)
         Pe*y(2,:)+Pe*Da*y(1,:)];


function Rand = frandbedlin(ya,yb,Pe,Da)  % boundary condition, first order DE
Rand = [ yb(2)
         ya(2)-Pe*ya(1)+Pe];


function dydx = frhsquad(x,y,Pe,Da)       % Right hand side of second order DE
dydx = [ y(2,:)
         Pe*y(2,:)+Pe*Da*y(1,:).^2];


function Rand = frandbedquad(ya,yb,Pe,Da) % boundary condition, second order DE
Rand = [ yb(2)
         ya(2)-Pe*ya(1)+Pe];


function f = initial(x)                   % initial guess
   f = [-0.6*x+0.9;1.2*x-1.2];
```

The plotting output of `linquadbvp` depends on our choice of Peclet number P_e and Damköhler number D_a. If we know the theoretical solution to the BVP, as we do for $P_e = 15$ and $D_a = 1.4$, for example, from p. 266, we start the program by plotting the theoretical solution using + signs. Then we initialize the BVP solver `bvp4cfsinghouseqr.m` and call it to obtain the numerical solution and its derivative for the first-order reaction. This is followed by solving the BVP for the second-order reaction and by general plotting commands. The boundary value problem solver `bvp4cfsinghouseqr` uses the right-hand side of the DE, called `frhslin` and `frhsquad` by us, respectively, as well as the boundary condition functions `frandbedlin` or `frandbedquad` depending on the order of the reaction. These two functions are problem specific and have to be altered according to the actual BVP that needs to be solved. `bvp4cfsinghouseqr.m` furthermore uses our initial guess `sol` of the solution, that is given by the function `initial`, as well as various technical option parameters and the two relevant system parameters P_e and D_a. More details on the derivation and use of our MATLAB BVP solver `bvp4cfsinghouseqr.m` is given in Section 5.1.3.

Figure 5.9 gives the graphical output of a call of `linquadbvp(15,1.4)`.

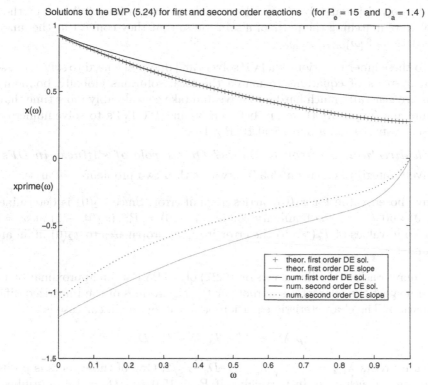

Solutions to the BVP (5.24) for first and second order reactions   (for $P_e = 15$ and $D_a = 1.4$)

Solution curves for the BVP (5.24) and first- and second-order reactions
Figure 5.9

In Figure 5.9 we observe that the theoretical solution (+) and the numerical solution (−) of the DE (5.24) coincide for the first-order reaction. But for the nonlinear second-order reaction, there is no known theoretical solution, hence we cannot compare.
Note that in most applications, BVPs have no known theoretical solutions and one has to rely on numerical analysis when solving BVPs.

First Numerical Considerations

The axial dispersion model has led to the two-point boundary value problem (5.37) for ω from $\omega = \omega_{start} = 0$ to $\omega = \omega_{end} = 1$. DEs are standardly solved by numerical integration over subintervals of the desired interval $[\omega_{start}, \omega_{end}]$. For more on the process of solving BVPs, see Section 1.2.4 or click on the Help line under the View icon on the MATLAB desktop, followed by a click on the Search tab in the Help window and searching for "BVP".
For initial value problems such as the ones we have encountered in Chapter 4, there is only one possible direction of integration, namely from the initial value ω_{start} onwards. But for two-point BVPs we have function information at both ends $\omega_{start} < \omega_{end}$ and we could **integrate forwards** from ω_{start} to ω_{end}, or **backwards** from ω_{end} to ω_{start}. Regardless of the direction of integration, each boundary value problem on $[\omega_{start}, \omega_{end}]$

is generally solved iteratively via small sub-interval integrations, followed by the nonlinear problem of matching the parts of a solution so that they connect at the inner points of the partition of $[\omega_{start}, \omega_{end}]$.

Due to these inner iterations via IVP solvers and due to the need to solve an associated nonlinear systems of equations to match the local solutions globally, boundary value problems are generally much harder to solve and take considerably more time than initial value problems. Typically there are between 30 and 120 IVPs to solve numerous times in each successful run of a numerical BVP solver.

In which direction to integrate BVPs? Or the role of stiffness in DEs.

If we naively integrate equation (5.37) forwards then two problems will arise.

1. Any chosen value for $x_1(0)$ carries a small error. Since $x_2(0)$ is determined from $x_1(0)$ and P_e via the boundary condition $x_2(0) = P_e \cdot (x_1(0) - 1)$ at $\omega = 0$, then for large values of $P_e(\gg 5)$ the error in $x_1(0)$ propagates to $x_2(0)$ in an amplified manner.

2. We may consider the linear version (5.27) of the BVP as an approximation to (5.37) and inspect the characteristic roots or the eigenvalues of the linearized differential equation. The characteristic equation $p(\lambda) = 0$ for the linear case is

$$p(\lambda) = \lambda^2 - P_e \cdot \lambda - P_e \cdot D_a = 0$$

with its roots $\lambda_{1,2} = P_e \pm \sqrt{P_e^2 + 4 \cdot D_a \cdot P_e}/2$. One of these roots is positive and the other is negative. In particular, if $P_e = 15.0$ and $D_a = 1.4$ as studied earlier for 72.649% conversion, then the two roots were computes as $\lambda_1 = 16.29$ and $\lambda_2 = -1.29$. And the solution of the DE has the general form $x(\omega) = C_1 e^{16.29 \cdot \omega} + C_2 e^{-1.29 \cdot \omega}$.

Note that $|\lambda_1| \gg |\lambda_2|$ by a factor exceeding 12 here and that the $C_1 e^{16.29 \cdot \omega}$ part of the solution will grow rapidly in the positive ω direction, while the other $C_2 e^{-1.29 \cdot \omega}$ part of the solution will decrease slowly as ω increases. We say that $e^{16.29 \cdot \omega}$ exhibits instability for growing ω, while $e^{-1.29 \cdot \omega}$ is quite stable as ω increases. DEs with two largely disproportionate eigenvalues $\lambda_{1,2}$ that satisfy

   (a) $\lambda_1 \cdot \lambda_2 < 0$,

   (b) $\lambda_2 < 0$,   and

   (c) $|\lambda_1| \gg |\lambda_2|$

are considered to be **stiff differential equations**. Such stiff DEs need special care in their numerical treatment. Stiff DEs carry inherent instabilities with them due to a potentially devastating mixture of slow decline and rapid growth in different parts of their solutions.

[Refer to subsection (H) of Appendix 1 for more on linear DEs.]

3. Given the Peclet number $P_e = 0.1$ and the Damkoehler number $D_a = 1.5$ instead, the roots of $\lambda^2 - P_e\lambda - P_e \cdot D_a = 0$ are 0.44051 and -0.34051. These roots satisfy

two of the conditions for stiffness, but not the third that $|\lambda_1| \gg |\lambda_2|$. Hence for this case the linearized DE is not stiff and its integration creates fewer problems.

Thus due to potential stiffness, integrating a BVP such as the one in equation (5.37) in the positive ω direction from 0 to 1 may not be wise in all cases. In fact, backward integration is much more stable for our simplified model since in backward integration, the eigenvalues switch signs and then the problem is no longer stiff according to the definition.

Actually a theoretical proof of whether a given DE is stiff or not is rather complicated if not impossible for most DE problems. Therefore it is best to use numerical codes that switch internally to backward integration if forward integration encounters troubles with numerical convergence and vice versa, giving us the best method for either situation.

5.1.3 Numerical Solution of Nonlinear BVPs. The Non-Isothermal Case

In the previous subsection we have studied two simplified single DEs, linear and nonlinear, that were derived from our system of coupled BVPs (5.20) and (5.19). We now want to solve the system of coupled BVPs (5.20) and (5.19) numerically with all its complexities.

Following our numerical insights for the linear single simplified DE above, we first show how to rewrite (5.20) and (5.19) and the associated boundary conditions

$$\frac{dy(1)}{d\omega} = \frac{dx_A(1)}{d\omega} = 0 , \quad \text{and} \quad \frac{1}{P_{\partial H}}\frac{dy(0)}{d\omega} = y(0) - 1 ,$$

$$\text{as well as} \quad \frac{1}{P_{eM}}\frac{dx_A(0)}{d\omega} = x_A(0) - 1$$

for backwards integration. We can do so by introducing the backwards variable $\omega' = 1-\omega$. Then the first derivative terms in (5.20) and (5.19) change signs since $d\omega = -d\omega'$, but the second derivative terms do not. A simple calculus explanation for this is as follows: if we reverse direction by setting $\omega' = 1 - \omega$ in the DEs, then function increases become function decreases as we change our direction and thus the first derivative terms change signs in the DEs. As the first derivative measures the slope of a function, the second derivative measures curvature and convexity. This does not change, however, when we reverse our directional variable from ω to ω'. What was convex up stays so upon a change of direction. Moreover, the boundary conditions for ω' flip from those for ω so that the BVP with the DEs (5.20) and (5.19) expressed in terms of ω' for the adiabatic case with $Q' = 0$ becomes

$$\text{for the mass} \quad \frac{1}{P_{eM}}\frac{d^2 x_A}{d\omega'^2} + \frac{dx_A}{d\omega'} = D_a e^{-\gamma/y} \cdot x_A , \tag{5.38}$$

and

$$\text{for the heat} \quad \frac{1}{P_{eH}}\frac{d^2 y}{d\omega'^2} + \frac{dy}{d\omega'} = -\beta \cdot D_a e^{-\gamma/y} \cdot x_A \tag{5.39}$$

with the "flipped" boundary conditions with reversed signs on the right-hand sides, namely

$$\text{at } \omega' = 0 \quad : \quad \frac{dx_A(0)}{d\omega'} = \frac{dy(0)}{d\omega'} = 0$$

and   (5.40)

$$\text{at } \omega' = 1 \quad : \quad \frac{1}{P_{eM}}\frac{dx_A(1)}{d\omega'} = 1 - x_A(1) \text{ and } \frac{1}{P_{eH}}\frac{dy(1)}{d\omega'} = 1 - y(1) \, .$$

Recall that $w' = 1 - w$ here.

Our next task is to rewrite both, the second-order two-DE systems (5.20), (5.19) used in forward integration and the DE system (5.38), (5.39) that is used for backwards integration, in the form of a first-order system of four DEs each. These have the form $u' = f(u)$ or $v' = g(v)$, respectively with u, $v \in \mathbb{R}^4$. We refer to Section 1.2.2 for the formal reduction of high-order DEs to first-order enlarged systems of DEs.

In the following we work on the second-order system of DEs (5.38) and (5.39). To obtain an equivalent first-order system of DEs, we set $u = u(\omega') = (u_1(\omega'), \ldots, u_4(\omega'))^T$ for $u_1 = x_A(\omega')$, $u_2 = u_1'$, $u_3 = y(\omega')$, and $u_4 = u_3'$. From the differential equations (5.38) and (5.39) we can readily see that the following system of DEs holds for $u(\omega')$:

$$u' = \begin{pmatrix} u_1' \\ u_2' \\ u_3' \\ u_4' \end{pmatrix} = \begin{pmatrix} u_2 \\ P_{eM} \cdot (D_a \cdot e^{-\gamma/u_3} \cdot u_1 - u_2) \\ u_4 \\ -P_{eH} \cdot (\beta \cdot D_a \cdot e^{-\gamma/u_3} \cdot u_1 + u_4) \end{pmatrix} = f(u) \, . \qquad (5.41)$$

Note that the differential equation $u' = f(u)$ in (5.41) is a set of first-order DEs since it contains only first derivatives, and it is nonlinear since its right-hand-side function f is highly nonlinear in the component functions of u. According to (5.40) its boundary conditions are

$$\text{at } \omega' = 0 \quad : \quad u_2(0) = u_4(0) = 0$$

and   (5.42)

$$\text{at } \omega' = 1 \quad : \quad \frac{u_2(1)}{P_{eM}} + u_1(1) - 1 = 0 \text{ and } \frac{u_4(1)}{P_{eH}} + u_3(1) - 1 = 0 \, .$$

Similarly we obtain the following first-order system $v' = g(v)$ for the BVP (5.20) and (5.19) in the variable ω with $v = v(\omega) = (v_1, \ldots, v_4(\omega))^T$ where we have set $v_1 = x_A(\omega)$, $v_2 = v_1'$, $v_3 = y(\omega)$, and $v_4 = v_3'$.

$$v' = \begin{pmatrix} v_1' \\ v_2' \\ v_3' \\ v_4' \end{pmatrix} = \begin{pmatrix} v_2 \\ P_{eM} \cdot (D_a \cdot e^{-\gamma/v_3} \cdot v_1 + v_2) \\ v_4 \\ P_{eH} \cdot (-\beta \cdot D_a \cdot e^{-\gamma/v_3} \cdot v_1 + v_4) \end{pmatrix} = g(v) \qquad (5.43)$$

with the boundary conditions

$$\text{at } \omega = 0 \ : \quad \frac{v_2(0)}{P_{eM}} - v_1(0) + 1 \ = \ 0 \ \text{ and } \ \frac{v_4(0)}{P_{eH}} - v_3(0) + 1 \ = \ 0$$

and $\hspace{10cm}$ (5.44)

$$\text{at } \omega = 1 \ : \quad v_2(1) \ = \ v_4(1) = 0 \ .$$

We leave the reduction of the two coupled DEs in (5.20) and (5.19) to this first-order DE system as an exercise.

Here are our MATLAB implementations of the four dimensional first-order DE (5.41) in dydx, and of its boundary conditions (5.42) in Rand. These two auxiliary functions dydx and Rand are called upon in MATLAB's built-in BVP solver bvp4c and in our modified BVP solver bvp4cfsinghouseqr.m.

```
function dydx = frhsDaPe(x,y,Da,bt,ga,PeH,PeM)      % right hand side of DE
dydx = [ y(2,:)                                     % backward integration
         PeM*(Da*exp(-ga./y(3,:)).*y(1,:)-y(2,:))
         y(4,:)
         PeH*(-bt*Da*exp(-ga./y(3,:)).*y(1,:)-y(4,:)) ];

function Rand = frandbedDaPe(ya,yb,Da,bt,ga,PeH,PeM)  % boundary condition
Rand = [ ya(2)                                        %    reversed also
         ya(4)
         yb(2)/PeM+yb(1)-1
         yb(4)/PeH+yb(3)-1 ] ;
```

Note that we use $u_i = y(i,:)$ in the derivative function dydt and that the symbol ya(i) denotes $u_i(\omega' = 0)$ in Rand while yb(i) denotes $u_i(\omega' = 1)$, all for $i = 1, ..., 4$.

Before delving into the numerics of solving the BVP (5.41) with (5.42) or the BVP (5.43) with (5.44) we shall first draw a few profiles of x_A and y in terms of ω for the Peclet numbers $P_{eH} = 4$ and $P_{eM} = 8$ and the Dahmköhler numbers $D_a = 10^5$, $10^{6.5}$, and 10^7, for example. We do this in MATLAB by calling the following commands in sequence:

```
axialdisp4DaPerunbackw(10^5,1,20,4,8,1,1,1,1); pause,
axialdisp4DaPerunbackw(10^6.5,1,20,4,8,1,1,1,1); pause,
axialdisp4DaPerunbackw(10^7,1,20,4,8,1,1,1,1); .
```

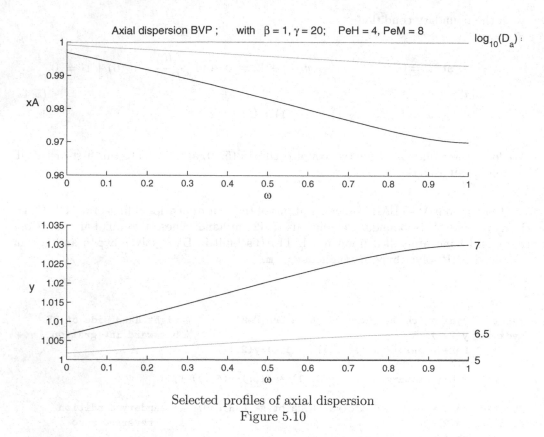

Selected profiles of axial dispersion
Figure 5.10

Note that the three x_A plots of Figure 5.10 appear like upside-down versions of the respective graphs for y. This is physically correct since for an exothermic reaction ($\beta = 1$), the temperature y increases as the concentration x_A decreases. The printed graph density of each pair of solutions $x_A(\omega)$ and $y(\omega)$ that solve the BVP for the same value of D_a is the same in the top and bottom part of Figure 5.10. We indicate the Dahmköhler number by its logarithm to base 10 on the right side of the graphs: For $D_a = 10^7$, the solutions in Figure 5.10 appear as a solid curve, for $D_a = 6.5$ as a wide dotted curve, and for $D_a = 5$ as a closely dotted curve. Generally the curve color of the computer screen plot is assigned by selecting one integer between 1 and 128 at random and then defining the color in each subsequent `axialdisp4DaPerunbackw` call randomly in the MATLAB code line `SS = colormap(hsv(128)); r = ceil(128*rand(1)); S1 = SS(r,:);` below.

```
function ...
[sol,sp,err] = axialdisp4DaPerunbackw(Da,bt,ga,PeH,PeM,a,p,halten,setDa,sol)
% [sol,sp,err] = axialdisp4DaPerunbackw(Da,bt,ga,PeH,PeM,a,p,halten,setDa,sol)
% Sample call : (without using a nearby solution "sol" for continuation)
%               axialdisp4DaPerunbackw(10^5,1,20,4,8,1,1,1,1); pause,
%               axialdisp4DaPerunbackw(10^6.5,1,20,4,8,1,1,1,1); pause,
%               axialdisp4DaPerunbackw(10^7,1,20,4,8,1,1,1,1);
```

```
%                       (click "return" after each graph has appeared)
%    THIS version uses backward integration, best in case of
%          Stiffness of the DE.
% Input : Da : Damkoehler number (typically 10^6, not its log10 !)
%          bt, ga : usual inputs; bt = 1, ga = 20 for example
%          PeH, PeM : Peclet numbers for heat and mass;
%                   Use 4 and 8  OR  8 and 4  for example
%          a : parameter to set up initial guess for the solution
%          p : of p = 1 : we plot; else we only keep the solution "sol",
%              sp, and err.
%          halten: if halten = 1 : we hold previous plots.
%          setDa : if setDa = 1 : we indicate log10(Da) on the right margin
%                   for each curve.
%          sol : starting solution (if known and supplied, otherwise "a" sets
%                   up the initial guess.
%          The color of the graph is chosen at random.
% Output : sol: solution structure;
%              If sp = 0, solution was found; if sp = 1, solution is
%              spurious and useless; data is discarded;
%          err : err set to pi (= 3.14...), if bvp integration was
%              unsuccessful.
% m-file used: bvp4cfsinghouseqr.m  to solve the BVP.

warning off % because of MATLAB: divideByZero

if nargin < 5, 'Too few inputs, we stop!', return, end
if nargin == 5, a = 1 + bt, p = 0; halten = 0; setDa = 0; end,
if nargin <= 7, halten = 0; setDa = 0;   end,
if nargin == 8, setDa = 0;   end
if nargin == 10, oldsol = sol; end
err = pi; sp = 0;
SS = colormap(hsv(128)); r = ceil(128*rand(1)); S1 = SS(r,:);

if PeH == 4 & PeM == 8, guess = [1;0;a;0];
    sol = bvpinit(linspace(0,1,15),guess); end,
if PeH == 8 & PeM == 4,              % best initial guesses below and above
    if a == 1, sol = bvpinit(linspace(0,1,24),@initialt);   end,
    if a == .5, sol = bvpinit(linspace(0,1,24),@initialm);   end,
    if a == .7, sol = bvpinit(linspace(0,1,24),@initialmt);   end,
    if a == 0, sol = bvpinit(linspace(0,1,24),@initialb);   end,   end

options = bvpset('RelTol',10^-2,'AbsTol',10^-4,'NMax',1000,'Vectorized','on');

try                       % if BVP solves all right
    sol = bvp4cfsinghouseqr(@frhsDaPe,@frandbedDaPe,sol,options,Da,bt,ga,...
        PeH,PeM);
catch                       % if BVP could not be solved
    if nargin == 11, sol = oldsol; end,
```

```
    sp = 1; return
end

if p == 1, xint = linspace(0,1); Sxint = deval(sol,xint);
  figure(1), subplot(2,1,1),                      % plot in reverse later
  if halten == 0, hold off, else hold on,
    if setDa == 1, text(1,1,[' log_{10}(D_a) : '],'FontSize',12), end, end
  plot(xint,fliplr(Sxint(1,:)),'Color',S1);
  figure(1), subplot(2,1,2),
  if halten == 0, hold off, else hold on,
    if setDa == 1,      % backwards integration looks at the starting point
      text(1,Sxint(3,1),['  ', num2str(log10(Da),'%7.4g')],'FontSize',12),
  end, end
  plot(xint,fliplr(Sxint(3,:)),'Color',S1); end,

if p == 1,
  if halten == 0,
    figure(1),   subplot(2,1,1),
    title(['Axial dispersion BVP ;       with  \beta = ',...
      num2str(bt,'%5.3g'),', \gamma = ',num2str(ga,'%5.3g'),...
      ';    PeH = ',num2str(PeH,'%5.3g'),', PeM = ',...
    num2str(PeM,'%5.3g')],'FontSize',12),
  else
    figure(1), subplot(2,1,1),
    title(['Axial dispersion BVP ;       with   \beta = ',...
        num2str(bt,'%5.3g'),', \gamma = ',num2str(ga,'%5.3g'),...
      ';    PeH = ',num2str(PeH,'%5.3g'),', PeM = ',num2str(PeM,'%5.3g')],...
        'FontSize',12), end
  xlabel('\omega','FontSize',12),
  ylabel('xA    ','Rotation',0,'FontSize',12),
  subplot(2,1,2), xlabel('\omega','FontSize',12),
  ylabel('y   ','Rotation',0,'FontSize',12), end
warning on, subplot(2,1,1), hold off, subplot(2,1,2), hold off,

function dydx = frhsDaPe(x,y,Da,bt,ga,PeH,PeM)     % right hand side of DE
dydx = [ y(2,:)                                    % backward integration
       PeM*(Da*exp(-ga./y(3,:)).*y(1,:)-y(2,:))
       y(4,:)
       PeH*(-bt*Da*exp(-ga./y(3,:)).*y(1,:)-y(4,:)) ];

function Rand = frandbedDaPe(ya,yb,Da,bt,ga,PeH,PeM)  % boundary condition
Rand = [ ya(2)                                    %   reversed also
       ya(4)
       yb(2)/PeM+yb(1)-1
       yb(4)/PeH+yb(3)-1 ]  ;

function f = initialt(x)                           % initial guesses
  f = [1;0;1;0];
```

```
function f = initialm(x)    % for 8/4 for lower middle
x = 1 - x;                  % reverse for ...backw
  if x < .7, f = [1;0;1;0]; end
  if x >= .7 & x < .95,  f = [-2*x+2.3;-2;2*x-.3;2]; end
  if x >= 0.95, f = [.4;0;1.6;0]; end

function f = initialmt(x)   % for 8/4 for top middle
x = 1 - x;                  % reverse for ...backw
  if x < .75,  f = [-.2*x+1;-.2;1;0]; end
  if x > 0.75, f = [-x+1.4;-1;1.6*x-.2;1.6]; end

function f = initialb(x)
  f = [0;0;2;0];
```

The code `axialdisp4DaPerunbackw.m` consists of a comments block, followed by the general default and initialization block, as well as the BVP initialization and solver block. The actual solution is found inside the central `try ... catch ... end` lines of code. This is followed by two blocks of plotting code and the coded DE in `dydx` with its boundary conditions in `Rand`, both expressed as vector valued functions. Finally four initial guess functions for the shape of the solution are given that are used inside the BVP solver for different parameter data.

Note that `axialdisp4DaPerunbackw.m` uses backwards integration and plots the reversed function data as commanded in the lines `plot(xint,fliplr(Sxint(1,:)),...)` and `plot(xint,fliplr(Sxint(3,:)),...)` of the above code so that we view the graphs of $x_A(\omega)$ and $y(\omega)$ in their natural orientation in Figure 5.10. We could have used `axialdisp4DaPerunfwd.m` from the CD instead without plot reversal and we would have generated the same graphs, indicating that the above data does not lead to a stiff system of DEs.

The BVP is solved in `axialdisp4DaPerunbackw.m` by our modified boundary value code `bvp4cfsinghouseqr.m`. This code is modified from MATLAB's standard BVP solver `bvp4c`. MATLAB's `bvp4c` is a finite difference code that uses a collocation method built on cubic splines. In its MATLAB version, an initial guess such as our four `initial...` functions at the end of the above code serves as a starting guess. The task of `bvp4c` is to match the endpoint values of the computed piecewise graphs so that they agree from left and right up to within a small negligible tolerance. To do this, the previous free IVP parameter data is altered and improved internally by a Newton search method to achieve a better match at the subinterval ends, see Section 1.2.1. In `bvp4c` this is implemented via Gaussian elimination to find the corrective Newton term $u_{change} = Df^{-1}(u_{old})f(u_{old})$ in (1.5). Unfortunately in our chemical engineering problems, the Jacobian $Df(u_{old})$ occasionally turns out to be singular or nearly singular, so that Gaussian elimination breaks down. This is rather unfortunate, for then our main task cannot be accomplished. In the modified BVP code `bvp4cfsinghouseqr.m` we have repaired this restriction in `bvp4c` of working only for nonsingular Jacobians by solving the least squares problem

$$\min \| Df(u_{old}) \cdot u_{change} - f(u_{old}) \|$$

for u_{change} rather than solving the linear equation

$$Df(u_{old}) \cdot u_{change} = f(u_{old})$$

via Gaussian elimination. In `bvp4cfsinghouseqr.m` we solve all linear equations that involve the Jacobian matrix Df by forcing MATLAB to find the least squares solution u_{change} of $\min \| Df(u_{old}) \cdot u_{change} - f(u_{old}) \|$ by appending a zero row to the (otherwise square) Jacobian matrix Df and a zero entry to the right-hand-side vector $f(u_{old})$ and then using the built-in \ MATLAB command. Type $\gg$ `help` \ to verify our use of \ for this purpose. The remainder of `bvp4cfsinghouseqr.m` is identical to MATLAB's `bvp4c` code.

Our reasoning for this change is as follows: If $n = 1$ and we are looking for a Newton correction of a 1-dimensional function $f : \mathbb{R} \to \mathbb{R}$ but encounter $f'(u_{old}) = 0 \in \mathbb{R}$, then the tangent to f at u_{old} is horizontal, there is no indication of where to look for a better root approximation, and Newton iteration correctly ends without result. If $n > 1$, however, as is the case with our DEs (5.41) or (5.43) for $n = 4$, and if $Df(u_{old})$ is singular, then the linear system $Df(u_{old}) \cdot u_{change} = f(u_{old})$ cannot be solved uniquely. But it can be solved in the least squares sense. Taking the Newton correction vector u_{change} as the least squares solution in this case and setting $u_{i+1} = u_i + u_{change}$ generally works and improves the boundary value guesses on each subinterval to become better, more agreeable estimates. And this often allows our algorithm to finish with an accurate solution of the BVP on a limited number of subintervals, where the original MATLAB code `bvp4c` fails by creating ever finer partitions due to its reliance on Gaussian elimination and is not converging.

The remainder of this section uses our BVP solver code and our experience with BVP stiffness to try and plot families of solution profiles for fixed values of the Peclet numbers P_{eH} and P_{eM} and changing Damköhler numbers D_a. In fact we compute two double plots: the first set shows the solution profiles for the concentration x_A and the temperature y of the BVP in (5.38) to (5.40) or equivalently in (5.43) to (5.44) depending on ω in Figure 5.11, while the second set in Figure 5.12 shows the corresponding values of x_A and y at the end $\omega = 1$, i. e., the concentration rate and temperature at the end of the reactor, this time in terms of the Damköhler number. There is a range of D_a values with multiple steady states. This is shown in the second set of plots in Figure 5.12 which show the bifurcation phenomenon most clearly. Namely, in the middle range of Damköhler numbers around $10^{6.5}$ there are three possible concentration rates x_A and three possible temperatures y at the end of the reactor, corresponding to three different x_A and y profiles along the reactor for the same set of parameters.

This bifurcation analysis is very valuable for every chemical, biochemical, and biomedical process that involves axial dispersion in a tubular reactor.

Figure 5.11 plots our first set of profile for the Peclet numbers $P_{eH} = 4$ and $P_{eM} = 8$. it is obtained using `runaxialdispDa.m`.

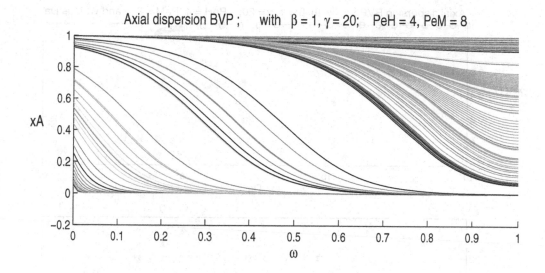

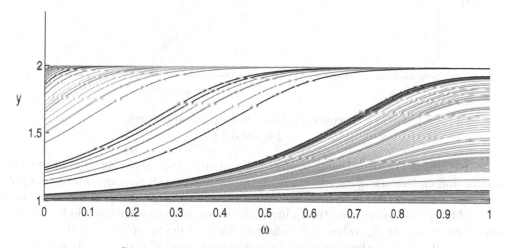

Concentration x_A and temperature y profiles
Figure 5.11

Note that the profiles in Figure 5.11 are drawn for Damköhler numbers between 10^5 and 10^8. The solution curve densities in Figure 5.11 and the colors[5] on the computer screen are chosen at random for high and low Damköhler numbers as explained earlier with Figure 5.10. The middle profiles, corresponding to the middle gray circles in Figure 5.12 and displayed in green on the computer screen, are printed in Figure 5.11 in uniformly thin dotted curves that occupy the center portion of the family of solutions.

---

[5]References to color always refer to a computer generated color graph of the figure in question.

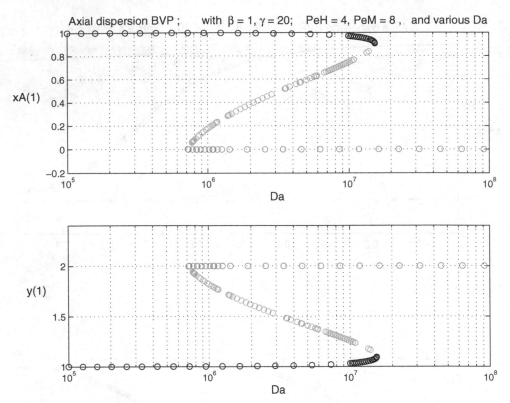

Exit concentrations and exit temperatures
Figure 5.12

In Figures 5.11 and 5.12 our readers will notice certain gaps (six in number) in the profiles and end-values plots, especially on the middle branch near the bifurcation points of Figure 5.12. This seems to be caused by our inability to compute reliable solutions for the middle steady state near the bifurcation points as well as in the middle branch at several intermediate D_a values. This effect is due to a *double stiffness* of the underlying system of DEs. This is the case if in a linearization of the DE system, there are two pairs of stiff eigenvalues for the linearized system matrix A: one pair λ_1, λ_2 is stiff for increasing values of the parameter ω in the sense of our conditions (a), (b), and (c) on p. 276 and this pair affects the forward integration from left to right. For example, an eigenvalue pair such as $\lambda_1 = 1$ and $\lambda_2 = -1/100$ may cause stiffness when integrating forward. And in case of double stiffness, besides such a pair there is another pair of eigenvalues λ_3, λ_4 for A with $\lambda_3 \cdot \lambda_4 < 0$ and the altered conditions (b') $\lambda_3 > 0$, and (c') $|\lambda_4| \gg |\lambda_3|$, such as $\lambda_3 = 1/100$ and $\lambda_4 = -2$. The latter pair of "backwards stiff" eigenvalues of A makes the backwards integration of the DEs stiff at this point of the data as well. This is the reason for our gaps in the graphs.

In a sense, the numerical results, i. e., the high level of instability in computing the middle solution to our BVP near the bifurcation points and elsewhere mimics the instability of the reactor. The middle steady state is generally highly unstable as a saddle-type unstable steady state. This despite the fact that it is highly desirable to operate many industrial processes there, as shown in Chapter 7 since operating on the middle, the unstable branch is generally associated with more manageable temperatures and higher yields. In a plant, the middle steady state needs to be stabilized through controls.

The program to generate these roughly 200 plots takes approximately 2 seconds of CPU time per plot on a 450 MHz SunBlade 100 with 1 Gig of RAM, or 400 seconds overall. To run this program involves much computing time. The multiple BVP solution plots of Sections 5.1 and 5.2 use more computing resources than any other program in this book. Here is the MATLAB program runaxialdispDa.m that plots Figures 5.10 through 5.14 and gives the user a running on-screen output of $x_A(1)$ and $y(1)$ for the D_a data of Figure 5.12. In order to save this screen output to a file, the user should learn how to use the diary function of MATLAB by entering >> help diary.

```
function [s0,s1] = runaxialdispDa(bt,ga,PeH,PeM,d)
%         runaxialdispDa(bt,ga,PeH,PeM,d)
%  Sample call :
%            d = clock; [s0,s1] = runaxialdispDa(1,20,4,8,1); etime(clock,d)
%  Output : two plots:
%               the first one shows the solution profiles for the input
%               variables, while the second one shows the corresponding end
%               values of xA and y in dependence of Da, the Damkoehler number.
%            screen output of xA(1), y(1), error of solution for each Da.
%               (if d = 1)
%            s0 : number of successful runs of BVP solver
%            s1 : number of unsuccessful runs of BVP solver
%               Note: the algorithm works with parameter continuation from a
%               known solution for a nearby Da value to the next Da value,
%               up or down as specified.
%               The algorithm will stop in its continuation from a
%               previous successful run when 3 BVP solver failures have
%               occurred in sequence. After the first and second such
%               failure, parameter continuation from the last previous
%               successful solution will be attempted with closer steps of Da.
%  Input: bt, ga   : such as bt = 1; ga = 20.
%         PeH, PeM : the Peclet numbers, such as 4 and 8 or 8 and 4.
%         d = 1    : display stepping data to the screen;
%         d = 0    : no screen display.
%
%  functions called internally : Damoveupbwd.m , Damoveupfwd.m
%                                 Damovedownbwd.m , Damovedownfwd.m
%  advantages for using "backwards" integration in case of DE stiffness;
%  else use ...fwd for best results.

if nargin == 4, d = 0; end    % for short list of inputs, no display
if d == 1, format short g, format compact, end
s1 = 0; s0 = 0;
```

```
if PeH == 4 & PeM == 8,          % one specifically fleshed out example

 if d == 1, disp('top xA, going up :'), end             % blue for top
   [s00,s11] = Damoveupbwd(5,7.2,.02,bt,ga,PeH,PeM,1,d,'ob');
   s0 = s0+s00; s1 = s1+s11;
 if d == 1, disp('top xA, going up in small steps:'), end    % blue for top
   [s00,s11] = Damoveupbwd(7,7.2,.002,bt,ga,PeH,PeM,1,d,'ob');
   s0 = s0+s00; s1 = s1+s11;
 if d == 1, disp('upper upper middle xA, going up (in 4 sections):'), end
   [s00,s11] = Damoveupfwd(6.6,7,.004,bt,ga,PeH,PeM,20.2,d,'og');  % green
   s0 = s0+s00; s1 = s1+s11;
   [s00,s11] = Damoveupfwd(6.7,7,.004,bt,ga,PeH,PeM,20,d,'og');
   s0 = s0+s00; s1 = s1+s11;
   [s00,s11] = Damoveupfwd(6.8,7.3,.002,bt,ga,PeH,PeM,20,d,'og');
   s0 = s0+s00; s1 = s1+s11;
   [s00,s11] = Damoveupbwd(7.1,7.3,.002,bt,ga,PeH,PeM,19.9,d,'og');
   s0 = s0+s00; s1 = s1+s11;
 if d == 1, disp('upper upper middle xA, going down :'), end
   [s00,s11] = Damovedownbwd(6.58,6.4,.004,bt,ga,PeH,PeM,20.2,d,'og');
   s0 = s0+s00; s1 = s1+s11;
 if d == 1, disp('upper middle xA, going up :'), end
   [s00,s11] = Damoveupfwd(6.2,6.9,.005,bt,ga,PeH,PeM,21,d,'og');
   s0 = s0+s00; s1 = s1+s11;
 if d == 1, disp('upper middle xA, going down :'), end
   [s00,s11] =  Damovedownbwd(6.2,6,.005,bt,ga,PeH,PeM,21,d,'og');
   s0 = s0+s00; s1 = s1+s11;
 if d == 1, disp('lower middle xA, going up :'), end
   [s00,s11] = Damoveupfwd(5.9,6.3,.005,bt,ga,PeH,PeM,22,d,'og');
   s0 = s0+s00; s1 = s1+s11;
 if d == 1, disp('lower middle xA, going down :'), end
   [s00,s11] =  Damovedownbwd(6,5.7,.005,bt,ga,PeH,PeM,22,d,'og');
   s0 = s0+s00; s1 = s1+s11;
 if d == 1, disp('lower middle xA, going down in small steps :'), end
   [s00,s11] =  Damovedownbwd(5.9,5.885,.0002,bt,ga,PeH,PeM,22,d,'og');
   s0 = s0+s00; s1 = s1+s11;
 if d == 1, disp('bottom xA, going up :'), end            % red for bottom
   [s00,s11] =  Damoveupbwd(5.8,7.96,.02,bt,ga,PeH,PeM,1000,d,'or');
   s0 = s0+s00; s1 = s1+s11;
 if d == 1, disp('bottom xA, going down :'), end
   [s00,s11] =  Damovedownbwd(6.1,5.6,.005,bt,ga,PeH,PeM,1000,d,'or');
   s0 = s0+s00; s1 = s1+s11;
 if d == 1, disp('bottom xA, going down in small steps :'), end
   [s00,s11] =  Damovedownbwd(5.86,5.6,.0002,bt,ga,PeH,PeM,1000,d,'or');
   s0 = s0+s00; s1 = s1+s11;
end

if PeH == 8 & PeM == 4,          % another fleshed out example
```

```
if d == 1, disp('top xA, going up :'), end
   [s00,s11] = Damoveupbwd(5,7.3,.05,bt,ga,PeH,PeM,1,d,'ob');
   s0 = s0+s00; s1 = s1+s11;
if d == 1, disp('top xA, going up in small steps:'), end
   [s00,s11] = Damoveupbwd(7.2,7.3,.005,bt,ga,PeH,PeM,1,d,'ob');
   s0 = s0+s00; s1 = s1+s11;
if d == 1, disp('middle top xA, going up :'), end
   [s00,s11] = Damoveupbwd(7,7.3,.005,bt,ga,PeH,PeM,0.7,d,'og');
   s0 = s0+s00; s1 = s1+s11;
if d == 1, disp('middle top xA, going down :'), end   % was 6.97
   [s00,s11] = Damovedownbwd(7.1,6.975,.002,bt,ga,PeH,PeM,0.7,d,'og');
   s0 = s0+s00; s1 = s1+s11;
if d == 1, disp('middle xA, going up :'), end   % was 6.5
   [s00,s11] = Damoveupbwd(6.7,7.3,.005,bt,ga,PeH,PeM,0.5,d,'og');
   s0 = s0+s00; s1 = s1+s11;
if d == 1, disp('middle xA, going up in small steps'), end
   [s00,s11] = Damoveupbwd(6.8,7,.002,bt,ga,PeH,PeM,0.5,d,'og');
   s0 = s0+s00; s1 = s1+s11;
if d == 1, disp('middle xA, going down :'), end   % was 6.6
   [s00,s11] = Damovedownbwd(6.7,6.48,.005,bt,ga,PeH,PeM,0.5,d,'og');
   s0 = s0+s00; s1 = s1+s11;
if d == 1, disp('middle xA, going down in small steps :'), end
   [s00,s11] = Damovedownbwd(6.5,6.48,.002,bt,ga,PeH,PeM,0.5,d,'og');
   s0 = s0+s00; s1 = s1+s11;
if d == 1, disp('bottom xA, going up :'), end
   [s00,s11] = Damoveupbwd(6.55,8.75,.05,bt,ga,PeH,PeM,0,d,'or');
   s0 = s0+s00; s1 = s1+s11;
if d == 1, disp('bottom xA, going up in small steps:'), end
   [s00,s11] = Damoveupbwd(6.5,6.56,.0005,bt,ga,PeH,PeM,0,d,'or');
   s0 = s0+s00; s1 = s1+s11;
if d == 1, disp('bottom xA, going down in small (and very small) steps :'), end
   [s00,s11] = Damovedownbwd(6.50,6.47,.0005,bt,ga,PeH,PeM,0,d,'or'); %was 6.52
   s0 = s0+s00; s1 = s1+s11;
end , format
```

Note that our program runaxialdispDa is specialized to run for just two sets of values for P_{eH} and P_{eM} as 4 and 8 or as 8 and 4, respectively. We have not attempted to make runaxialdispDa truly global by internal parameterizations for other Peclet data. It is obvious, however, how our readers can modify the program with a bit of trial and error to work for other sets of Peclet numbers P_{eH} and P_{eM}. The main task for this consists of figuring out - by trial and error - how to set up the various integration subintervals differently and to experiment with the direction of integration and the step size used for different Damköhler numbers.

runaxialdispDa uses four further programs Damoveupbwd, Damovedownbwd, Damoveupfwd, and Damovedownfwd, all available on the CD. The names of these programs indicate the direction of change (up or down) for D_a, as well as the direction of integration in their "fwd" or "bwd" endings. Here is the listing of Damovedownbwd, for example.

```
function ...
[s0,s1,P,Df,Dl,solf,soll] = Damovedownbwd(Ds,De,step,bt,ga,PeH,PeM,a,d,S,sols)
% Auxiliary function for    runaxialdispDa(1,20,4,8,1), which calls
%            Damovedownbwd(5,7.3,.02,bt,ga,PeH,PeM,1,1,'b'); for example
% Input : Ds = start of log10(Da) values;
%         De = end of log10(Da) values;
%         step = step size suggested;
%         bt, ga, PeH, PeM = given parameters;
%         a = starting guess [a = 1, 21, 1000, 22000, ... for example]
%             (not used if a starting solution sols is supplied)
%         d = display indicator. If d is set to 1, we screen-print log10(Da),
%                          function values at 1, error, timing data etc.
%         S = color and symbol entry for plots; sample S = 'oy', or S = '+y'.
%         sols = starting solution (if known and supplied, otherwise a sets
%                 up an initial guess solution automatically).
% Output : P = k by 3 matrix with Da, xA(1), y(1) at the computed nodes.
%          Df = first log10(Da) value for which a profile, called solf, was
%               successfully computed in the log10(Da) range.
%          Dl = last such log10(Da) value with the solution soll.
% m-files used: axialdisp4DaPecurvespots.m

if nargin <= 10, sp = 1; end
if nargin <= 9, d = 0; S = 'ob'; end
if nargin <= 7, disp('a AND solstart sols unspecified'), return, end
good = 1; turn = -1; first = 0; twice = 0;   % turn = -1 to integrate backwards
s0 = 0; s1 = 0;                              % counters of success with BVP

if De > Ds | step <= 0,
  disp('wrong order in Ds, De, or nonpositve step: NOT going down'),
  return, end

if d == 1,                       % when screen display is activated
disp('    log10(Da)        xA(1)         y(1)  error of DE     sp          time')
    end

step = min((Ds - De)/5,step); Step = 1 - step; D = Ds/Step;   % set up steps
tries = log(Ds/De)/log(Step); P = zeros(floor(tries)+10,3); k = 0;
while D >= De & good == 1,                   % loop up in steps as prescribed
  if twice == 1, Step = 1 - step/2; end, D = D*Step; Da = 10^D; tic;
  if sp == 1 | first == 0,
    [sol,sp,p,err]=axialdisp4DaPecurvespots(Da,bt,ga,PeH,PeM,a,S,turn);
  else,
    [sol,sp,p,err]=axialdisp4DaPecurvespots(Da,bt,ga,PeH,PeM,a,S,turn,sol); end
  s0 = s0+1-sp; s1 = s1+sp;
  if d == 1, t = toc; disp([log10(p(1)),p(2),p(3),err,sp,t]), end
  if sp == 0, k = k + 1; P(k,:) = p';
    if first == 0, Df = D; solf = sol; first = 1;
    else, Dl = D; soll = sol; end, end
```

```
if sp == 1 & first == 1  % look closer if unsuccessful after orig. success
  D = D/Step;
  if twice == 0, halfstep = 1 - step/2;
  else, halfstep = 1 - step/4; end, D = D*halfstep;  Da = 10^D; tic;
  [sol,sp,p,err]=axialdisp4DaPecurvespots(Da,bt,ga,PeH,PeM,a,S,turn,sol);
  s0 = s0+1-sp; s1 = s1+sp;
  if d == 1, t = toc; disp([log10(p(1)),p(2),p(3),err,sp,t]), end
  if sp == 0, k = k + 1; P(k,:) = p';
    if first == 0, Df = D; solf = sol; first = 1;
    else Dl = D; soll = sol; end, end, twice = 1;
  if sp == 1;                    % look closer once more
    D = D/halfstep;
    if twice == 0, quartstep = 1 - step/4;
    else, quartstep = 1 - step/8; end, D = D*quartstep; Da = 10^D; tic;
    [sol,sp,p,err]=axialdisp4DaPecurvespots(Da,bt,ga,PeH,PeM,a,S,turn,sol);
    s0 = s0+1-sp; s1 = s1+sp;
    if d == 1, t = toc; disp([log10(p(1)),p(2),p(3),err,sp,t]), end
    if sp == 0, k = k + 1; P(k,:) = p';
      if first == 0, Df = D; solf = sol; first = 1;
      else Dl = D; soll = sol; end, end
    if sp == 1, good = 0; end, end, end, end

i = find(P(:,1) ~=0); P = P(i,:);         % clean off zero rows in P
```

This program's central part computes solutions to the BVP for decreasing values of D_a (as specified) from a previous nearby solution via **parameter continuation**, i.e., it takes a nearby known solution as the initial guess for the next nearby problem. If this fails for one step, the program halves the step size. And it gives up after a lack of success documented by two subsequent failures in a row. On screen the user can monitor the following data: the value of D_a used, the computed values of $x_A(1)$ and $y(1)$ if successful for D_a, as well as the error committed by the solution for the BVP, a success indicator sp, and the CPU time taken for each individual BVP problem in case the parameter d in the call of Damovedownbwd is set equal to 1.

We close with the plots for $P_{eH} = 8$ and $P_{eM} = 4$ and D_a ranging from 10^5 to 10^8. These are similar in nature to the earlier Figures 5.11 and 5.12 for $P_{eH} = 4$ and $P_{eM} = 8$ in Figure 5.13.

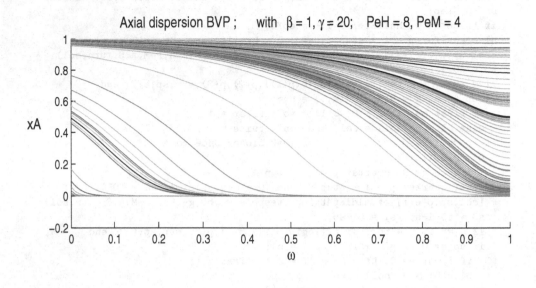

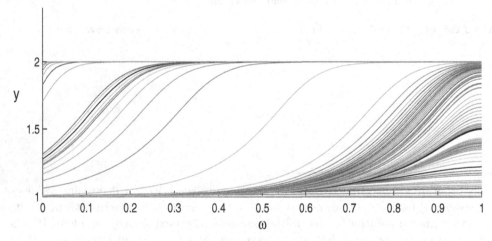

Concentration x_A and temperature y profiles
Figure 5.13

In Figure 5.14 note that for this set of Peclet numbers the middle uniformly dotted branch steady-state instabilities seem not to be as severe as before. There is only one gap in the family of solutions where double stiffness is an issue. To corroborate, note that the middle branch of the plot takes only 135 individual BVPs, each indicated by a pair of green circles on a computer screen and a pair of gray circles in Figure 5.14. This represents about two thirds of our earlier effort for $P_{eH} = 4$ and $P_{eM} = 8$ for obtaining these graphs. Each BVP now takes about 3 seconds on average.

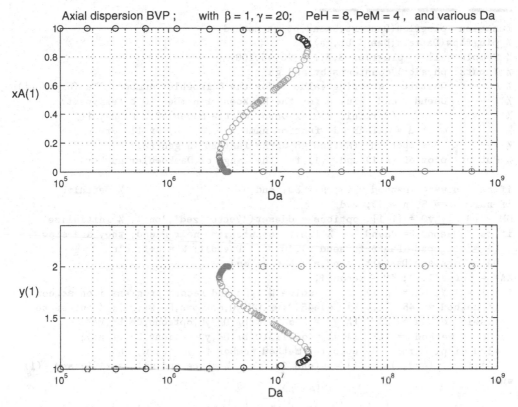

End concentrations and end temperatures
Figure 5.14

In general, smaller values of P_{eH} give wider ranges of bifurcation as the above two sets of Figures exemplify. And very large values for the Peclet numbers P_{eH} and P_{eM} transform the BVP problem in equations (5.19) and (5.20) into an IVP for plug flow. We illustrate this behavior for the adiabatic case when $\hat{Q} = 0$ in (5.19) and when $P_{eH} = P_{eM} = \infty$ in both (5.19) and (5.20). In this case the leading second derivative terms of (5.19) and (5.20) become zero and we are left with the two DEs

$$-\frac{dy}{d\omega} = -\beta \cdot D_a \cdot e^{-\gamma/y} \cdot x_A \tag{5.45}$$

and

$$-\frac{dx_A}{d\omega} = D_a \cdot e^{-\gamma/y} \cdot x_A . \tag{5.46}$$

The associated boundary conditions (5.40) for this case reduce to

$$x_A(0) - 1 = 0 \quad \text{and} \quad y(0) - 1 = 0 .$$

I.e., (5.45) and (5.46) describe an IVP which we can easily solve in MATLAB. For this we introduce the following MATLAB code that is built exactly as our IVP solvers of Chapter 4.

```
function plugflow(Dastart,Daend,bt,ga,method,n)
% plugflow(Dastart,Daend,bt,ga)
% Sample call :  plugflow(6,7.7,1,20,45,100)
% Inputs: physical constants bt and ga
%         Dastart : starting value a for the Damkoehler number 10^a (exponent)
%         Daend : end value b for the Damkoehler number 10^b (exponent)
%                  (Careful, use exponent of the power of ten only !!)
%         method = MATLAB ode funcion number;  23, 45, or 113 are
%             supported. (Method 45 works fastest in general)
% Output: plot of xA(1) and y(1) for the range of Damkoehler numbers

if nargin == 4, method = 45; n = 50; end,                % defaults
if nargin == 5, n = 50; end
OM = [0 1]; y0 = [1;1]; options = odeset('Vectorized','on'); % initialize
if ceil(Daend) ~= Daend      % limit computations near Daend; they are slow
  DA = [logspace(Dastart,Daend,n),10^ceil(Daend)]; N = n+1; else
  DA = logspace(Dastart,Daend,n); N = n; end
XA1 = zeros(1,n); Y1 = zeros(1,n);
for i = 1:N, Da = DA(i); i    % solve N IVPs and record progress i on screen
  if method == 45, [A,y] =  ode45(@fbr,OM,y0,options,Da,bt,ga); % integrate
  elseif method == 23,  [A,y] =  ode23(@fbr,OM,y0,options,Da,bt,ga);
  elseif method == 113, [A,y] = ode113(@fbr,OM,y0,options,Da,bt,ga);
  else disp('Error : Unsupported method number !'), return, end
  XA1(i) = y(end,1); Y1(i) = y(end,2);       % store end values xA(1) and y(1)
end

subplot(2,1,1);             % plot solution value at 1 to dxA/dom   versus Da
  semilogx(DA,XA1), v = axis; axis([v(1),v(2),-0.09,1.09]); % widen vert area
  title([{'Limiting plug flow end concentration and end temperature '},...
           {' '},{['for \beta = ',...
           num2str(bt,'%5.4g'),', \gamma = ', num2str(ga,'%5.4g'),...
           ' and varying values of D_a']}],'FontSize',13)
  xlabel('D_a','FontSize',12),
  ylabel('xA(1)       ','Rotation',0,'FontSize',12),

subplot(2,1,2); % plot solution value at 1 to dy/dom    versus Da
  semilogx(DA,Y1), v = axis; axis([v(1),v(2),0.92,2.08]);   % widen vert area
  xlabel('D_a','FontSize',12), ylabel('y(1)       ','Rotation',0,'FontSize',12),

function dydt = fbr(om,y,Da,bt,ga)          % right hand side of ODE IVP
dydt = [-Da * exp(-ga./y(2,:)) .* y(1,:);
        bt * Da * exp(-ga./y(2,:)) .* y(1,:)];
```

A call of `plugflow(6,7.7,1,20,15,100)` creates the following plot of values for $x_A(1)$ and $y(1)$ with the same set of parameters ($\beta = 1$, $\gamma = 20$) and Dahmköhler numbers around 10^7 as was used in Figures 5.12 and 5.14.

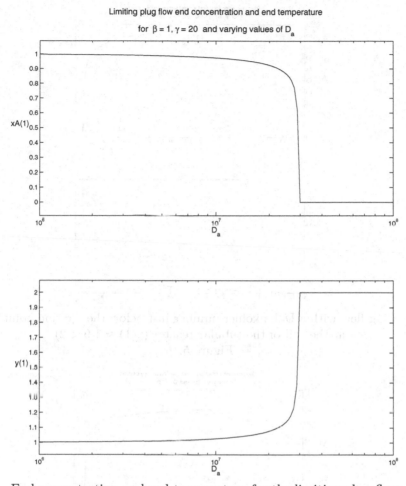

Limiting plug flow end concentration and end temperature

for $\beta = 1, \gamma = 20$ and varying values of D_a

End concentration and end temperature for the limiting plug flow
when P_{eH} and $P_{eM} \rightarrow \infty$
Figure 5.15

Note the very steep slope and the sharp turn of the curves when $x_A(1)$ reaches 0 and $y(1)$ reaches 2. This is called an *ignition point*, which is the critical value of D_a where the reactor is ignited to maximum temperature $y(1) = 2$ and zero reactants $x_A(1) = 0$ reach the exit of the reactor.

We illustrate this reaction behavior further using our concentration and temperature profile plotting code `plugflowxAy.m` for a Damköhler number such as $10^{7.456}$ just below ignition in Figure 5.16 and then for $D_a = 10^{7.463}$ which indicates ignition and $y(\omega) = 2$ near the end of the tube for $\omega < 1$ in Figure 5.17.

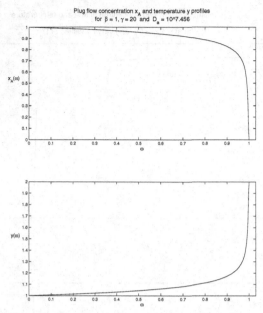

Plug flow with a Dahmköhler number just below the ignition point
at the end of the tubular reactor $(y(1) \approx 1.9 < 2)$
Figure 5.16

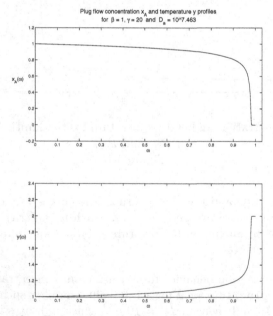

Plug flow with a Dahmköhler number just above the ignition point
at the end of the tubular reactor $(y(\omega) = 2$ for $\omega < 1)$
Figure 5.17

Exercises for 5.1

1. **Project**
   The equation (5.35) can be interpreted by a function $F_3(D_a, P_e, x_A(1)) = 0$ in the three variables D_a, P_e, and $x_A(1)$ since $x_A(1) = 1 - x(1)$.

   (a) For a fixed value $P_e = P$, adapt the surface plotting routines of Chapter 3 to plot the surface $z = F_3(D_a, P, x_A(1)) = f(D_a, x_A(1))$ and its level zero contour in order to obtain the curve $D_a(x_A(1))$ that was plotted via `fzero` in Figure 5.7.

   (b) Repeat Problem 1(a) for a constant value of $D_a = D$ and plot the surface $z = F_3(D, P_e, x_A(1)) = g(P_e, x_A(1))$ and its level zero contour in order to obtain the curve $P_e(x_A(1))$ that was plotted via `fzero` in Figure 5.8.

   (c) How do the curves obtained via `fzero` and via the level set method compare? Which method is simpler to implement in code? Explain, please.

2. Give details how to transform the second-order BVP (5.20) and (5.19) with the associated boundary conditions $dy(1)/d\omega = dx_A(1)/d\omega = 0$ and $(1/P_{eM}) \cdot (dy(0)/d\omega) = y(0) - 1$ into a system of four linear DEs and find its associated boundary conditions using forward integration in terms of ω.

3. Modify the stepping inside `runaxialdispDa.m` for $P_{eH} = 6$ and $P_{eM} = 12$ or $P_{eH} = 12$ and $P_{eM} = 6$ and an appropriate range of Damköhler numbers D_a that shows bifurcation, if possible.

4. Repeat the previous problem for $P_{eH} = P_{eM} = 12$ and $P_{eH} = P_{eM} = 4$. Do the profiles for these Peclet numbers look different from those of Problem 3 and from our two sets of profiles in this section? How would you describe their differences and similarities?

5. Replace our BVP solver `bvp4cfsinghouseqr.m` in any of our plotting routines of this section by MATLAB's original BVP solver `bvp4c` and observe the differences in output.

6. Solve the same nonisothermal axial dispersion problem when the reaction is second-order with $P_{eH} = 4$ and $P_{eM} = 8$.

7. Repeat the previous problem when the reaction is consecutive

$$A \rightarrow B \rightarrow C .$$

   Here B is the desired product and the preexponential factor for the second reaction is assumed to be half that of the first reaction, while the activation energies are the same and the heat of reaction of the second reaction $B \rightarrow C$ is twice that of the first reaction $A \rightarrow B$.
   When $P_{eM} = 4$, $P_{eH} = 8$, $\gamma = 10$, and $\beta = 0.8$, what is the value of D_{a_1} that maximizes the yield of the desired product B. (Notice that $D_{a_2} = 0.5 \cdot D_{a_1}$, $\gamma_1 = \gamma_2$, and $\beta_2 = 2\beta_1$.)

8. Develop the boundary conditions for the heat balance design equation of axial dispersion in the nonisothermal case as suggested on p. 262.

9. Convert the two second-order differential equations given by (5.20) and (5.19) to one coupled four dimensional first-order system of DEs as suggested on p. 279.

Conclusions

In this section we have presented the first example of two-point boundary value problems that occur in chemical/biological engineering. The axial dispersion model for tubular reactors is a generalization of the plug flow model for tubular reactors which removes some of the limiting assumptions of plug flow. Our model includes additional axial diffusion terms that are based on the simple physics laws of Fick for mass and of Fourier for heat dispersion.

The most simple isothermal case was studied first by finding the analytical solution to the BVP. Possible numerical difficulties of this case were analyzed via the eigenvalues of the corresponding constant coefficient linear system of ODEs in order to make our readers aware of the potential harm of stiffness when trying to solve DEs. The isothermal case with nonlinear kinetics was later solved numerically in a MATLAB program.

Building on the isothermal problem and its analytical solution, we then studied the nonisothermal adiabatic case and developed an efficient numerical algorithm based on a modified version of the built-in MATLAB BVP solver bvp4c.m. We have learnt that the nonisothermal adiabatic axial dispersion problem for tubular reactors may have multiple steady-state solutions over a certain range of parameters. The associated bifurcation diagrams were drawn for selected Peclet numbers via MATLAB. Our numerical adaptation of stock MATLAB functions is very efficient and manages to solve this type of coupled and highly nonlinear BVPs accurately in record time. Some of the MATLAB codes were printed and explained in this section. All programs are available on the CD. Our readers should be able to adapt the codes for use with other parameters and other BVP problems in chemical/biological engineering.

5.2   The Porous Catalyst Pellet BVP

In this section we are interested in the boundary value problem that is associated with catalyst pellets in a reactor.

5.2.1   Diffusion and Reaction in a Porous Structure (Porous Catalyst Pellet)

We consider a spherical catalyst particle and a simple reaction

$$A \to B$$

with the rate of reaction equal to

$$r = k \cdot C_A \ \ in \ \ moles/(g \cdot min) \,.$$

The considered particle is spherical in shape and it is symmetrical around its center. The concentration profile inside the catalyst pellet is shown in Figure 5.18, as well as the molar balance over an element Δr.

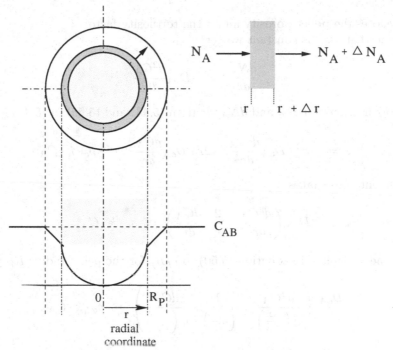

Elemental molar balance inside a catalyst pellet

Figure 5.18

The molar balance for component A gives us

$$4\pi \cdot r^2 \cdot N_A \;=\; 4\pi(r + \Delta r)^2(N_A + \Delta N_A) + (4\pi \cdot r^2 \cdot \Delta r)\rho_c \cdot k \cdot C_A \; .$$

This can be rewritten as

$$r^2 \cdot N_A \;=\; (r^2 + 2r\Delta r + (\Delta r)^2)(N_A + \Delta N_A) + r^2 \cdot \Delta r \rho_c \cdot k \cdot C_A \; .$$

By further simplification and by neglecting higher powers of Δr we obtain

$$0 \;=\; r^2 \Delta N_A + 2r \cdot N_A \cdot \Delta r + r^2 \cdot \Delta r \rho_c \cdot k \cdot C_A \; .$$

On dividing the equation by Δr and taking the limit of $\Delta N_A / \Delta r$ as $\Delta r \to 0$ we get

$$0 \;=\; r^2 \frac{dN_A}{dr} + 2r \cdot N_A + r^2 \cdot \rho_c \cdot k \cdot C_A \; . \tag{5.47}$$

By Fick's law we have

$$N_A \;=\; -D_{eA} \frac{dC_A}{dr} \; . \tag{5.48}$$

For diffusion in porous structures, the diffusion coefficient D_{eA} has the form

$$D_{eA} = \frac{D_A \cdot \epsilon}{\tau} \,,$$

where ϵ denotes the pellet porosity and τ the tortuosity factor.
If we assume that D_{eA} is constant we get

$$\frac{dN_A}{dr} = -D_{eA}\frac{d^2C_A}{dr^2} \,. \tag{5.49}$$

Using the expressions for N_A and dN_A/dr from (5.48) and (5.49) in (5.47) gives us

$$0 = -r^2 \cdot D_{eA}\frac{d^2C_A}{dr^2} - 2r \cdot D_{eA}\frac{dC_A}{dr} + r^2 \cdot \rho_c \cdot k \cdot C_A \cdot$$

Rearrangement then makes

$$D_{eA}\left(\frac{d^2C_A}{dr^2} + \frac{2}{r} \cdot \frac{dC_A}{dr}\right) = \rho_c \cdot k \cdot C_A \,. \tag{5.50}$$

Changing the variable r in equation (5.50) to r/R_p for the pellet radius R_P yields

$$\frac{D_{eA}}{R_P^2}\left(\frac{d^2C_A}{d\left(\frac{r^2}{R_P^2}\right)} + \frac{2}{\left(\frac{r}{R_P}\right)} \cdot \frac{dC_A}{d\left(\frac{r}{R_P}\right)}\right) = \rho_c \cdot k \cdot C_A \,. \tag{5.51}$$

If we introduce the dimensionless terms

$$\omega = \frac{r}{R_P} \quad \text{and} \quad x_A = \frac{C_A}{C_{Aref}} \,,$$

then equation (5.51) can be written as

$$\frac{D_{eA}}{R_P^2}\left(\frac{d^2x_A}{d\omega^2} + \frac{2}{\omega} \cdot \frac{dx_A}{d\omega}\right) = \rho_c \cdot k \cdot x_A \,. \tag{5.52}$$

This can be expressed as

$$\nabla^2 x_A = \Phi^2 x_A \,, \tag{5.53}$$

where

$$\nabla^2 = \frac{d}{d\omega^2} + \frac{a}{\omega} \cdot \frac{d}{d\omega} \quad \text{and} \tag{5.54}$$

$$\Phi^2 = \frac{\rho_c \cdot k \cdot R_P^2}{D_{eA}} \,.$$

Φ is called the Thiele[6] modulus.

[6]Ernest W. Thiele, US chemical engineer, 1895-1993

The value of a depends on the shape of the particle as follows. For a sphere we have $a = 2$, while $a = 1$ for a cylinder, and $a = 0$ for a slab.

Boundary Conditions :
Due to symmetry at the center we have $dx_A/d\omega = 0$ at $\omega = 0$.

At the surface where $r = R_P$, the mass transfer is shown in Figure 5.19.

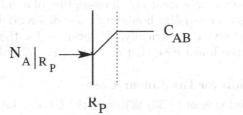

Mass transfer across the boundary at the surface of the catalyst pellet
Figure 5.19

In Figure 5.19 we observe that

$$4\pi R_P^2 \cdot k_{gA}(C_{AB} - C_A|_{r=R_P}) + 4\pi R_P^2 \cdot N_A|_{r=R_P} = 0 \,.$$

Simplification makes

$$k_{gA}(C_{AB} - C_A|_{r=R_P}) = -N_A|_{r=R_P} \,.$$

Since $N_A = -D_{eA} \cdot dC_A/dr$ by equation (5.48), we thus obtain

$$k_{gA}(C_{AB} - C_A|_{r=R_P}) = \left(\frac{D_{eA}}{R_P}\right) \frac{dC_A}{d(r/R_P)}|_{r=R_P} \,.$$

Here we define $x_{AB} = C_{AB}/C_{ref}$ to get

$$\frac{dx_A}{d\omega}|_{\omega=1.0} = \underbrace{\left(\frac{R_P \cdot k_{gA}}{D_{eA}}\right)}_{Sh_A}(x_{AB} - x_A|_{\omega=1.0}) \,,$$

where

$$\left(\frac{R_P \cdot k_{gA}}{D_{eA}}\right) = Sh_A$$

is the Sherwood[7] number for component A.
Thus the second boundary condition is $dx_A/d\omega = Sh_A(x_{AB} - x_A)$ at $\omega = 1.0$.

---

[7]Thomas Kilgore Sherwood, US chemical engineer, 1903-1976

The Limiting Case

When the external mass transfer resistance is negligible and k_{gA} is large, leading to $Sh_A \to \infty$, then the boundary condition at $\omega = 1.0$ can be written as

$$\frac{1}{Sh_A}\frac{dx_A}{d\omega} = (x_{AB} - x_A) .$$

Besides, if $Sh_A \to \infty$, then $x_A \mid_{\omega=1.0} = x_{AB}$. This also corresponds to a negligible external mass transfer resistance. In both cases, that of a finite Sherwood number Sh_A or for $Sh_A \to \infty$, we get a two-point boundary value differential equation. For the nonlinear case this has to be solved numerically. However, as for the axial dispersion model, we will start out with the linear case that can be solved analytically.

Analytical Solution for the Linear Case

When the right-hand side of (5.53) is linear, the DE can be solved analytically. The second-order differential equation for the spherical catalyst pellet is

$$\frac{d^2x_A}{d\omega^2} + \frac{2}{\omega}\frac{dx_A}{d\omega} = \Phi^2 x_A$$

for the Thiele modulus Φ. We rewrite this equation as

$$\omega\frac{d^2x_A}{d\omega^2} + 2\frac{dx_A}{d\omega} = \Phi^2 x_A \omega \tag{5.55}$$

and define the new variable

$$y = x_A \cdot \omega . \tag{5.56}$$

Differentiating equation (5.56) with respect to ω gives us

$$\frac{dy}{d\omega} = \omega\frac{dx_A}{d\omega} + x_A . \tag{5.57}$$

And the second derivative of (5.56) yields

$$\frac{d^2y}{d\omega^2} = \omega\frac{d^2x_A}{d\omega^2} + \frac{dx_A}{d\omega} + \frac{dx_A}{d\omega} = \omega\frac{d^2x_A}{d\omega^2} + 2\frac{dx_A}{d\omega} . \tag{5.58}$$

From (5.55) and (5.58) we thus get

$$\frac{d^2y}{d\omega^2} = \Phi^2 y . \tag{5.59}$$

This we can write as

$$D^2 y - \Phi^2 y = 0 .$$

The characteristic equation for the above equation is the second-degree polynomial

$$\lambda^2 - \Phi^2 = 0$$

with the roots $\lambda_1 = \Phi$ and $\lambda_2 = -\Phi$.
Therefore the solution of equation (5.59) is given analytically by

$$y(\omega) = C_1 e^{\Phi \omega} + C_2 e^{-\Phi \omega} \qquad (5.60)$$

for arbitrary constants C_1 and C_2. These two constants are determined by the boundary conditions and thus we have obtained the complete analytical solution in the linear case.

For the nonlinear case, the nonlinear two-point boundary value differential equation(s) for the catalyst pellet can be solved using the same method as used for the axial dispersion model in Section 5.1, i.e., by the orthogonal collocation technique of MATLAB's `bvp4c.m` boundary value solver.

The Non-Isothermal Catalyst Pellet

For the nonisothermal catalyst pellet with negligible external mass and heat transfer resistances, i.e., with $Sh \to \infty$ and $Nu \to \infty$ and for a first-order reaction, the dimensionless concentration and temperature are governed by the following couple of boundary value differential equations

$$\nabla^2 x_A = \bar{\alpha} \cdot e^{\gamma(1-1/y)} \cdot x_A \qquad (5.61)$$

and

$$\nabla^2 y = -\bar{\alpha} \cdot \beta \cdot e^{\gamma(1-1/y)} \cdot x_A , \qquad (5.62)$$

with ∇^2 as defined in formula (5.54) and with the boundary conditions

$$\frac{dx_A}{d\omega} = \frac{dy}{d\omega} = 0 \quad \text{at } \omega = 0 \qquad (5.63)$$

and

$$x_A(1) = 1 = y(1) \quad \text{at } \omega = 1 . \qquad (5.64)$$

Here we use

$$x_A = \frac{C_A}{C_{AB}}, \ y = \frac{T}{T_B}, \gamma = \frac{E}{R \cdot T_B}, \ \text{and} \ \beta = \frac{(-\Delta H) \cdot C_{AB}}{\lambda_e \cdot T_B}$$

for the effective thermal conductivity λ_e of the catalyst pellet.

5.2.2   Numerical Solution for the Catalytic Pellet BVP

Here we treat the more realistic case that the right-hand side of equation (5.53) is nonlinear. Such a nonlinearity is generally introduced by a nonconstant reaction rate that depends upon the temperature as exemplified by the Arrhenius dependence (2.1).
First we study a model made up of one second-order BVP, followed by a more advanced model that relies on two coupled second-order BVPs.

The Heat Balance Model Equations of the Catalytic Pellet BVP

From equations (5.61) and (5.62) and the boundary conditions for $Sh \to \infty$ and $Nu \to \infty$, i.e., by assuming (5.63) and (5.64), we obtain the relation $x_A = (1 + \beta - y)/y$. If we replace x_A in equation (5.62) by $x_A = (1 + \beta - y)/y$ then we obtain the following DE in terms of y :

$$\frac{d^2y}{d\omega^2} + \frac{2}{\omega}\frac{dy}{d\omega} = -\bar{\alpha} \cdot e^{(\gamma(1-1/y))} \cdot (1 + \beta - y) \tag{5.65}$$

with $\bar{\alpha} = \rho_c \cdot R_P^2/D_{eA} = \Phi^2$ for the Thiele modulus Φ. The boundary conditions for y are

$$\frac{dy}{d\omega} = 0 \quad \text{at} \quad \omega = 0, \quad \text{and} \quad y = 1 \quad \text{at} \quad \omega = 1 .$$

Note that the DE is singular at $\omega = 0$ due to the term $+(2/\omega) \cdot (dy/d\omega)$ on the left-hand side of (5.65). The term 'singular' refers to the indeterminate value $(2/0) \cdot 0$ in $(2/\omega) \cdot (dy/d\omega)$ when $\omega = 0$ according to the required boundary condition. MATLAB can handle singular boundary problems easily. To do so, the user has to set up the differential operator **dydt** in a specific way. First we must rewrite (5.65) as a first-order two-dimensional system of DEs.
By setting

$$Y(\omega) = \begin{pmatrix} y(\omega) \\ \dfrac{dy(\omega)}{d\omega} \end{pmatrix} = \begin{pmatrix} y_1 \\ y_2 \end{pmatrix}$$

we can rewrite (5.65) as a first-order system consisting of two DEs, namely

$$Y'(\omega) = \begin{pmatrix} \dfrac{dy_1}{d\omega} \\ \dfrac{dy_2}{d\omega} \end{pmatrix} = \begin{pmatrix} \dfrac{dy}{d\omega} \\ \dfrac{dy^2}{d^2\omega} \end{pmatrix} = \begin{pmatrix} y_2(\omega) \\ -\dfrac{2}{\omega}y_2(\omega) - \bar{\alpha} \cdot e^{(\gamma(1-1/y_1(\omega)))} \cdot (1 + \beta - y_1(\omega)) \end{pmatrix} .$$

$$\tag{5.66}$$

Equation (5.66) contains the singular term $-2 \cdot (y_2(\omega)/\omega)$. This is singular for $\omega = 0$ since it asks for the division of $y_2(0) = 0$ by zero there. To account for this in MATLAB and to avoid dividing by zero, we define the **SingularTerm** vector used in MATLAB's BVP solver as the coefficient matrix

$$S = \begin{pmatrix} 0 & 0 \\ 0 & -2 \end{pmatrix} \quad \text{for the singular part affecting} \quad Y = \begin{pmatrix} y_1 \\ y_2 \end{pmatrix} .$$

We can verify that this choice of S is correct by evaluating

$$S \cdot \left(\frac{Y}{\omega}\right) = \begin{pmatrix} 0 & 0 \\ 0 & -2 \end{pmatrix}\begin{pmatrix} y_1 \\ y_2 \end{pmatrix} \cdot \frac{1}{\omega} = \begin{pmatrix} 0 \\ -\dfrac{2}{\omega}y_2(\omega) \end{pmatrix}$$

which agrees with (5.66). A look at the **function dydx** differential operator block inside the MATLAB code for **pelletrunfwd.m** and at the line S=[0 0;0 -2]; options = bvpset('NMax', 1000,'SingularTerm',S,'Vectorized','on'); that precedes the

call of `bvp4cfsinghouseqr` in the code that follows shows how to prepare MATLAB for
this singular BVP.
Once prepared, this singular BVP can be solved as before via our modified version
`bvp4cfsinghouseqr.m` of MATLAB's `bvp4c.m` boundary value problem solver.

But instead of the solution curve $y(\omega)$ itself, we are interested here in finding the effectiveness factor η of the catalyst pellet. This is defined by the integral

$$
\eta \; = \; \frac{\int_0^1 \omega^2 e^{\gamma(1-1/y)} x_A \, d\omega}{\int_0^1 \omega^2 \, d\omega} \; = \; 3 \int_0^1 \omega^2 e^{\gamma(1-1/y)} x_A \, d\omega \; =
$$

$$
= \; 3 \int_0^1 \omega^2 \cdot e^{(\gamma(1-1/y(\omega)))} \cdot \frac{(1+\beta-y(\omega))}{\beta} \, d\omega \; . \tag{5.67}
$$

The system parameters are $\bar{\alpha}$, β, and γ. For a fixed set of $\beta = 1$ and $\gamma = 15$, for example,
our task is to find the range of $\bar{\alpha}$ with multiplicity. We shall plot the multiplicity regions
for η in terms of the Thiele modulus $\Phi = \sqrt{\bar{\alpha}}$.

Here is an auxiliary program for this task. It computes the solution curve $y(\omega)$ from
an initial guess of the solution to the BVP and evaluates η using MATLAB's definite
integral evaluator `quad` once the BVP has been solved successfully.

```
function [sol,eta,sp] = pelletrunfwd(phi,bt,ga,a,p,halten,sol)
%       [sol,eta,sp] = pelletrunfwd(phi,bt,ga,a,p,halten,sol)
% Sample call :  (without using a nearby solution "sol" for continuation)
%  pelletrunfwd(.5,.8,15,1,1,1); pause, pelletrunfwd(.5,.8,15,1.3,1,1); pause,
%       pelletrunfwd(.5,.8,15,1.8,1,1);
% For the one second order function BVP; porous catalyst pellet.
% Input : 0.001 <= phi <= 100; -1 <= bt <= 1.2 (bt ~= 0 needed);
%         4 <= ga <= 25;
%         a is an estimate of the solution, a is chosen between 1 and 1+bt;
%         If a = 1 or a = 1+ bt, we start with a horizontal line at a as the
%         initial guess of the solution;
%         If a differs from 1 and 1+bt, we use the join of a horizontal line,
%         followed by a (downward) slope and a horizontal line again. (See the
%         function "initial2" or "initial3" at the bottom.)
%         p = 1: draw a plot; p = 0 : no plot.
%         halten = 0: do not hold the graph; if halten = 1: hold on.
%         if sol is specified in the call: we have a nearby solution and
%         continue from it.
%         If sol is not given, we start with our initial guess.
% Output : solution structure sol; value for eta if sp = 0;
%          if sp = 1, the solution was spurious and therefore useless;
%                     data should be discarded.

if nargin <= 5, halten = 0; end,                % default settings
```

```
if nargin == 3, a = 1 + bt; p = 0; end, t = 0; sp = 0; eta = 0; ret = 0;
if nargin >= 4 & nargin ~= 7,
    if a == 1+bt | a == 1 | abs(bt*ga) < 10 | bt < .5,
      guess = a*[1;0]; toprint = guess; sol = bvpinit(linspace(0,1,5),guess);
    else if bt >= 15, toprint = a*[1;0];
                      sol = bvpinit(linspace(0,1,8),@initial2, [], a); end,
        if bt < 15, toprint = a*[1;0];
                      sol = bvpinit(linspace(0,1,8),@initial3, [], a, bt); end,
                end, end
if nargin >= 5, if p ~=0, p = 1; end, else, p = 0; end

S=[0 0;0 -2]; options = bvpset('NMax',1000,'SingularTerm',S,'Vectorized','on');

try                                            % solve BVP
    sol = bvp4cfsinghouseqr(@frhs,@frandbed,sol,options,phi,bt,ga);
catch                                          % if no solution found
    sp = 1; return
end

eta = 3/bt * quad(@bwsol,0,1,[],[],sol,bt,ga);    % find value of eta

d0f = deval(sol,0.5);                          % initialize annotations
if p == 1, xint = linspace(0,1); Sxint = deval(sol,xint);
    if halten == 0, hold off,
    else hold on,
        text(1,Sxint(1,1),['  [ ',num2str(a,'%5.3g'),',',...
                 num2str(eta,'%5.3g'),']']), end,
    plot(xint,Sxint(1,:)); hold on, end,

if p == 1,
 if halten == 0,
  title([{'Pellet BVP 1 DE model solution with'},...
      {['\phi = ',num2str(phi,'%9.4g'),'; \beta = ',num2str(bt,'%9.4g'),...
      ', \gamma = ',num2str(ga,'%9.4g'),'   (start a = ',...
      num2str(a,'%5.3g'),'; \eta = ',num2str(eta,'%5.3g'),' )']}],...
      'FontSize',12),
  else,
   title([{'Pellet BVP 1 DE model solution with'},{['\phi = ',...
          num2str(phi,'%9.4g'),'; \beta = ',num2str(bt,'%9.4g'),...
          ', \gamma = ',num2str(ga,'%9.4g'),' ;',...
'                                    for   [a, \eta] =']}],'FontSize',12), end
if nargin <= 5,
    xlabel(['\omega             ( for initial guess [ ',...
          num2str(toprint(1),'%9.4g'),'; ',num2str(toprint(2),'%9.4g'),...
          ' ] )'],'FontSize',12),
else, xlabel('\omega','FontSize',12), end,
ylabel('y   ','Rotation',0,'FontSize',12), end,
```

```
function dydx = frhs(x,y,phi,bt,ga)              % right hand side of the DE
dydx = [ y(2,:)
         -phi^2*exp(ga*(1-1./y(1,:))).*(1+bt-y(1,:)) ];

function Rand = frandbed(ya,yb,phi,bt,ga)        % boundary conditions
Rand = [ ya(2)
         yb(1)-1 ];

function f = bwsol(t,sol,bt,ga)                  % evaluation of the eta integrand
temp = deval(sol,t); temp = temp(1,:);
f = t.^2.*exp(ga*(1-1./temp)).*(1+bt-temp);

function f = initial2(x,a)                       % initial solution guesses
if x <= 1/8, f = [a;0]; end
if x > 1/8 & x < 1/2, f = [ (8*(1-a)*x+4*a-1)/3; 8/3*(1-a) ]; end
if x >= 1/2, f = [1;0]; end

function f = initial3(x,a,bt)
b = .1*a+.9*(1+bt);
if x <= 1/8, f = [b;0]; end
if x > 1/8 & x < 1/2, f = [ 8/3*(1-b)*(x-1/8)+b; 8/3*(1-b) ]; end
if x >= 1/2, f = [1;0]; end
```

The data and graphics output of calling `pelletrunfwd(.5,.8,15,1,1,1)`; pause, `pelletrunfwd(.5,.8,15,1.3,1,1)`; pause, `pelletrunfwd(.5,.8,15,1.8,1,1)`; is shown in Figure 5.20 for the parameters $\Phi = 0.5 (= \sqrt{\alpha})$, $\beta = 0.8$ and $\gamma = 15$. Figure 5.20 shows three separate solution curves $y(\omega)$ to the BVP that are obtained numerically by starting from different initial guesses $a = 1$, $a = 1.3$, and $a = 1.8$ for the solution. On the right margin of the graph we indicate the starting guess a, followed by the value of the Pellet efficiency η for this particular solutions. Note that for our parameter values there are multiple solutions for the BVP, i.e, this is a problem with static bifurcation.

Our program `pelletrunfwd.m` relies on our modified BVP solver `bvp4cfsinghouseqr.m` and the built-in MATLAB definite integral evaluator **quad** to evaluate η for each solution y of (5.65).

The CD contains a multi-input version, called `multipelletrunfwdxAy.m` of `pelletrunfwd.m` which draws the plots of the temperature $y(\omega)$ and the concentration $x_A(\omega)$ profiles for up to six values of the starting parameter $a \approx y(0)$.

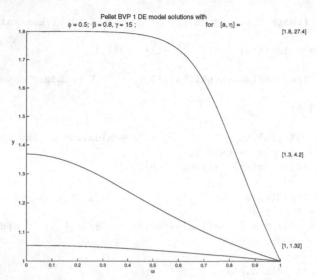

Multiple BVP solutions with varying pellet efficiencies

Figure 5.20

Figure 5.21 gives the graph of the solution profiles for y and x_A produced by calling `multipelletrunfwdxAy(.5,.8,15,[1,1.3,1.8])`; for the same data as was used in Figure 5.20.

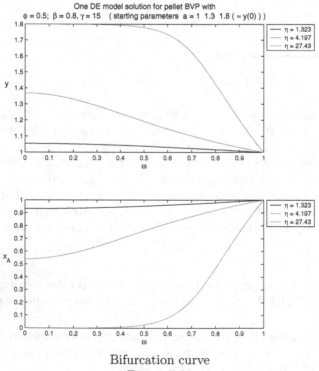

Bifurcation curve

Figure 5.21

The auxiliary function `pelletrunfwd.m` is used in `pelletetacurve.m` to evaluate η repeatedly for an interval of Φ values in order to draw the bifurcation curve of η with respect to $\Phi = \sqrt{\bar{\alpha}}$. Below is a plot of the inverted S shape bifurcation curve $\eta(\Phi)$ for $\beta = 1$ and $\gamma = 10$. This curve is plotted for $0.01 \leq \Phi \leq 3$ in about 72 seconds on a 2002 vintage SunBlade 100 by solving 104 BVPs by calling `pelletetacurve(.01,3,1,10,1,1);`.

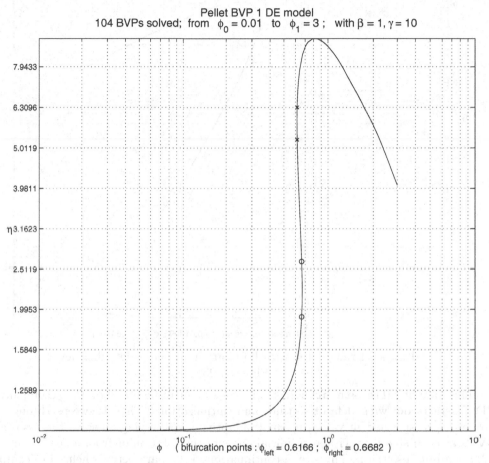

Very small range of Φ with bifurcation for the pellet efficiency

Figure 5.22

If we increase β and γ to 1.2 and 25, respectively, we find a much larger interval of Φ values with multiple pellet efficiencies. The curve shows a nascent 5-fold bifurcation kink in the lower part of the middle branch and it requires us to solve nearly 2.7 times as many BVPs in 285 seconds.

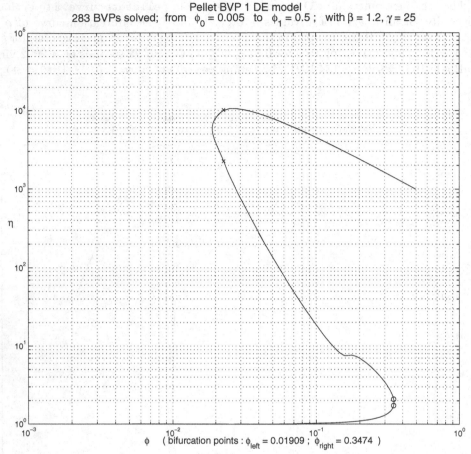

Extended range of Φ with bifurcation for the pellet efficiency
Figure 5.23

The code of `pelletetacurve.m` is quite long and can be found on the accompanying CD. It starts out with default settings and initializations. Then it sweeps through the range of pre specified Φ values in small adaptive steps, trying to solve the associated BVPs at each step from either an initial guess of the solution or from a nearby successfully computed solution. This use of continuation in the parameter is helpful to compute points of the $\eta(\Phi)$ curve near the bifurcation limits and specifically on the middle branch where computations are otherwise rather unstable. The computational difficulties that one generally encounters on the middle branch with numerical instabilities mimic the corresponding unstable saddle-type characteristics of the system in this region very well in their behavior.

Throughout the Φ range, the curve is drawn by interpolating the discrete computed values of η. This is done in the last part of `pelletetacurve.m` where we select and sort the useful computed data before plotting an interpolating curve. The two graphs of Figures 5.22 and 5.23 contain two small x and o marks, each near the bifurcation limits.

These indicate the limit of our successful numerical BVP integrations near the bifurcation points. In between the x and o marks on the middle branch, the curve is drawn using interpolation of our successful BVP solutions data, while in between two adjacent x or two adjacent o marks, the curve is drawn by extrapolating nearby computed function data. This is done automatically by MATLAB's plot commands.

Next we list the program of `pelletetacurvemulti.m`. It uses `pelletetacurve.m` to draw multiple $\eta(\Phi)$ curves for a fixed interval of Φ values and a vector of either β or γ values for one fixed value of γ or β, respectively.

```
function pelletetacurvemulti(phi0,phi1,vbt,vga,d)
%   pelletetacurvemulti(phi0,phi1,vbt,vga,d)
% Sample calls:
%   a) pelletetacurvemulti(0.05,10,[1 .8 .6 .4 .2 .1 -.1 -.3 -.6],15,1);
%   b) pellototacurvemulti(.01,100,2,2:2:12,1);
% For the one function BVP; porous catalyst pellet.
% Computes and plots multiple Pellet run curves for varying values of beta
% or gamma % between phi0 and phi1 in one figure.
% Only one of vbt or vga can be a vector of length exceeding 1, the other
% must be a scalar.
% Uses pelletetacurve.m program.
% d = 1 makes pelletetacurve display its location data on screen.

format short g, format compact
lvbt = length(vbt); lvga = length(vga); hold off,
if (lvbt == 1 & lvga == 1) | (lvbt > 1 & lvga >1), % check: improper vbt, vga
    disp(' ERROR : either vbeta or vgamma must have length 1 '), return, end
if nargin == 4, d = 0; end    % no running log, unless d = 1 has been set

if lvbt > 1,                   % running for several betas and one gamma
  for i = 1:lvbt, runningbeta = vbt(i),
    [PPP,bwp,Ptop,Pr,Pbot,Pl,Pmid] = pelletetacurve(phi0,phi1,vbt(i),vga,0,d);
    hold on,    [t,pt] = size(Ptop);
    text(phi1,Ptop(2,pt),[' \beta = ',num2str(vbt(i),'%5.3g'),' (',...
            num2str(bwp,'%5.3g'),')')],'Fontsize',12), end

title([{' Pellet BVP 1 DE model'},{[' for   ',...
        num2str(phi0,'%9.4g'),' \leq  \phi \leq ',num2str(phi1,'%9.4g'),...
        ' ;   with \beta = ',num2str(vbt,'%5.3g'),' ;  \gamma = ',...
        num2str(vga,'%5.3g')]}],'Fontsize',12),
xlabel('\phi','Fontsize',12),
ylabel('\eta   ','Rotation',0,'Fontsize',12), end

if lvga > 1,                   % running for several gammas  and one beta
  for j = 1:lvga, runninggamma = vga(j),
    [PPP,bwp,Ptop,Pr,Pbot,Pl,Pmid] = pelletetacurve(phi0,phi1,vbt,vga(j),0,d);
    hold on, [t,pt] = size(Ptop);
    text(phi1,Ptop(2,pt),[' \gamma = ',num2str(vga(j),'%5.3g'),' (',...
```

```
                num2str(bwp,'%5.3g'),')'],'Fontsize',12), end

title([{' Pellet BVP 1 DE model'},{[' for   ',num2str(phi0,'%9.4g'),...
     ' \leq  \phi \leq ',num2str(phi1,'%9.4g'),' ;   with \beta = ',...
  num2str(vbt,'%5.3g'),' ; \gamma = ',num2str(vga,'%5.3g')]}],'Fontsize',12),
xlabel('\phi','Fontsize',12), ylabel('\eta','Rotation',0,'Fontsize',12), end
```

The call of `pelletetacurvemulti(.01,100,2,2:2:12,1);` produces the following plot
with multiple curves $\eta(\Phi)$ for varying values of $\gamma = 2,\ 4, ..., 10,\ 12$ as indicated by the
vector [2:2:12] of γ values in the fourth position of the calling sequence. On the right
margin of Figure 5.24, the program automatically indicates the value of γ that was used
for the respective curve, appended by the number of BVPs (in parentheses) needed to
solve the problem adaptively for each individual curve.

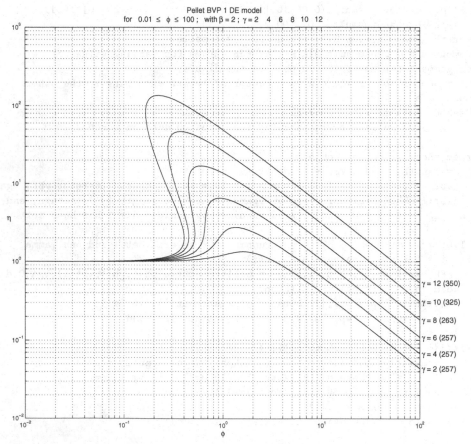

Bifurcation curves for various values of γ
Figure 5.24

To draw Figure 5.24 on a SunBlade 100 takes about 1,000 seconds for 1709 solved BVPs.
To verify this number 1709 simply add the numbers in parentheses at the right margin

of Figure 5.24. Note that for one fixed β, increasing the values of γ requires increasing numerical work, and that only the top three values of $\gamma = 8$, 10, and 12 give rise to bifurcation when $\beta = 2$.

Figure 5.25 gives another plot of 9 curves for $0.01 \leq \Phi \leq 10$, this time for one fixed $\gamma = 15$ and varying $\beta = 1$, 0.8, ..., -0.6 obtained by calling `pelletetacurvemulti(.01,10,`
`[-.6,-.3,-.1,.1,.2,.4,.6,.8,1],15,1);`.

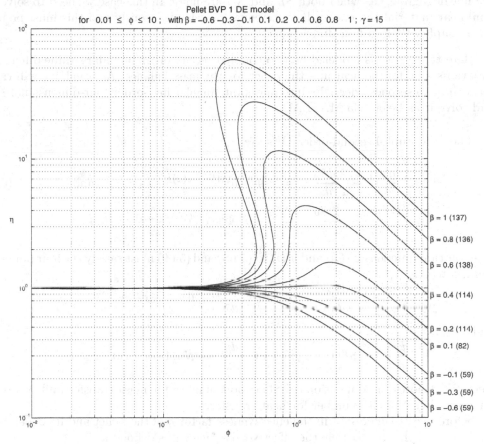

Bifurcation curves for various values of β
Figure 5.25

Here we note that increasing values of β for a fixed γ lead to increasing computational costs, and that only the top values of $\beta \geq 0.6$ lead to bifurcation when $\gamma = 15$. The above plot of nine $\eta(\Phi)$ curves was drawn from the data of 898 solved BVPs that took about 600 seconds to compute.

Overall, for the two Figures 5.24 and 5.25 we have solved 2607 BVPs in around 1600 seconds, or in about 27 minutes of CPU time. The average time for one BVP solution thus comes to about 0.6 seconds on the 450 Mhz, 1Gb RAM SunBlade 100 of 2002. The speed achieved by the reader's computer may vary when attempting to draw these or similar plots.

Non-Isothermal Mass and Heat Balance Model of the Catalytic Pellet BVP with Finite External Mass and Heat Transfer Resistance

In this section we consider the case where external mass and heat transfer resistances are not negligible, i.e., when both Sh and Nu are finite. In this case we need to solve the nonlinear material and energy balance equations simultaneously and this must be done for a coupled BVP in four dimensions.

Here the system is modeled by two coupled second-order boundary value differential equations, one for the heat and the other for the mass balance. As usual, we will transform the two second-order DEs into one joint singular first-order four-dimensional BVP and solve it as before via MATLAB.

The two coupled second-order DEs are

$$\frac{d^2 x_A}{d\omega^2} + \frac{2}{\omega}\frac{dx_A}{d\omega} = \Phi^2 \cdot e^{(\gamma(1-1/y(\omega)))} \cdot x_A(\omega) , \quad \text{and} \tag{5.68}$$

$$\frac{d^2 y}{d\omega^2} + \frac{2}{\omega}\frac{dy}{d\omega} = -\beta \cdot \Phi^2 \cdot e^{(\gamma(1-1/y(\omega)))} \cdot x_A(\omega) . \tag{5.69}$$

The solution functions $x_A(\omega)$ and $y(\omega)$ to (5.68) and (5.69) must satisfy the four boundary conditions

$$\frac{dx_A}{d\omega} = 0 \quad \text{and} \quad \frac{dy}{d\omega} = 0 \quad \text{at} \quad \omega = 0$$

and

$$\frac{dx_A}{d\omega} = Sh(1 - x_A) \quad \text{and} \quad \frac{dy}{d\omega} = Nu(1 - y) \quad \text{at} \quad \omega = 1 .$$

Here Sh denotes the dimensionless Sherwood number for mass transfer and Nu is the dimensionless Nusselt[8] number for heat transfer.

As before, we are interested in the effectiveness factor η of the pellet and its dependence on Φ. For a spherical particle the effectiveness factor η is defined as

$$\eta = 3 \int_0^1 \omega^2 \cdot e^{(\gamma(1-1/y(\omega)))} \cdot x_A(\omega) \, d\omega . \tag{5.70}$$

The useful range of Φ is from around 0.01 to 100, that of Sh from around 50 to 500, and Nu varies between 5 and 50.

As usual we first transform the coupled two second-order DEs in (5.68) and (5.69) into one first-degree system with 4 first-order differential equations.

---

[8]Wilhelm Nusselt, German engineer, 1882-1957

For

$$Y = \begin{pmatrix} x_A \\ \dfrac{dx_A}{d\omega} \\ y \\ \dfrac{dy}{d\omega} \end{pmatrix} = \begin{pmatrix} y_1 \\ y_2 \\ y_3 \\ y_4 \end{pmatrix}$$

the two DEs (5.68) and (5.69) combine to the 4 dimensional first-order system

$$Y'(\omega) = \begin{pmatrix} \dfrac{dy_1}{d\omega} \\ \dfrac{dy_2}{d\omega} \\ \dfrac{dy_3}{d\omega} \\ \dfrac{dy_4}{d\omega} \end{pmatrix} = \begin{pmatrix} y_2(\omega) \\ -\dfrac{2}{\omega}y_2(\omega) + \Phi^2 \cdot e^{(\gamma(1-1/y_3(\omega)))} \cdot y_1(\omega) \\ y_4(\omega) \\ -\dfrac{2}{\omega}y_4(\omega) - \beta \cdot \Phi^2 \cdot e^{(\gamma(1-1/y_3(\omega)))} \cdot y_1(\omega) \end{pmatrix} . \quad (5.71)$$

Equation (5.71) contains two singularities in the second and fourth components in the terms $-(2/\omega) \cdot y_2(\omega)$ and $-(2/\omega) \cdot y_4(\omega)$ since for $\omega = 0$ these indicate a division by zero. As explained earlier, for MATLAB we have to define the 4 by 4 **SingularTerm** matrix

$$S - \begin{pmatrix} 0 & 0 & 0 & 0 \\ 0 & -2 & 0 & 0 \\ 0 & 0 & 0 & 0 \\ 0 & 0 & 0 & -2 \end{pmatrix} \text{ with } S \cdot \left(\dfrac{Y}{\omega}\right) = \begin{pmatrix} 0 \\ -\dfrac{2}{\omega}y_2(\omega) \\ 0 \\ -\dfrac{2}{\omega}y_4(\omega) \end{pmatrix} \text{ as can be readily checked.}$$

We have implemented the 4 dimensional first-order singular BVP system (5.71) in the MATLAB code pellet4etarunfwd.m.

```
function [sol,eta,sp] = pellet4runfwd(phi,bt,ga,Sh,Nu,a,p,halten,sol)
%       [sol,eta,sp] = pellet4runfwd(phi,bt,ga,Sh,Nu,a,p,halten,sol)
% Sample call   :  (without using a nearby solution "sol" for continuation)
% pellet4runfwd(.5,.8,15,5000,1000,1,1); pause,
%    pellet4runfwd(.5,.8,15,5000,1000,1.4,1);
%       pause, pellet4runfwd(.5,.8,15,5000,1000,1.8,1);
%       (click "return" after each graph has appeared)
% For the one function BWP; porous catalyst pellet.
% Input : 0.001 <= phi <= 100; -1 <= bt <= 1.2 (bt ~= 0 needed); 4 <= ga <= 25;
%         a is an guess for the solution, a is chosen between 1 and 1+bt;
%         If a = 1 or a = 1+ bt, we start with a horizontal line at a as the
%         initial guess;
%         If a differs from 1 and 1+bt, we use the join of a horizontal line,
%         followed by a (downward) slope and a horizontal again. (See the
%         function "initial2" or "initial3" at the bottom.)
%         p = 1: draw a plot; p = 0 : no plot.
```

```
%           halten = 0: do not hold the graphs;
%           if halten = 1: hold on for both graphs.
%           if sol is specified in the call: we have a nearby solution and
%           continue from it. If sol is not specified, we start with our
%           initial guess.
% Output : solution structure sol; value for eta and this data if sp = 0;
%           if sp = 1, solution is spurious and useless; data is discarded,

warning off %MATLAB:divideByZero

if nargin < 5, 'Too few inputs, we stop!', return, end
if nargin == 5, a = 1 + bt; p = 0; halten = 0; end,
t = 0; sp = 0; eta = 0; ret = 0;
if nargin >= 5 & nargin < 9,
    if nargin == 6, p = 0; halten = 0; end,
    if nargin == 7, halten = 0; end,
    if a == 1, guess = [1;0;1;0]; sol = bvpinit(linspace(0,1,5),guess); end
 if a == 1+bt, sol = bvpinit(linspace(0,1,32),@initialtop,[],bt,ga,Sh,Nu); end
    if a > 1 & a < 1+bt,
        sol = bvpinit(linspace(0,1,8),@initialmiddle, [], bt, ga,Sh,Nu); end,
    if a > 1+bt & a <= 1+3*bt,
  sol = bvpinit(linspace(0,1,32),@initialmiddlemiddle, [], bt,ga,Sh,Nu); end,
    if a > 1+3*bt,
  sol = bvpinit(linspace(0,1,40),@initialtoptop, [], bt, ga,Sh,Nu); end, end,
if nargin >= 7, if p ~=0, p = 1; end, else, p = 0; end,

S=[0 0 0 0;0 -2 0 0;0 0 0 0;0 0 0 -2];
options = bvpset('NMax',1000,'SingularTerm',S,'Vectorized','on');

try
   sol = bvp4cfsinghouseqr(@frhs4,@frandbed4,sol,options,phi,bt,ga,Sh,Nu);
catch
   sp = 1; return
end

eta = 3*quad(@bwsol4,0,1,[],[],sol,bt,ga,Sh,Nu);

dOf = deval(sol,0.5);
if p == 1, xint = linspace(0,1); Sxint = deval(sol,xint);
  subplot(2,1,1),
  if halten == 0, hold off, else hold on, text(1,1,['   [a, \eta] = ']),  end,
  plot(xint,Sxint(1,:));
  subplot(2,1,2),
  if halten == 0, hold off, else hold on,
     text(1,Sxint(3,end),...
         ['   [ ',num2str(a,'%5.3g'),',',num2str(eta,'%5.3g'),']']), end,
  plot(xint,Sxint(3,:)); end,
```

```
if p == 1,
   if halten == 0,
     subplot(2,1,1),
     title([' Pellet 4 BVP ;    with \phi = ',num2str(phi,'%5.3g'),...
       '; \beta = ',num2str(bt,'%5.3g'),', \gamma = ',num2str(ga,'%5.3g'),...
       ';    Sh = ',num2str(Sh,'%5.3g'),', Nu = ',num2str(Nu,'%5.3g'),...
 '    (start a = ',num2str(a,'%5.3g'),'; \eta = ',num2str(eta,'%5.3g'),' )']),
   else
     subplot(2,1,1),
     title([' Pellet 4 BVP ;    with \phi = ',num2str(phi,'%5.3g'),...
       '; \beta = ',num2str(bt,'%5.3g'),', \gamma = ',num2str(ga,'%5.3g'),...
       ';    Sh = ',num2str(Sh,'%5.3g'),', Nu = ',num2str(Nu,'%5.3g')]), end
   xlabel('\omega'), ylabel('xA ','Rotation',0),
   subplot(2,1,2), xlabel('\omega'), ylabel('y ','Rotation',0), end
warning on

function dydx = frhs4(x,y,phi,bt,ga,Sh,Nu)
dydx = [ y(2,:);
         phi^2*exp(ga*(1-1./y(3,:))).*y(1,:);
         y(4,:);
         -bt*phi^2*exp(ga*(1-1./y(3,:))).*y(1,:) ];

function Rand = frandbed4(ya,yb,phi,bt,ga,Sh,Nu)
Rand = [ ya(2)
         ya(4)
         yb(2)-Sh*(1-yb(1))
         yb(4)-Nu*(1-yb(3)) ] ;

function f = bwsol4(t,sol,bt,ga,Sh,Nu)
temp = deval(sol,t); temp1 = temp(1,:); temp3 = temp(3,:);
f = t.^2.*exp(ga*(1-1./temp3)).*temp1;

function f = initialtop(x,bt,ga,Sh,Nu)   % various initial guesses follow below
if x <= .8, f = [0;0;1+3*bt;0]; end
if x > .8,  f = [ 5*x-4; 5; -5*bt*x+6.5; -5*bt]; end,
if [bt,ga,Sh,Nu] == [1,15,200,20] ,
   if x <= .3, f = [.05;0;1+1.05*bt;0]; end
   if x > .3 & x <= .5, f = [ x-.25; 1; -x+2.35; -1 ]; end
   if x > .5,   f = [ 1.6*x-.6; 1.6; -1.6*x+2.7; -1.6]; end
elseif [bt,ga,Sh,Nu] == [1,15,400,20],
   if x <= .7, f = [0;0;1+1.2*bt;0]; end
   if x > .7,
      f = [ 3.5*x-2.5; 3; 3.5*(1-x)+1.2; -3]; end,
elseif [bt,ga,Sh,Nu] == [1,15,100,18] ,
   if x <= .5, f = [0;0;1+1.1*bt;0]; end
   if x > .5,
    f = [ 2*x-1; 2; -2*x+3; -2]; end,
elseif [bt,ga,Sh,Nu] == [1,15,500,5],
```

```
if x <= 0.9, f = [0;0;1+7*bt;0]; end
if x > .9 & x <= .93, f = [.5*x-.4; .5; -.5*x+7; -.5]; end
if x > 0.93,
    f = [35*x-34; 34; -35*x+43; -34];  end, end

function f = initialmiddle(x,bt,ga,Sh,Nu)
if x <= .3, f = [0.05;0;1+1.1*bt;0]; end
if x > 0.3 & x <= 0.6,
    f = [ .7*x-.17; .7; -1*bt*x+2.4; -1*bt]; end,
if x > .6,
    f = [ 1.8*x-.8; 1.8; -1.5*bt*x+2.7; -1.5*bt]; end,
if [bt,ga,Sh,Nu] == [1,15,200,20] | [bt,ga,Sh,Nu] == [1,15,400,20],,
    f = [ (1-bt/2)*x+bt/2; 1-bt/2; -bt/2*x+1+bt/2; -bt/2 ];
elseif [bt,ga,Sh,Nu] == [1,15,100,10],
    if x <= .3, f = [0.05;0;1+1.1*bt;0]; end
    if x > 0.3 & x <= 0.6,
        f = [ .7*x-.17; .7; -1*x+2.4; -1]; end,
    if x > .6,
     f = [ 1.8*x-.8; 1.8; -1.5*bt*x+2.7; -1.5*bt]; end,
elseif [bt,ga,Sh,Nu] == [1,15,500,5]
if x <= .8, f = [0;0;1+6*bt;0]; end
if x > .8,  f = [ 20*x-19; 15; -20*x+24; -15]; end, end

function f = initialtoptop(x,bt,ga,Sh,Nu)
if x <= 0.99, f = [0;0;1+15*bt;0]; end
if x > 0.99,
    f = [150*x-149; 150; -150*x+155; -150];  end,
if [bt,ga,Sh,Nu] == [1,15,200,20],
    if x <= 0.97, f = [0;0;1+6*bt;0]; end
    if x > 0.97,
        f = [34*x-34.5; 30; -34*bt*x+40; -30]; end
elseif [bt,ga,Sh,Nu] == [1,15,400,20],
    if x <= 0.96, f = [0;0;1+5*bt;0]; end
    if x > .96 & x <=.98, f = [10*x-9.6; 10; -10*x+15.7; -10]; end
    if x > 0.98,  f = [30*x-29; 30; -30*x+35; -30];  end,
elseif [bt,ga,Sh,Nu] == [1,15,100,18]
    if x <= 0.99, f = [0;0;1+9*bt;0]; end
    if x > 0.99,  f = [70*x-73; 70; -70*x+1+75; -70]; end, end

function f = initialmiddlemiddle(x,bt,ga,Sh,Nu)
if x <= 0.92, f = [0;0;1+2.5*bt;0]; end
if x > 0.92, f = [  13*x-12; 12; -13*x+1+15.5; -12 ];     end,
if [bt,ga,Sh,Nu] == [1,15,200,20],
  if x <= 0.9, f = [0;0;1+5*bt;0]; end
  if x > 0.9,  f = [ 10*x-11; 10; -9*x+1+11; -9];  end
elseif [bt,ga,Sh,Nu] == [1,15,400,20],
    if x <= 0.92, f = [0;0;1+2.5*bt;0]; end
    if x > 0.92,  f = [13*x-12; 12; -13*x+1+15.5; -12];  end,
```

```
elseif [bt,ga,Sh,Nu] == [1,15,100,18],
    if x <= 0.93, f = [0;0;1+2.2*bt;0]; end
    if x > 0.93,  f = [12*x-11.2; 10; -12*x+14.5; -10]; end, end
```

Note specifically the differential operator function `function dydx` which is central to the program, as well as the boundary condition settings in `function Rand`, the singularity matrix S, and the various starting solution guesses at the end of the program.

The m file `pellet4runfwd.m` acts as an auxiliary program for `pellet4etacurve.m` which draws an interpolation curve for the $\eta(\Phi)$ data that is computed in `pellet4runfwd` for a multitude of input parameters. `pellet4etacurve.m` again uses continuation in the parameter and is adaptive. Here the term "parameter continuation" refers to our use of a nearby known solutions as the initial guess for solving the BVP with a new parameter Φ and our approach is "adaptive" in the sense that if a BVP computation fails to converge for a predetermined step size in Φ then we reduce the step size and try again. The program consists of over 450 lines of code and is available on the CD.

A call of `pellet4etacurve(0.1,0.7,1,15,100,18,1,1);` draws the following plot with fivefold bifurcation for the parameters as indicated in Figure 5.26.

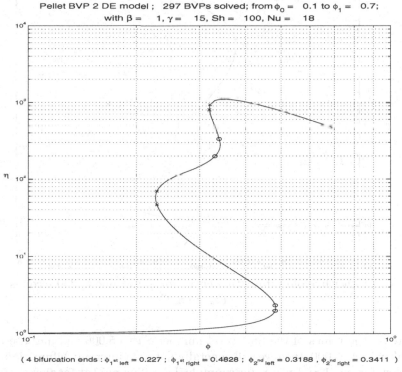

Pellet BVP 2 DE model ;   297 BVPs solved; from $\phi_0 =$ 0.1 to $\phi_1 =$ 0.7; with $\beta =$ 1, $\gamma =$ 15, Sh =   100, Nu =   18

(4 bifurcation ends : $\phi_{1^{st} \text{ left}} = 0.227$; $\phi_{1^{st} \text{ right}} = 0.4828$; $\phi_{2^{nd} \text{ left}} = 0.3188$, $\phi_{2^{nd} \text{ right}} = 0.3411$)

Fivefold bifurcating curve for η; runtime $\approx$ 290 seconds

Figure 5.26

The x and o marks in the plot are used here exactly as explained earlier in Section 5.2 on p. 310 and depict the closest computed data near the bifurcation ends.

Figure 5.27 shows another fivefold bifurcating $\eta(\Phi)$ curve for different Sherwood and Nusselt numbers.

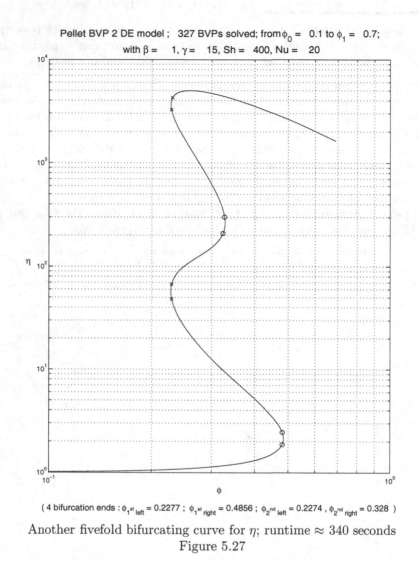

Pellet BVP 2 DE model ;   327 BVPs solved; from $\phi_0 =$ 0.1 to $\phi_1 =$ 0.7;
with $\beta =$ 1, $\gamma =$ 15, Sh =   400, Nu =   20

(4 bifurcation ends : $\phi_{1^{st}\text{left}}$ = 0.2277 ; $\phi_{1^{st}\text{right}}$ = 0.4856 ; $\phi_{2^{nd}\text{left}}$ = 0.2274 , $\phi_{2^{nd}\text{right}}$ = 0.328)

Another fivefold bifurcating curve for η; runtime $\approx$ 340 seconds
Figure 5.27

If we increase the values of the Sherwood number Sh to 5,000 and increase the Nusselt number likewise to 5,000, then we will generally not encounter any bifurcation. This is depicted below for $\beta = 1$ and $\gamma = 8$, for example, by calling `pellet4etacurve(0.01,10,1, 8,5000,5000,1,1,2)`; In this call, the very last parameter, called `tt` in `pellet4etacurve` is set to 2 in order to plot the nonbifurcating curve $\eta(\Phi)$ correctly.

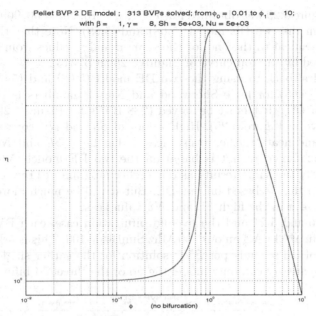

Pellet BVP 2 DE model ;   313 BVPs solved; from ϕ_0 = 0.01 to ϕ_1 =   10;
with β =   1, γ =   8, Sh = 5e+03, Nu = 5e+03

No bifurcation for η; runtime $\approx$ 110 seconds
Figure 5.28

Figure 5.29 draws the output of the call `pellet4etacurve(0.001,3,1,15,50000,50000,
1,1,0);` with `tt` set to zero for threefold bifurcation.

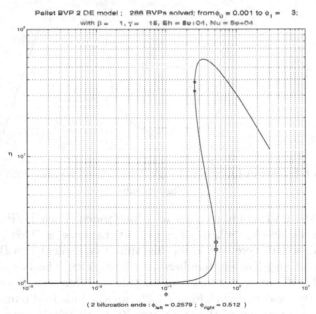

Pellet BVP 2 DE model ;   288 BVPs solved; from ϕ_0 = 0.001 to ϕ_1 =   3;
with β =   1, γ =   15, Sh = 5e+04, Nu = 5e+04

(2 bifurcation ends : ϕ_{left} = 0.2579 ; ϕ_{right} = 0.512)

Bifurcation for η; runtime $\approx$ 200 seconds
Figure 5.29

The curve in Figure 5.29 was obtained for $\beta = 1$, $\gamma = 15$, $Sh = 50,000$, and $Nu = 50,000$ numerically from the two DE model (5.68) and (5.69). Note that the graph in Figure 5.29 is nearly identical to the one for the same β and γ values from the one DE model (5.65) as depicted by the top curve of Figure 5.25 on p. 313.

One practical drawback to using the two DE model (5.68) and (5.69) rather than the one DE model (5.65) for huge Sherwood and Nusselt numbers is the extra time and extra number of DEs that must be solved (288 BVPs for Figure 5.29 versus 137 BVPs for the top curve in Figure 5.25). High values of Sh and Nu correspond to negligible external mass and heat transfer resistances. For large Sh and Nu values, the program `pellet4etacurve.m` that is based on the two DE model (5.68) and (5.69) will always give nearly the same results as our earlier program `pelletetacurve.m` does for the model (5.65) that is based on one DE. But it will be much more expensive to run `pellet4etacurve.m` in the high Sh and Nu values case.

For Figures 5.26 and 5.27 and the fivefold bifurcation case, each BVP takes around 1 second to solve in MATLAB on our 450 Mhz SunBlade 100. This nearly doubles the time of our average of 0.6 seconds per BVP solution for the earlier single second-order DE model which apparently can only model the no or the threefold bifurcation cases.

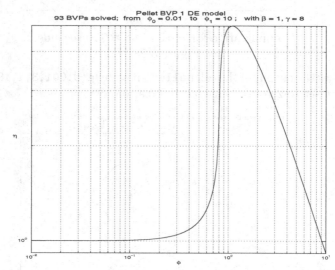

No bifurcation for η; single BVP model; 45 seconds run time

Figure 5.30

For Figure 5.30, 93 BVP computations were performed. Each BVP problem took less than 0.5 seconds on average when using `pelletetacurve.m`. This is much less effort than the 313 BVPs that were solved for the almost equivalent two BVP problem with two high Sherwood and Nusselt numbers in Figure 5.28. The two resulting curves of Figures 5.28 and 5.30 are nearly identical for this nonbifurcating $\eta - \Phi$ case. The single second-order BVPs associated with Figure 5.30 can also be solved numerically faster with fewer nodes for the partition of the ω interval [0, 1], thereby requiring fewer parametric IVP solutions and smaller sized nonlinear equations that match the solutions at the nodes of the partition for ω.

We have mentioned these runtimes comparisons only to indicate the complexity of our computations and the level and amount of work that BVPs generally require. Industrial problems with pellet effectiveness multiplicities will also be studied in Chapter 7.

Exercises for 5.2

1. Derive the porous pellet diffusion reaction equations for a consecutive reaction network
$$A \longrightarrow B \longrightarrow C \ .$$

2. Calculate the isothermal effectiveness factor η for the porous catalyst pellet in problem 1 as a function of the Thiele modulus Φ for the first reaction $A \longrightarrow B$ utilizing the fact that the rate constant of the second reaction $B \longrightarrow C$ is half the rate constant of $A \longrightarrow B$, the pellet is isothermal, and the external mass transfer resistance is negligible.

   Also compute the yield and selectivity η of the desired product B and the range of Φ' values from 0.01 to 10, where Φ' is the isothermal Thiele modulus.

3. For problem 2 above compute the effect of $10 \leq Sh \leq 1,000$ on the $\eta - \Phi$ diagram and on the yield of B versus Φ diagram.

4. For problem 2 assume that the particle is nonisothermal, that the nonisothermal Thiele moduli Φ_i of both reactions are equal, that the activation energy of the second reaction is double that of the first reaction, and that its heat of reaction is also double the heat of reaction of the first reaction.

   Compute the $\eta - \Phi$ diagram for negligible external mass and heat transfer resistances. Also compute the yield y_B of B versus Φ_1 diagram.

5. For problem 4 compute the effect of Sh and Nu on the $\eta-\Phi_1$ and the $y_B-\Phi_1$ diagrams.

Express your results for each of the above problems in graphs, please.

Conclusions

In this section we have presented and solved the BVPs associated with the diffusion and reaction that take place in the pores of a porous catalyst pellet. The results were expressed graphically in terms of the effectiveness factor η versus the Thiele modulus Φ for two cases: One with negligible external mass and heat transfer resistances, i.e., when Sh and $Nu \to \infty$, and another with finite Sh and Nu values. This problem is very important in the design of fixed-bed catalytic reactors. The sample results presented here have shown that for exothermal reactions multiple steady states may occur over a range of Thiele moduli Φ. Efficient numerical techniques have been presented as MATLAB programs that solve singular two-point boundary value problems.

Exercises for Chapter 5

1. Develop the model equations and MATLAB code for solving the problem of a packed bed reactor which is packed with porous catalyst pellets that catalyze a first-order exothermic reaction with $\Phi^2 = 1.8$, $\gamma = 1.1$, and $\beta = 1.1$. Use a dimensionless feed concentration and reactor length, as well as $Sh = 250.0$ and $Nu = 10.0$.

2. Construct the diagrams of the effectiveness factor and the desired yield versus the Thiele modulus Φ_A of the reactant A for a first-order consecutive exothermic catalytic reaction

$$A \longrightarrow B \longrightarrow C$$

in a porous catalyst pellet for $\Phi_B = 0.2 \cdot \Phi_A$, $\gamma_B = 1.5 \cdot \gamma_A$, $\gamma_A = 15$, β_A(exothermic factor for $A \to B$) $= 1.1$, and β_B(exothermic factor for $B \to C$) $= 1.8$.

3. Construct the diagrams of the effectiveness factor and the desired yield versus the Thiele modulus Φ_A of the reactant A for the first-order parallel exothermic catalytic reactions

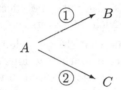

Here B is the desired product, while C is undesired. Use $\Phi_2 = 0.3 \cdot \Phi_1$, $\gamma_2 = 2.0 \cdot \gamma_1$, $\gamma_1 = 12.0$, $\beta_1 = 1.0$, and $\beta_2 = 2.0$.

Flower tower

Chapter 6

Heterogeneous and Multistage Systems

Many chemical and biological systems include multistage processes rather than only continuous contact ones. The most common multistage systems are absorption and distillation columns. Most of these systems involve more than one phase and they therefore fall under the category of heterogeneous multistage systems. Multistage systems can be cocurrent or countercurrent.

6.1 Heterogeneous Systems

Most real world chemical and biological systems are heterogeneous. What does *heterogeneous* mean? In our context it means that the system contains matter in more than one phase and that there is a strong interaction between the various phases, gaseous, liquid, or solid.

A **distillation** process involves a heterogeneous system formed of at least two phases, gas and liquid. Distillation is impossible unless at least two phases are present. Packed bed distillation columns are formed of three phases that also take the solid packing phase into consideration. Most distillation processes are not only heterogeneous, but also multistage.

Absorption is another gas-liquid process with two phases. Packed bed absorbers have three phases when also taking the solid packing of the column into consideration. Both continuous, i.e., packed bed, and multistage absorption are common in the chemical and biological industry.

Adsorption involves a solid-gas (or solid-liquid) two-phase system.

Fermentation gives rise to at least a two-phase system including the fermentation liquid mixture and the solid microorganisms that catalyze the fermentation process. Fermentation can also have three phases such as for aerobic fermentation where gaseous oxygen is bubbled through the fermentor. And immobilized packed-bed aerobic fermentors are formed of four phases, two of them are solid, plus one liquid and one gaseous phase.

Gas solid catalytic systems are two-phase systems involving a solid catalyst with reactants and products that occur in the gas phase. Most gas-solid catalytic systems are continuous rather than multistage systems.

In the previous chapters of this book we have dealt only with one phase systems. There the emphasis was mainly on reacting systems. Single-phase nonreacting systems such as mixers and splitters are almost trivial and shall not be dealt with at all. However, in the present section we deal with heterogeneous systems and here nonreacting systems are generally as nontrivial as reacting systems.

6.1.1   Material Balance and Design Equations for Heterogeneous Systems

Generalized Mass-Balance and Design Equations for a Single-Phase Reactor

We first recall the generalized mass-balance equations for one-phase systems with M components.

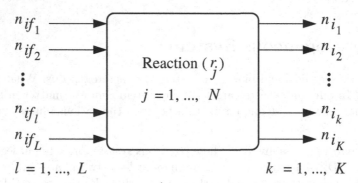

Material balance for homogeneous systems and component i
Figure 6.1

The generalized mass-balance equations are given by

$$\sum_{k=1}^{K} n_{ik} = \sum_{l=1}^{L} n_{if_l} + \sum_{j=1}^{N} \sigma_{ij} \cdot r_j \tag{6.1}$$

for $i = 1, ..., M$, where $r_j = R_{ij}/\sigma_{ij}$ is the overall generalized rate of reaction for reaction j, R_{ij} is the rate of production for component i in reaction j, σ_{ij} is the stoichiometric number of component i in reaction j, n_{if_l} is the input (feed) molar flow rate of component

i, and n_{ik} is the output (product) molar flow rate of component i.

Equation (6.1) is the most general mass-balance equation for single phase systems. It applies to all possible mass-balance cases.

If there are no reactions then equation (6.1) reduces to the form

$$\sum_{k=1}^{K} n_{ik} = \sum_{l=1}^{L} n_{if_l} \ \ \text{for each} \ \ i = 1, ..., M .$$ (6.2)

If there are N reactions but only a single input and a single output, then equation (6.1) reduces to

$$n_i = n_{if} + \sum_{j=1}^{N} \sigma_{ij} \cdot r_j \ \ \text{for} \ \ i = 1, ..., M$$ (6.3)

and any number M of components i and any number of reactions N.

If we have a single reaction and a single input single output system, then equation (6.3) becomes

$$n_i = n_{if} + \sigma_i r .$$

For no reactions, i.e., a nonreacting system and single in- and output, the equations reduce to the trivial form $n_i = n_{if}$ for each component i.

For each component i the material-balance equation (6.1) is correct regardless of whether the system is lumped or distributed. However, when turning the material-balance equations (6.1) into design equations given in (6.4) below, the situation differs for lumped and distributed systems. The design equations for a lumped system such as depicted in Figure 6.1 is given by

$$\sum_{k=1}^{K} n_{ik} = \sum_{l=1}^{L} n_{if_l} + \sum_{j=1}^{N} \sigma_{ij} \cdot r'_j \cdot V$$ (6.4)

for each component i, where r'_j is the generalized rate of reaction of the jth reaction per unit volume of the reactor. As explained before, only the relations $r_j = r'_j \cdot V$ are needed to turn the mass-balance equations into design equations for lumped systems.

We start with the simple single-input, single-output, single-reaction mass-balance equation as represented in Figure 6.2. This is a one-phase system.

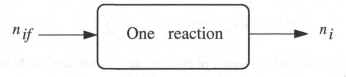

$$n_i = n_{if} + \sigma_i \, r$$

Mass-balance and design equations for a single-input, single-output, one-reaction system
Figure 6.2

We use the very simple case of a first-order irreversible liquid-phase reaction $A \rightarrow B$ where the rate of reaction is given by $r' = k \cdot C_A$ in $mol/(l \cdot sec)$, k is the reaction rate constant in sec^{-1} and C_A is the concentration of component A in mol/l. Later we will show how the same principles can be applied to distributed system and also for other rates like rate of mass transfer for heterogeneous systems with multiple phases.

The design equation for this single input single output system for the single reaction case becomes $n_i = n_{if} + \sigma \cdot r' \cdot V_R$. Applied to component A we get

$$n_A = n_{Af} - k \cdot C_A \cdot V_R . \tag{6.5}$$

This is a simple problem since C_A and n_A occur in the same equation. For liquid-phase systems when there is no change in volume, this problem is trivial. For gas-phase systems with a change in the number of moles accompanying the reaction, it is simple enough, but not as trivial as will be shown later.

For liquid-phase system we have $n_A = q \cdot C_A$ and $n_{Af} = q \cdot C_{Af}$. Hence in this case equation (6.5) becomes

$$q \cdot C_A = q \cdot C_{Af} - k \cdot C_A \cdot V_R . \tag{6.6}$$

Thus for a specific volume associated with the flow rate q, and for the rate of reaction constant k we can obtain the volume of the reactor that is needed to achieve a certain conversion. Rearranging equation (6.6) gives us

$$V_R = \frac{q \cdot (C_{Af} - C_A)}{k \cdot C_A} .$$

Thus if we want 90% conversion for example, then $C_A = 0.1 \cdot C_{Af}$ and V_R can be obtained as $V_R = 9 \cdot q/k$.

This is how simply a mass-balance equation can be turned into a design equation and become a mathematical model for a lumped isothermal system.

The same very simple principles apply to the heat-balance equations for nonisothermal system and also to distributed systems as will be shown in the following subsections. The same principles also apply to heterogeneous systems.

Next we consider distributed systems.

To illustrate we consider a homogeneous tubular reactor. The simplest model is given by plug flow and the design equations are obtained from the mass-balance equations by taking the mass balance over an element of length Δl. This is expressed in the formula

$$n_i(l + \Delta l) = n_i(l) + A_t \cdot \Delta l \cdot \sum_{j=1}^{N} \sigma_{ij} \cdot r'_j ,$$

where A_t is the cross-sectional area of the tubular reactor and Δl is the length increment. This difference equation can be written as

$$(\Delta n_i / \Delta l) = A_t \cdot \sum_{j=1}^{N} \sigma_{ij} \cdot r'_j .$$

Taking the limit of $\Delta l \rightarrow 0$, the difference equation changes into the differential equation

$$\frac{dn_i}{dl} = A_t \cdot \sum_{j=1}^{N} \sigma_{ij} \cdot r'_j .$$

More specifically we consider the consecutive reaction network

$$3A \xrightarrow{k_1} 2B \xrightarrow{k_1} 5C .$$

Rearrangement and taking the limit of $\Delta l \rightarrow 0$ gives us the following differential equation for the change of the molar flow rate of component A along the length of the tubular reactor as

$$\frac{dn_A}{dl} = -3A_t \cdot k_1 \cdot C_A . \tag{6.7}$$

This equation can be written in terms of the volume increment $dV = A_t \cdot \Delta l$ as

$$\frac{dn_A}{dV} = -3k_1 \cdot C_A .$$

With the initial conditions $n_A = n_{Af}$ at $l = 0$ or $C_A = C_{Af}$ when $V = 0$, the-balance equations for component B are

$$\frac{dn_B}{dl} = A_t \cdot (2k_1 \cdot C_A - 2k_2 \cdot C_B) \tag{6.8}$$

or

$$\frac{dn_B}{dV} = 2k_1 \cdot C_A - 2k_2 \cdot C_B ,$$

respectively. For component C we obtain the analogous equation

$$\frac{dn_C}{dl} = 5A_t \cdot k_2 \cdot C_B . \tag{6.9}$$

From the equations (6.7) to (6.9) we deduce that

$$\frac{n_A}{3} + \frac{n_B}{2} + \frac{n_C}{5} = \frac{n_{Af}}{3} + \frac{N_{Bf}}{2} + \frac{n_{Cf}}{5} . \tag{6.10}$$

Therefore only the two differential equations (6.7) and (6.8) need to be solved for n_A and n_b as n_C can be computed from equation (6.10) once n_A and n_B are known. However, these two equations contain expression for the components A and B in two different forms; they use n_A and n_B as well as C_A and C_B. To simplify this situation we can write the equations in terms of n_A and n_B alone, or in terms of C_A and C_B alone, or in terms of x_A, the conversion of A, and Y_B, the yield of B.
In terms of n_A and n_B we have

$$C_A = n_A/q \quad \text{and} \quad C_B = n_B/q .$$

For the gas phase we use $P \cdot q = n_T \cdot R \cdot T$ and $q = n_T \cdot R \cdot T / P$. Therefore $q = \alpha \cdot n_T$ with $n_T = n_A + n_B + n_C$. Equation (6.10) gives us

$$n_C = 5 \left(\sum_i \frac{n_{if}}{|\sigma_i|} - \frac{n_A}{3} - \frac{n_B}{2} \right) .$$

Thus

$$n_T = n_A + n_B + 5 \left(\sum_i \frac{n_{if}}{|\sigma_i|} - \frac{n_A}{3} - \frac{n_B}{2} \right) ,$$

or written out

$$n_T = 5\frac{n_{Af}}{3} + 5\frac{n_{Bf}}{3} + n_{Cf} - \frac{2}{3}n_A - \frac{3}{2}n_B = n_{Cf} + \frac{1}{3}(5n_{Af} - 2n_A) + \frac{1}{2}(5n_{Bf} - 3n_B) .$$

$$(6.11)$$

Equation (6.11) expresses n_T as a function of n_A and n_B, i.e.,

$$n_T = f_1(n_A, n_B) .$$

Thus $q = \alpha \cdot n_T = \alpha \cdot f_1(n_A, n_B)$ and

$$C_A = \frac{n_A}{\alpha f_1(n_A, n_B)} = g_A(n_A, n_B) , \quad \text{while} \qquad (6.12)$$

$$C_B = \frac{n_B}{\alpha f_1(n_A, n_B)} = g_B(n_A, n_B) \qquad (6.13)$$

for two functions g_A and g_B of n_A and n_B. Substituting (6.12) and (6.12) into the equations (6.7) and (6.8) gives us

$$\frac{dn_A}{dl} = -3A_t \cdot k_1 \cdot g_A(n_A, n_B) \text{ and} \qquad (6.14)$$

$$\frac{dn_B}{dl} = A_t \cdot [3k_1 \cdot g_A(n_A, n_B) - 2k_2 \cdot g_B(n_A, n_B)] . \qquad (6.15)$$

These are two differential equations that can be solved simultaneously using one of the standard MATLAB IVP solvers `ode...` with the initial conditions $n_A = n_{Af}$ and $n_B = n_{Bf}$ at $l = 0$.

Once $n_A(l)$ and $n_B(l)$ have been computed numerically for every position l along the length of the tubular reactor, all other variables x_A, Y_B, C_A, C_B, q, etc. can also be computed.

Next we formulate the DEs (6.7) and (6.8) in terms of C_A and C_B instead. Here we write equation (6.7) as

$$\frac{d(qC_A)}{dl} = -3A_t \cdot k_1 \cdot C_A .$$

And the product rule of differentiation gives us

$$q\frac{dC_A}{dl} + C_A\frac{dq}{dl} = -3A_t \cdot k_1 \cdot C_A . \qquad (6.16)$$

From our earlier formulation in terms of n_A and n_B we now that $q = \alpha(n_A + n_B + n_C)$. Thus

$$\frac{dq}{dl} = \alpha \left[\frac{dn_A}{dl} + \frac{dn_B}{dl} + \frac{dn_C}{dl} \right] . \qquad (6.17)$$

By using the equations (6.7) to (6.9) inside equation (6.17) we obtain the DE

$$q\frac{dC_A}{dl} + C_A \cdot \alpha \cdot A_t \cdot (-k_1 \cdot C_A + 3k_2 \cdot C_B) = -3k_1 \cdot CA .$$

Rearrangement finally yields

$$\frac{dC_A}{dl} = \frac{A_t}{q} \cdot (-\alpha \cdot C_A(-k_1 \cdot C_A + 3k_2 \cdot C_B) - 3k_1 \cdot C_A) . \qquad (6.18)$$

Equation (6.8) can similarly be written in term of C_B and q. The resulting differential equation the has to be solved simultaneously with the equations (6.17) and (6.18). It is obvious that the formulation in terms of n_A and n_B is more straight forward than the formulation in terms of C_A and C_B.

Overall Heat-Balance and Design Equations with Single and Multiple Reactions

We start with the following diagram

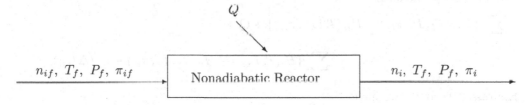

Figure 6.3

Here i is an index used for numbering all components involved in the reaction, i.e., the reactants and products. If any product i is not included in the feed, then we put $n_{if} = 0$ and if a reactant j does not exist in the output then we put $n_j = 0$.

The enthalpy balance is

$$\sum_i n_{if} \cdot H_{if}(T_f, P_f, \pi_{if}) + Q = \sum_i n_i \cdot H_i(T, P, \pi) \qquad (6.19)$$

where Q is the external heat added to the system or reactor, T_f and T are the feed and output temperatures, P_f and P are the feed and output pressures, and π_{if} and π_i are the phases of component i in the feed and the output, each respectively.

Note that the above equation does not contain any heat of reaction terms. These are automatically included as will become clear in next few lines.

If we define a reference state r and we add and subtract the terms $\sum_i n_{if} \cdot H_{ir}(T_r, P_r, \pi_{ir})$ and $\sum_i n_i \cdot H_{ir}(T_r, P_r, \pi_{ir})$, respectively, twice to equation (6.19), then after further rearrangement of the terms, we obtain the heat-balance equation in the form

$$\sum_i \{H_{if}(T_f, P_f, \pi_{if}) - H_{ir}(T_r, P_r, \pi_{ir})\} + Q = \tag{6.20}$$

$$= \sum_i \{H_i(T, P, \pi) - H_{ir}(T_r, P_r, \pi_{ir}\} + \sum_i (n_i - n_{if}) \cdot H_{ir}(T_r, P_r, \pi_{ir}) .$$

From the mass balance of a single reaction we have

$$n_i = n_{if} + \sigma_i \cdot r \tag{6.21}$$

for $r = R_i/\sigma_i$. Equation (6.21) is equivalent to $n_i - n_{if} = \sigma_i \cdot r$ and therefore the heat-balance equation becomes

$$\sum_i \{H_{if}(T_f, P_f, \pi_{if}) - H_{ir}(T_r, P_r, \pi_{ir})\} + Q =$$

$$= \sum_i \{H_i(T, P, \pi) - H_{ir}(T_r, P_r, \pi_{ir}\} + r \cdot \sum_i \sigma_i \cdot H_{ir}(T_r, P_r, \pi_{ir}) .$$

By definition

$$\sum_i \sigma_i \cdot H_{ir}(T_r, P_r, \pi_{ir}) = (\Delta H_R)_r$$

for the heat of reaction $(\Delta H_R)_r$ at the reference conditions. With this, the heat-balance equation takes the form

$$\sum_i \{H_{if}(T_f, P_f, \pi_{if}) - H_{ir}(T_r, P_r, \pi_{ir})\} + Q =$$

$$= \sum_i \{H_i(T, P, \pi) - H_{ir}(T_r, P_r, \pi_{ir}\} + r \cdot (\Delta H_R)_r .$$

Expressed verbally, we have

{*Enthalpy in (exceeding the reference condition) plus Heat added*} =
$$= \{Enthalpy\ out\ (exceeding\ the\ reference\ condition)\ plus$$
$$Heat\ absorbed\ by\ the\ reaction\ (at\ reference\ conditions)\} .$$

For a single reaction, the term "heat absorbed by the reaction" above equals $(\Delta H_R)_r$.

The most general heat-balance equation for multiple-reactions, multiple-input and multiple-output system is obtained by summing the enthalpies of the input streams and the output streams. If we have F input streams and we use l as the counter for the input streams and if we have K output streams and use k as their counter with M components and N reactions, then we obtain the most general heat-balance equation in the form

$$\sum_{l=1}^{F} \left(\sum_{i=1}^{M} n_{if_l}(H_{if_l} - H_{ir}) \right) + Q = \sum_{k=1}^{K} \left(\sum_{i=1}^{M} n_{ik}(H_{ik} - H_{ir}) \right) + \sum_{j=1}^{N} r_j (\Delta H_R)_{r_j} .$$
$$\tag{6.22}$$

This is the most general heat-balance equation for a multiple input, multiple output, multiple reactions (and of course multi-components) system.

In order to turn this heat-balance equation into a part of the design equations we simply replace r_j by $V \cdot r'_j$ as done before for the mass balance, namely

$$\sum_{l=1}^{L} \left(\sum_{i=1}^{M} n_{if_l}(H_{if_l} - H_{ir}) \right) + Q = \sum_{k=1}^{K} \left(\sum_{i=1}^{M} n_{ik}(H_{ik} - H_{ir}) \right) + V \cdot \sum_{j=1}^{N} r'_j \cdot (\Delta H_j) \ .$$

(6.23)

One-phase systems have been investigated in detail in the earlier chapters of this book. In the following sections we extend these studies to two-phase systems.

Two Phase Systems

If a system is formed of two phases, we draw the mass-balance schematic diagram as follows.

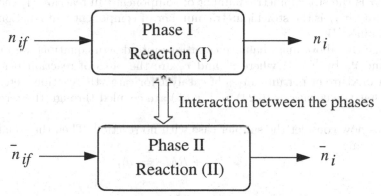

Interaction between two phases of a system
Figure 6.4

The interaction between the two phases usually includes mass transfer from phase I to phase II or vise-versa. This transfer can be in one direction for one component and in the opposite direction for another component. The easiest way is to view the balance in one direction as shown in Figure 6.5. The sign of the mass-transfer term RM_i of each component i depends on the direction of the mass transfer for the specific component, which in turn depends on the concentration driving force.

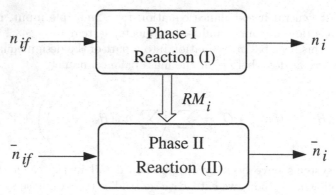

Interaction acting in one direction between different phases
Figure 6.5

Here the mass-balance equations can be written as

$$n_i + RM_i = n_{if} + \sigma_{iI} \cdot r_I \tag{6.24}$$

and

$$\bar{n}_i - RM_i = \bar{n}_{if} + \sigma_{iII} \cdot r_{II} \, , \tag{6.25}$$

where σ_{iI} is the stoichiometric number of component i in reaction (I) that takes place in phase I and σ_{iII} is the stoichiometric number of component i in reaction (II) that takes place in phase II.

To change the above mass-balance equations into design equations we just replace r'_j by $V \cdot r'_j$ and $\bar{r}_j$ by $V \cdot \bar{r}'_j$, where r'_j and $\bar{r}'_j$ are the rates of reaction per unit volume of reaction mixture or per unit mass of catalyst for catalytic reactions, etc.

Notice that the equations (6.24) and (6.25) are coupled through the term RM_i.

Let us now consider the simpler case with no reaction. Then the equations (6.24) and (6.25) become

$$n_i + RM_i = n_{if}$$

and

$$\bar{n}_i - RM_i = \bar{n}_{if} \, .$$

For constant flow rates q_I and q_{II}, we can write

$$q_I \cdot C_i + RM_i = q_I \cdot C_{if}$$

and

$$q_{II} \cdot \bar{C}_i + RM_i = q_{II} \cdot \bar{C}_{if} \, .$$

Here RM_i can be expressed in terms of the concentrations C_i and $\bar{C}_i$ as

$$RM_i = a_m \cdot K_{gi} \cdot (C_i - \bar{C}_i) \, , \tag{6.26}$$

where C_i denotes the concentration of component i in phase I, $\bar{C}_i$ denotes the concentration of component i in phase II, a_m is the total area for mass transfer between the two phases, and K_{gi} is the mass-transfer coefficient for component i. The expression in formula (6.26) is based on the assumption that the equilibrium is established when $C_i = \bar{C}_i$. Otherwise we need to replace formula (6.26) with the expression

$$RM_i = a_m \cdot K_{gi} \cdot (C_i - F_i(\bar{C}_i)) \,, \tag{6.27}$$

where $F_i(\bar{C}_i)$ is the concentration in phase I at equilibrium with $\bar{C}_i$. Thus in the general case the term $\bar{C}_i$ in equation (6.26) is replaced by $F_i(\bar{C}_i)$. Note that if the concentration C_i in phase I is equal to the equilibrium concentration $\bar{C}_i$ in phase I, i.e., if $C_i = F_i(\bar{C}_i)$, then there is no mass transfer between the phases, i.e., we observe equilibrium. Assuming (6.26) to be true, we have

$$q_I \cdot C_i + a_m \cdot K_{gi} \cdot (C_i - \bar{C}_i) \;=\; q_I \cdot C_{if} \tag{6.28}$$

and

$$q_{II} \cdot \bar{C}_i - a_m \cdot K_{gi} \cdot (C_i - \bar{C}_i) \;=\; q_{II} \cdot \bar{C}_{if} \,. \tag{6.29}$$

And therefore for given values of q_I, q_{II}, a_m, K_{gi}, C_{if}, and $\bar{C}_{if}$ we can compute the values of C_i and $\bar{C}_i$ from equations (6.28) and (6.29).

The Co- and Countercurrent Cases

The diagram below describes cocurrent flow.

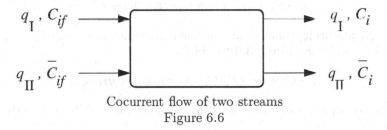

Cocurrent flow of two streams
Figure 6.6

For countercurrent-flow streams we have the situation of Figure 6.7.

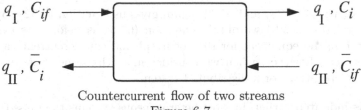

Countercurrent flow of two streams
Figure 6.7

The two mass-balance equations (6.28) and (6.29) remain the same for both cases. This is so because the difference between cocurrent and countercurrent flows does not appear in a single lumped stage. It appears in a sequence of stages, or takes the form of a distributed system as will be shown later.

The Equilibrium Case

For a system in which the contact between the two phases is long and/or the rate of mass transfer is relatively high, the concentrations of different components in the two phases generally reach a state of equilibrium, i.e., a state where no further mass transfer is possible. More details regarding equilibrium states are usually covered in thermodynamics courses. Equilibrium relations relate the concentrations of the phases to each other such as in

$$\bar{C}_i^* = F_i'(C_i) , \quad \text{or} \quad C_i^* = F_i(\bar{C}_i) . \tag{6.30}$$

For narrow regions of concentration values, we can replace (6.30) by linear relations such as

$$\bar{C}_i^* = K_i \cdot C_i , \quad \text{or} \quad C_i^* = K_i \cdot \bar{C}_i \tag{6.31}$$

where * refers to the equilibrium concentration and K_i is the equilibrium constant for component i.

In the equilibrium case we neither know, nor need to know the rates of mass transfer. The simple and systematic approach is to add equations (6.28) and (6.29) for both the cocurrent and the countercurrent cases and thereby use only one equation instead of two, coupled with the mass-balance equations which are the same for both flow cases.

The Cocurrent Case

Adding equations (6.28) and (6.29) gives us

$$q_I \cdot C_i + q_{II} \cdot \bar{C}_i = q_I \cdot C_{if} + q_{II} \cdot \bar{C}_{if} . \tag{6.32}$$

This is a single algebraic equation linking the two unknowns C_i and $\bar{C}_i$. The equilibrium relation (6.30) can be used in (6.32) to obtain

$$q_I \cdot C_i + q_{II} \cdot F_i'(C_i) = q_I \cdot C_{if} + q_{II} \cdot \bar{C}_{if} . \tag{6.33}$$

Equation (6.33) can be solved for C_i and the value of $\bar{C}_i$ then follows from the equilibrium relation (6.30).

The Countercurrent Case

Adding the equations (6.28) and (6.29) again gives us equation (6.32). By replacing $\bar{C}_i$ according to relation (6.30) we obtain equation (6.33) as before. As evidenced, there is no difference in the equations for the cocurrent and countercurrent cases. There will be case differences, however, when we consider more than one stage of cocurrent and countercurrent operations or a distributed system.

Remark: As indicated earlier, in most heterogeneous systems the mass-transfer driving force between phase I and phase II includes the concentration in phase I minus the concentration in phase I which is in equilibrium with the concentration in phase II (or vice versa), i.e.,

$$RM_i = a_m \cdot K_{gi} \cdot (C_i - F_i(\bar{C}_i)) \quad \text{or} \quad RM_i = a_m \cdot K_{gi} \cdot (\bar{C}_i - F_i'(C_i))$$

where RM_i represents the transfer of component i from phase I to phase II and $F_i(\bar{C}_i)$ is the concentration in phase I which is at equilibrium with the concentration in phase II. Obviously, if $C_i > F_i(\bar{C}_i)$ then the transfer is from phase I to phase II. Note that the transfer is in the opposite direction if $C_i < F_i(\bar{C}_i)$. And when $C_i = F_i(\bar{C}_i)$ then the system is at equilibrium with no further mass transfer between the phases.

This situation can be expressed in the opposite sense if we write the driving force RM_i as

$$RM_i = -a_m \cdot K_{gi} \cdot (\bar{C}_i - F_i'(C_i))$$

as shown on the previous page. Here $F_i'(C_i)$ denotes the concentration in phase II which is in equilibrium with the concentration in phase I and $\bar{C}_i$ is the concentration in phase II.

Stage Efficiency

In contrast to continuous packed bed columns, each stage, whether cocurrent or countercurrent, can be considered to be at equilibrium for many multi-phase mass-transfer processes such as distillation, absorption, extraction etc. Such stages are usually called "ideal stages".

The deviation of the actual system from this "ideality" or "equilibrium" is compensated for by using a "stage efficiency" coefficient, which varies between 1.0 for an "ideal stage" and 0 for a "useless stage". Usually the stage efficiencies range between 0.6 and 0.8. They differ widely for different processes, for different components, and for different designs (especially with respect to retention time) that are involved in a particular process.

However, in some cases the concept of "stage efficiency" can cause unacceptable inaccuracies. The rate of mass transfer between the phases should be used in both mass-balance and the design equations.

It is clear from the above that the main concepts used in the mass balance for one-phase systems can be extended to heterogeneous system by simply writing mass-balance equations for each phase and taking the interaction (mass transfer) between the phases into account. The same applies when there are reactions in both phases. Then the rate of reaction terms must be included in the phase mass balances as shown earlier.

Generalized Mass-Balance for Two Phase Systems

For the sake of generality, we now develop most general mass-balance equations for a two phase system in which each phase has multiple inputs and multiple outputs and in which each phase is undergoing reactions within its boundaries.

Refer to Figure 6.8 for a representation of the generalized mass-balance expression in a heterogeneous system.

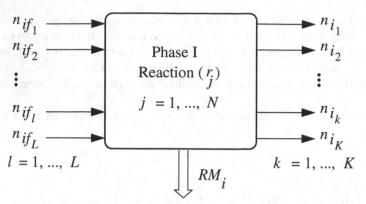

A two-phase heterogeneous system with multiple inputs, multiple outputs,
and multiple reactions in each phase and with mass transfer between the two phases
Figure 6.8

In Figure 6.8 the index i refers to component i in the two phases, namely to the reactants,
the products, and the inerts.

The mass-balance equations for phase I are

$$\sum_{k=1}^{K} n_{ik} + RM_i = \sum_{l=1}^{L} n_{if_l} + \sum_{j=1}^{N} \sigma_{ij} \cdot r_j , \qquad (6.34)$$

where RM_i is the overall rate of mass transfer from phase I to phase II, r_j is the overall
generalized rate of reaction for reaction j, and σ_{ij} is the stoichiometric number of component i in reaction j.

Likewise, for phase II:

$$\sum_{\bar{k}=1}^{\bar{K}} \bar{n}_{i\bar{k}} - RM_i = \sum_{\bar{l}=1}^{\bar{L}} \bar{n}_{if_{\bar{l}}} + \sum_{\bar{j}=1}^{\bar{N}} \sigma_{i\bar{j}} \cdot \bar{r}_{\bar{j}} , \qquad (6.35)$$

where $\bar{K}$ is the number of output streams from phase II, $\bar{L}$ is the number of input streams
to phase (II) other than the mass transfer between the two phases, and $\bar{N}$ is the number
of reactions in phase II.

To change the mass-balance equations (6.34) and (6.35) into design equations we replace
r_j by $V \cdot r_j'$ and $\bar{r}_{\bar{j}}$ by $V \cdot \bar{r}_{\bar{j}}'$, where r_j' and $\bar{r}_{\bar{j}}'$ denote the rates of reaction per unit volume
of the reaction mixture, or per unit mass of catalyst for catalytic reactions for example.
More details of the design equations are given below.

6.1.2   The Design Equations (Steady-State Models) for Isother-
mal Heterogeneous Lumped Systems

Design equations are used in sizing the equipment, while mass-balance equations are
components inventory equations.

For simplicity and clarity, let us now consider a two-phase system where each phase has a single input and a single output with a single reaction taking place in each phase.

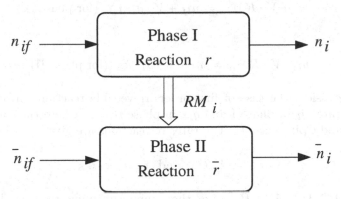

A two-phase heterogeneous system with single input, single output,
and a single reaction in each phase
Figure 6.9

The mass-balance equations are

$$n_i + RM_i = n_{if} + \sigma_i \cdot r \quad \text{(for phase I)} \tag{6.36}$$

and

$$\bar{n}_i - RM_i = \bar{n}_i + \bar{\sigma}_i \cdot \bar{r} \quad \text{(for phase II)} , \tag{6.37}$$

where

n_i	=	molar flow rate of component i out of phase I (mol/min),
$\bar{n}_i$	=	molar flow rate of component i out of phase II (mol/min),
n_{if}	=	molar flow rate of component i fed to phase I (mol/min),
n_{if}	=	molar flow rate of component i fed to phase II (mol/min),
r	=	generalized rate of the single reaction in phase I (mol/min),
σ_i	=	stoichiometric number of component i in the reaction in phase I ,
$\bar{r}$	=	generalized rate of the single reaction in phase II (mol/min),
$\bar{\sigma}_i$	=	stoichiometric number of component i in the reaction in phase II, and
RM_i	=	overall mass-transfer rate of component i from phase I to phase II (mol/min).

Here σ_i and $\bar{\sigma}_i$ are both dimensionless numbers.

In order to turn these mass-balance equations into design equations, we turn all the rate processes r, $\bar{r}$, and RM_i into rates per unit volume of the process for the specific phase, namely

r'	=	rate of reaction in phase I per unit volume of the process $(mol/(min \cdot cm^3))$,
$\bar{r}'$	=	rate of reaction in phase II per unit volume of the process $(mol/(min \cdot cm^3))$,
RM_i'	=	rate of mass transfer of component from phase I to phase II per unit volume of the process $(mol/(min \cdot cm^3))$, and
V	=	volume of the process (cm^3).

We note that RM_i' can also be expressed per unit of mass-transfer area between the two phases.

Then the equations (6.36) and (6.37) can be rewritten as

$$n_i + V \cdot RM_i^{'} = n_{if} + V \cdot \sigma_i \cdot r^{'} \quad \text{(for phase I)} \tag{6.38}$$

and

$$\bar{n}_i - V \cdot RM_i^{'} = \bar{n}_i + V \cdot \bar{\sigma}_i \cdot \bar{r}^{'} \quad \text{(for phase II)} . \tag{6.39}$$

For example, consider the case of first-order irreversible reactions in both phases with constant flow rates q_I in phase I and q_{II} in phase II, and where the concentrations are C_i in phase I and $\bar{C}_i$ in phase II. The rates of reactions are given by

$$r^{'} = k \cdot C_A \quad \text{and} \quad \bar{r}^{'} = \bar{k} \cdot \bar{C}_B .$$

Note that $i = 1, 2, 3, ..., A, ..., B, ...$, i.e., the component in question may be labeled by an integer number or referred to by a letter.

If we assume that the equilibrium between the phases is established when $C_i = \bar{C}_i$, then the equations (6.38) and (6.39) become

$$q_I \cdot C_i + V \cdot a_m^{'} \cdot K_{gi}(C_i - \bar{C}_i) = q_I \cdot C_{if} + V \cdot \sigma_i \cdot k \cdot C_A \tag{6.40}$$

and

$$q_{II} \cdot \bar{C}_i - V \cdot a_m^{'} \cdot K_{gi}(C_i - \bar{C}_i) = q_{II} \cdot \bar{C}_{if} + V \cdot \bar{\sigma}_i \cdot \bar{k} \cdot \bar{C}_B , \tag{6.41}$$

where $a_m^{'}$ is the area of mass transfer per unit volume of the process unit and K_{gi} is the coefficient of mass transfer of component i between the two phases. Note that we again assume here that the equilibrium between the two phases occurs when the concentrations in both phases are equal.

For many two-phase systems these equations can be written as

$$q_I \cdot C_i + V \cdot a_m^{'} \cdot K_{gi}(C_i - F(\bar{C}_i)) = q_I \cdot C_{if} + V \cdot \sigma_i \cdot k \cdot C_A \tag{6.42}$$

and

$$q_{II} \cdot \bar{C}_i - V \cdot a_m^{'} \cdot K_{gi}(C_i - F(\bar{C}_i)) = q_{II} \cdot \bar{C}_{if} + V \cdot \bar{\sigma}_i \cdot \bar{k} \cdot \bar{C}_B , \tag{6.43}$$

where $F(\bar{C}_i)$ is the concentration in phase I which is in equilibrium with the concentration in phase II.

A Simple Illustrative Example

Let us consider an example whose flow rates $q_{..}$ and rates of reactions $r_{..}$ are shown in Figure 6.10.

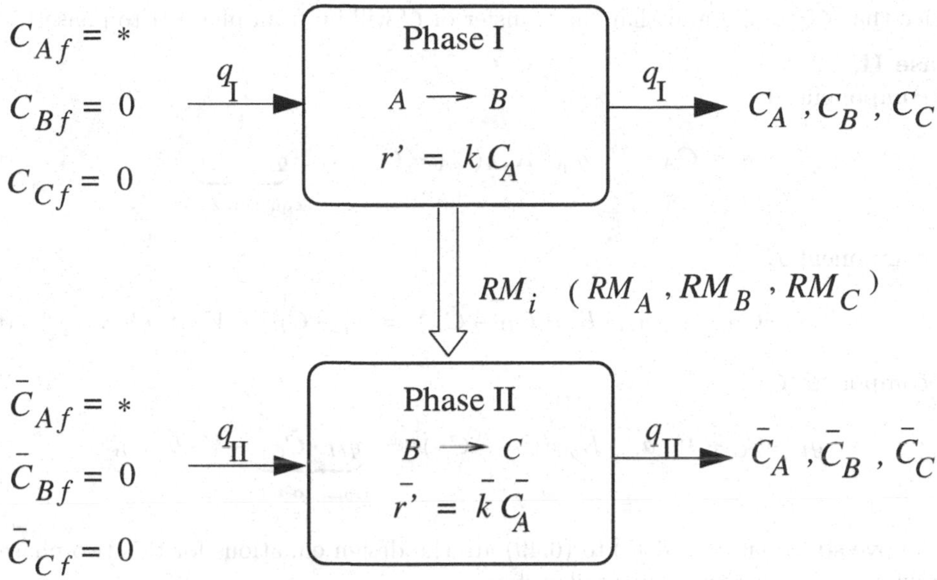

$C_{Af} = *$

$C_{Bf} = 0$

$C_{Cf} = 0$

q_I →

Phase I

$A \longrightarrow B$

$r' = k\,C_A$

q_I → C_A , C_B , C_C

RM_i (RM_A , RM_B , RM_C)

$\bar{C}_{Af} = *$

$\bar{C}_{Bf} = 0$

$\bar{C}_{Cf} = 0$

q_{II} →

Phase II

$B \longrightarrow C$

$\bar{r}' = \bar{k}\,\bar{C}_A$

q_{II} → $\bar{C}_A$, $\bar{C}_B$, $\bar{C}_C$

Mass flow diagram
Figure 6.10

Notice in this example that C_{Af} is fed to the first phase, while the feed concentrations C_{Bf} and C_{Cf} are equal to zero. For phase II, $\bar{C}_{Bf}$ is fed to this second phase, while both feed rates $\bar{C}_{Af}$ and $\bar{C}_{Cf}$ are equal to zero.

Design equations (for the simplified RM_i driving force $C_i - \bar{C}_i$ instead of $C_i - F(\bar{C}_i)$):

Phase I:

For component A:

$$q_I \cdot C_A + V \cdot a'_m \cdot K_{gA} \cdot (C_A - \bar{C}_A) = q_I \cdot C_{Af} - V \cdot k \cdot C_A . \tag{6.44}$$

Here the term $V \cdot k \cdot C_A$ carries a negative sign since A is a reactant and $\sigma_A = -1$. Component A is only available in phase I, $C_A > \bar{C}_A$, and any transfer of A will be from phase I to phase II.

For component B:

$$q_I \cdot C_B + V \cdot a'_m \cdot K_{gB}(C_B - \bar{C}_B) = \underbrace{q_I \cdot C_{Bf}}_{\text{equal to zero}} + V \cdot k \cdot C_B , \tag{6.45}$$

since B is produced in phase I and consumed in phase II. Thus $C_B > \bar{C}_B$ and the transfer of B is from phase I to phase II.

For component C:

$$q_I \cdot C_C + V \cdot a'_m \cdot K_{gC}(C_C - \bar{C}_C) = \underbrace{q_I \cdot C_{Cf}}_{\text{equal to zero}} . \tag{6.46}$$

Notice that $\bar{C}_C > C_C$ and that the transfer of C will be from phase II to phase I.

Phase II:

For component A:

$$q_{II} \cdot \bar{C}_A - V \cdot a'_m \cdot K_{gA}(C_A - \bar{C}_A) \; = \; \underbrace{q_I \cdot \bar{C}_{Af}}_{\text{equal to zero}} \; . \qquad (6.47)$$

For component B:

$$q_{II} \cdot \bar{C}_B - V \cdot a'_m \cdot K_{gB}(C_B - \bar{C}_B) \; = \; q_{II} \cdot \bar{C}_{Bf} - V \cdot \bar{k} \cdot \bar{C}_B \; . \qquad (6.48)$$

For component C:

$$q_{II} \cdot \bar{C}_C - V \cdot a'_m \cdot K_{gC}(C_C - \bar{C}_C) \; = \; \underbrace{q_{II} \cdot \bar{C}_{Cf}}_{\text{equal to zero}} + V \cdot \bar{k} \cdot \bar{C}_B \; . \qquad (6.49)$$

The above six equations (6.44) to (6.49) are the design equations for this two-phase system when both phases are lumped systems.

For more accuracy in the above six equations, we could have used the driving force terms $(C_{...} - f(\bar{C}_{...}))$ instead in the above equations, with $f(\bar{C}_{...})$ more accurately denoting the concentration of the component in play in phase I which is at equilibrium with the concentration in phase II.

6.1.3   The Design Equations (Steady-State Models) for Isothermal Heterogeneous Distributed Systems

As explained in detail for homogeneous systems in Chapter 4, a distributed system includes variations in the space direction. Therefore for a distributed system we can not use the overall rate of reaction, i.e., the rate of reaction per unit volume multiplied by the total volume, nor the overall rate of mass transfer, i.e., the mass transfer per unit area multiplied by the area of mass transfer. For lumped systems the area of mass transfer is treated as the multiple of the area per unit volume of the process multiplied by the volume of the process for the design equations.

In the case of a distributed system we need to establish the mass-balance for a small element ΔV and then take $\lim_{\Delta V \to 0}$ to arrive at a DE, as detailed for distributed homogeneous systems earlier in Chapter 4 and depicted in Figure 6.11.

As an illustration, let us consider a system with the single reaction

$$A \longrightarrow B \; ,$$

taking place completely in phase I with no reaction taking place in phase II, but with mass transfer between the two phases.

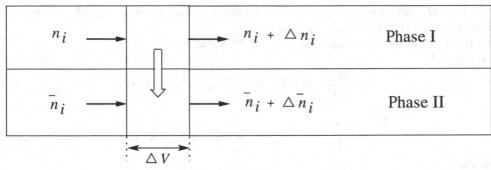

A distributed heterogeneous system
Figure 6.11

The Cocurrent Case

Here the-balance equation for phase I, or the first design equation is

$$n_i + \Delta n_i + \Delta V \cdot RM_i' = n_i + \Delta V \cdot \sigma_i \cdot r' \ . \tag{6.50}$$

And for phase II the second design equation is

$$\bar{n}_i + \Delta \bar{n}_i - \Delta V \cdot RM_i' = \bar{n}_i + 0 \tag{6.51}$$

because there is no reaction in phase II.
Rearranging equation (6.50) and taking the difference quotient limit as $\Delta V \to 0$ gives us

$$\frac{dn_i}{dV} + RM_i' = \sigma_i \ r' \ . \tag{6.52}$$

Furthermore, rearranging equation (6.51) makes

$$\frac{d\bar{n}_i}{dV} - RM_i' = 0 \ . \tag{6.53}$$

Both (6.52) and (6.53) are differential equations with the initial conditions of $n_i = n_{if}$ and $\bar{n}_i = \bar{n}_{if}$ at $V = 0$.

If the flow rates q in phase I and $\bar{q}$ in phase II are constant and the reaction is first-order with $r' = k \cdot C_A$ and if we assume that the equilibrium between the phases is established when $C_i = \bar{C}_i$, then we can rewrite the two equations (6.52) and (6.53) in the following form

$$q \cdot \frac{dC_i}{dV} + a_m' \cdot K_{gi}(C_i - \bar{C}_i) = \sigma_i \cdot k \cdot C_A \tag{6.54}$$

and

$$\bar{q} \cdot \frac{d\bar{C}_i}{dV} - a_m' \cdot K_{gi}(C_i - \bar{C}_i) = 0 \ , \tag{6.55}$$

where C_i and $\bar{C}_i$ are the concentrations of component i in phase I and II, respectively. The initial conditions for the cocurrent case are $C_i = C_{if}$ and $\bar{C}_i = \bar{C}_{if}$ at $V = 0$, where

$i = A$ or B. Here we again assume that the two phases are at equilibrium when the concentrations in both phases are equal.

Specifically for the two components A and B, we write:

For phase I:

$$q \cdot \frac{dC_A}{dV} \;=\; -a'_m \cdot K_{gA}(C_A - \bar{C}_A) - k \cdot C_A \tag{6.56}$$

and

$$q \cdot \frac{dC_B}{dV} \;=\; -a'_m \cdot K_{gB}(C_B - \bar{C}_B) + k \cdot C_A \; . \tag{6.57}$$

For phase II:

$$\bar{q} \cdot \frac{d\bar{C}_A}{dV} \;=\; a'_m \cdot K_{gA}(C_A - \bar{C}_A) \tag{6.58}$$

and

$$\bar{q} \cdot \frac{d\bar{C}_B}{dV} \;=\; a'_m \cdot K_{gB}(C_B - \bar{C}_B) \; . \tag{6.59}$$

For the cocurrent case both sets of DEs are subject to the initial conditions $C_A = C_{Af}$, $C_B = C_{Bf}$, $\bar{C}_A = \bar{C}_{Af}$, and $\bar{C}_B = \bar{C}_{Bf}$ at $V = 0$.

Generalized Form of Multiple Reactions in Both Phases I and II

We now develop general design equations for N reactions (counted by the index j) taking place in phase I and $\bar{N}$ reactions (counted in $\bar{j}$) taking place in phase II as depicted in Figure 6.12.

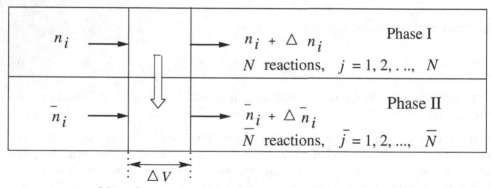

Mass flow for the heterogeneous distributed system
Figure 6.12

Here the design equations for the cocurrent case are as follows:
For phase I:

$$n_i + \Delta n_i + \Delta V \cdot RM'_i \;=\; n_i + \Delta V \sum_{j=1}^{N} \sigma_{ij} \cdot r'_j \; .$$

This equation can be turned into a DE, namely

$$\frac{dn_i}{dV} + RM_i' = \sum_{j=1}^{N} \sigma_{ij} \cdot r_j' . \tag{6.60}$$

Similarly for phase II, the design equation DE becomes

$$\frac{d\bar{n}_i}{dV} - RM_i' = \sum_{\bar{j}=1}^{\bar{N}} \bar{\sigma}_{i\bar{j}} \cdot \bar{r}_{\bar{j}}' . \tag{6.61}$$

Both DEs (6.60) and (6.61) are subject to the initial conditions $n_i = n_{if}$ and $\bar{n}_i = \bar{n}_{if}$ at $V = 0$ for the cocurrent case.

Notice that for a distributed system, the multiple input problem changes into an artificial idle stage mass balance with no reaction or mass transfer at the input for each phase as shown in Figure 6.13.

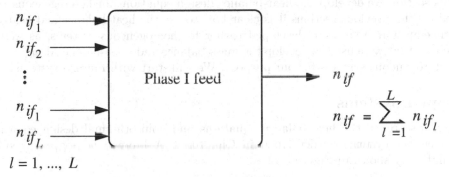

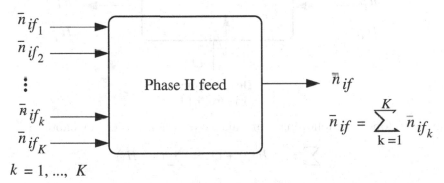

Multiple inputs can be combined into a single feed
Figure 6.13

This simplifies the problem of multiple inputs. One can proceed similarly for multiple outputs.

In some instances one of the phases is distributed and the other phase is lumped. In such a case the distributed phase will contribute its mass and/or heat-transfer term to the lumped phase through an integral as shown in the model of a bubbling fluidized bed in Chapter 4.

The Countercurrent Case

For the countercurrent case, the differential equation of phase II is given by

$$-\frac{d\bar{n}_i}{dV} - RM_i' = \sum_{\bar{j}=1}^{\bar{N}} \bar{\sigma}_{i\bar{j}} \cdot \bar{r}_{\bar{j}}' \tag{6.62}$$

with the boundary conditions $\bar{n}_i = \bar{n}_{if}$ at $V = V_t$, where V_t is the total volume of the process unit.

6.1.4   Nonisothermal Heterogeneous Systems

In this section we develop the heat-balance design equations for heterogeneous systems. Based on the previous sections it is clear how to use the heat-balance and heat-balance design equations that were developed earlier for homogeneous systems, as well as the principles that were used to develop the mass-balance and mass-balance design equations for heterogeneous systems for our purpose. We will start with lumped systems.

Lumped Systems

Let us first recall the heat-balance equations and nonisothermal design equations for homogeneous systems as developed in Chapter 3. A lumped homogeneous system is schematically shown in Figure 6.14.

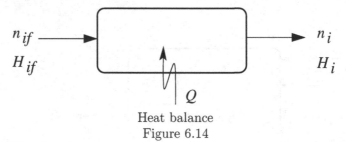

Heat balance
Figure 6.14

We have developed the following heat-balance equation earlier in Chapter 3:

$$\sum_i (n_{if} \cdot Hif) + Q = \sum_i (n_i \cdot Hi) . \tag{6.63}$$

Recall how to transform this enthalpy equation into an enthalpy difference equation using enthalpy of each component at the reference condition H_{ir}. Thus

$$\sum_i (n_{if} \cdot Hif - n_{if} \cdot Hir) + \sum_i (n_{if} \cdot Hir) + Q = \sum_i (n_i \cdot Hi - n_i \cdot Hir) + \sum_i (n_i \cdot Hir) .$$

We can rearrange this equation to obtain

$$\sum_i n_{if}(Hif - Hir) + Q = \sum_i n_i(Hi - Hir) + \sum_i (n_i \cdot Hir - n_{if} \cdot Hir) \ . \qquad (6.64)$$

The last term in equation (6.64) can be written as

$$\sum_i (n_i - n_{if})Hir \ . \qquad (6.65)$$

This term can be replaced by using the mass-balance relation

$$n_i = n_{if} + \sigma_i \cdot r \ , \quad \text{or} \quad n_i - n_{if} = \sigma_i \cdot r$$

as

$$\sum_i \sigma_i \cdot r \cdot \sum_i (n_i - n_{if})Hir = H_{ir} = r \sum_i \sigma_i \cdot H_{ir} = r \cdot \Delta H_r \ .$$

Thus the heat-balance equation (6.63) has become

$$\sum_i n_{if}(Hif - Hir) + Q = \sum_i n_i(Hi - Hir) + r \cdot \Delta H_r \ . \qquad (6.66)$$

In order to convert equation (6.66) into a nonisothermal heat-balance design equation, we replace r with $r' \cdot V$, where r' is the rate of reaction per unit volume V and V is the volume of the reactor. Then equation (6.66) becomes

$$\sum_i n_{if}(Hif - Hir) + Q = \sum_i n_i(Hi - Hir) + V \cdot r' \cdot \Delta H_r \ . \qquad (6.67)$$

Notice that the units of r' and V depend upon the type of the process. For example, if we are dealing with a gas-solid catalytic system, we will usually define r' as per unit mass of the catalyst and replace V with W_S which is the weight of the catalyst.
And for N reactions (counted in j), equation (6.67) becomes

$$\sum_i n_{if}(Hif - Hir) + Q = \sum_i n_i(Hi - Hir) + V \sum_{j=1}^{N} r_j' \cdot \Delta H_{r_j} \ . \qquad (6.68)$$

The most suitable reference conditions r are the standard conditions. The standard conditions for any component are a temperature of $25°\ C$, a pressure of $1\ atm$ and the true natural phase of the component under these conditions, e.g., H_2O is liquid, H_2 is a gas, and C is solid under the standard conditions.

Heterogeneous Lumped Systems

For a heterogeneous system we write equation (6.67) out for each phase and account for the heat transfer Q between the two phases. For a nonadiabatic system the heat $Q_{external}$

that is added from the outside is added to the phase that receives it. Which phase receives which amount of heat depends on the configuration and on our knowledge of the physical system as we have seen for example for the nonadiabatic bubbling fluidized bed catalytic reactor in Chapter 4.

Now we consider an **adiabatic** two-phase system in which each phase has one reaction as shown schematically in Figure 6.15.

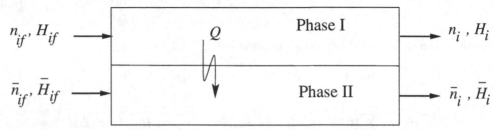

Heat balance for a two-phase system
Figure 6.15

In Figure 6.15, Q denotes the heat transfer between the two phases.
For phase I, the heat-balance equation is

$$\sum_i n_{if}(Hif - Hir) - Q = \sum_i n_i(Hi - Hir) + r_I \cdot \Delta H_{rI} , \qquad (6.69)$$

while for phase II it is

$$\sum_i \bar{n}_{if}(\bar{H}if - \bar{H}ir) + Q = \sum_i \bar{n}_i(\bar{H}i - \bar{H}ir) + r_{II} \cdot \Delta H_{rII} . \qquad (6.70)$$

To turn these heat-balance equations into nonisothermal heat balance design equations, we define the rate of reaction per unit volume (or per unit mass of catalyst, depending on the system) and the heat transfer per unit volume of the process unit (or per unit length), whichever is more convenient.
Then the equations (6.69) and (6.70) become:
For phase I:

$$\sum_i n_{if}(Hif - Hir) - V \cdot Q' = \sum_i n_i(Hi - Hir) + V \cdot r' \cdot \Delta H_{rI} . \qquad (6.71)$$

And for phase II:

$$\sum_i \bar{n}_{if}(\bar{H}if - \bar{H}ir) + V \cdot Q' = \sum_i \bar{n}_i(\bar{H}i - \bar{H}ir) + V \cdot \bar{r}' \cdot \Delta H_{rI} . \qquad (6.72)$$

The rate of heat transfer per unit volume of the process unit is given by

$$Q' = a_h' \cdot h(T - \bar{T}) , \qquad (6.73)$$

where

a'_h =   the area of heat transfer per unit volume of the system,
h =   the heat-transfer coefficient between the two phases,
$\bar{T}$ =   the temperature of phase II, and
T =   the temperature of phase I.

Obviously for multiple reactions in each phase, we must change the equations accordingly. For N reactions (counted in j) in phase I and $\bar{N}$ reactions (counted in $\bar{j}$) in phase II, the equation for phase I is

$$\sum_i n_{if}(Hif - Hir) + V \cdot a'_h \cdot h(\bar{T} - T) \;=\; \sum_i n_i(Hi - Hir) + V \sum_{j=1}^{N}(r'_j \cdot \Delta H_{rj}) \;, \quad (6.74)$$

and that for phase II is

$$\sum_i \bar{n}_{if}(\bar{H}if - \bar{H}ir) - V \cdot a'_h \cdot h(\bar{T} - T) \;=\; \sum_i \bar{n}_i(\bar{H}i - \bar{H}ir) + V \sum_{\bar{j}=1}^{N}(\bar{r}'_{\bar{j}} \cdot \Delta \bar{H}_{r\bar{j}}) \;. \quad (6.75)$$

Now our picture is almost complete and we can determine the heat-balance design equation for our distributed two phase system.

Distributed Systems

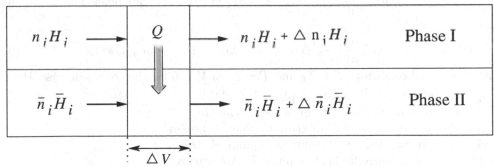

Heat balance for a heterogeneous distributed system
Figure 6.16

The Cocurrent Case

Phase I with one reaction gives rise to the equation

$$\sum_i n_i H_i - \Delta V \sum_i (RM'_i \cdot H_i) - \Delta V \cdot Q' \;=\; \sum_i (n_i Hi + \Delta n_i H_i) \;. \quad (6.76)$$

The equivalent DE for equation (6.76) is

$$\sum_i \frac{d(n_i H_i)}{dV} \;=\; -Q' - \sum_i (RM'_i \cdot H_i)$$

which can be written as

$$\sum_i n_i \frac{dHi}{dV} + \sum_i H_i \frac{dni}{dV} = -Q' - \sum_i (RM_i' \cdot H_i) . \tag{6.77}$$

From the mass-balance design equation (6.60), used here for a single reaction, we obtain

$$\frac{dni}{dV} + RM_i' = \sigma_i \cdot r' . \tag{6.78}$$

Substituting dn_i/dV from equation (6.78) into equation (6.77) gives us

$$\sum_i n_i \frac{dHi}{dV} + \sum_i H_i [\sigma_i \cdot r' - RM_i'] = -Q' - \sum_i (RM_i' \cdot H_i) ,$$

or

$$\sum_i n_i \frac{dHi}{dV} + r' \cdot \Delta H = -Q' . \tag{6.79}$$

For phase II and cocurrent operation, the corresponding DE is

$$\sum_i \bar{n}_i \frac{d\bar{H}i}{dV} + \bar{r}' \cdot \Delta \bar{H} = Q' . \tag{6.80}$$

If the change is only sensible heat and if we use an average constant C_p for the mixture, then we obtain the following more popular approximate heat-balance design equations

$$q \cdot \rho \cdot C_p \frac{dT}{dV} + r' \Delta H = -a_h' \cdot h(T - \bar{T}) \qquad \text{for phase I} \tag{6.81}$$

and

$$\bar{q} \cdot \bar{\rho} \cdot \bar{C}_p \frac{d\bar{T}}{dV} + \bar{r}' \Delta \bar{H} = a_h' \cdot h(T - \bar{T}) \qquad \text{for phase II} , \tag{6.82}$$

with the initial conditions $T = T_f$ and $\bar{T} = \bar{T}_f$ at $V = 0$ for the cocurrent case. Here

q = volumetric flow rate in phase I (m^3/h),
$\bar{q}$ = volumetric flow rate in phase II (m^3/h),
ρ = average density of mixture in phase I (kg/m^3),
$\bar{\rho}$ = average density of mixture in phase II (kg/m^3),
C_p = average specific heat of phase I $(kJ/(kg \cdot K)$,
$\bar{C}_p$ = average specific heat of phase II $(kJ/(kg \cdot K)$.

The Countercurrent Case

In this case we have

$$q \cdot \rho \cdot C_p \frac{dT}{dV} + r' \Delta H = -a_h' \cdot h(T - \bar{T}) \qquad \text{for phase I} . \tag{6.83}$$

Note that the equation for phase I for the countercurrent case is the same as the one for the cocurrent case. And

$$-\bar{q} \cdot \bar{\rho} \cdot \bar{C}_p \frac{d\bar{T}}{dV} + \bar{r}' \Delta \bar{H} = -a_h' \cdot h(T - \bar{T}) \qquad \text{for phase II} \tag{6.84}$$

with the two-point boundary conditions $T = T_f$ at $V = 0$ and $\bar{T} = \bar{T}_f$ at $V = V_t$.

Exercises for 6.1

1. Show that the steady- state and dynamic models for a double-pipe, counter-current heat exchanger can have the same form as the model of a packed bed absorber. Discuss the assumptions inherent in both the heat exchanger and the absorber models which might lead to significant differences in the kinds of model equations used to describe each system.

2. An exothermic reaction

$$A \longrightarrow B \quad \text{with} \quad r' = k_0 \cdot e^{-E/(R \cdot T)} \cdot C_A$$

takes place in a plug flow tubular reactor. The reactor is cooled by a cooling jacket. Derive the steady-state model for both countercurrent and cocurrent flow in the cooling jacket. Structure a solution algorithm for both cases.

6.2   Nonreacting Multistage Isothermal Systems and High Dimensional Linear and Nonlinear Systems

Many chemical and biological processes are multistage. Multistage processes include absorption towers, distillation columns, and batteries of continuous stirred tank reactors (CSTRs). These processes may be either cocurrent or countercurrent. The steady state of a multistage process is usually described by a set of linear equations that can be treated via matrices. On the other hand, the unsteady-state dynamic behavior of a multistage process is usually described by a set of ordinary differential equations that gives rise to a matrix differential equation.

We start with isothermal processes with fixed temperatures in all stages, so that the model simply consists of the mass-balance design equations.

6.2.1   Absorption Columns or High Dimensional Lumped, Steady State and Equilibrium Stages Systems

Absorption is a process in which a component in a gas stream is absorbed by a liquid stream, such as the CO_2 absorption in ammonia and methanol production lines.

We study absorption columns or towers, and multistage or multitray systems in which all stages are equilibrium stages, also called ideal stages. In an equilibrium or ideal stage, the liquid and gas streams at the exit of each tray or stage are at equilibrium with each other. This occurs if the mass transfer rate of the stage and the residence time in the stage are both relatively high, so that the gas and liquid streams reach equilibrium during their contact in each stage.

We assume that in our system the component A is absorbed from the gas phase into the liquid phase. Figure 6.17 describes the jth stage (or tray) of a multistage system in terms of the following variables:

Y_j, the mole fraction of component A in the gas phase leaving the jth tray;

n_j, the molar flow rate of component A in the gas phase leaving the jth tray;

X_j, the mole fraction of component A in the liquid phase leaving the jth tray; and

$\bar{n}_j$, the molar flow rate of component A in the liquid phase leaving the jth tray.

Here $n_j = n_{t_j} \cdot Y_j$, where n_{t_j} is the total molar flow rate of the gas phase leaving the jth tray and $\bar{n}_j = \bar{n}_{t_j} \cdot X_j$, where $\bar{n}_{t_j}$ is the total molar flow rate of the liquid phase leaving the jth tray.

The mass flow streams at tray j are shown in Figure 6.1 for a countercurrent process of a dilute system. Here the total molar flow rate of the **rising gas** is constant at V *moles/min* while the total molar flow rate of the **sinking liquid** is constant at L *moles/min*. Both L and V are molar flow rates as long as X_j and Y_j are molar fractions. If, however, X_j and Y_j describe concentrations, measured in $mole/m^3$, then L and V are volumetric flow rates measured in m^3/min.

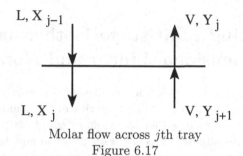

Molar flow across jth tray
Figure 6.17

By looking at the molar flow balance for the jth tray in Figure 6.17 we find that

$$\bar{n}_{j-1}X_{j-1} + n_{j+1}Y_{j+1} = \bar{n}_j X_j + n_j Y_j . \tag{6.85}$$

For simplicity, we assume that the total molar flow rates of the liquid and gas phases remain constant, that is $\bar{n}_j = L$ and $n_j = V$ for all j as described above.
With this assumption equation (6.85) becomes

$$LX_{j-1} + VY_{j+1} = LX_j + VY_j . \tag{6.86}$$

When we assume equilibrium stages, we write $Y_j = F(X_j)$ for an equilibrium function $F(X_j)$ that depends on the variable X_j only.

The Case of a Linear Equilibrium Relation

Specifically for a linear equilibrium function $F(x) = ax + b$, we have $Y_j = aX_j + b$ for two process constants a and b.
By using an equilibrium function F of this linear form in equation (6.86), we obtain

$$LX_{j-1} + V(aX_{j+1} + b) = LX_j + V(aX_j + b) .$$

This simplifies to

$$LX_{j-1} + (Va)X_{j+1} = (L + Va)X_j .$$

And further rearrangement yields

$$LX_{j-1} - \underbrace{(L + Va)}_{\alpha}X_j + \underbrace{(Va)}_{\beta}X_{j+1} = 0 .$$

Thus we have

$$LX_{j-1} - \alpha X_j + \beta X_{j+1} = 0 . \tag{6.87}$$

Figure 6.18 shows the schematic diagram of a countercurrent absorption column with the molar flow rates shown symbolically at each of its N trays.

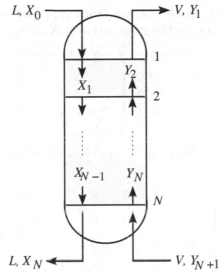

Absorption column with N stages

Figure 6.18

Note that the order of the labeling of the trays or stages from 1 to N or in reverse order from N to 1 does not affect the problem.

Next we look at equation (6.87) for each $j = 1, ..., N$.

For the first tray with $j = 1$, we have

$$LX_0 - \alpha X_1 + \beta X_2 = 0$$

which we can rewrite as

$$-\alpha X_1 + \beta X_2 + 0X_3 + 0X_4 + ... + 0X_N = -LX_0 .$$

For $j = 2$ we get

$$LX_1 - \alpha X_2 + \beta X_3 + 0X_4 + 0X_5 + ... + 0X_N = 0 .$$

For $j = 3$ we obtain

$$0X_1 + LX_2 - \alpha X_3 + \beta X_4 + 0X_5 + \ldots + 0X_N = 0 .$$

Similarly for $j = N - 1$ equation (6.87) reads as

$$0X_1 + 0X_2 + 0X_3 + \ldots + LX_{N-2} - \alpha X_{N-1} + \beta X_N = 0 .$$

And finally for the tray with number $j = N$ we have

$$0X_1 + 0X_2 + 0X_3 + \ldots + 0X_{N-2} + LX_{N-1} - \alpha X_N + \beta X_{N+1} = 0 ,$$

or after rearranging

$$0X_1 + 0X_2 + 0X_3 + \ldots + 0X_{N-2} + LX_{N-1} - \alpha X_N = -\beta X_{N+1} .$$

These are N equations that can be written in matrix form. For this purpose we define a vector X whose components are the state variables X_j, namely

$$X = \begin{pmatrix} X_1 \\ X_2 \\ \vdots \\ X_{N-1} \\ X_N \end{pmatrix} . \tag{6.88}$$

The matrix of coefficients of X for the N equations given by (6.87) for $j = 1, \ldots, N$ is the N by N matrix

$$A = \begin{pmatrix} -\alpha & \beta & 0 & 0 & \ldots & \ldots & 0 \\ L & -\alpha & \beta & 0 & & & \vdots \\ 0 & L & -\alpha & \beta & \ddots & & \vdots \\ \vdots & \ddots & \ddots & \ddots & \ddots & \ddots & \vdots \\ \vdots & & \ddots & L & -\alpha & \beta & 0 \\ \vdots & & & \ddots & L & -\alpha & \beta \\ 0 & \ldots & \ldots & \ldots & 0 & L & -\alpha \end{pmatrix} . \tag{6.89}$$

The matrix A is a tridiagonal **Toeplitz**[1] **matrix** with constant main diagonal and co-diagonal entries.

The right-hand sides of the N equations can be combined into the constant right-hand-side vector

$$B = \begin{pmatrix} -L\,X_0 \\ 0 \\ \vdots \\ 0 \\ -\beta\,X_{N+1} \end{pmatrix} , \tag{6.90}$$

[1]Otto Toeplitz, German mathematician, 1881 - 1940

since X_0 and X_{N+1} are not state variables, but rather known input variables. Thus, the matrix form of the mass-balance design equations for the absorption column is the linear system of equations

$$A\,X\ =\ B\,.\tag{6.91}$$

If A is invertible, then equation (6.91) has a unique solution, theoretically written in the form

$$X\ =\ A^{-1}\,B\,.\tag{6.92}$$

Numerically, matrix inversion is at least twice as costly as solving the corresponding linear system of equations. Therefore for numerical purposes there is no need to go beyond equation (6.91) to the algebraically more satisfying, but numerically meaningless closed form solution X of (6.92).

Unfortunately the first simplifying assumption of a linear equilibrium relation in the mass balance model is not very accurate for practical chemical/biological systems. Therefore we will also present numerical solutions for linear high-dimensional systems with nonlinear equilibrium relations. A model that accounts for mass transfer in each tray will be presented in the next subsection.

For the moment we continue with our very simple case.

Example of a linear absorption column

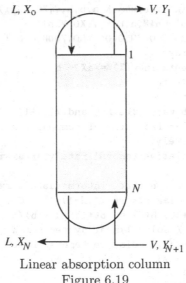

Linear absorption column
Figure 6.19

The matrix equation for the mole fractions X_j and Y_j in Figure 6.19 is

$$A\,X\ =\ B$$

where A is the tridiagonal matrix

$$
A = \begin{pmatrix}
-\alpha & \beta & & & 0 \\
L & -\alpha & \beta & & \\
 & L & -\alpha & \beta & \\
 & & \ddots & \ddots & \ddots \\
0 & & & L & -\alpha
\end{pmatrix}.
$$

Here L is the liquid flow rate (in l/min), $\beta = V \cdot a$, where V is the gas flow rate (in l/min) and a is the coefficient in the equilibrium relation $Y_j = aX_j + b$, $\alpha = L + V \cdot a$ $(= L + \beta)$, $B = (-L\,X_0,\, 0,\, \ldots ,\, 0,\, -\beta\,X_{N+1})^T$, and $X_{N+1} = (Y_{N+1} - b)/a$.
We use the following data for our simulation:
The number of trays is $N = 6$, making A a 6 by 6 matrix. Moreover,
$L = 750\ l/min$, $V = 7000\ l/min$, $a = 0.72$, $b = 0.01$, $X_0 = 0.001\ g/l$,
$Y_{N+1} = 1.4\ g/l$, and $X_{N+1} = (Y_{N+1} - b)/a = (1.4 - 0.01)/0.72 = 1.943\ g/l$.
Notice that we are using volumetric flow rates and concentrations here, rather than the total molar flow rate and mole fractions.

Our task is to find all X and Y values from the linear system $A\,X = B$. For this it is sufficient to find all X values first and then to compute the corresponding Y values from the equations $Y_j = aX_j + b$ for $j = 1, \ldots, N$.
Here is a simple MATLAB function that performs this task by solving $AX = B$ via MATLAB's \ built-in backslash linear equations solver.

```
function [relrem,X,Y] = linearcolumn(N,a,b,L,V,X0,YNp1)
% [relrem,X,Y] = linearcolumn(N,a,b,L,V,X0,YNp1)
% Sample call: linearcolumn(6,0.72,0.01,750,7000,0.001,1.4)
% Input : N = number of trays
%         a, b = coefficients in F(X) = aX + b
%         L = liquid flow rate
%         V = gas flow rate
%         X0, YNp1 = input variables X_0 and Y_{N+1}
% Output: X and Y, the molar fractions of component A in the liquid and gas
%         phases, respectively.
%         relrem is the relative removal rate (for example 0.14 means 14 %)

bt = V*a; al = L + bt;    % construct intermediate terms and system matrices
A = -diag(al*ones(1,N)) + diag(bt*ones(1,N-1),1) + diag(L*ones(1,N-1),-1);
B = zeros(N,1); B(1) = -L*X0; B(N) = -bt*(YNp1 - b)/a;
X = A\B; Y = a*X + b;     % solve linear system for X and compute Y from X
relrem = (YNp1 - Y(1))/YNp1; % relative removal rate
disp('         Xj              Yj                j'),
disp([X,Y,[1:length(X)]'])
```

For our specific data the computed concentrations X_j and Y_j of component A in the jth tray appear in the first column of the on-screen output matrix for the liquid phase and in the second column for the gaseous phase, respectively, after calling `linearcolumn(6,0.72, 0.01,`**750**`,7000,0.001,1.4)`.

Our output data includes the relative removal rate $(Y_{N+1} - Y_1)/Y_{N+1}$, called **ans** below, of the absorbable component.

```
>> linearcolumn(6,0.72,0.01,750,7000,0.001,1.4)
        Xj              Yj              j
      1.6434          1.1933           1
      1.8878          1.3692           2
      1.9242          1.3954           3
      1.9296          1.3993           4
      1.9304          1.3999           5
      1.9305           1.4             6
ans =
      0.14767
```

Note that the mole concentrations X_j of the liquid phase, collected in the first column of the screen output is increasing from tray to tray, while the mole concentrations Y_j of the gaseous phase in the second on-screen column decrease from tray N to tray 1. For our data there is a sharp initial jump from $X_0 = 0.001$ to $X_1 = 1.6434$. This is related to the characteristics of the equilibrium relation, the number of trays, and the liquid and gas flow rates.

If we change the number N of trays to 2, 4, 8, or 10 (see Problem 1(a) of the Chapter Exercises), we notice that the removal rate **ans** changes and if $N = 10$ for example, then there is no removal in trays 6 to 10, since the liquid has apparently become saturated. The reader should verify this assertion, see Problem 1(a). Therefore we conclude that this specific removal process with the specified flow rates and characteristics needs only 6 trays to reach full saturation of the absorbing liquid. Its relative removal rate, however, is only about 15%.

On the other hand, several more trays may be needed if L is much higher such as $L = 7500$ instead of 750. Here is the output for a call of `linearcolumn(20,0.72,0.01,7500, 7000,0.001,1.4)` where we have increased the liquid flow rate L ten-fold from 750 l/min in the previous case to 7,500 l/min.

```
        Xj              Yj              j
     0.0012233       0.010881          1
     0.0015555       0.01112           2
     0.0020499       0.011476          3
     0.0027856       0.012006          4
     0.0038804       0.012794          5
     0.0055096       0.013967          6
     0.007934        0.015713          7
     0.011542        0.01831           8
     0.01691         0.022176          9
     0.024899        0.027928         10
     0.036788        0.036487         11
     0.054479        0.049225         12
     0.080805        0.06818          13
     0.11998         0.096387         14
     0.17828         0.13836          15
```

0.26503	0.20082	16
0.39413	0.29377	17
0.58623	0.43209	18
0.87211	0.63792	19
1.2975	0.94421	20

```
ans =
     0.99223
```

The relative removal rate now exceeds 99% when we use 20 trays and $X_1 = 0.0012233 \approx X_0 = 0.001$ here. For differing tray numbers N, see problem 1(b). The large liquid flow rate of 7500 l/min has increased the removal rate to almost 100% and there is only a very small increase in the liquid concentration from the input to tray 1.

In many application the least number N of trays or stages that achieves a certain mole fraction or concentration in the liquid phase of the last tray must be found for a chemical/biological process. We can adapt our previous program to find this number. N depends on the desired value of X_N for an unknown number N of trays, i.e., the dimension N of the matrix A in (6.91) is unknown and has to be found for the given system parameters and for the desired value of X_N. Note that $X_N \leq X_{N+1} = (Y_{N+1} - b)/a$ due to the equilibrium limitations of the unit. We can compute the number N of trays that are needed to achieve a certain value for X_N and the given data by solving the problem for increasing dimensions $n = 1, 2, 3, \ldots$ until the computed concentration X_n exceeds the desired value X_N for the first time. This simple algorithm solves increasingly larger linear systems of equations $AX = B$ (6.91) with A, X, and B as specified in (6.89), (6.88), and (6.90) in terms of the system parameters. Here is our MATLAB code for this problem based on a linear equilibrium relation.

```
function [N,X,Y] = findNlinearcolumn(XN,a,b,L,V,X0,YNp1)
%   [N,X,Y] = findNlinearcolumn(XN,a,b,L,V,X0,YNp1)
% Sample call: [N,X,Y] = findNlinearcolumn(1.8,0.72,0.01,750,7000,0.001,1.4)
% Input : XN = desired mole fraction in liquid phase in tray N
%         a, b = coefficients in F(X) = aY + b
%         L = liquid flow rate
%         V = gas flow rate
%         X0, YNp1 = input variables X_0 and Y_{N+1}
% Output: N = number of trays for desired value of X_N
%         X and Y, the molar fractions of component A in the liquid and gas
%         phases, respectively.

bt = V*a; al = L + bt; N = 1; X = zeros(2,1);   % intermediate terms
while N == 1 | (X(N-1) < XN & Y(N-1) < YNp1 & N <=200)    % if X(N) < XN :
  A = -diag(al*ones(1,N)) + diag(bt*ones(1,N-1),1) + diag(L*ones(1,N-1),-1);
  B = zeros(N,1); B(1) = -L*X0; B(N) = -bt*(YNp1 - b)/a; % construct A and RHS
  X = A\B; Y = a*X + b;        % solve linear system for X and compute Y from X
  N = N + 1; end, N = N - 1; % reduce N to last used value
```

A call of [N,X,Y] = findNlinearcolumn(1.9305,0.72,0.01,**750**,7000,0.001,1.4) with $L = 750$ gives us $N = 6$ and the same output of X_j and Y_j values as obtained

on p. 359 where we started with $N = 6$ trays. The readers should change the value of XN slightly from our setting of 1.9305 to see how the new desired value for X_N increases or decreases the number of trays N that are necessary.
Note in your experiments that a higher value than $(Y_{N+1} - b)/a$ for XN cannot be reached for the given parameters, due to the equilibrium limitations.

A call of findNlinearcolumn(1.2975,0.72,0.01,7500,7000,0.001,1.4) with $L = 7,500$ gives us the answer ans = 20 for the number N of needed trays. Note that we have used the computed value of XN = 1.2975 from p. 359 in this call. The value of $X_N = 1.2975$ is near the limit for feasible X_N. If we increase the desired value for XN slightly to 1.3 for example, the algorithm computes up to the code specific maximally allowed dimension $N = 200$ without reaching this value for X_N. If we increase XN a bit less less to 1.297661 for example and use the same other data to call findNlinearcolumn(1.297661,0.72,0.01,7500,7000, 0.001,1.4), then our algorithm computes that $N = 36$ trays are needed.
Notice that for our test data $(Y_{N+1} - b)/a = (1.4 - 0.01)/0.72 = 1.9306$ and the gap between the largest feasible value (≈ 1.2975) obtained numerically for X_N and $X_{N+1} = (Y_{N+1} - b)/a = 1.9306$ is rather large.
This low value for X_N is caused by the large value of L. Notice that Y_N in turn drops very sharply to 0.9442 from the feed $Y_{N+1} = 1.4$ for $L = 7500$, while for $L = 750$ it drops relatively little from the feed value Y_{N+1} in the bottom stage.

Multistage Absorption with a Nonlinear Equilibrium Relation

Here we start over with equation (6.86) in the form

$$LX_{j-1} + VF(X_{j+1}) = LX_j + VF(X_j) , \qquad (6.93)$$

where $Y_j = F(X_j)$ now is a nonlinear function of X_j.
This gives rise to a set of nonlinear equations that must be solved numerically. This more realistic and more accurate model of the nonlinear equilibrium case is much more interesting.
Below we explain the steps involved with finding the solution in case the equilibrium relation F is quadratic, or

$$Y_j = F(X_j) = a X_j^2 + b . \qquad (6.94)$$

All other nonlinear equilibrium relations $Y = F(X)$ can be treated analogously.

By combining (6.93) and (6.94) we have to solve the equation

$$LX_{j-1} + V(a X_{j+1}^2 + b) - LX_j - V(a X_j^2 + b) = 0 , \qquad (6.95)$$

which is in standard form $f(x) = 0$ for $x = (X_1, \ldots, X_N)^T$. In order to solve this problem for given values of a, b, L, and V and the input concentrations X_0 and Y_{n+1}, we can apply Newton's iterative method (1.5) to equation (6.95). Newton's method solves nonlinear equations $f(x) = 0$ iteratively. It involves the N by N Jacobian matrix $Df = (\partial f_\ell / \partial X_k)$ of f and uses the iteration sequence

$$x^{(i+1)} = x^{(i)} - \left(Df\left(x^{(i)}\right)\right)^{-1} \cdot f\left(x^{(i)}\right) \tag{6.96}$$

in which the upper index (i) in $x^{(i)}$ indicates the ith Newton iterate. See p. 26 in Section 1.2 for more details.

To be able to apply Newton's method we must determine what $f : \mathbb{R}^N \to \mathbb{R}^N$ and its Jacobian matrix $Df(x)$ are in equation (6.96).

For $j = 1$ equation (6.95) states that

$$V \cdot a\, X_2^2 - LX_1 - V \cdot aX_1^2 + LX_0 = 0\ ;$$

while for $1 < j < N$ we have

$$LX_{j-1} + V \cdot aX_{j+1}^2 - LX_j - V \cdot aX_j^2 = 0\ ;$$

and for $j = N$

$$LX_{N-1} - LX_N - V \cdot aX_N^2 + V \cdot a\frac{Y_{N+1} - b}{a} = 0$$

since $Y_{N+1} = aX_{N+1}^2 + b$. Note that X_0 and Y_{N+1}, as well as a, b, L, and V are given input parameters and $X_1, ..., X_N$ are unknown.

By first separating all terms that involve $V \cdot a$ in the above set of formulas, then gathering those terms that involve L, and finally those that remain, we arrive at the expanded left-hand side $f(x)$ of equation (6.95) in matrix/vector form

$$f(x) = V \cdot a \left(\begin{pmatrix} X_2^2 \\ \vdots \\ X_N^2 \\ 0 \end{pmatrix} - \begin{pmatrix} X_1^2 \\ \vdots \\ \vdots \\ X_N^2 \end{pmatrix} \right) + L \left(\begin{pmatrix} 0 \\ X_1 \\ \vdots \\ X_{N-1} \end{pmatrix} - \begin{pmatrix} X_1 \\ \vdots \\ \vdots \\ X_N \end{pmatrix} \right) + \begin{pmatrix} LX_0 \\ 0 \\ \vdots \\ 0 \\ V \cdot a \cdot \frac{Y_{N+1} - b}{a} \end{pmatrix}. \tag{6.97}$$

Here the function f maps a vector $x = (X_1, \ldots, X_N)^T$ to $f(x) = (f_1(x), \ldots, f_N(x))^T \in \mathbb{R}^N$. The Jacobian of f is the matrix composed of the gradients of each component function f_ℓ of f for $\ell = 1, ..., N$. It has the matrix form $Df(x) =$

$$= \begin{pmatrix} -2VaX_1 - L & 2VaX_2 & 0 & \cdots & \cdots & 0 \\ L & -2VaX_2 - L & 2VaX_3 & 0 & & \vdots \\ 0 & \ddots & \ddots & \ddots & \ddots & \vdots \\ \vdots & \ddots & \ddots & \ddots & \ddots & 0 \\ 0 & \cdots & 0 & L & -2VaX_{N-1} - L & 2VaX_N \\ 0 & \cdots & \cdots & 0 & L & -2VaX_N - L \end{pmatrix}. \tag{6.98}$$

This form for $Df(x)$ can be verified by finding the gradient $grad\ f_2\left((X_1,\ \dots\ ,\ X_N)^T\right)$ for the second component function $f_2\left((X_1,\ \dots\ ,\ X_N)^T\right) = V\cdot a\cdot(X_3^2-X_2^2)+L(X_1-X_2)$ as

$$grad\ f_2\ =\ \left(\frac{\partial f_2}{\partial X_1},\ \frac{\partial f_2}{\partial X_2},\ \dots\ ,\frac{\partial f_2}{\partial X_N}\right)\ =\ (L,\ -L-2\cdot V\cdot a\cdot X_2,\ 2\cdot V\cdot a\cdot X_3,\ 0,\ \dots\ ,\ 0)\ .$$

The readers should practice computing partial derivatives and gradients on their own for this extended example in order to verify and understand formula (6.98).

When using Newton's method to solve (6.95) we start from a guess x_{start} of the vector x of X_j values for $j = 1, ..., N$ and iterate

$$x_{new} = x_{start} - Df(x_{start})^{-1}\cdot x_{start}\ .$$

The linear system $Df(x_{start})y = x_{start}$ that needs to be solved to find $x_{new} = x_{start} - y$ from x_{start} changes its system matrix and its right-hand side in each iteration. Our code quadcolumn.m iterates until the relative error of the iterates falls to below 1%. This accuracy limit is arbitrary and can be changed by the reader to higher or smaller values depending on the sensitivity of the specific problem by modifying the bound of 0.01 in the while line of the quadcolumn.m code accordingly.

```
function [relrem,X,Y] = quadcolumn(N,a,b,L,V,X0,YNp1)
% [relrem,X,Y] = quadcolumn(N,a,b,L,V,X0,YNp1)
% Sample call: quadcolumn(12,0.72,0.01,750,7000,0.001,1.4)
% Input : N = number of trays
%         a, b = coefficients in quadratic equilibrium relation F(X) = aX^2 + b
%         L = liquid flow rate
%         V = gas flow rate
%         X0, YNp1 = input variables X_0 and Y_{N+1}
% Output: X and Y, the molar fractions of component A in the liquid and gas
%         phases, respectively, in trays 1 to N.
%         relrem = (YNp1 - Y1)/YNp1 = relative removal rate

bt = V*a; Lx0 = L* X0; VaXNp1 = bt*(YNp1-b)/a;  % convenience constant settings
xstart = linspace(X0,((YNp1-b)/a)^0.5,N)';       % Newton starting value
xmod = dh(xstart,bt,L)\h(xstart,bt,L,Lx0,VaXNp1); % Newton modification
xlast = xstart - xmod; i=1;                       % next Newton iterate
while norm(xmod)/norm(xlast) > 0.01 & i < 100,  % Further Newton steps
    xstart = xlast;                               %  (at most 100)
    xmod = dh(xstart,bt,L)\h(xstart,bt,L,Lx0,VaXNp1);
    xlast = xstart - xmod; i=i+1; end             % until < 2% error
X = xlast; Y = a*X.^2 + b;                        % prepare output data
relrem = (YNp1 - Y(1))/YNp1;                      % relative removal rate
disp('        Xj              Yj              j'),
disp([X,Y,[1:length(X)]'])

function H = h(x,bt,L,Lx0,VaXNp1)                 % quadratic function H(x)
```

```
H = bt*([x(2:end);0].^2 - x.^2) + L*([0;x(1:end-1)] - x);
H(1) = H(1) + Lx0; H(end) = H(end) + VaXNp1;
```

```
function DH = dh(x,bt,L)                 % Jacobian matrix of H(x)
  lx = length(x);
DH = 2*bt*(diag(x(2:lx),1) - diag(x)) + L*(diag(ones(1,lx-1),-1) - eye(lx));
```

Let us compare the results for a quadratic equilibrium relation with those for linear equilibrium relations that we have obtained earlier. Here is the output of quadcolumn(6, 0.72,0.01,**750**,7000,0.001,1.4).

Xj	Yj	j
1.313	1.2513	1
1.3853	1.3918	2
1.3892	1.3996	3
1.3894	1.4	4
1.3894	1.4	5
1.3894	1.4	6

```
ans =
   0.10625
```

Apparently, with a quadratic equilibrium relation we need to use only 3 stages, but achieve only about 10% removal.

For $L \approx V$, the same quadratic equilibrium relation achieves 99% removal in 8 trays as computed by calling quadcolumn(8,0.72,0.01,**7500**,7000,0.001,1.4)

Xj	Yj	j
0.0010004	0.010001	1
0.0012884	0.010001	2
0.020744	0.01031	3
0.17141	0.031156	4
0.50358	0.19259	5
0.86481	0.54848	6
1.1338	0.93551	7
1.2983	1.2237	8

```
ans =
   0.99286
```

For the relatively low value of $L = 750$ we still notice a sudden jump from $X_0 = 0.001$ to $X_1 = 1.313$, while for $L = 7,500$ there is no such jump. In the quadratic equilibrium relation case, the largest possible value for X_N is bounded by $X_{N+1} = \sqrt{(Y_{N+1} - b)/a}$. For our test data we compute $\sqrt{(Y_{N+1} - b)/a} = \sqrt{(1.4 - 0.01)/0.72} = 1.3894$ and the gap between the numerically obtained value of $X_N = 1.2983$ above and $X_{N+1} = 1.3894$ is much smaller than in the previous subsection for a linear equilibrium relation.

This difference is due to the different characteristics of linear and nonlinear equilibrium relations.

Multistage Absorption with Nonequilibrium Stages

When the stages can be considered to be nonequilibrium stages, then the model should also account for the rate of mass transfer between the phases. In this case the overall molar balance that we have previously used will not be adequate, because the equilibrium relation between the two phases can not be used. To remedy this, we must introduce the mass-transfer rate between the two phases as shown in Figure 6.20.

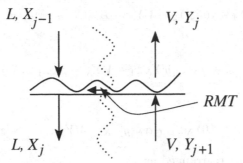

Mass transfer in a nonequilibrium stage
Figure 6.20

In Figure 6.20 the term RMT denotes the **R**ate of **M**ass **T**ransfer of component A from the gas phase to the liquid phase. It can be expressed in the relation

$$RMT = K_{gj} \cdot \bar{a}_{mj} \cdot \text{(driving force)} \,,$$

where the "driving force" means the concentration difference, K_{gj} is the mass-transfer coefficient, and $\bar{a}_{mj}$ is the area of mass transfer for the tray.

Assuming that the tray is perfectly mixed as regards both the liquid and the gas phases and that we have a linear equilibrium relation, we get

$$RMT = K_{gj}\bar{a}_{mj}(Y_j - (aX_j + b))$$

where $aX_j + b = Y_j^*$ is the gas-phase mole fraction at equilibrium with X_j.
The molar balance for the gas phase is given by

$$VY_{j+1} = VY_j + K_{gj}\bar{a}_{mj}(Y_j - (aX_j + b)) \,. \tag{6.99}$$

By rearranging we get

$$\underbrace{(V + K_{gj}\bar{a}_{mj})}_{\alpha_j[y]}Y_j - VY_{j+1} - \underbrace{(K_{gj}\bar{a}_{mj}a)}_{\beta_j[x]}X_j - \underbrace{K_{gj}\bar{a}_{mj}b}_{\gamma_j[y]} = 0 \,.$$

This gives us

$$\alpha_j[y]Y_j - VY_{j+1} - \beta_j[x]X_j - \gamma_j[y] = 0 \,, \tag{6.100}$$

where $\alpha_j[y] = (V + K_{gj}\bar{a}_{mj})$, $\beta_j[x] = K_{gj}\bar{a}_{mj}a$, and $\gamma_j[y] = K_{gj}\bar{a}_{mj}b$ and the symbols $[x]$ and $[y]$ are used to denote the associated variables and serve to distinguish the coefficients of equation (6.100) from those of the molar balance equation (6.102) for the liquid phase.

Now we write out equation (6.100) for a number of indices j.

For $j = 1$:
$$\alpha_1[y]Y_1 - VY_2 - \beta_1[x]X_1 - \gamma_1[y] = 0 .$$

For $j = 2$:
$$0Y_1 + \alpha_2[y]Y_2 - VY_3 - \beta_2[x]X_2 - \gamma_2[y] = 0 .$$

For $j = N - 1$:
$$0Y_1 + 0Y_2 + 0Y_3 + \ldots + \alpha_{N-1}[y]Y_{N-1} - VY_N - \beta_{N-1}[x]X_{N-1} - \gamma_{N-1}[y] = 0 .$$

And for $j = N$:
$$0Y_1 + 0Y_2 + 0Y_3 + \ldots + 0Y_{N-1} + \alpha_N[y]Y_N - VY_{N+1} - \beta_N[x]X_N - \gamma_N[y] = 0 .$$

The last equation can be rearranged as
$$0Y_1 + 0Y_2 + 0Y_3 + \ldots + 0Y_{N-1} + \alpha_N[y]Y_N - \beta_N[x]X_N - \gamma_N[y] = VY_{N+1} .$$

In order to write these N equations in matrix form, we define the matrix
$$A_Y = \begin{pmatrix} \alpha_1[y] & -V & 0 & 0 & 0 \\ 0 & \alpha_2[y] & -V & 0 & 0 \\ \vdots & \vdots & \ddots & \ddots & \vdots \\ 0 & 0 & 0 & \alpha_{N-1}[y] & -V \\ 0 & 0 & 0 & 0 & \alpha_N[y] \end{pmatrix} .$$

This matrix is an upper bidiagonal matrix with varying diagonal entries $\alpha_j[y]$ and constant upper diagonal entries $-V$.

As before, we introduce the state variable vectors X and Y, comprised of the X_j and Y_j values, namely
$$X = \begin{pmatrix} X_1 \\ X_2 \\ \vdots \\ X_N \end{pmatrix} \quad \text{and} \quad Y = \begin{pmatrix} Y_1 \\ Y_2 \\ \vdots \\ Y_N \end{pmatrix} .$$

Moreover, we define the diagonal matrix
$$\beta_X = \begin{pmatrix} \beta_1[x] & & & 0 \\ & \beta_2[x] & & \\ & & \ddots & \\ 0 & & & \beta_N[x] \end{pmatrix}, \; \gamma_Y = \begin{pmatrix} \gamma_1[y] \\ \gamma_2[y] \\ \vdots \\ \gamma_N[y] \end{pmatrix}, \; \text{and } B_Y = \begin{pmatrix} 0 \\ \vdots \\ 0 \\ V Y_{N+1} \end{pmatrix} .$$

With these settings, equation (6.100) becomes
$$A_Y Y - \beta_X X - \gamma_Y = B_Y . \tag{6.101}$$

[*Equation (6.101) should be worked out with pencil and paper by our readers.*]
Similarly we obtain the equation

$$LX_{j-1} + K_{gj}\bar{a}_{mj}(Y_j - (aX_j + b)) = LX_j \tag{6.102}$$

for the molar balance of the liquid phase. Putting equation (6.102) in analogous matrix form gives us

$$A_X X + \beta_Y Y - \gamma_X = B_X . \tag{6.103}$$

Here are the specific matrices and vectors as derived from equation (6.102), that occur in equation (6.103):

$$A_X = \begin{pmatrix} -\alpha_1[x] & 0 & \cdots & \cdots & 0 \\ L & -\alpha_2[x] & \ddots & & \vdots \\ 0 & \ddots & \ddots & \ddots & \vdots \\ \vdots & & \ddots & \ddots & 0 \\ 0 & \cdots & 0 & L & -\alpha_N[x] \end{pmatrix} , \quad \beta_Y = \begin{pmatrix} \beta_1[y] & & \\ & \ddots & \\ & & \beta_N[y] \end{pmatrix} ,$$

$$\gamma_X = \begin{pmatrix} \gamma_1[x] \\ \vdots \\ \gamma_N[x] \end{pmatrix} , \quad \text{and} \quad B_X = \begin{pmatrix} -L\,X_0 \\ 0 \\ \vdots \\ 0 \end{pmatrix} .$$

Note that in the liquid-phase molar-balance equation (6.103), the matrix A_X is lower bi-diagonal and, more specifically, equation (6.102) makes $\alpha_1[x] = K_{gj}\bar{a}_{mj}a + L$, $\beta_1[y] = K_{gj}\bar{a}_{mj}$, and $\gamma_j[x] = K_{gj}\bar{a}_{mj}b$.

The two matrix equations (6.101) and (6.103) can be solved simultaneously to find the two solution vectors X and Y.
First we try to simplify the two equations by algebraic matrix manipulations.
We can formally solve equation (6.103) for X, namely

$$X = A_X^{-1}[\gamma_X + B_X - \beta_Y Y] . \tag{6.104}$$

By plugging X in (6.104) into equation (6.101) we obtain

$$A_Y Y - \beta_X \left\{ A_X^{-1}[\gamma_X + B_X - \beta_Y Y] \right\} - \gamma_Y = B_Y .$$

Some simple manipulation gives us

$$\underbrace{[A_Y + \beta_X A_X^{-1} \beta_Y]}_{C} Y = \underbrace{B_Y + \gamma_Y + \beta_X \left\{ A_X^{-1}[\gamma_X + B_X] \right\}}_{M} . \tag{6.105}$$

Thus we have obtained the linear equation

$$C\,Y = M \tag{6.106}$$

for the unknown vector Y. The matrix equation (6.106) can be solved for Y (in theory) by matrix inversion, namely

$$Y = C^{-1}M .\tag{6.107}$$

We thus can obtain the solution vector Y from (6.107) explicitly. Then we can use Y in (6.104) to compute the vector X via another matrix inversion using formula (6.104). Here

$$C = A_Y + \beta_X A_X^{-1}\beta_Y$$

and

$$M = B_Y + \gamma_Y + \beta_X \left\{A_X^{-1}[\gamma_X + B_X]\right\} .$$

What is wrong with this approach?

In theory there is nothing wrong; but for numerical computations, the above formulas are rather costly and inaccurate: They involve inverting the upper triangular bidiagonal matrix A_X, whose inverse will be a dense upper triangular matrix. Moreover, formula (6.107) for Y involves the inverse of C which itself is given by a triple and complicated matrix product correction term subtracted from A_X in (6.105).

Some of the golden rules of modern matrix computations are:

> *Do not compute or use a matrix inverse unless absolutely needed for theoretical purposes; do not **ever** destroy the sparsity of a matrix, nor increase the density of an intermediate matrix; be mindful of the structure of each matrix and try to take advantage of it and preserve it during numerical computations. Rely on matrix factorizations, and view coupled matrix equations as block matrix equations before actually trying to solve them.*

Let us thus revisit our two coupled matrix equations (6.101) and (6.103) while heeding the above matrix numerics caveats and let us see how to extract X and Y more accurately and economically than using the equations (6.104) and (6.107).

Our first step is to represent (6.101) and (6.103) by one joint $2N$ by $2N$ system of linear equations in block matrix form as

$$\begin{pmatrix} -\beta_X & A_Y \\ A_X & \beta_Y \end{pmatrix} \begin{pmatrix} X \\ Y \end{pmatrix} = \begin{pmatrix} B_Y + \gamma_Y \\ B_X + \gamma_X \end{pmatrix} .\tag{6.108}$$

Our readers should verify - using pencil and paper - that the block matrix equation (6.108) is identical to the two joint matrix equations (6.101) and (6.103).

The system matrix of equation (6.108) contains the two diagonal blocks β_X and β_Y and the two offdiagonal blocks A_X and A_Y which are both banded tridiagonal. Rather than inverting a tridiagonal block as naively suggested earlier, it is much less costly to multiply the two sides of the block matrix equation (6.108) from the left by the nonsingular block matrix

$$\begin{pmatrix} -\beta_X^{-1} & O \\ O & I_N \end{pmatrix}$$

as follows:

$$\left\{ \begin{pmatrix} -\beta_X^{-1} & O \\ O & I_N \end{pmatrix} \cdot \begin{pmatrix} -\beta_X & A_Y \\ A_X & \beta_Y \end{pmatrix} \right\} \begin{pmatrix} X \\ Y \end{pmatrix} = \begin{pmatrix} -\beta_X^{-1} & O \\ O & I_N \end{pmatrix} \cdot \begin{pmatrix} B_Y + \gamma_Y \\ B_X + \gamma_X \end{pmatrix} .\tag{6.109}$$

Multiplying this equation out we obtain the equivalent block matrix system

$$\begin{pmatrix} I_N & -\beta_X^{-1}A_Y \\ A_X & \beta_Y \end{pmatrix} \begin{pmatrix} X \\ Y \end{pmatrix} = \begin{pmatrix} -\beta_X^{-1}(B_Y + \gamma_Y) \\ B_X + \gamma_X \end{pmatrix}. \qquad (6.110)$$

Here $I = I_N$ is the N by N identity matrix and $O = O_N$ is the N by N zero matrix.

The block system (6.110) can be solved more easily now. To do so we suggest to eliminate the (1,1) block I_N of the system matrix in (6.110) by left multiplication of the two sides of equation (6.110) with

$$\begin{pmatrix} A_X & -I \\ O & I \end{pmatrix}.$$

Then we obtain

$$\begin{pmatrix} O_N & A_X\beta_X^{-1}A_Y - \beta_Y \\ A_X & \beta_Y \end{pmatrix} \begin{pmatrix} X \\ Y \end{pmatrix} = \begin{pmatrix} -A_X\beta_X^{-1}(B_Y + \gamma_Y) - (B_X + \gamma_X) \\ B_X + \gamma_X \end{pmatrix}.$$

$$(6.111)$$

This system amounts to two separate linear systems of equations. The first block row in (6.111) involves only the unknown vector Y, namely

$$\left(A_X\beta_X^{-1}A_Y + \beta_Y\right) Y = A_X\beta_X^{-1}(B_Y + \gamma_Y) + (B_X + \gamma_X) \qquad (6.112)$$

where we have switched all signs to become positive. This system of N linear equations can be solved easily by Gaussian elimination and the backslash, i.e., the \ command in MATLAB for Y. Care should be taken to implement the left multiplication by β_X^{-1} as row scalings since β_X is a diagonal matrix. To find X we then use Y to solve the second block row equation

$$A_X X = B_X + \gamma_X - \beta_Y Y. \qquad (6.113)$$

Let us test our two-phase model numerically. The model is captured in the two equations (6.99) and (6.102) that account for the gas and liquid molar balances. We shall compare these results for nonequilibrium stages with those of the one-phase equilibrium model given in equation (6.85) earlier.

The MATLAB code `linearnoneqicol.m` implements and solves the two linear equations (6.112) and (6.113) under the assumption that the term $K_{gj} \cdot \bar{a}_{mj} = const = K_a$ is constant for all stages j. It is easy for the reader to change our program below to accommodate varying mass transfer coefficients K_{gj} or varying areas of mass transfer $\bar{a}_{mj}$ from tray to tray, if needed.

```
function [relrem,X,Y] = linearnoneqicol(N,Ka,a,b,L,V,X0,YNp1)
% [X,Y,relrem] = linearnoneqicol(N,Ka,a,b,L,V,X0,YNp1)
% Sample call: linearnonequicol(6,100,0.72,0.01,750,7000,0.001,1.4)
% Input : N = number of trays
%         Ka = product of mass transfer coeff and area of transfer
%         a, b = coefficients in F(X) = aX + b
%         L = liquid flow rate
%         V = gas flow rate
```

```
%           X0, YNp1 = input variables X_0 and Y_{N+1}
% Output: X and Y, the molar fractions of component A in the liquid and gas
%           phases, respectively.
%           relrem is the relative removal rate (for example 0.06 means 6 %)

Ax = -(Ka*a + L)*eye(N) + L*diag(ones(1,N-1),-1);   % setting up data
Ay = (V + Ka)*eye(N) - V*diag(ones(1,N-1),1); bty = (Ka)*eye(N);
gax = Ka*b*ones(N,1); Bx = zeros(N,1); Bx(1) = -L*X0; gay = gax;
By = zeros(N,1); By(N) = V*YNp1;

A = 1/(Ka*a) * Ax * Ay + bty;                      % solving for Y
b = 1/(Ka*a) * Ax * (By + gay) + (Bx + gax); Y = A\b;
b = Bx + gax - Ka *Y; X = Ax\b;                    % solving for X

relrem = (YNp1 - Y(1))/YNp1;  % relative removal rate
disp('        Xj              Yj              j'),
disp([X,Y,[1:length(X)]'])
```

First we repeat our earlier computations with $N = 6$ trays and varying L. A call of
linearnonequicol(6,**100**,0.72,0.01,**750**,7000,0.001,1.4) with $K_a = 100$ generates the output

Xj	Yj	j
0.1597	1.3152	1
0.30656	1.3322	2
0.44248	1.3479	3
0.56826	1.3625	4
0.68466	1.376	5
0.79239	1.3885	6

ans =
    0.060565

This indicates a 6.06% removal rate for the relatively low value of $K_a = 100$.
If we increase K_a to 1,000, we obtain a removal rate of 14.3% when calling
linearnonequicol(6,**1000**,0.72,0.01,**750**,7000,0.001, 1.4).

Xj	Yj	j
0.81018	1.2002	1
1.282	1.2869	2
1.5571	1.3375	3
1.7175	1.3669	4
1.8111	1.3841	5
1.8656	1.3942	6

ans =
    0.1427

If we increase K_a to 5000, we reach almost the same results as on p. 359 when calling
linearnonequicol(6,**5000**,0.72,0.01,**750**,7000,0.001,1.4).

Xj	Yj	j

1.3604	1.1934	1
1.7622	1.339	2
1.8809	1.3821	3
1.916	1.3948	4
1.9264	1.3986	5
1.9295	1.3997	6

ans =

    0.14759

Further increases to K_a values such as 50,000 or 500,000 show that for 6 trays the upper limit of X_6 is around 1.9305 for the given parameters and model.

For $L = 7500$ the corresponding results are as follows, first for $K_a = 100$ by calling `linearnonequicol(6,100,0.72,0.01,7500,7000,0.001,1.4)`.

Xj	Yj	j
0.017896	1.2901	1
0.034871	1.3082	2
0.051925	1.3264	3
0.069057	1.3447	4
0.08627	1.363	5
0.10356	1.3815	6

ans =

    0.078491

We note that if K_a is small, then the removal rate is small at 7.85%. For $K_a = 1,000$, the removal rate grows to around 46.2% as computed below by calling `linearnonequicol(6,1000,0.72,0.01,7500,7000,0.001,1.4)`.

Xj	Yj	j
0.091353	0.75342	1
0.18557	0.85023	2
0.28381	0.95118	3
0.38626	1.0564	4
0.49308	1.1662	5
0.60447	1.2807	6

ans =

    0.46184

When we increase the number of trays N from 6 to 20, we get even better removal rates ranging from 22% to 95%, depending on the magnitude of the factor K_a in RMT.
The first computed data comes from calling `linearnonequicol(20,100,0.72,0.01,7500,7000,0.001,1.4)`.

Xj	Yj	j
0.015159	1.0828	1
0.029384	1.098	2
0.043674	1.1132	3
0.058031	1.1286	4
0.072455	1.1439	5
0.086946	1.1594	6

0.1015	1.1749	7
0.11613	1.1905	8
0.13082	1.2062	9
0.14558	1.2219	10
0.16041	1.2377	11
0.17531	1.2536	12
0.19028	1.2696	13
0.20532	1.2856	14
0.22043	1.3017	15
0.2356	1.3179	16
0.25085	1.3342	17
0.26617	1.3505	18
0.28156	1.3669	19
0.29702	1.3834	20

ans =

   0.22655

If we increase K_a to 1000 and call `linearnonequicol(20,1000,0.72,0.01,7500,7000, 0.001,1.4)` instead, we create the following data.

Xj	Yj	j
0.034842	0.2889	1
0.070131	0.32516	2
0.10693	0.36297	3
0.1453	0.4024	4
0.18531	0.44351	5
0.22703	0.48638	6
0.27054	0.53108	7
0.3159	0.57769	8
0.36321	0.6263	9
0.41254	0.67698	10
0.46397	0.72983	11
0.51761	0.78494	12
0.57354	0.84241	13
0.63185	0.90233	14
0.69267	0.96482	15
0.75608	1.03	16
0.8222	1.0979	17
0.89115	1.1688	18
0.96305	1.2426	19
1.038	1.3197	20

ans =

   0.79364

And with $K_a = 200,000$ we reach the same value for $X_{20} = 1.2975$ and a 99.2 % removal rate using this model when calling `linearnonequicol(20,200000,0.72,0.01,7500, 7000,0.001,1.4)` as we did on p. 359 for a linear equilibrium relation and nonequilibrium stages.

Xj	Yj	j

0.0012944	0.010943	1
0.0017255	0.011259	2
0.0023565	0.01172	3
0.0032803	0.012396	4
0.0046326	0.013386	5
0.0066124	0.014835	6
0.0095106	0.016956	7
0.013753	0.020062	8
0.019965	0.024607	9
0.029057	0.031262	10
0.042368	0.041004	11
0.061855	0.055266	12
0.090382	0.076145	13
0.13214	0.10671	14
0.19328	0.15145	15
0.28278	0.21696	16
0.4138	0.31285	17
0.60561	0.45323	18
0.8864	0.65873	19
1.2975	0.95958	20

ans =

0.99218

The reciprocal $1/K_a$ of K_a measures how far the system is from equilibrium. The equilibrium stages are naturally the most efficient ones and they occur for high K_a values. For lower K_a values the system efficiency decreases. Here the tray efficiency is defined as the ratio between the separation efficiency for specific low K_a values and that for high K_a values. Note that high K_a values indicate that the system is very close to equilibrium.

6.2.2   Nonequilibrium Multistages with Nonlinear Equilibrium Relations

Next we consider the case where the equilibrium relation is not linear and the stages are not in equilibrium. Then the *rate of mass transfer* between the phases RMT becomes nonlinear, i.e.,

$$RMT = K_{gj} \cdot a_{mj} \cdot (Y_j - F(X_j))$$

for a nonlinear function $F(X_j)$ of X_j and $j = 1, 2, ..., N$. The transfer is from the gas phase to the liquid phase when RMT is positive. $F(X_j)$ is the equivalent gas-phase concentration $\bar{Y}_j$ that is in equilibrium with X_j.

RMT can also be expressed as

$$RMT = -K_{gj} \cdot a_{mj} \cdot (X_j - F'(Y_j))$$

where $F'(Y_j)$ is equal to $\bar{X}_j$, the equivalent liquid-phase concentration in equilibrium with Y_j. The transfer is from the liquid phase to the gas phase if $X_j > F'(Y_j)$, i.e., if RMT is negative.

This model leads to a set of nonlinear equations that can be solved numerically by using the multi-dimensional Newton method of Section 1.2. It is this more realistic and more accurate nonlinear case which we concentrate on in this book. We now present an efficient MATLAB program to solve this nonlinear and more general case.

We start with a modified version of the gas-phase molar-balance equation (6.99) and a modified version of the liquid-phase molar-balance equation (6.102) with a quadratic equilibrium relation $F(X_j) = Y_j = aX_j^2 + b$. This modification affects the term RMT by a nonlinearly, namely

$$RMT \; = \; K_a \cdot (Y_j - (aX_j^2 + b)) \,.$$

Here for simplicity we assume a constant mass-transfer coefficient times area of mass transfer K_a for all trays, i.e., $K_{gj}\bar{a}_{mj} = const = K_a$.

Thus for the gas phase we now have

$$VY_{j+1} \; = \; VY_j + K_a(Y_j - (aX_j^2 + b)) \,, \tag{6.114}$$

for $j = 1, 2, ..., N$. And for the liquid phase

$$LX_{j-1} + K_a(Y_j - (aX_j^2 + b)) \; = \; LX_j, \tag{6.115}$$

again for $j = 1, 2, ..., N$.

Our task is to solve these two joint nonlinear sets of equations in the unknowns

$$x = \begin{pmatrix} X_1 \\ \vdots \\ X_N \end{pmatrix} \text{ and } y = \begin{pmatrix} Y_1 \\ \vdots \\ Y_N \end{pmatrix} .$$

The two vectors x and $y \in \mathbb{R}^N$ help us to view and solve the $2N$ nonlinear equations defined in (6.114) and (6.115). For this purpose we rewrite the problem given in (6.114) and (6.115) as follows in matrix and vector form.

First we want to express the problem in terms of

$$z = \begin{pmatrix} x \\ y \end{pmatrix} \in \mathbb{R}^{2N} \text{ in standard form } f(z) = f\begin{pmatrix} x \\ y \end{pmatrix} = 0$$

for x, $y \in \mathbb{R}^N$ as defined above.

By collecting the squared terms of X_j, the unshifted and shifted Y_j terms, and the terms in (6.114), the N equations that are given to us in (6.114) for $j = 1, ..., N$ can be written jointly in vector form as

$$f_1\begin{pmatrix} x \\ y \end{pmatrix} = K_a{\cdot}a \begin{pmatrix} X_1^2 \\ \vdots \\ X_N^2 \end{pmatrix} - (V+K_a)\begin{pmatrix} Y_1 \\ \vdots \\ Y_N \end{pmatrix} + V\begin{pmatrix} Y_2 \\ \vdots \\ Y_N \\ Y_{N+1} \end{pmatrix} + K_a{\cdot}b \begin{pmatrix} 1 \\ \vdots \\ 1 \end{pmatrix} = \begin{pmatrix} 0 \\ \vdots \\ 0 \end{pmatrix} \in \mathbb{R}^N.$$

$$\tag{6.116}$$

Likewise, the N equations given by (6.115) for $j = 1, ..., N$ can be written jointly as

$$f_2 \begin{pmatrix} x \\ y \end{pmatrix} = K_a \cdot a \begin{pmatrix} X_1^2 \\ \vdots \\ X_N^2 \end{pmatrix} - L \begin{pmatrix} X_0 - X_1 \\ \vdots \\ X_{N-1} - X_N \end{pmatrix} - K_a \begin{pmatrix} Y_1 \\ \vdots \\ Y_N \end{pmatrix} + K_a \cdot b \begin{pmatrix} 1 \\ \vdots \\ 1 \end{pmatrix} = \begin{pmatrix} 0 \\ \vdots \\ 0 \end{pmatrix} \in \mathbb{R}^N.$$

(6.117)

[Our readers should work this out with pencil and paper in case $N = 3$ in order to fully understand the role of vectors in this context.]

Finally we define the function

$$f \begin{pmatrix} x \\ y \end{pmatrix} = \begin{pmatrix} f_1 \begin{Bmatrix} x \\ y \end{Bmatrix} \\ f_2 \begin{Bmatrix} x \\ y \end{Bmatrix} \end{pmatrix} \in \mathbb{R}^{2N}$$

whose zeros in $\mathbb{R}^{2N}$ we seek, because they solve the twice N nonlinear equations given in (6.114) and (6.115).

Our solution method of choice is Newton's iterative method, see Section 1.2. Newton's method requires us to evaluate the Jacobian matrix Df of

$$f = \begin{pmatrix} f_1 \\ f_2 \end{pmatrix} : \mathbb{R}^{2N} \to \mathbb{R}^{2N}$$

for the component functions f_i given in the equations (6.116) and (6.117). It is best to visualize DF as a block matrix comprised of 4 blocks , namely

$$Df \begin{pmatrix} x \\ y \end{pmatrix} = \begin{pmatrix} Df_1x & Df_1y \\ Df_2x & Df_2y \end{pmatrix}$$

(6.118)

where

$$Df_1x = \begin{pmatrix} \dfrac{\partial f_1 \begin{pmatrix} x \\ y \end{pmatrix}}{\partial X_j} \end{pmatrix}, \; Df_1y = \begin{pmatrix} \dfrac{\partial f_1 \begin{pmatrix} x \\ y \end{pmatrix}}{\partial Y_j} \end{pmatrix},$$

$$Df_2x = \begin{pmatrix} \dfrac{\partial f_2 \begin{pmatrix} x \\ y \end{pmatrix}}{\partial X_j} \end{pmatrix}, \; \text{and} \; Df_2y = \begin{pmatrix} \dfrac{\partial f_2 \begin{pmatrix} x \\ y \end{pmatrix}}{\partial Y_j} \end{pmatrix}$$

are the respective Jacobian matrices Df_i of f_i with respect to x or y.

Specifically for our f, we note that Df_1x is the diagonal N by N matrix with the entries $2K_a \cdot a \cdot X_j$ in position (j, j) and zeros everywhere else since each X_j occurs only once in equation (6.116), namely in the jth component function of f_1 in squared form with the coefficient $K_a \cdot a$.

Similarly Df_1y has constant diagonal entries $-(V + K_a)$ since each Y_j occurs in the jth component function with the coefficient $-(V + K_a)$ and in the $(j-1)^{st}$ component function with the coefficient V, giving rise to the first superdiagonal entries equal to V

and zeros everywhere else in Df_1y.

Likewise, Df_2x has the diagonal entries $2K_a \cdot a \cdot X_j + L$ from equation (6.117), its first subdiagonal entries are equal to $-L$, and the rest of the entries are zero. Finally Df_2y is $-K_a \cdot I_N$.

[*These statements should be worked out for $N = 3$ by our readers with paper and pencil again.*]

To iterate according to Newton's method, we have to update each approximate solution z_{old} by the modification term

$$ z_{mod} = -Df(z_{old})^{-1} f(z_{old}) \text{ with } z_{old} = \begin{pmatrix} x_{old} \\ y_{old} \end{pmatrix} \in \mathbb{R}^{2N} $$

according to formula (1.5) of Section 1.2.

One can avoid solving the $2N$ by $2N$ linear system

$$ Df(z_{old}) \, w = f(z_{old}) \tag{6.119} $$

directly at $O((2N)^3) = O(8N^3)$ cost by recognizing the **structure** of the system matrix $Df(z_{old})$ and simplifying it first. Note that the entries of the system matrix $Df(z_{old})$ are defined in formula (6.118) and that we have to use $z_{old} = (x_{old}, \, y_{old})^T$ in place of $(x, \, y)^T$ here:

If we multiply both the system matrix $Df(z_{old})$ and the right-hand side $b = f(z_{old})$ of (6.119) by the matrix product

$$ \begin{pmatrix} I_N & O \\ O & \frac{-1}{K_a} I_N \end{pmatrix} \begin{pmatrix} -I_N & \frac{-1}{K_a} Df_1y \\ O & I_N \end{pmatrix} $$

from the left, we obtain the updated system matrix

$$ \widetilde{Df}(z_{old}) = \begin{pmatrix} I_N & O \\ O & \frac{-1}{K_a} I_N \end{pmatrix} \cdot \begin{pmatrix} -I_N & \frac{-1}{K_a} Df_1y \\ O & I_N \end{pmatrix} \cdot \begin{pmatrix} Df_1x & Df_1y \\ Df_2x & Df_2y \end{pmatrix} = $$

$$ = \begin{pmatrix} -Df_1x - \frac{1}{K_a} Df_1y \cdot Df_2x & O \\ \frac{-1}{K_a} Df_2x & I_N \end{pmatrix} \tag{6.120} $$

and the updated right-hand side

$$ \widetilde{b} = \begin{pmatrix} \frac{-1}{K_a} Df_1y \cdot f_2(z_{old}) - f_1(z_{old}) \\ \frac{-1}{K_a} f_2(z_{old}) \end{pmatrix}. \tag{6.121} $$

[*Readers: please evaluate the triple matrix product in the reduction formula (6.120) with pencil and paper.*]

The first block row of the updated system $\widetilde{Df}(z_{old})w = \widetilde{b}$ with $w = (w_x, \, w_y(^T \in \mathbb{R}^{2N}$ can be solved for $w_x = (w_{x1}, \, \ldots, \, w_{xN})^T \in \mathbb{R}^N$ in $O(N)$ time since all intermediate matrices are at most bidiagonal matrices. Then the values of w_{xj} can be used to solve the second

block row of $\widetilde{Df}(z_{old})w = \widetilde{b}$ for $w_y = (w_{y1}, \; \ldots \; , \; w_{yN})^T$, also in $O(N)$ operations; see
problem 2 at the end of this chapter. Finally $z_{new} = z_{old} + (w_x, \; w_y)^T \in \mathbb{R}^{2N}$ and all this
can be achieved in $O(N)$ operations.

The following MATLAB code `quadnonequicol.m` performs these steps naively without implementing the above labor reducing operations. We leave this modification to the
reader as a valuable exercise.

```
function [relrem,X,Y] = quadnonequicol(N,Ka,a,b,L,V,X0,YNp1)
% [relrem,X,Y] = quadnonequicol(N,Ka,a,b,L,V,X0,YNp1)
% Sample call: quadnonequicol(6,1000,0.72,0.01,7500,7000,0.001,1.4)
% Input : N = number of trays
%         Ka = constant product of mass transfer coeff and area of transfer
%         a, b = coefficients in quadratic equilibrium relation F(X) = aX^2 + b
%         L = liquid flow rate
%         V = gas flow rate
%         X0, YNp1 = input variables X_0 and Y_{N+1}
% Output: X and Y, the molar fractions of component A in the liquid and gas
%         phases, respectively, in trays 1 to N.
%         relrem = (YNp1 - Y1)/YNp1 = relative removal rate

xstart = linspace(X0,(X0+YNp1)/2,N)';        % starting values for [X;Y] :
ystart = linspace((X0+YNp1)/2,YNp1,N)'; xystart = [xstart;ystart];
xymod = df(xstart,Ka,a,L,V)\f(xstart,ystart,Ka,a,b,L,V,X0,YNp1); % Newton mod
xylast = xystart - xymod; i=1;               % next Newton iterate
while norm(xymod)/norm(xylast) > 0.01 & i < 100,  % Further Newton steps
    xystart = xylast; xstart = xystart(1:N); ystart = xystart(N+1:end);
    xymod = df(xstart,Ka,a,L,V)\f(xstart,ystart,Ka,a,b,L,V,X0,YNp1);
    xylast = xystart - xymod; i=i+1; end,    % until < 1% error
relrem = (YNp1 - xylast(N+1))/YNp1;          % relative removal rate
disp('        Xj              Yj                j'),
disp([xylast(1:N),xylast(N+1:end),[1:N]'])

function F = f(x,y,Ka,a,b,L,V,X0,YNp1)       % quadratic RMT function f(x,y)
  lx = length(x);
  F = [Ka*a*x.^2 - (V+Ka)*y + V*[y(2:lx);YNp1] + Ka*b*ones(lx,1);
       Ka*a*x.^2 - L*([X0;x(1:lx-1)] - x) - Ka*y + Ka*b*ones(lx,1)];

function DF = df(x,Ka,a,L,V)                 % Jacobian matrix of f(x,y)
  lx = length(x);
  DF = [2*Ka*a*diag(x), -(V + Ka)*eye(lx) + V*diag(ones(1,lx-1),1);
        2*Ka*a*diag(x) + L*eye(lx) - L*diag(ones(1,lx-1),-1), - Ka*eye(lx)];
```

Here is the output of various calls of `quadnonequicol`. Note that the final screen output `ans` of each call gives the percentage of removal, i.e., if `ans` = 0.066486, as in the
first call `quadnonequicol(6,100,0.72,0.01,750,7000,0.001,1.4)` for $K_a = 100$ and
$L = 750$ below, then 6.65% is removed.

Xj	Yj	j
0.17111	1.3069	1

0.33565	1.3251	2
0.49029	1.3428	3
0.63189	1.3593	4
0.7586	1.3745	5
0.86974	1.3881	6

ans =

0.066486

Next we change K_a to 1000 and call `quadnonequicol(6,1000,0.72,0.01,750,7000,` `0.001,1.4)` to get the following results that show an increase in the removal rate to 10.615 %.

Xj	Yj	j
0.89213	1.2514	1
1.2277	1.3469	2
1.3384	1.3828	3
1.3735	1.3947	4
1.3845	1.3985	5
1.388	1.3996	6

ans =

0.10615

Next we show the results of two runs with $K_a = 100$ and 1000, but with $L = 7500$ instead of 750 and for 6 trays:

```
>> quadnonequicol(6,100,0.72,0.01,7500,7000,0.001,1.4)
```

Xj	Yj	j
0.018022	1.2869	1
0.035278	1.3051	2
0.052766	1.3236	3
0.070483	1.3423	4
0.088425	1.3613	5
0.10659	1.3805	6

ans =

0.080809

Notice that the above increase in efficiency is quite limited from 6.65% to 8.1% for the change of L from 750 to 7500 as long as K_a is low at 100 and the number of trays remains at 6.

If we increase K_a to 1000 as promised, then we obtain a much larger efficiency increase:

```
>> quadnonequicol(6,1000,0.72,0.01,7500,7000,0.001,1.4)
```

Xj	Yj	j
0.090842	0.68976	1
0.19082	0.78602	2
0.29993	0.89313	3
0.41661	1.01	4
0.53875	1.1351	5
0.66389	1.2659	6

ans =

0.50732

Because the value of $K_a = 1000$ is quite large, we observe a considerable increase from 10.615% removal for $L = 750$ to 50.7% removal for $L = 7500$.

Now we change the number of trays to 20 and vary K_a as before while $L = 7500$ and we have 6 trays.

```
>> quadnonequicol(20,100,0.72,0.01,7500,7000,0.001,1.4)
```

Xj	Yj	j
0.015029	1.0623	1
0.029252	1.0773	2
0.043668	1.0926	3
0.058276	1.108	4
0.073074	1.1237	5
0.08806	1.1395	6
0.10323	1.1556	7
0.11859	1.1719	8
0.13413	1.1883	9
0.14984	1.205	10
0.16574	1.2218	11
0.1818	1.2388	12
0.19804	1.256	13
0.21445	1.2734	14
0.23101	1.291	15
0.24774	1.3088	16
0.26462	1.3267	17
0.28166	1.3448	18
0.29884	1.363	19
0.31617	1.3814	20

```
ans =
    0.2412
```

For this data the removal rate of 24.12% is much higher than the 8.1% for the previous corresponding case.

Next we try increasing K_a to 1000 for 20 trays.

```
>> quadnonequicol(20,1000,0.72,0.01,7500,7000,0.001,1.4)
```

Xj	Yj	j
0.027771	0.21134	1
0.058116	0.24002	2
0.092303	0.27253	3
0.13055	0.30916	4
0.17303	0.35015	5
0.21982	0.39566	6
0.27088	0.44578	7
0.32607	0.50049	8
0.38511	0.55962	9
0.4476	0.62289	10
0.51298	0.68984	11
0.5806	0.75989	12
0.64972	0.83234	13
0.71954	0.9064	14

0.78924	0.9812	15
0.85801	1.0559	16
0.92513	1.1296	17
0.98992	1.2015	18
1.0518	1.2709	19
1.1104	1.3372	20

ans =

0.84904

Here the removal rate has increased again from 50.7% with 6 trays to around 85 % when using 20 trays.

If we increase K_a further to 200,000 and leave L unchanged, we can replicate the removal percentage of 99.2% obtained for our earlier simpler model on p. 359 with this more sophisticated model as seen below.

```
>> linearnonequicol(20,200000,0.72,0.01,7500,7000,0.001,1.4)
```

Xj	Yj	j
0.0012944	0.010943	1
0.0017255	0.011259	2
0.0023565	0.01172	3
0.0032803	0.012396	4
0.0046326	0.013386	5
0.0066124	0.014835	6
0.0095106	0.016956	7
0.013753	0.020062	8
0.019965	0.024607	9
0.029057	0.031262	10
0.042368	0.041004	11
0.061855	0.055266	12
0.090382	0.076145	13
0.13214	0.10671	14
0.19328	0.15145	15
0.28278	0.21696	16
0.4138	0.31285	17
0.60561	0.45323	18
0.8864	0.65873	19
1.2975	0.95958	20

ans =

0.99218

For this very high value of K_a we approach 100% removal (99.218%) and the result is close to the earlier equilibrium case.

To clarify the differences in the removal efficiencies of absorption columns, we look at the linear and quadratic equilibrium relations in more detail now.

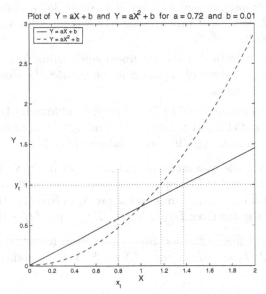

Graph of the **linear** and quadratic equilibrium relations
Figure 6.21

We first use the mass-transfer relation between the phases in terms of X and Y, namely

$$RMT = K_{gj} \cdot a_{mj} \cdot (Y_j - F(X_j)) .$$

For a fixed Y value, such as $Y = y_t = 1$ in Figure 6.21 and the linear equilibrium relation $Y = a \cdot X + b$, the mass transfer will be from the gas phase to the liquid phase when $RMT > 0$. Specifically for $a = 0.72$ and $b = 0.01$, this mass transfer from the gas to the liquid phase will occur from $X = 0$ until $1 = 0.72 \cdot X + 0.01$, or until $X = 1.375$. This value of X can be obtained by solving the linear equation for X. For $X > 1.375$, the transfer will be from the liquid phase to the gas phase since then $RMT < 0$.
Repeating this process for the quadratic equilibrium relation, we have to solve $1 = 0.72 \cdot X^2 + 0.01$ for X instead, making $X = \sqrt{1.375} = 1.1726$. These calculations corroborate the plot in Figure 6.21. Thus for a quadratic equilibrium relation, the mass transfer is from the gas phase to the liquid phase for $0 < X < 1.1726$ with our a and b values, and it is in the opposite direction if $X > 1.1726$.
Vice versa, if we use the relation

$$RMT = -K_{gj} \cdot a_{mj} \cdot (X_j - F'(Y_j))$$

and fix $X = x_t = 0.8$ for example, then we can obtain the same mass-transfer direction analysis by looking for the Y values for which RMT is negative.

Exercises for 6.2

1. (a) Run **linearcolumn.m** for our data on p. 358 and $N = 2, 3.4, 8$, and 10. Compare the removal rates for various N and interpret the results. Vary L and V as well.

(b) Run `linearcolumn.m` for our data on p. 359 and $N = 5, 10.15, 25$, and 50 with $L = 7500$. Compare the removal rates for various N and interpret the results.

(c) Repeat part (a) for the nonlinear equilibrium function $Y_j = aX_j^2 + b$ with the same values of a and b as on p. 358. Use our MATLAB program `quadcolumn.m`.

(d) Develop and test a MATLAB code `findNquadcolumn.m` that mimics our code `findNlinearcolumn.m` for finding the number of required trays, but for a quadratic equilibrium relation instead.

2. Rewrite the code `quadnonequicol.m` to run in $O(N)$ time as detailed on p. 376. To achieve this fast operations count, you need to implement all matrix times matrix or matrix times vector products in $O(N)$ time by recognizing the sparsity of the matrices Df_1x, Df_1y, Df_2x, and Df_2y that make up $Df(z_{old})$.

3. Complete the mass-transfer direction analysis for a two-phase system from the relation $RMT = -K_{gj} \cdot a_{mj} \cdot (X_j - F'(Y_j))$ as indicated at the end of this section.

6.3   Isothermal Packed Bed Absorption Towers

A packed bed absorption tower is a distributed system and Figure 6.22 shows the molar flow rates across a small element of this tower.

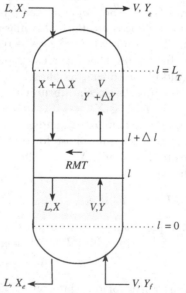

Schematic diagram of a packed absorption tower
Figure 6.22

6.3.1   Model Development

We consider the absorption of component A from the gas (vapor) phase to the liquid phase. The rate of mass transfer RMT from the vapor phase to the liquid phase is given by

$$RMT = a_m K_g (Y - (aX + b)) , \qquad (6.122)$$

where a_m is the specific area, i.e., the mass-transfer area per unit volume of the column and $(aX + b) = Y^*$ is the gas-phase mole fraction at equilibrium with X in the liquid phase. The molar-balance equation for the gas-phase element gives us

$$VY = V(Y + \Delta Y) + A_c \cdot \Delta l \cdot a_m \cdot K_g (Y - (aX + b)) . \qquad (6.123)$$

Here V is the total molar flow rate of the gas phase (assumed to be constant), Y is the mole fraction of component A in the gas phase, X is the mole fraction of component A in the liquid phase, and A_c is the cross sectional area of the tower.
By simple manipulation and rearrangement of (6.123) we obtain

$$V \frac{\Delta Y}{\Delta l} = -A_c \cdot a_m \cdot K_g (Y - (aX + b)) .$$

After taking the limit of $\Delta l \to 0$ we get

$$V \frac{dY}{dl} = -A_c \; a_m \cdot K_g (Y - (aX + b)) . \qquad (6.124)$$

With the initial condition $Y = Y_f$ at $l = 0$, the molar balance for the liquid phase satisfies

$$LX = L(X + \Delta X) + A_c \cdot \Delta l \cdot a_m \cdot K_g (Y - (aX + b)) ,$$

where L is the total molar flow rate of the liquid phase (also assumed constant).
By simple manipulation and rearrangement we obtain

$$L \frac{\Delta X}{\Delta l} = -A_c \cdot a_m \cdot K_g (Y - (aX + b)) . \qquad (6.125)$$

Taking the limit as $\Delta l \to 0$ in (6.125) we get

$$L \frac{dX}{dl} = -A_c \cdot a_m \cdot K_g (Y - (aX + b)) \qquad (6.126)$$

with the boundary condition $X = X_f$ at $l = H_t$. Here H_t is the total height of the absorption column.
Notice that the set of joint differential equations

$$V \frac{dY}{dl} = -A_c \cdot a_m \cdot K_g (Y - (aX + b)) , \quad \text{with the initial condition} \quad Y = Y_f \text{ at } l = 0,$$

and $\qquad\qquad\qquad\qquad\qquad\qquad\qquad\qquad\qquad\qquad\qquad\qquad\qquad (6.127)$

$$L \frac{dX}{dl} = -A_c \cdot a_m \cdot K_g (Y - (aX + b)) , \text{ for the boundary condition } X = X_f \text{ at } l = H_t$$

forms a system of two-point BVPs.

6.3.2   Manipulation of the Model Equations

Due to the simplicity of this illustrative problem and of the linear equilibrium isotherm $Y = aX + b$ we can reduce the two coupled two point boundary value differential equations (6.127) to one differential equation as follows.

Subtracting equation (6.126) from (6.124) we get

$$V\frac{dY}{dl} - L\frac{dX}{dl} = 0 \ .$$

Thus by the product rule of differentiation, used in reverse, we have

$$\frac{d(VY - LX)}{dl} = 0 \quad \text{or} \quad VY - LX = C_1 \ . \tag{6.128}$$

At $l = 0$ we have $Y = Y_f$ and $X = X_e$, where X_e is the mole fraction of component A in the liquid phase at the exit or the bottom of the column.

Using this initial condition in equation (6.128) we find the value of C_1 as

$$VY_f - LX_e = C_1 \ .$$

Substituting this value of C_1 into equation (6.128) we then have

$$VY - LX = VY_f - LX_e \quad \text{or} \quad VY = L(X - X_e) + VY_f \ .$$

By rearrangement we see that

$$Y = \frac{L}{V}(X - X_e) + Y_f \ . \tag{6.129}$$

Substituting the value of Y in equation (6.129) into equation (6.126) we finally get

$$L\frac{dX}{dl} = -A_c \cdot a_m \cdot K_g \left(\frac{L}{V}(X - X_e) + Y_f - (aX + b) \right) \tag{6.130}$$

with the boundary condition $X(H_t) = X_f$ at $l = H_t$.

6.3.3   Discussion and Results for both the Simulation and the Design Problem

The two coupled equations (6.129) and (6.130) can be used to solve two different problems:

(a) *a simulation problem*, and

(b) *a design problem* for packed bed absorption towers.

(a) *The Simulation Problem*:

If the system parameters L, V, A_c, and $a_m \cdot K_g$ are known, as well as the feed concentrations X_f and Y_f of the liquid and gas, respectively, then the two equations (6.129) and (6.130) become an initial value problem that we need to solve in order to find the unknown liquid output concentration X_e of the given absorption column. The gaseous output concentration Y_e can then be determined from X_e using equation (6.129).

Notice that equation (6.130) will be solved from the top to the bottom of the column, i.e., from $l = H_t$ where we know $X(H_t) = x_f$ down to the bottom at $l = 0$. We also note that the value of X_e used in equation (6.130) is an assumed or approximate value and as equation (6.130) is integrated from $l = H_t$ to $l = 0$, at $l = 0$ the solution provides us with a value of X_e that will generally be different from the assumed or approximate value of X_e at the start of the integration process for equation (6.130). We can find the correct value for X_e quickly by an iterative process: Take two guesses X_{e1} and X_{e2} for X_e and find the two values $X_1(0)$ and $X_2(0)$ of X at height zero of the corresponding IVP solutions. Note that when X_{e1} and X_{e2} are chosen correctly, then $X_1(0) = X_{e1}$ and $X_2(0) = X_{e2}$. Our iterative process involves solving two initial value problems (6.130) with X_{e1} and X_{e2}, respectively, as parameters and evaluating $X_1(0)$ and $X_2(0)$. The computed values will generally differ from each other. Our task is to find a value of X_e that we can use inside the IVP (6.130) at $l = H_t$ so that the IVP solution $X(l)$ gives us the equal exit X concentration $X_{X_e}(0) = X_e$ back at height zero, i.e., at the lower exit of the tower as depicted in Figure 6.22.

In Figure 6.23 we plot the typical scenario for this **fixed-point iteration** to illustrate our algorithm.

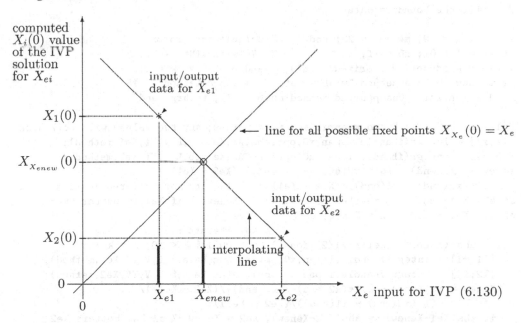

Fixed point iteration scheme for the simulation problem
Figure 6.23

Figure 6.23 plots the computed IVP solution $X_i(0)$ at height zero versus the DE parameter X_{ei} for two initial guesses X_{ei} at the points marked by $*$. This data is interpolated by a solid black line. We also draw the "fixed point" line in the plot as the angular bisector of the first quadrant where $X_{X_{e}}(0) = X_e$. A new approximation X_{enew} for the DE parameter X_e is taken where the two lines intersect. This intersection is denoted by a small circle in Figure 6.23.

Here is our MATLAB routine absorbtoweriterXe.m that finds the output liquid concentration X_e for the simulation problem (a) according to Figure 6.23.

```
function [Xenew, check] = absorbtoweriterXe(Ac,Ht,Ka,a,b,L,V,Xf,Yf,method)
%     Xenew = absorbtoweriterXe(Ac,Ht,Ka,a,b,L,V,Xf,Yf,method)
% Sample call : absorbtoweriterXe(2.5,20,68,.72,0.01,720,7000,0.001,1.4,23)
% Input : Ac (= cross section area of tower),
%         Ht (= height of tower),
%         Ka (= K_g * a_m), a, b, L, V,
%         Xf (= liquid feed at top input),
%         Yf (= gas feed at bottom entrance); These are the system parameters
%         method : 15, 23, 45, or 113;
%                    method of MATLAB integrator: ode15s, ode23, ode45, or ode113
% Output: Plot of  solution X and Y curves from l = 0 to l = Ht.
%         Xenew is the liquid output
% to solve:
%     dy/dx = 1/L [-A*Ka* (L/V (y-Xe) + Yf -(a*y + b))]
%     choose the initial condition: x(0) = Xe and iteratively adjust Xe to
%     fit the boundary data

if nargin == 9, method = 23; end      % default integrator
lspan = [Ht 0]; x0 = Xf;              % IVP with X(Ht) = Xf;
options = odeset('Vectorized','on'); fhandle = @tow; k = 0;
if method ~= 23 & method ~= 45 & method ~= 113 & method ~= 15
   disp('Error : Unsupported method number !'), return, end

Xe1 = 2; Xe2 = 1; % two generic starting points; any two values will work here
[l1,x1] = integr(fhandle,lspan,x0,options,Ac,Ka,a,b,L,V,Yf,Xe1,method);
[l2,x2] = integr(fhandle,lspan,x0,options,Ac,Ka,a,b,L,V,Yf,Xe2,method);
Xenew = (x1(end) - Xe1*(x2(end) - x1(end))/(Xe2 - Xe1))/...
   (1-(x2(end) - x1(end))/(Xe2 - Xe1));  % looking for self-replicating Xenew
if abs(Xe1-Xenew) <= abs(Xe2-Xenew), Xe2 = Xenew; % if Xe1 is better than Xe2
else,  Xe1 = Xe2; Xe2 = Xenew; end
                              % Now iterate :
while abs((Xenew - (Xe1+Xe2)/2)/Xenew) >= 0.0001 & k < 10, k = k +1;
   [l1,x1] = integr(fhandle,lspan,x0,options,Ac,Ka,a,b,L,V,Yf,Xe1,method);
   [l2,x2] = integr(fhandle,lspan,x0,options,Ac,Ka,a,b,L,V,Yf,Xe2,method);
   Xenew = (x1(end) - Xe1*(x2(end) - x1(end))/(Xe2 - Xe1))/...
           (1-(x2(end) - x1(end))/(Xe2 - Xe1));
   if abs(Xe1-Xenew) <= abs(Xe2-Xenew), Xe2 = Xenew; % if Xe1 betters Xe2
   else,  Xe1 = Xe2; Xe2 = Xenew; end, end
                              % Find best X and Y  from Xenew
[l,x] = integr(fhandle,lspan,x0,options,Ac,Ka,a,b,L,V,Yf,Xenew,method);
y = L/V * (x - Xenew) + Yf; rr = (max(y) - min(y))/max(y);
```

```
check = L*Xf + V*Yf - L*x(end) - V*y(1); % mass balance check

subplot(2,1,1);                     % plot graph of solution h(X) to BVP
  plot(l,x,'Linewidth',1), hold on, v = axis; plot(v(1),Xenew,'+r'),
  xlabel(['      h          (with computed X_e = ',num2str(Xenew,'%9.6g'),' )'],...
      'FontSize',12), ylabel('X    ','Rotation',0,'FontSize',12),
  title([{'Absorption Tower (using an IVP iteration) : '},{'   '},{[' A_c = ',...
    num2str(Ac,'%5.4g'),', H_t = ',num2str(Ht,'%5.4g'),', K_a = ',...
    num2str(Ka,'%5.4g'),', a = ',num2str(a,'%5.4g'),', b = ',...
    num2str(b,'%5.4g'),', L = ',num2str(L,'%5.4g'),', V = ',...
    num2str(V,'%5.4g'),', X_f = ',num2str(Xf,'%5.4g'),...
    ', Y_f = ',num2str(Yf,'%5.4g')]}],'FontSize',12), hold off
subplot(2,1,2);                     % plot graph of associated solution h(Y)
  plot(l,y,'Linewidth',1), hold on, v = axis; plot(v(1),y(1),'+g'),
  xlabel(['              h    (rel. removal rate = ',num2str(rr,'%5.4g'),...
      ' ; Y_e = ',num2str(y(1),'%9.6g'),')'],'FontSize',12),
  ylabel('Y    ','Rotation',0,'FontSize',12), hold off

function [l,x] = integr(fhandle,lspan,x0,options,Ac,Ka,a,b,L,V,Yf,Xe,method)
  ghandle = @tow;          % module for the integration
  if method == 15,
    [l,x] = ode15s(ghandle,lspan,x0,options,Ac,Ka,a,b,L,V,Yf,Xe); end
  if method == 45,
    [l,x] = ode45(ghandle,lspan,x0,options,Ac,Ka,a,b,L,V,Yf,Xe); end
  if method == 23,
    [l,x] = ode23(ghandle,lspan,x0,options,Ac,Ka,a,b,L,V,Yf,Xe); end
  if method == 113,
    [l,x] = ode113(ghandle,lspan,x0,options,Ac,Ka,a,b,L,V,Yf,Xe); end

function dydt = tow(l,x,A,Ka,a,b,L,V,Yf,Xe)          % r.h.s of the DE
  dydt = 1/L * (-A*Ka*(L/V * (x-Xe)+Yf-(a*x+b)));    %  use all inputs
```

A close look at the intermediate output values of Xenew in a call of absorbtoweriterXe reveals that convergence is almost instantaneous. This is so because the role of X_e in the IVP (6.130) is linear, indicating that one linear interpolation will give the exact self-replicating value of X_e for the simulation problem in case of a linear equilibrium relation. For a quadratic equilibrium relation, refer to problem 3 in the Chapter exercise section.

Equation (6.130) is a single DE which can be solved as an IVP once $X(0) = X_e$ is specified. Its solution gives the value of $X(l)$ for every value of l and we can use X to evaluate the corresponding Y values using the simple relation (6.129).
Notice that this reduction of the two DEs into one has not destroyed the two point boundary value nature of the problem because we have to find a solution with the known boundary values of Y_f at $l = 0$ and of X_f at $l = H_f$ while X_e is unknown. We can solve this problem iteratively by assuming a value for X_e, integrating (6.130) from $l = 0$ to $l = H_f$, checking the value of X_f that was found against the desired value, and then varying X_e to better approximate X_f.

Alternately, the problem can be viewed as an initial value differential equation with the boundary condition $X(0) = X_e$ on the right-hand side of equation (6.130), once the boundary value $X(H_t) = X_f$ is specified.

We want to compare the solution for this continuous packed bed model with our earlier computations of the values of X_j and Y_j. For this purpose we first determine the value of $K_a = K_g \cdot a_m$ via quadnonequicol.m. A call of quadnonequicol(6,68,0.72,0.01,720, 7000,0.001,1.4) indicates that we have about 5% removal for our original system data with 6 trays, $K_a = 68$, and $L = 720$:

```
      Xj             Yj            j
    0.12461         1.33          1
    0.24634         1.3427        2
    0.36436         1.3552        3
    0.47708         1.3673        4
    0.58324         1.3789        5
    0.68193         1.3898        6
ans =

    0.050028
```

Thus we settle on 68 for the value of $K_a = K_g \cdot a_m$ used in equation (6.130).

We can solve the corresponding IVP (6.130) for a set value of $X_e = 0.001$ or a set value of $X_f = 1.9$ via our MATLAB program absorbtowerivp.m on the CD by alternately setting the input parameter called zero equal to 0 and starting the integration from $(0, X_e)$ or by setting zero not equal to 0 and thereby starting the integration from (H_t, X_f), as depicted in Figure 6.22.

Figures 6.24 and 6.25 show the graphical and numerical output of the X and Y profiles, as well as X_e, Y_e and the removal rate when calling absorbtoweriterXe(2.5,20,68, .72,0.01,720,7000,0.001,1.4) and absorbtoweriterXe(2.5,20,600,.72,0.01,720, 7000,0.001,1.4), respectively.

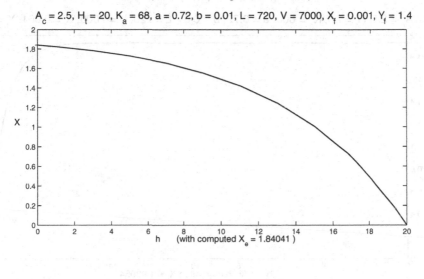

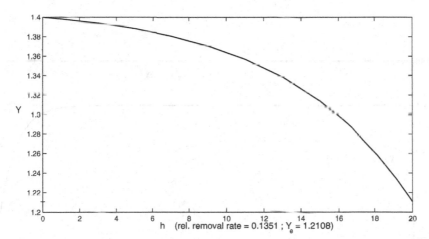

X and Y profiles for the simulation problem (6.129) and (6.130) with $K_a = 68$

Figure 6.24

The increase in the value of K_a from 68 in Figure 6.24 to 600 in Figure 6.25 shows that the increase in the rate of mass transfer increases the efficiency per unit height of the column. Thus a much smaller column can be used to achieve the same separation for $K_a = 600$.

Absorption Tower (using an IVP iteration) :

$A_c = 2.5$, $H_t = 20$, $K_a = 600$, $a = 0.72$, $b = 0.01$, $L = 720$, $V = 7000$, $X_f = 0.001$, $Y_f = 1.4$

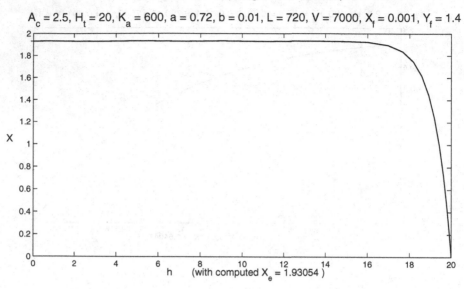

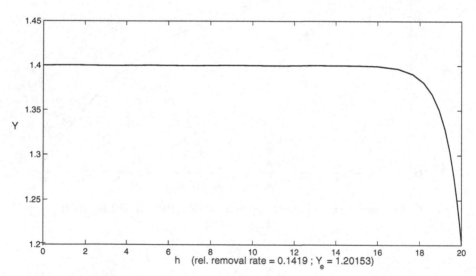

X and Y profiles for the simulation problem (6.129) and (6.130) with $K_a = 600$

Figure 6.25

We include two more graphs for increasing values of $L = 7200$ and $72{,}000$ and $K_a = 600$.

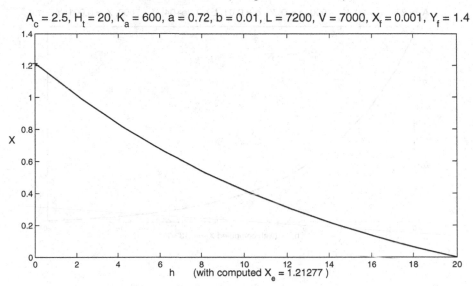

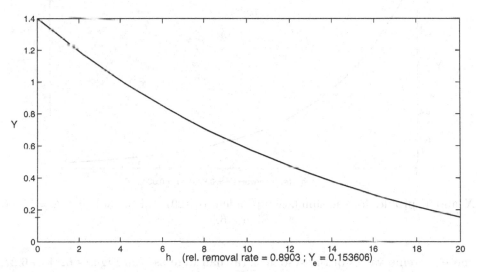

X and Y profiles for the simulation problem (6.129) and (6.130) with $L = 7200$

Figure 6.26

Absorption Tower (using an IVP iteration) :

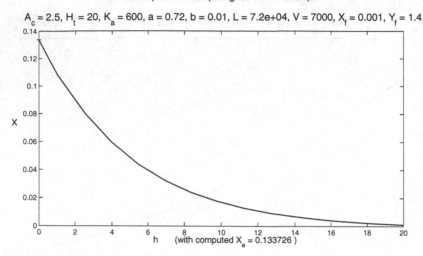

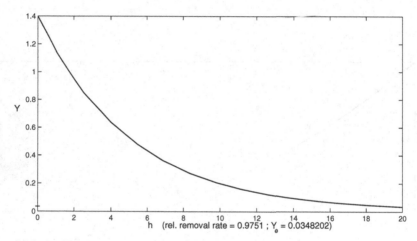

X and Y profiles for the simulation problem (6.129) and (6.130) with $L = 72000$
Figure 6.27

For more extreme values of the system data than were used in Figures 6.24 to 6.27, we note that the $+$ mark on the vertical axis of the top X plot may not coincide with the left, i.e., the bottom end of the X profile as it should. In this case we advise to use the extended call [Xe, check] = absorbtoweriterXe(... ... ...) that displays the error called check to equation (6.129) at height zero on screen. This error can be improved by tightening the allowed relative error bound 0.0001 of the solution X to (6.130) used in the while abs((Xenew - (Xe1+Xe2)/2)/Xenew) >= 0.0001 & k < 10, ... code line near the upper third of our absorbtoweriterXe.m code.

Notice that an increase of the liquid flow rate together with a high K_a value give us an efficient separation with a low exit-flow concentration, see Figure 6.26. As the flow rate increases further (for the same K_a), the efficiency of the column increases also and the liquid exit concentration becomes lower as depicted in Figure 6.27.

(b) *The Design Problem*:

Differing from the previous simulation problem in which the apparatus and system parameters, as well as the feed rates were given to us and we needed to find the exit liquid concentration X_e, we now consider the task of designing an absorption tower with a specified cross-sectional area A_c, the given system parameter values L, V, and $K_a = a_m \cdot K_g$, and the known feed rates X_f and Y_f for a linear equilibrium relation $Y = aX + b$.

In the design problem, our task is to determine H_t so that the liquid output X_e at height zero is as desired. Note that not all X_e values are possible here due to the equilibrium relation $Y = aX + b$ which links X_e to Y_f according to the formula

$$X_e = \frac{Y_f - b}{a} \, . \tag{6.131}$$

This limits the possible liquid output concentrations to $X_e \leq (Y_f - b)/a$.

To solve this problem we again proceed iteratively. For two initial guesses H_{t1} and H_{t2} of the tower height, we solve (6.130) and record the computed output concentrations X_{e1} and X_{e2}. We use linear interpolation between the two data points to find an H value H_{tnew} where the line connecting (H_{t1}, X_{e1}) and (H_{t2}, X_{e2}) reaches the value of the desired X_e and repeat the process until the relative difference between the computed $X_{e..}$ value and the desired X_e is negligibly small.

If the desired X_e exceeds what can be reached according to the equilibrium relation at height zero, our program automatically computes H_t for the maximal possible $X_e = (Y_f - b)/a$ and annotates the graph accordingly.

Here is our MATLAB code `absorbtoweriterHt.m` that solves the design problem numerically.

```
function [Htnew, check] = absorbtoweriterHt(Ac,Ka,a,b,L,V,Xf,Xe,Yf,method)
%      Xenew = absorbtoweriterHt(Ac,Ka,a,b,L,V,Xf,Xe,Yf,method)
% Sample call :
%   absorbtoweriterHt(2.5,68,.72,0.01,720,7000,0.001,1.8404,1.4,23)
% Input : Ac (= cross section area of tower),
%         Ka (= K_g * a_m), a, b, L, V,
%         Xf (= liquid feed at top input),
%         Xe (= desired liquid output),
%         Yf (= gas feed at bottom entrance); These are the system parameters
%         method : 15, 23, 45, or 113;
%                  method of MATLAB integrator: ode 15s, ode23, ode45, or ode113
% Output: Plot of  solution X and Y curves from l = 0 to l = Ht.
%         Ht = necessary height of column
% to solve:
%    dy/dx = 1/L [-A*Ka* (L/V (y-Xe) + Yf -(a*y + b))]
%    choose an initial height and continue integrating until desired Xe is
```

```
%    reached

if nargin == 9, method = 23; end     % default integrator and preparations
Ht1 = 200/Ka; Ht2 = 2*Ht1; Xemax = (Yf - b)/a; toolarge = 0;
                                     % compute max possible Xe
options = odeset('Vectorized','on'); fhandle = @tow; k = 0; Hmax = 1000;
if method ~= 23 & method ~= 45 & method ~= 113 & method ~= 15
  disp('Error : Unsupported method number !'), return, end
if Xe >= Xemax,                      % reduce Xe to Xemax iff too large
    disp(['  desired Xe exceeds the physically possible; ',...
          ' X_e is set to max possible value ',num2str(Xemax,'%10.8g')]),
    Xe = Xemax; toolarge = 1; end

lspan1 = [Ht1 0]; lspan2 = [Ht2 0]; x0 = Xf;   % prepare for two integrations
[l1,x1] = integr(fhandle,lspan1,x0,options,Ac,Ka,a,b,L,V,Yf,Xe,method);
[l2,x2] = integr(fhandle,lspan2,x0,options,Ac,Ka,a,b,L,V,Yf,Xe,method);
Htnew = (Xe - x1(end))/(x2(end) - x1(end)) * (Ht2 - Ht1) + Ht1; % find Htnew
if abs(Xe-x1(end)) <= abs(Xe-x2(end)), Ht2 = Htnew; % if Ht1 is better than Ht2
else,  Ht1 = Ht2; x1 = x2; Ht2 = Htnew; end,
if abs(Htnew) > Hmax,
    disp(['  bad data: Ht exceeds ',num2str(Hmax,'%6.5g'),' m']), return, end
                                    % Now iterate :
while abs((x1(end) - Xe)/Xe) >= 10^-5 & k < 10 & abs(Htnew) <= Hmax
  k = k +1; lspan2 = [Ht2 0];             % evaluate X for new Ht value only
  [l2,x2] = integr(fhandle,lspan2,x0,options,Ac,Ka,a,b,L,V,Yf,Xe,method);
  Htnew = (Xe - x1(end))/(x2(end) - x1(end)) * (Ht2 - Ht1) + Ht1; %interpolate
  if abs(Xe-x1(end)) <= abs(Xe-x2(end)), Ht2 = Htnew; % if Ht1 is better
  else,  Ht1 = Ht2; x1 = x2; Ht2 = Htnew; end, end,

                                    % Find best X and Y  for Htnew
if abs(Htnew) <= Hmax
  [l,x] = integr(fhandle,[Htnew 0],x0,options,Ac,Ka,a,b,L,V,Yf,Xe,method);
  y = L/V * (x - Xe) + Yf; rr = (max(y) - min(y))/max(y);
  check = L*Xf + V*Yf - L*x(end) - V*y(1); % mass balance check
else
  disp(['  bad data: Ht exceeds ',num2str(Hmax,'%6.5g'),' m']), return, end

subplot(2,1,1);                          % plot graph of solution h(X) to BVP
  plot(l,x,'Linewidth',1), hold on, %v = axis; plot(v(1),Xenew,'+r'),
  xlabel(['  h       (with computed H_t = ',num2str(Htnew,'%9.5g'),' )'],...
      'FontSize',12), ylabel('X    ','Rotation',0,'FontSize',12),
  if toolarge == 0,
    title([{'Absorption Tower (using an IVP iteration) : '},{'  '},...
    {[' Ac = ',num2str(Ac,'%5.4g'),', K_a = ',num2str(Ka,'%5.4g'),...
      ' a = ',num2str(a,'%5.4g'),', b = ',num2str(b,'%5.4g'),', L = ',...
    num2str(L,'%5.4g'),', V = ',num2str(V,'%5.4g'),', X_f = ',...
    num2str(Xf,'%5.4g'),', X_e = ',num2str(Xe,'%7.6g'),...
      ', Y_f = ',num2str(Yf,'%5.4g')]}],'FontSize',12), hold off,
```

```
else
   title([{'Absorption Tower (using an IVP iteration) : '},{' '},{[' A_c =
',...
      num2str(Ac,'%5.4g'),', K_a = ',num2str(Ka,'%5.4g'),', a = ',...
      num2str(a,'%5.4g'),', b = ',num2str(b,'%5.4g'),', L = ',...
      num2str(L,'%5.4g'),', V = ',num2str(V,'%5.4g'),', X_f = ',...
      num2str(Xf,'%5.4g'),', X_e = ',num2str(Xe,'%7.6g'),' (= max X_e)',...
      ', Y_f = ',num2str(Yf,'%5.4g')]}],'FontSize',12), hold off, end

subplot(2,1,2);                    % plot graph of associated solution h(Y)
   plot(l,y,'Linewidth',1), hold on, %v = axis; plot(v(1),y(1),'+g'),
   xlabel(['                 h   (rel. removal rate = ',num2str(rr,'%5.4g'),...
        ' ; Y_e = ',num2str(y(1),'%9.5g'),')'],'FontSize',12),
   ylabel('Y     ','Rotation',0,'FontSize',12), hold off

function [l,x] = integr(fhandle,lspan,x0,options,Ac,Ka,a,b,L,V,Yf,Xe,method)
   ghandle = @tow;            % module for the integration
   if method == 15,
     [l,x] = ode15s(ghandle,lspan,x0,options,Ac,Ka,a,b,L,V,Yf,Xe); end
   if method == 45,
     [l,x] = ode45(ghandle,lspan,x0,options,Ac,Ka,a,b,L,V,Yf,Xe); end
   if method == 23,
     [l,x] =  ode23(ghandle,lspan,x0,options,Ac,Ka,a,b,L,V,Yf,Xe); end
   if method == 113,
     [l,x] = ode113(ghandle,lspan,x0,options,Ac,Ka,a,b,L,V,Yf,Xe); end

function dydt = tow(l,x,A,Ka,a,b,L,V,Yf,Xe)         % r.h.s of the DE
   dydt = 1/L * (-A*Ka*(L/V * (x-Xe)+Yf-(a*x+b)));   % use all inputs
```

A call of absorbtoweriterHt(2.5,68,.72,0.01,720,7000,0.001,1.8404,1.4) with the same data used in Figure 6.24 and that computed $X_e = 1.8404$ for a given height of $20\ m$ in the simulation problem now finds $H_t = 19.999\ m$ in the design problem when we specify the simulated output value $X_e = 1.8404$ instead. This validates our algorithms.

Absorption Tower (using an IVP iteration) :

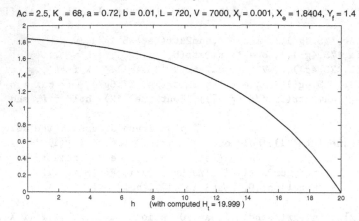

$Ac = 2.5, K_a = 68, a = 0.72, b = 0.01, L = 720, V = 7000, X_f = 0.001, X_e = 1.8404, Y_f = 1.4$

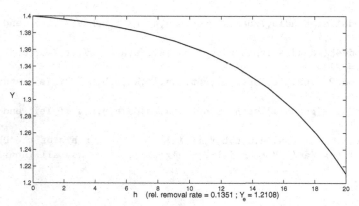

X and Y profiles for the design problem (6.129) and (6.130)

Figure 6.28

Calling   `[H,check] = absorbtoweriterHt(25,68,.72,0.01,7200,70000,0.001,3,`
`1.4)` for much larger parameter values creates the screen following output:

```
   desired Xe exceeds the physically possible;  X_e is set to max possible value
1.9305556
H =
      57.31
check =
      2.5125
```

Particularly note that the error `check` is sizable for this example and not nearly zero as it should be. We will comment on this numerically ill-conditioned problem on p. 398 after we have displayed and discussed the graphics output of this particular call and following the on-screen output of several other similar calls.

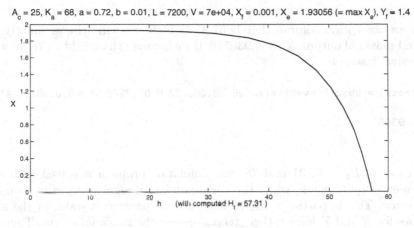

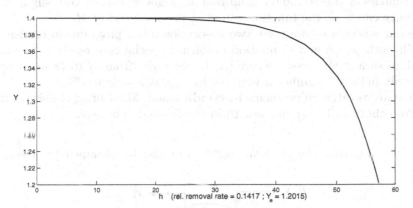

X and Y profiles for the design problem (6.129) and (6.130)
Figure 6.29

Note the physical size differences for the cross-sectional area A_c and the liquid and gas feed volumes L and V in this example when compared to the ones of Figure 6.28. Our original quest to achieve $X_e = 3$ as stipulated in the above call turns out to be unreasonable. The program recognizes this and gives us the height of the tower needed for achieving the maximally possible $X_e = 1.93056$ instead on the title line.

For this data our computations are, however, not very accurate in absolute terms, as the mass-balance check `check = 2.5125` $\gg 0$ indicates. But for the given size of L and V, the IVP solver is still relatively accurate.

A cross check for the same data with `absorbtoweriterXe` gives us the following screen output

```
>> [Xe,check] = absorbtoweriterXe(25,57.31,68,.72,0.01,7200,70000,0.001,1.4)
Xe =
        1.9303
```

```
check =
     0
```

Thus for an absorption column that is 57.31 m high, we will achieve slightly less than the desired maximal output $X_e = 1.930556$. If we increase the height to 78 m, we achieve our objective, however:

```
>> [Xe,check] = absorbtoweriterXe(25,78,68,.72,0.01,7200,70000,0.001,1.4)
Xe =
        1.9306
check =
     0
```

In both cases of $H_t = 57.31$ and 78, the simulation problem is solved with zero error in the screen output `check`, while the corresponding design problem is fraught with a sizeable error. This is partially due to the extreme parameter data, in the sense that the profiles for X and Y have rather steep slopes at the maximal height. Therefore even miniscule changes of the boundary condition at height zero (i.e., very slight changes of X_e) have huge effects on the function values at the other end at H_t.

We advise our students to use these two absorption tower programs in tandem for best results with both design and simulation problems. Special care needs to be taken if we compute data such as the above, where the height varies from 57 to 78 m, or by 37% as X_e varies only in its 5th significant digit, or by approximately 0.02%.

Numerical analysts call such problems **ill-conditioned**. Monitoring the size of the mass-balance error `check` will alert the user to ill-conditioned data here.

Finally we note that the relation (6.129) can also be obtained by a simple mass balance.

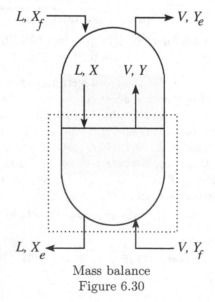

Mass balance
Figure 6.30

A mass balance across the dotted boundary in Figure 6.30 gives us

$$L \cdot X + V \cdot Y_f \; = \; L \cdot X_e + V \cdot Y \; .$$

This yields

$$V \cdot Y \; = \; L(X - X_e) + V \cdot Y_f \; ,$$

and consequently

$$Y \; = \; \frac{L}{V}(X - X_e) + Y_f \; .$$

The above equation is identical to equation (6.129) that was obtained earlier by subtracting the two differential equations (6.124) and (6.126) and integrating the resulting DE.

Exercises for 6.3

1. Rewrite the code quadnonequicol.m to run in $O(N)$ time as detailed on p. 376. To achieve this fast operations count, you need to implement all matrix times matrix or matrix times vector products in $O(N)$ time by recognizing the sparsity of the matrices Df_1x, Df_1y, Df_2x, and Df_2y that make up $Df(z_{old})$.

2. Derive the design equations for multi-tray absorption towers taking into account the change of the liquid and vapor flow rates L and V due to absorption.

3. Develop a numerical algorithm for solving problem 2 above. Use one set of parameters from this section in order to find out the effect of changing L and V on the results.

4. Repeat problems 2 and 3 above for a packed bed absorption tower.

6.4   The Nonisothermal Case: a Battery of CSTRs

In this section we formulate a high-dimensional multi-stage problem with consecutive exothermic reactions that take place in three consecutive adiabatic CSTRs with no recycle. This is done for a variable base set of parameters α_1, α_2, β_1, β_2, γ_1, and γ_2 for the two reactions. The dynamic characteristics of this system are obtained by numerical simulation.

6.4.1   Model Development

Specifically, we consider the consecutive reaction

$$A \xrightarrow{r_1} B \xrightarrow{r_2} C$$

with $r_1 = k_1 \cdot C_A$ and $r_2 = k_2 \cdot C_B$ for $k_i = k_{i,o} \cdot e^{-E_i/(R \cdot T)}$ and $i = 1, 2$. Here B is the desired product and the process takes place in three adiabatic CSTRs in series as depicted in Figure 6.31.

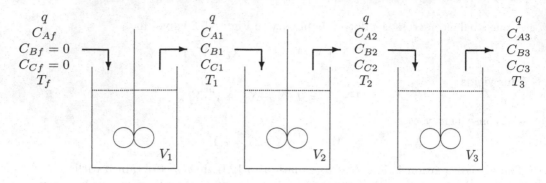

Battery of three adiabatic CSTRs
Figure 6.31

If the three CSTRs have the same constant volume $V = V_i$ for $i = 1, 2, 3$, then the reaction can be described by the following set of DEs.

For the first tank:

$$V\frac{dC_{A1}}{dt} + q \cdot C_{A1} = q \cdot C_{Af} - V \cdot k_{1,0} \cdot e^{-E_1/(R \cdot T_1)} \cdot C_A \, , \qquad (6.132)$$

$$V\frac{dC_{B1}}{dt} + q \cdot C_{B1} = 0 + V \cdot k_{1,0} \cdot e^{-E_1/(R \cdot T_1)} \cdot C_A - V \cdot k_{2,0} \cdot e^{-E_2/(R \cdot T_1)} \cdot C_B \, , \qquad (6.133)$$

and

$$V \cdot \rho \cdot C_p \frac{dT_1}{dt} + q \cdot \rho \cdot C_p \cdot (T_1 - T_f) = \qquad (6.134)$$
$$= V \left(k_{1,0} \cdot e^{-E_1/(R \cdot T_1)} \cdot C_A \cdot (-\Delta H_1) + k_{2,0} \cdot e^{-E_2/(R \cdot T_1)} \cdot C_B \cdot (-\Delta H_2) \right) \, .$$

These three differential equations can be put into dimensionless form to become the system of DEs (6.135), (6.136), and (6.137) below.

$$\frac{dx_{A1}}{d\tau} = 1 - x_{A1} - \alpha_1 \cdot e^{-\gamma_1/y_1} \cdot x_{A1} \, , \qquad (6.135)$$

where $x_{A1} = C_{A1}/C_{Af}$, $\alpha_1 = V \cdot k_{1,0}/q$, $\gamma_1 = E_1/(R \cdot T_f)$, $y_1 = T_1/T_f$, and $\tau = t \cdot q/V$ denotes dimensionless time in the first DE (6.132) for the first tank. And for (6.133) :

$$\frac{dx_{B1}}{d\tau} = -x_{B1} + \alpha_1 \cdot e^{-\gamma_1/y_1} \cdot x_{A1} - \alpha_2 \cdot e^{-\gamma_2/y_1} \cdot x_{B1} \, , \qquad (6.136)$$

where $x_{B1} = C_{B1}/C_{Af}$, $\alpha_2 = V \cdot k_{2,0}/q$, and $\gamma_2 = E_2/(R \cdot T_f)$ according to the second DE (6.133) for the first tank.
And finally

$$\frac{dy_1}{d\tau} = 1 - y_1 + \alpha_1 \cdot \beta_1 \cdot e^{-\gamma_1/y_1} \cdot x_{A1} + \alpha_2 \cdot \beta_2 \cdot e^{-\gamma_2/y_1} \cdot x_{B1} \, , \qquad (6.137)$$

where $\beta_1 = (-\Delta H_1) \cdot C_{Af}/(\rho \cdot C_P \cdot T_f)$ and $\beta_2 = (-\Delta H_2) \cdot C_{Af}/(\rho \cdot C_P \cdot T_f)$ for the third DE (6.134) above.

Similarly we can derive two sets of three dimensionless DEs that govern the second and the third tank in Figure 6.31, respectively.

For the second tank:

$$\frac{dx_{A2}}{d\tau} = x_{A1} - x_{A2} - \alpha_1 \cdot e^{-\gamma_1/y_2} \cdot x_{A2} ,$$ (6.138)

where $x_{A2} = C_{A2}/C_{Af}$ and $y_2 = T_2/T_f$.

$$\frac{dx_{B2}}{d\tau} = x_{B1} - x_{B2} + \alpha_1 \cdot e^{-\gamma_1/y_2} \cdot x_{A2} - \alpha_2 \cdot e^{-\gamma_2/y_2} \cdot x_{B2} ,$$ (6.139)

where $x_{B2} = C_{B2}/C_{Af}$.
And

$$\frac{dy_2}{d\tau} = y_1 - y_2 + \alpha_1 \cdot \beta_1 \cdot e^{-\gamma_1/y_2} \cdot x_{A2} + \alpha_2 \cdot \beta_2 \cdot e^{-\gamma_2/y_2} \cdot x_{B2} .$$ (6.140)

And similarly *for the third tank:*

$$\frac{dx_{A3}}{d\tau} = x_{A2} - x_{A3} - \alpha_1 \cdot e^{-\gamma_1/y_3} \cdot x_{A3} ,$$ (6.141)

where $x_{A3} = C_{A3}/C_{Af}$ and $y_3 = T_3/T_f$.

$$\frac{dx_{B3}}{d\tau} = x_{B2} - x_{B3} + \alpha_1 \cdot e^{-\gamma_1/y_3} \cdot x_{A3} - \alpha_2 \cdot e^{-\gamma_2/y_3} \cdot x_{B3} ,$$ (6.142)

where $x_{B3} = C_{B3}/C_{Af}$.
And

$$\frac{dy_3}{d\tau} = y_2 - y_3 + \alpha_1 \cdot \beta_1 \cdot e^{-\gamma_1/y_3} \cdot x_{A3} + \alpha_2 \cdot \beta_2 \cdot e^{-\gamma_2/y_3} \cdot x_{B3} .$$ (6.143)

The above nine DEs (6.135) to (6.143) describe our system of three CSTRs depicted in Figure 6.31. It has six dimensionless parameters α_i, β_i, and γ_i for $i = 1, 2$. This system can be visualized as a matrix DE, namely

$$\frac{dX}{d\tau} = F(X) \text{ for the state variable vector } X = \begin{pmatrix} x_{A1} \\ x_{B1} \\ y_1 \\ x_{A2} \\ x_{B2} \\ y_2 \\ x_{A3} \\ x_{B3} \\ y_3 \end{pmatrix} \in \mathbb{R}^9 .$$ (6.144)

In particular, we are interested in the solution for the parameters $\alpha_1 = 10^{6.426}$, $\alpha_2 = 10^{6.85}$; $\beta_1 = 0.4$, $\beta_2 = 0.6$; $\gamma_1 = 18$, and $\gamma_2 = 27$. Moreover, we want to vary the appropriate design and operating parameters that are contained inside the dimensional parameters in order to maximize the yield of the desired intermediate product B at the output of the third tank.

6.4.2   Numerical Solutions

The system of DEs (6.144) describes an initial value problem in nine dimensions, once we have chosen the initial values for the dimensionless temperatures $y_i(0)$ and the dimensionless concentrations $x_{Ai}(0)$ and $x_{Bi}(0)$ of the components A and B, respectively, in each of the three tanks numbered $i = 1, 2, 3$ at the dimensionless starting time $\tau = 0$. We study various sets initial conditions for the problem (6.144) that lead to different steady-state output concentrations x_{B3} of the desired component B in the three CSTR system.

The MATLAB ODE solver `ode15s` is most appropriate for solving the nonlinear system (6.144) of nine DEs for the nine state variables quickly. This multistep integrator is built on Gear's[2] method. We have experimented with using simpler ODE solvers such as `ode23` or `ode45`, but these simple Runge-Kutta embedding formulas experience difficulties right from the start $\tau = 0$ and do not produce a reliable solution in any reasonable amount of time. This, combined with the success of `ode15s` (and of `ode23s`, which is much slower) indicates that the system of DEs (6.144) is in fact stiff with slowly increasing *and*, at the same time, slowly decreasing fundamental solutions to the DE. This fact can be verified theoretically with great effort, but stiffness of a system of DEs is much more easily verified by numerical experimentation, as we have just described: try various DE solvers and the most successful one asserts the DEs system's stiffness or nonstiffness numerically.

See also the earlier remark on stiffness and ODE solver validation in section 4.3.5 on p. 201.

We start with a multi-density (or multi-colored on the computer screen) plot of six phase plots for the solution of (6.144) starting from various initial conditions. Each phase plot begins at a small o mark and ends at a specified time $\tau = $ `Tend` at a small □ square. The plots in Figure 6.32 are generated by our MATLAB code `threeCSTRrun.m` which in turn relies on `threeCSTR.m` to solve each IVP of the form (6.144).

```
function threeCSTRrun(a1,a2,y1,y2,y3,Tend,S,tl,colorswitch)
%  threeCSTRrun(a1,a2,y1,y2,y3,Tend,S,tl)
% Sample call : threeCSTRrun(10^6.426,10^6.85,1.3,1.3,1.3,100,'b',0,1)
% Input : a1, a2: system parameters;
%         y1, y2, y3: initial temperature values in tank i = 1,2,3;
%         Tend = time end of integration;
%         S = color choice
%         tl = 1: we plot with pluses after each time step
%         tl = 0: no plot of time steps
%         colorswitch = 1: we switch color for each new initial condition
% Output: 5 sets of phase plots using threeCSTR.m, overlaid in one figure

if nargin == 6, S = 'b'; tl = 0; colorswitch = 0; end   % defaults
if nargin == 7, tl = 0; colorswitch = 0; end

clf, multi = 0;
```

---

[2]Charles William Gear, British mathematician, 1935-

```
disp('start at zero, zero')  % on screen display of progress :   % blue
[t,y,multi] = ...
    threeCSTR(a1,a2,Tend,[.0;.0;y1; .0;.0;y2; .0;.0;y3],S,1,multi,tl); drawnow
disp('start at 0.95, zero'), if colorswitch == 1, S = 'k'; end, % black
[t,y,multi] = ...
  threeCSTR(a1,a2,Tend,[.95;.0;y1; .95;.0;y2; .95;.0;y3],S,1,multi,tl); drawnow
disp('start at 0.45, 0.45'), if colorswitch == 1, S = 'r'; end, % red
[t,y,multi] = ...
    threeCSTR(a1,a2,Tend,[.45;.45;y1; .45;.45;y2; .45;.45;y3],S,1,multi,tl);
    drawnow
disp('start at 0, 0.95'), if colorswitch == 1, S = 'g'; end,    % green
[t,y,multi] = ...
    threeCSTR(a1,a2,Tend,[0;.95;y1; 0;.95;y2; 0;.95;y3],S,1,multi,tl);   drawnow
disp('start at 0.4, 0.15'), if colorswitch == 1, S = 'c'; end, % cyan
[t,y,multi] = ...
    threeCSTR(a1,a2,Tend,[0.4;.15;y1; 0.4;.15;y2; 0.4;.15;y3],S,1,multi,tl);
    drawnow
```

The specific graphs in Figure 6.32 are generated by the call of $\texttt{threeCSTRrun(10^6.426,}$ $\texttt{10^6.85,1.3,1.3,1.3,100,'b',0,1)}$. Here the system parameters α_i are chosen as $\alpha_1 = 10^{6.426}$ and $\alpha_2 = 10^{6.85}$. The next three inputs in this call of $\texttt{threeCSTRrun}$ denote the dimensionless temperatures $y_1 = y_2 = y_3 = 1.3$ for each of the three tanks. These are followed by the end $T_{end} = 100$ of the dimensionless time of integration, the plot color[3] specification of blue for the computer screen, as well as two output toggles, explained fully in the comments preamble to $\texttt{threeCSTRrun.m}$.

Note that the complete set of initial value inputs can be retrieved from the six phase plots by reading the location of each starting point from the o marks in the plots for each tank. On a computer screen for example, in tank 1 the green trajectory starts from the initial values $x_{A1} = 0$, $x_{B1} = 0.95$ and $y_1 = 1.3$ as can be seen by looking at the red o marks that indicates the respective starting point for the top two graphs in Figure 6.32. Note that the dynamic trajectories for the second tank are depicted in the two middle plots of Figure 6.32, and those of the third tank in the two bottom plots. The specific initial values for x_{Ai} and x_{Bi} and $i = 1, 2, 3$ can be adjusted inside $\texttt{threeCSTRrun.m}$ if desired.

---

[3]References to color always refer to a computer generated color graph of the figure in question.

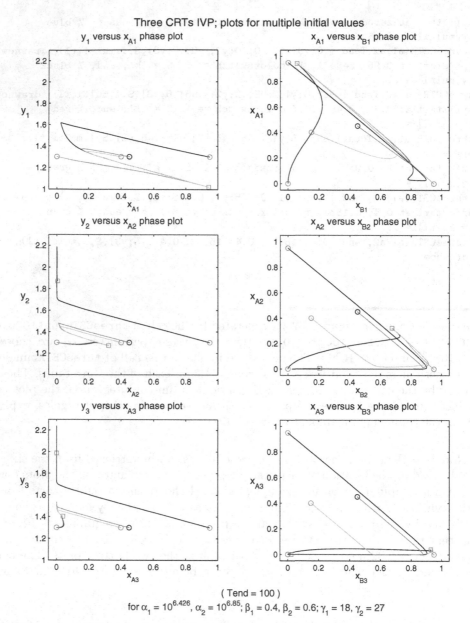

Three CRTs IVP; plots for multiple initial values

(Tend = 100)
for $\alpha_1 = 10^{6.426}$, $\alpha_2 = 10^{6.85}$; $\beta_1 = 0.4$, $\beta_2 = 0.6$; $\gamma_1 = 18$, $\gamma_2 = 27$

Multiple phaseplots of solutions to (6.144) from four different sets of initial values

Figure 6.32

The top two graphs depict the phase trajectories of y_1 versus x_{A1} and of x_{A1} versus x_{B1} for the first tank. The middle two graphs do the same for the second tank, while the bottom two plots describe the dynamics in tank 3 for the chosen initial conditions $y_3 = 1.3$, and various $x_{A3}(0)$ and $x_{B3}(0)$ values .

In the three double plots of Figure 6.32 we have chosen the following five sets of initial values $(x_{Ai}(0),\ x_{Bi}(0),\ y_i(0))$ as $(0, 0, 1.3)$, $(0.95, 0, 1.3)$, $(0.45, 0.45, 1.3)$, $(0, 0.95, 1.3)$, and $(0.4, 0.15, 1.3)$ for each tank $i = 1, 2, 3$, i.e., we have chosen a feasible set of initial conditions for the concentrations and the constant initial dimensionless temperatures of 1.3 in each tank. Note that always $1 \leq y_i = 1.3 \leq 1 + \beta_1 = 1 + 0.4 = 1.4$ and $x_{Ai}(0) + x_{Bi}(0) \leq 1\ (= 100\%)$ for each i.

More specifically in each horizontal pair of plots, all realistic component concentrations x_{Ai} and x_{Bi} must satisfy the constraint $x_{Ai} + x_{Bi} \leq 1$. And the concentration of the component C in tank i is given by $x_{Ci} = 1 - x_{Ai} - x_{Bi}$ for each $i = 1, 2, 3$.

We note that in the top pair of phaseplots, all our initial values lead to a unique steady state with $(x_{A1}(Tend),\ x_{B1}(Tend),\ y_1(Tend)) \approx (0.92, 0.05, 1.03)$ as depicted by the red □ marks. Note further the large phaseplot swing of the black curve that starts at the initial value $(x_{A1}(0),\ x_{B1}(0)) = (0.95, 0)$. This curve nearly backs to its start in each of the topmost plots for tank 1.

The middle two plots show the dynamics of the reaction in the second tank. One steady state of tank 2 lies at $(x_{A1}(Tend),\ x_{B1}(Tend),\ y_1(Tend)) \approx (0.33, 0.67, 1.28)$ and another at $(x_{A1}(Tend),\ x_{B1}(Tend),\ y_1(Tend)) \approx (0, 0.2, 1.87)$. The latter gives the smaller yield of x_B and results from the initial second tank conditions $(x_{A2}(0),\ x_{B2}(0),\ y_2(0)) = (0.95, 0, 1.3)$ depicted in black. These two steady states are stable. There is another unstable steady state for this data, but our graphical method does obviously not allow us to find it because it is an unstable saddle-type steady state that will repel any profile that is near to it. It can be easily obtained from the steady-state equations, though. For a method to find all steady states of a three CSTR system, see Section 6.4.3.

The bottom two plots for tank 3 move the first steady state of the second tank to a higher dimensionless temperature that is almost equal to 2, the maximal dimensionless temperature, and reduce the concentrations of A and of B even further to nearly zero. Again this steady state is only reached from a relatively high concentration of $x_{A3}(0) = 0.95$ along the black curves. The second stable steady state of tank 2 moves to $(x_{A1}(Tend),\ x_{B1}(Tend),\ y_1(Tend)) \approx (0.05, 0.92, 1.4)$ in tank 3. Thus we can increase the production of the desired component B steadily from tank 1 to tank 3 using any of the five tested initial conditions, except for the one with high x_A initial values in tank 2 or 3. See Problem 2 of the Exercises.

Overall most of our five trial initial conditions will force the three CSTR unit to produce a yield of around 92% of component B at the output of tank 3. This is a very good result.

To draw Figures 6.33 to 6.35 below, we use the code **threeCSTR.m** printed on p. 411. We alter the initial component concentrations for A and B in the three tanks while keeping all tank temperatures at the same constant initial value $y_i = 1.3$. We start out with $(x_{A1}(0),\ x_{B1}(0)) = (0.4, 0)$, $(x_{A2}(0),\ x_{B2}(0)) = (0.1, 0.1)$, and $(x_{A3}(0),\ x_{B3}(0)) = (0.1, 0.5)$ to obtain Figure 6.33.

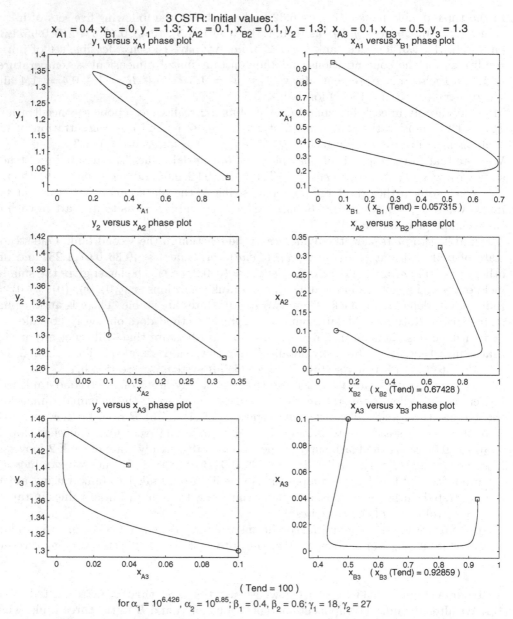

Phase plots of the solution to (6.144) from one set of initial values

Figure 6.33

Next we increase the initial concentration $x_{B3}(0)$ to become 1 and reduce both $x_{B2}(0)$ and $x_{A3}(0)$ to zero.

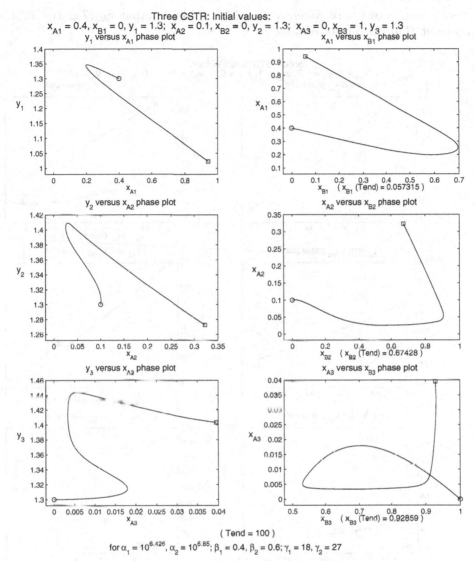

Phase plots of the solution to (6.144) from a different initial condition
Figure 6.34

Note that the top 4 phaseplots in Figure 6.34 for the reactors 1 and 2 look very similar to the ones in Figure 6.33, while the bottom two plots for the third CSTR differ. However, the output $x_{B3}(Tend) = 0.92859$ is nearly 93% in Figure 6.34 as it was in Figure 6.33.

But if we change the initial condition $(x_{A2}(0),\ x_{B2}(0))$ of tank 2 to become $(1,0)$ and use $(x_{A3}(0),\ x_{B3}(0)) = (0,1)$ for tank 3 in Figure 6.35, then the output of B at the exit of tank 3 becomes rather small at around 19% of the useful component B. In fact we could have done better for these initial conditions by reducing the three CSTR tank system to a two tank system with a higher useful output $x_{B2}(Tend) = 67.428\%$ of component B at the output of tank 2.

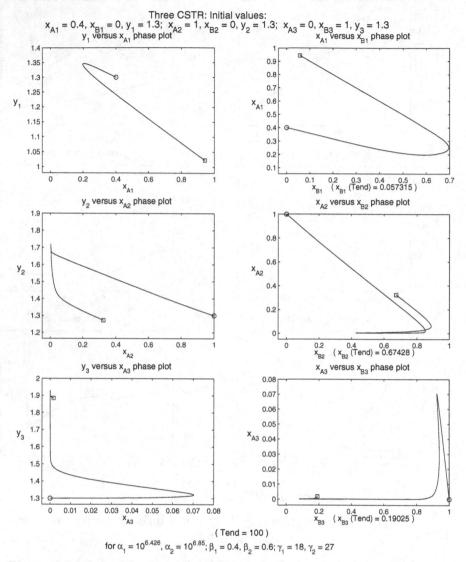

Phase plots of the solution to (6.144) with further altered initial conditions
Figure 6.35

Note that the reduction of useful output of B in tank 3 in Figure 6.35 is accompanied by a rather high steady-state temperature of around 1.9. This indicates that the reaction taking place in tank 3 mainly transforms the nearly 67% concentration of product B at its input into component C.

Figure 6.36 gives a single plot with another set of altered heat and concentration initial values spelled out in detail in the figure's title line. Recall that all phase trajectories start at a small o mark and end at a small □ mark when $\tau =$ **Tend**.

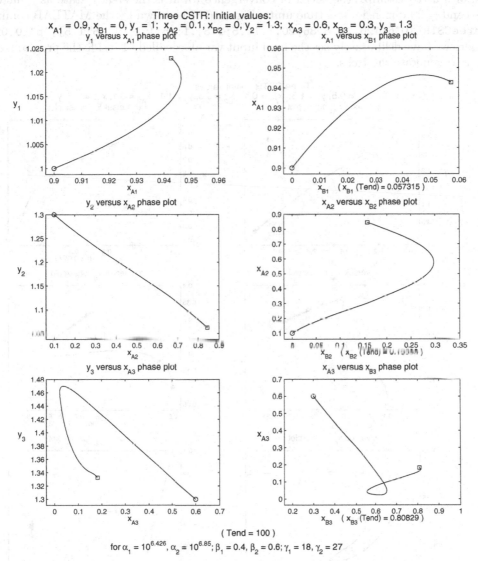

Phase plots of the solution to (6.144) with altered initial conditions
Figure 6.36

For this specific initial data the output of the desired product B increases from tray to tray from around 6% in tank 1, to about 15% in tank 2 and finally to 80% at the output of tank 3.

Before listing the ODE solver and phaseplot drawing MATLAB program `threeCSTR.m` for the previous four Figures 6.33 to 6.36, we show one more plot that marks the position reached by the solution at each full 1 unit time interval by a + plus mark. This enables us to visualize the speed of convergence towards the steady state as τ increases to `Tend` = 20 dimensionless time units. Figure 6.37 is drawn by the MATLAB command `threeCSTR(10^6.426,10^6.85,20,[.1;.5;1.3;.1;.5;1.1;.1;.5;1.5],'b',0,0,1)` ; where we have deliberately set the final input variable, called `tl` inside the program, equal to 1 to generate the ticks.

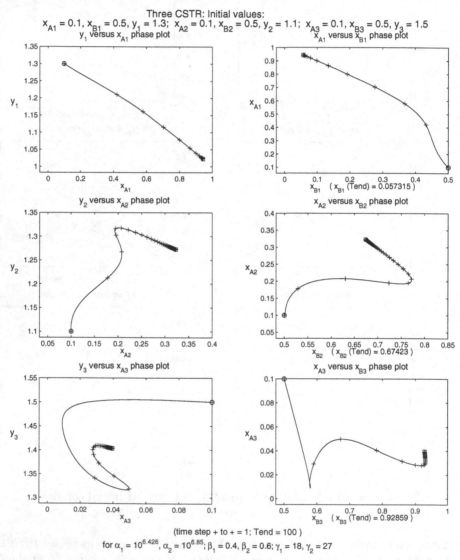

Phase plots with time tick marks of the solution to (6.144)

Figure 6.37

Note in Figure 6.37 that tank 1 reaches its steady state very quickly since we can only distinguish 8 or 9 of the $T_{end} = 20$ time marks. Tank 2 shows a slower convergence in close to $T_{end} = 20$ time steps, as does tank 3. It is fair to conclude from our graphs that the overall convergence to steady state takes about 30 dimensionless time units for this reactor. Our readers can verify this by extending T_{end} to 30 time units and counting the distinguishable tick marks.

The MATLAB program threeCSTR.m contains a few lines of numerical calculations at its top that generate the numerical solution for the IVP in equation (6.144), followed by more than two pages of plotting and graph annotation code, and it finishes with the right-hand side DE equation evaluator for the nine DEs collected in the equations (6.135) to (6.143).

```
function [t,y,multi] = threeCSTR(a1,a2,Tend,y0,S,halten,multi,tl)
%   [t,y,multi] = threeCSTR(a1,a2,Tend,y0,S,halten,multi,tl)
% Sample call :
%   threeCSTR(10^6.426,10^6.85,20,[.1;.5;1.3; .1;.5;1.1; .1;.5;1.5],'b',0,0,1);
% Input : a1, a2 system parameters;
%         Tend = time end of integration;
%         y0 = initial values for xAi, xBi, the initial concentrations of the
%              components A and B in tank i, and yi, the initial
%              temperature in tank i;
%         S = color specification for curve plots
%         halten = 1 for holding the graph for multiple runs;
%         halten = 0 for starting a new graph;
%         multi = 0 for a single run; multi > 0 for multiple runs
%                (size of multi (> 0) governs plot annotations)
%         tl = 0 for no intermediate "pluses" on the IVP solutions at
%              constant time intervals; (helps to judge the setting of Tend
%              for having reached a steady state)
%         tl = 0 for unmarked phase curves; tl = 1 for time ticks on profiles.
% Output: six phase plots of temperature versus concentration of component
%         A and of component A concentration versus component B
%         concentration in each of the tanks; with annotations etc.

if nargin == 4, S = 'b'; halten = 0; multi = 0; tl = 0; end % setting defaults
if nargin == 5, halten = 0; multi = 0; tl = 0; end
if nargin == 6, multi = 0; tl = 0; end
if nargin == 7, tl = 0; end

b1 = 0.4; b2 = 0.6; g1 = 18; g2 = 27;                  % fixed system parameters
step = 1; tstep = linspace(0,Tend,floor(Tend/step)); % starting step pluses
if tl ~=0, tl = length(tstep); end
if S ~= 'r', R = 'r'; else, R = 'b'; end  % different colors for start and end
tspan = [0 Tend];            % De IVP solver setup and execution
options = odeset('RelTol',10^-6,'AbsTol',10^-7,'Vectorized','on');
[t,y] = ode15s(@dXdT,tspan,y0,options,a1,a2,b1,b2,g1,g2);
        % using fast stiff DE integrator ode15s here for speed and accuracy
```

```
yi = interp1(t,y,tstep);        % Interpolating the solution at steps
if halten == 0, clf, end

subplot(3,2,1),                 % 6 plot routines for the phase plots
plot(y(:,1),y(:,3),S),
if halten == 0, v = axis; hold on,
    axis([v(1)-0.05*(v(2)-v(1)),v(2),v(3)-0.05*(v(4)-v(3)),v(4)]);
else hold on, v = [-0.05,1,1,2.3]; axis(v); end
plot(y(1,1),y(1,3),'Marker','o','Color',R),
for i = 1:tl, plot(yi(i,1),yi(i,3),'Marker','+','Color',S), end
plot(y(end,1),y(end,3),'Marker','s','Color',R),
xlabel('x_{A1}','FontSize',11); ylabel('y_1      ','FontSize',12,'Rotation',0);
title('y_1 versus x_{A1} phase plot ','FontSize',12);
if multi == 0 & halten == 0, ,
    text(v(1)-0.3*(v(2)-v(1)),v(4)+0.22*(v(4)-v(3)),...
 [{['                                                    Three CSTR:',...
 ' Initial values:']}, {['                   x_{A1} = ',num2str(y0(1),'%6.4g'),...
 ', x_{B1} = ',num2str(y0(2),'%6.4g'),', y_{1} = ',num2str(y0(3),'%6.4g'),...
 ';   x_{A2} = ',num2str(y0(4),'%6.4g'),', x_{B2} = ',num2str(y0(5),'%6.4g'),...
 ', y_{2} = ',num2str(y0(6),'%6.4g'),...
 ';   x_{A3} = ',num2str(y0(7),'%6.4g'),', x_{B3} = ',num2str(y0(8),'%6.4g'),...
 ', y_{3} = ',num2str(y0(9),'%6.4g')]}],'FontSize',14),
else, text(v(1)+0.5*(v(2)-v(1)),v(4)+0.2*(v(4)-v(3)),...
    ['Three CSTR IVP; plots for multiple initial values'],'FontSize',14), end
if halten == 1, multi = multi +1; end;
if halten == 0, hold off, end

subplot(3,2,2),
plot(y(:,2),y(:,1),S),
if halten == 0, v = axis; hold on,
    axis([v(1)-0.05*(v(2)-v(1)),v(2),v(3)-0.05*(v(4)-v(3)),v(4)]);
else hold on, v = [-0.05,1.05,-0.05,1.05]; axis(v); end
plot(y(1,2),y(1,1),'Marker','o','Color',R),
for i = 1:tl, plot(yi(i,2),yi(i,1),'Marker','+','Color',S), end
plot(y(end,2),y(end,1),'Marker','s','Color',R),
if halten == 0,
  xlabel(['x_{B1}     ( x_{B1} (Tend) = ',num2str(y(end,2),'%7.5g'),' )'],...
        'FontSize',11);
else xlabel('x_{B1}    ','FontSize',11); end
ylabel('x_{A1}      ','FontSize',12,'Rotation',0);
title('x_{A1} versus x_{B1} phase plot ','FontSize',12);
if halten == 1, hold on, end

subplot(3,2,3),
plot(y(:,4),y(:,6),S),
if halten == 0, v = axis; hold on,
    axis([v(1)-0.05*(v(2)-v(1)),v(2),v(3)-0.05*(v(4)-v(3)),v(4)]);
else hold on, v = [-0.05,1,1,2.3]; axis(v); end
```

```
plot(y(1,4),y(1,6),'Marker','o','Color',R),
for i = 1:tl, plot(yi(i,4),yi(i,6),'Marker','+','Color',S), end
plot(y(end,4),y(end,6),'Marker','s','Color',R),
xlabel('x_{A2}','FontSize',11); ylabel('y_2     ','FontSize',12,'Rotation',0);
title('y_2 versus x_{A2} phase plot ','FontSize',12);
if halten == 1, hold on, end

subplot(3,2,4),
plot(y(:,5),y(:,4),S), if halten == 0, v = axis; hold on,
    axis([v(1)-0.05*(v(2)-v(1)),v(2),v(3)-0.05*(v(4)-v(3)),v(4)]);
else hold on, v = [-0.05,1.05,-0.05,1.05]; axis(v); end
plot(y(1,5),y(1,4),'Marker','o','Color',R),
for i = 1:tl, plot(yi(i,5),yi(i,4),'Marker','+','Color',S), end
plot(y(end,5),y(end,4),'Marker','s','Color',R),
if halten == 0,
  xlabel(['x_{B2}    ( x_{B2} (Tend) = ',num2str(y(end,5),'%7.5g'),' )'],...
      'FontSize',11);
else xlabel('x_{B2}    ','FontSize',11); end
ylabel('x_{A2}     ','FontSize',12,'Rotation',0);
title('x_{A2} versus x_{B2} phase plot ','FontSize',12);
if halten == 1, hold on, end

subplot(3,2,5),
plot(y(:,7),y(:,9),S),
if halten == 0, v = axis; hold on,
    axis([v(1)-0.05*(v(2)-v(1)),v(2),v(3)-0.05*(v(4)-v(3)),v(4)]);
else hold on, v = [-0.05,1,1,2.3]; axis(v); end
plot(y(1,7),y(1,9),'Marker','o','Color',R),
for i = 1:tl, plot(yi(i,7),yi(i,9),'Marker','+','Color',S), end
plot(y(end,7),y(end,9),'Marker','s','Color',R),
xlabel('x_{A3}','FontSize',11); ylabel('y_3     ','FontSize',12,'Rotation',0);
title('y_3 versus x_{A3} phase plot ','FontSize',12);
if halten == 1, hold on, end

subplot(3,2,6),
plot(y(:,8),y(:,7),S),
if halten == 0, v = axis; hold on,
    axis([v(1)-0.05*(v(2)-v(1)),v(2),v(3)-0.05*(v(4)-v(3)),v(4)]);
else hold on, v = [-0.05,1.05,-0.05,1.05]; axis(v); end
plot(y(1,8),y(1,7),'Marker','o','Color',R),
for i = 1:tl, plot(yi(i,8),yi(i,7),'Marker','+','Color',S), end
plot(y(end,8),y(end,7),'Marker','s','Color',R),
if halten == 0,
  xlabel(['x_{B3}    ( x_{B3} (Tend) = ',num2str(y(end,8),'%7.5g'),' )'],...
      'FontSize',11);
else xlabel('x_{B3}    ','FontSize',11); end
ylabel('x_{A3}     ','FontSize',12,'Rotation',0);
title('x_{A3} versus x_{B3} phase plot ','FontSize',12);
```

```
if (halten == 0 & multi == 0) | (halten == 1 & multi == 1),
  if halten == 1, cv1 = .35; cv2 = .65; cv3 = .23; cv1m = .9; cv3m = 0.37;
  else cv1 = .43; cv2 = .67; cv3 = .28; cv1m = 1.1; cv3m = .44; end
  if tl ~= 0, text(v(1)-cv2*(v(2)-v(1)),v(3)-cv3*(v(4)-v(3)),...
    ['(time step + to + = ',num2str(step,'%6.4g'),'; Tend = ',...
    num2str(Tend,'%6.4g'),' )'],'FontSize',12);
  else, text(v(1)-cv1*(v(2)-v(1)),v(3)-cv3*(v(4)-v(3)),...
    ['( Tend = ',num2str(Tend,'%6.4g'),' )'],'FontSize',12); end
  text(v(1)-cv1m*(v(2)-v(1)),v(3)-cv3m*(v(4)-v(3)),...
    ['for \alpha_1 = 10^{',num2str(log10(a1),'%6.4g'),'}, \alpha_2 = 10^{',...
      num2str(log10(a2),'%6.4g'),'}; \beta_1 = ',num2str(b1,'%6.4g'),...
      ', \beta_2 = ',num2str(b2,'%6.4g'),'; \gamma_1 = ',...
      num2str(g1,'%6.4g'),', \gamma_2 = ',num2str(g2,'%6.4g')],...
    'FontSize',12); end
if halten == 1, hold on, end

function dydx = dXdT(x,y,a1,a2,b1,b2,g1,g2)   % right hand side of DE (9 eqs)
dydx = [1 - y(1,:) - a1*exp(-g1./y(3,:)).*y(1,:);
       -y(2,:) + a1*exp(-g1./y(3,:)).*y(1,:) - a2*exp(-g2./y(3,:)).*y(2,:);
     1-y(3,:) + a1*b1*exp(-g1./y(3,:)).*y(1,:) + a2*b2*exp(-g2./y(3,:)).*y(2,:);
         y(1,:) - y(4,:) - a1*exp(-g1./y(6,:)).*y(4,:);
     y(2,:) - y(5,:) + a1*exp(-g1./y(6,:)).*y(4,:) - a2*exp(-g2./y(6,:)).*y(5,:);
         y(3,:) - y(6,:) + a1*b1*exp(-g1./y(6,:)).*y(4,:) + ...
                                    a2*b2*exp(-g2./y(6,:)).*y(5,:);
         y(4,:) - y(7,:) - a1*exp(-g1./y(9,:)).*y(7,:);
     y(5,:) - y(8,:) + a1*exp(-g1./y(9,:)).*y(7,:) - a2*exp(-g2./y(9,:)).*y(8,:);
         y(6,:) - y(9,:) + a1*b1*exp(-g1./y(9,:)).*y(7,:) + ...
                                    a2*b2*exp(-g2./y(9,:)).*y(8,:)];
```

In threeCSTR.m, please note the use of the subplot MATLAB command that draws six separate graphs in our code, namely two for each tank.

The initial temperatures y_i of the three CSTRs that have so far been used were all relatively low compared to their possible maxima $1 \leq y_1$, y_2, $y_3 \leq 2$. If we keep $y_1 = 1.3$ and raise y_2 and y_3 to 1.6 and 1.8 respectively, the simulation shows multiple steady states in the second and the third tank reactors as depicted in Figure 6.38.

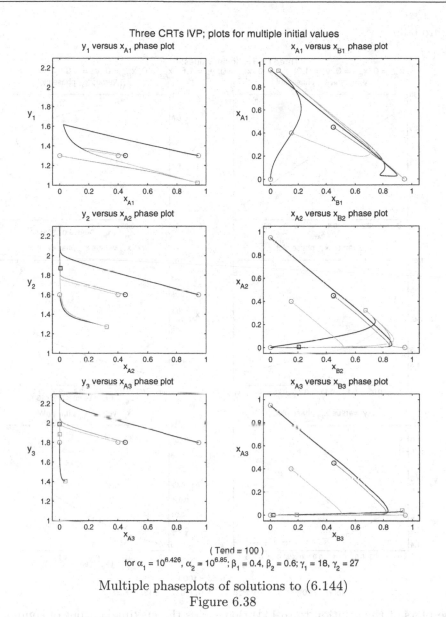

Multiple phaseplots of solutions to (6.144)
Figure 6.38

In Figure 6.38, best viewed in color on a monitor, all our chosen initial conditions lead to a single steady state in tank 1 with a concentration for component B of around 10%. For tank 2 there are at least two stable steady states with a maximal yield of component B of around 70%. And for tank 3 there are at least three stable steady states with yields for component B ranging from near zero at a high temperature to over 90% at $y_3 \approx 1.4$.

For the initial tank temperatures as chosen in Figure 6.38, the system reaches a concentration of around 93% for component B in tank 3 when starting all tanks without any trace of component A or B, i.e., by setting $x_{Ai} = 0 = x_{Bi}$ for all $i = 1, 2, 3$ as shown in Figure 6.39.

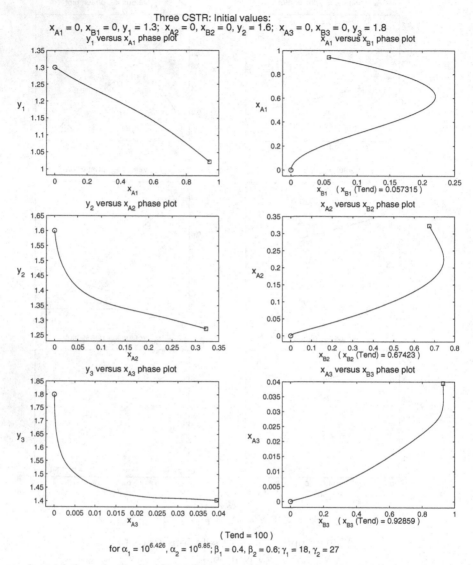

Phase plots of the solution to (6.144) that yields the maximal output of component B
in tank 3 for the initial values $x_{Ai} = 0 = x_{Bi}$ for all i
Figure 6.39

Thus for these initial temperatures y_i we need not prime any tank with the components A or B at start-up and the system will still produce the maximal yield of B in tank 3.

If we now lower the initial temperature y_1 of tank 1 to 1.0 and raise y_2 to 1.7 while keeping $y_3 = 1.8$ fixed, our simulation shows three (stable) steady states in tank 2, indicating a total of five steady states, including two unstable ones, and four (stable) steady states in tank 3, meaning a total of seven steady states, i.e., these four plus another three saddle-type unstable steady states. All of the stable steady states are reached in tank 3 from our standard set of five initial component concentrations x_{Ai} and x_{Bi} and our chosen tank temperatures, see Figure 6.40.

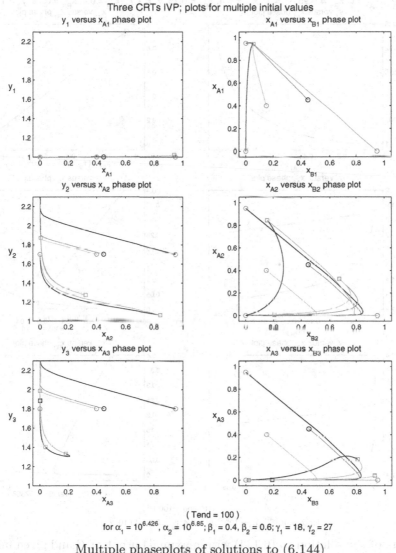

Multiple phaseplots of solutions to (6.144)
Figure 6.40

As before, Figure 6.40 is best created and viewed in color on a computer monitor.
Note that the yields of component B differ greatly between the various steady states of each tank reactor for the parameters of Figure 6.40, with different steady states and yields resulting from different initial conditions.

In Figure 6.41 we use one specific set of initial conditions that yields the maximal output of around 93% of component B in tank 3 for the same temperature data as used in Figure 6.40.

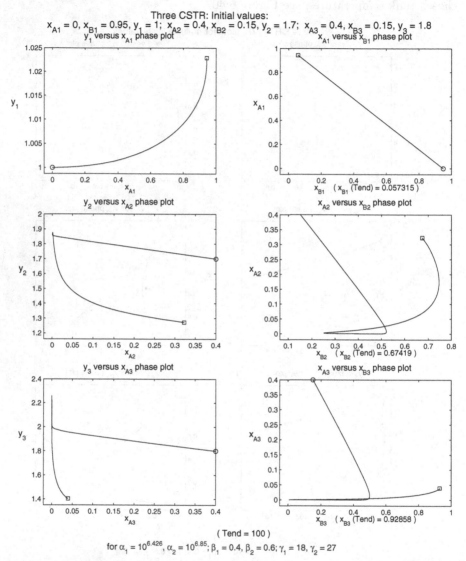

Phase plots of the solution to (6.144) with maximal output of B and given initial values for the temperatures and starting concentrations

Figure 6.41

For a problem on optimizing the start-up cost of a battery of three CSTRs, see the exercises for Section 6.4.

The numerical results of this section describe the dynamic behavior of a specific system that is modeled by an initial value problem in nine dimensions. The number of steady states of this system can be any odd number between 1 and $3^3 = 27$, because if tank 1 has one steady state, there may be three in the second tank and nine in the third tank, since there may be three steady states in a subsequent tank for each steady state in the preceding tank. Therefore, if tank 1 has three steady states, the second tank can have nine and the third tank 27 steady states. The maximal number 27 of possible steady states in tank 3 is achieved if there are three steady states in tank 1, with each of these spawning three in the next tank, giving us maximally 9 steady states in tank 2. If each of these steady states in tank 2 gives us three steady states in tank 3, then there is a maximum of 27 steady states in tank 3.

For certain ranges of the parameters it is possible that the first tank can have five steady states, then in the second tank there can be up to $5^2 = 25$ steady states and consequently in the third tank up to $5^3 = 125$ steady states. In our dynamic simulation examples for the three CSTRs with equal volume, the maximal number of steady states that we have obtained is seven.

The actual number of steady states of our three CSTR system can be found by solving the steady-state equations of Section 6.4.3.

For more details on multiplicity and bifurcation, see the Appendix with this title.

6.4.3   The Steady State Equations

The steady-state equations for (6.144) are given by the nine transcendental equations of the system of equations $F(X) = 0$. For the steady state, we can reduce each set of three equations for one tank to a single transcendental equation as explained below.

For example, for tank 1 the steady-state version of the DE (6.135) becomes

$$1 - x_{A1} - \alpha_1 \cdot e^{-\gamma_1/y_1} \cdot x_{A1} = 0 \, . \tag{6.145}$$

This equation obtained as usual by setting the derivative in (6.135) equal to zero. Equation (6.145) makes

$$x_{A1} = \frac{1}{1 + \alpha_1 \cdot e^{-\gamma_1/y_1}} \, . \tag{6.146}$$

Plugging this value of x_{A1} into the steady-state equation corresponding to the DE (6.136) for x_{B1} gives us

$$x_{B1} = \frac{\alpha_1 \cdot e^{-\gamma_1/y_1}}{(1 + \alpha_1 \cdot e^{-\gamma_1/y_1})(1 + \alpha_2 \cdot e^{-\gamma_2/y_1})} \tag{6.147}$$

for the feed concentration $x_{B_f} = 0$ of B.

Combining (6.146) and (6.147) with the steady-state version of (6.137), the heat balance finally gives us the following transcendental equation in the single variable y_1 that depends on the six parameters α_i, β_i, and γ_i for $i = 1, 2$, namely

$$y_1 - 1 = \frac{\alpha_1 \cdot \beta_1 \cdot e^{-\gamma_1/y_1}}{1 + \alpha_1 \cdot e^{-\gamma_1/y_1}} + \frac{\alpha_1 \cdot \alpha_2 \cdot \beta_2 \cdot e^{-(\gamma_1 + \gamma_2)/y_1}}{(1 + \alpha_1 \cdot e^{-\gamma_1/y_1})(1 + \alpha_2 \cdot e^{-\gamma_2/y_1})} \, . \tag{6.148}$$

Depending on the parameters, this equation can yield up to five steady state solutions.
For the second tank, the steady-state equations are

$$x_{A2} = \frac{x_{A1}}{1 + \alpha_1 \cdot e^{-\gamma_1/y_2}} \; ; \tag{6.149}$$

$$x_{B2} = \frac{x_{B1}}{1 + \alpha_1 \cdot e^{-\gamma_2/y_2}} + \frac{\alpha_1 \cdot e^{-\gamma_1/y_2} \cdot x_{A1}}{(1 + \alpha_1 \cdot e^{-\gamma_1/y_2})(1 + \alpha_2 \cdot e^{-\gamma_2/y_2})} \; ; \text{ and} \tag{6.150}$$

$$y_2 - y_1 = \frac{\alpha_1 \cdot \beta_1 \cdot e^{-\gamma_1/y_2} \cdot x_{A1}}{1 + \alpha_1 \cdot e^{-\gamma_1/y_2}} + \tag{6.151}$$

$$+ \alpha_2 \cdot \beta_2 e^{-\gamma_2/y_2} \cdot \left(\frac{x_{B1}}{1 + \alpha_2 \cdot e^{-\gamma_2/y_2}} + \frac{\alpha_1 \cdot e^{-\gamma_1/y_2} \cdot x_{A1}}{(1 + \alpha_1 \cdot e^{-\gamma_1/y_2})(1 + \alpha_2 \cdot e^{-\gamma_2/y_2})} \right) \; .$$

For each steady-state value for y_1 in tank 1, this can lead to five steady states in tank 2.
Thus for the maximal possible five values for y_1 in the first tank, this gives us up to 25
steady states in tank 2.
And for the third tank we have the steady-state equations

$$x_{A3} = \frac{x_{A2}}{1 + \alpha_1 \cdot e^{-\gamma_1/y_3}} \; ; \tag{6.152}$$

$$x_{B3} = \frac{x_{B2}}{1 + \alpha_2 \cdot e^{-\gamma_2/y_3}} + \frac{\alpha_1 \cdot e^{-\gamma_1/y_3} \cdot x_{A2}}{(1 + \alpha_1 \cdot e^{-\gamma_1/y_3})(1 + \alpha_2 \cdot e^{-\gamma_2/y_3})} \; ; \text{ and} \tag{6.153}$$

$$y_3 - y_2 = \frac{\alpha_1 \cdot \beta_1 \cdot e^{-\gamma_1/y_3} \cdot x_{A2}}{1 + \alpha_1 \cdot e^{-\gamma_1/y_3}} + \tag{6.154}$$

$$+ \alpha_2 \cdot \beta_2 \cdot e^{-\gamma_2/y_3} \cdot \left(\frac{x_{B2}}{1 + \alpha_2 \cdot e^{-\gamma_2/y_3}} + \frac{\alpha_1 \cdot e^{-\gamma_1/y_3} \cdot x_{A2}}{(1 + \alpha_1 \cdot e^{-\gamma_1/y_3})(1 + \alpha_2 \cdot e^{-\gamma_2/y_3})} \right) \; .$$

Here the process repeats and for each steady state (up to 25 in number) of tank 2, there
are five possible steady states in tank 3, making the maximal number of steady states in
tank 3 equal to $5^3 = 125$.

A suggested solution sequence for the above nine coupled equations (6.146) to (6.154)
and for given values of α_i, β_i, and γ_i for $i = 1, 2$ is as follows.

1. For the given α, β, and γ data, the only unknown in equation (6.148) is y_1. There
   may be multiple solutions for y_1 (up to five).

2. Using the fixed given parameter values of α_i, β_i, γ_i and the maximally five solutions
   for y_1, we can obtain the values of x_{A1} and x_{B1} from (6.146) and (6.147).

3. Now the only unknown in equation (6.151) is y_2, which may have multiple solutions
   for each value of the (possibly multiple) y_1 (and for the corresponding multiple x_{A1}
   and x_{B1} values) found in steps 1 and 2. Thus up to 25 steady states are possible
   for tank 2.
   For each of the found y_1 and y_2 value sets, we can compute the corresponding
   values of x_{A2} and x_{B2} using the equations (6.149) and (6.150), respectively.

4. Finally we need to solve (6.154) for y_3 and all previous subsets of solutions. Then we find the corresponding x_{A3} and x_{B3} values from the equations (6.152) and (6.153), respectively. There are up to 125 possible steady-state solutions for tank 3.

Exercises for 6.4

1. Assume that component C is very cheap, while component A is moderately expensive and component B is very expensive.

    (a) For the initial temperatures y_i chosen as in Figure 6.41, find out whether the three CSTR system can reach the maximal output of component B in tank 3 of about 93% without using any of component B in the tanks at start-up.

    (b) If problem (a) can be solved, how little of component A can be used to prime the tanks at start-up and still achieve the maximal almost 93% steady state output of component B in tank 3?
    (Hint: Experiment with `threeCSTR.m`.)

2. Find estimates for the minimal amounts of the components A and B that are needed to prime the three CSTR system at start-up in order to obtain near zero output of component B in tank 3. (Most wasteful case scenario)

3. Rewrite the MATLAB codes `threeCSTR.m` and `threeCSTRrun.m` to work for two CSTR and for four CSTR systems and test them on the plotted examples in Section 6.4.3.

4. **Project**

    (a) Implement the 4 steps given on p. 420 in order to solve the steady state equations (6.146) to (6.154) for any given parameter values of α_i, β_i, and γ_i and $i = 1, 2$ inside a MATLAB program that constructs the complete set of steady states for the desired data.
    Is it possible to obtain the maximal number of 125 steady states in tank 3? What is the maximal number of steady states in tank 3 that can be obtained while keeping the volume of the three tanks equal?

    (b) Try to obtain the maximal number of steady states in each tank by using different volumes for the three reactors.

5. Find all steady states for the IVPs with initial conditions as chosen for Figures 6.38 and 6.40 by using the algorithm developed in problem 4.
How can one distinguish between the stable and the unstable steady states found via the MATLAB code of problem 4?
[Hint: Linearize the equations and find the associated matrix eigenvalues by using the built-in MATLAB function `eig`.]

6. Repeat problems 1 to 5 for the consecutive reaction case:

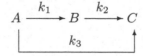

Conclusions

In this chapter we have presented multistage systems with special emphasis on absorption processes. We have studied multitray countercurrent absorption towers with equilibrium trays for both cases when the equilibrium relation is linear and when it is nonlinear. This study was accompanied by MATLAB codes that can solve either of the cases numerically. We have also introduced cases where the trays are not efficient enough to be treated as equilibrium stages. Using the rate of mass transfer *RMT* in this case, we have shown how the equilibrium case is the limit of the nonequilibrium cases when the rate of mass transfer becomes high. Both the linear and the nonlinear equilibrium relation were used to investigate the nonequilibrium case. We have developed MATLAB programs for the nonequilibrium cases as well.

Additionally, countercurrent packed bed absorption column has been studied and a MATLAB program involving IVPs for this problem has been developed and explained.

For all of the above configurations and situations, we have presented both the simulation problem, where the number of trays is given in the multitray case, or the height is given for the packed column, and the design problem, where the number of trays in the multitray case is not known, or the height is unknown for the packed column.

The last section of this chapter describes a rather complex multi-stage process involving several nonisothermal CSTRs in series and partially solves it numerically, showing the wide range of dynamic possibilities associated with this unit, including multiple steady states.

Exercises for Chapter 6

1. Choose a multistage absorption tower problem from industry. Collect all the necessary data, develop the model and the MATLAB code to simulate the industrial situation. Show how you can improve the performance of the chosen industrial unit.

2. Choose a packed bed absorption tower problem from industry. Collect all the necessary data, develop the model and the MATLAB code to simulate the industrial situation. Show how you can improve the performance of the chosen industrial unit.

Refinery

Chapter 7

Industrial Problems

In the previous chapters we have shown how to solve many types of problems that occur in Chemical and Biological Engineering through mathematical modeling, standard numerical methods, and MATLAB.

We have introduced many practical software based numerical procedures to solve physico-chemical models for simulation and design purposes. Therefore, we hope that our readers now feel comfortable and ready to handle more complex industrial problems from the modeling stage through the numerical solution and model validation stages on her/his own.

Industrial problems are usually more complicated than the earlier problems in this book. But their solutions generally require the same steps, tools and procedures. Therefore, an engineer needs to learn how to handle these problems in both a direct and an integrated way. A typical industrial problem might involve solving one system of differential equations and then solving an algebraic equation (or another DE) at each point of the solution profile.

We will start by introducing a relatively simple integrated example from industry to illustrate what this means. This example will be followed by a number of industrial problems in separate sections. In this chapter we will explain and model each industrial problem and we will develop strategies for a numerical solution. However, explicit solutions and MATLAB codes for these problems will be left to the reader throughout this chapter as exercises. We hope that we have given the necessary tools in the previous chapters to enable our readers to solve these industrial problems. Some results for the industrial problems of this chapter will be included so that the reader can check his or her own results. However, we emphasize that actual industrial problems take much effort, perseverance and labor for anyone who models and solves them.

7.1   A Simple Illustrative Example

We consider the following gas-solid catalytic reaction

$$A + B \quad \longrightarrow \quad C$$

inside a tubular cocurrent cooled reactor. The intrinsic rate of reaction is

$$r = k_0 \cdot e^{-E/(R \cdot T)} \cdot C_A \cdot C_B \quad \text{in } mol/(g \ (of \ catalyst) \cdot min) \ .$$

We assume that the change of the number of moles and the temperature change do not affect the volumetric flow rate q and that all physical properties (ρ, C_ρ, etc) are constant. Here the catalyst pellets are spherical and nonporous and the reaction is taking place in a nonadiabatic fixed-bed reactor with cocurrent cooling. The reaction is exothermic.

Statement of the Problem:

1. Derive the material and energy-balance design equations, including the equations of the catalyst pellets to calculate the effectiveness factor η.

2. Check whether you can reduce the number of equations. If possible, reduce them.

3. Suggest a sequence of steps for the solution of the design equations (for the reactor, jacket, catalytic pellet, η, etc).

Here is an outline for a solution. We start with two diagrams of our apparatus.

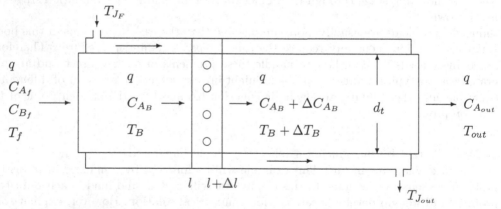

Schematic diagram of the packed-bed tubular cocurrent catalytic reactor of length L_t

Figure 7.1

Figure 7.1 is a schematic representation of the tubular reactor and its cooling jacket, together with the differential element from l to $l + \Delta l$.

An enlarged view of the central differential element in Figure 7.1 is drawn in Figure 7.2.

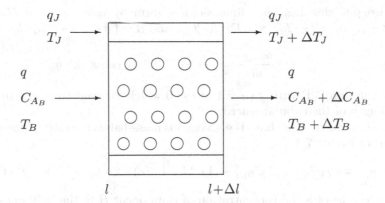

Enlarged view of the differential element of Figure 7.1
Figure 7.2

Note that our model assumes a constant volumetric flow rate q in the reactor, as well a constant flow rate q_J in the jacket.

7.1.1 Mass Balance of the Element of the Reactor

The rate of reaction per unit mass of the catalyst is given by

$$r - r_B \cdot \eta = k_0 \cdot e^{-E/(R \cdot T_B)} \cdot C_{A_B} \cdot C_{B_B} \cdot \eta .$$

Here r_B is the intrinsic rate of reaction, neglecting the mass and heat transfer resistances, r is the actual rate of reaction that takes the mass- and heat-transfer resistances into account, and η is the effectiveness factor that accounts for the effect of mass- and heat-transfer resistances between the bulk fluid and the catalyst pellet. By using η we are able to express the rate of reaction in terms of the bulk concentration and the temperature while the reaction is in fact taking place inside the pellet.

For component A we have the following balance over the differential element of Figure 7.2:

$$q \cdot C_{A_B} = q \left(C_{A_B} + \Delta C_{A_B} \right) + A \cdot \Delta l \cdot (1 - \epsilon) \cdot \rho_s \cdot k_o \cdot e^{-E/(R \cdot T_B)} \cdot C_{A_B} \cdot C_{B_B} \cdot \eta ,$$

where ϵ is the bed voidage, $A = \pi d_t^2 / 4$ is the cross-sectional area of the tubular reactor of diameter d_t in Figure 7.1, T_B is the temperature in the bulk phase, η is the effectiveness factor of the catalyst pellet, and C_{A_B} and C_{B_B} are the concentrations of component A and B in the bulk phase, respectively.

Replacing the difference equation by a differential equation by taking the limit $\lim \Delta l \to 0$ gives us

$$q \frac{dC_{A_B}}{dl} = -A \cdot (1 - \epsilon) \cdot \rho_s \cdot k_o \cdot e^{-E/(R \cdot T_B)} \cdot C_{A_B} \cdot C_{B_B} \cdot \eta .$$

Next we put this DE into dimensionless form by using $x_{AB} = C_{A_B}/C_{A_f}$, $x_{BB} = C_{B_B}/C_{B_f}$, $y_B = T_B/T_f$, $\gamma = E/(R \cdot T_f)$, and $\alpha = (A \cdot (1 - \epsilon) \cdot \rho_s \cdot k_0 \cdot C_{B_f})/q$, namely

$$\frac{dx_{AB}}{dl} = -\alpha \cdot e^{-\gamma/y_B} \cdot x_{AB} \cdot x_{BB} \cdot \eta \ . \qquad (7.1)$$

This defines an IVP with $x_{AB} = 1$ at $l = 0$ which we want to integrate until $l = L_t$, the total length of the tubular reactor.

For component B we have the analogous mass-balance equation across the differential element of Figure 7.1

$$q \cdot C_{B_B} = q \, (C_{B_B} + \Delta C_{B_B}) + A \cdot \Delta l \cdot (1 - \epsilon) \cdot \rho_s \cdot k_0 \cdot e^{-E/(R \cdot T_B)} \cdot C_{A_B} \cdot C_{B_B} \cdot \eta \ ,$$

where C_{B_B} denotes the concentration of component B in the bulk phase.
This leads to an analogous DE and finally to the dimensionless initial value problem

$$\frac{dx_{BB}}{dl} = -a \cdot \alpha \cdot e^{-\gamma/y_B} \cdot x_{AB} \cdot x_{BB} \cdot \eta \qquad (7.2)$$

with $x_{BB} = C_{B_B}/C_{B_f}$, the reactants ratio $a = C_{A_f}/C_{B_f}$, and the initial condition $x_{BB} = 1$ at $l = 0$.

7.1.2   Heat Balance for the Element of the Reactor

Here we assume for simplicity that there is no change of phase in the reactor and moreover we use the average density ρ and the average specific heat C_ρ for the mixture in our model formulation.

First we consider the heat balance inside the tubular reactor. Figure 7.2 gives rise to the incremental heat-balance equation

$$q \cdot \rho \cdot C_\rho \cdot T_B + A(1 - \epsilon)\Delta l \cdot \rho_s \cdot k_0 \cdot e^{-\gamma/y_B} \cdot C_{A_B} \cdot C_{B_B}(-\Delta H) \cdot \eta =$$
$$= q \cdot \rho \cdot C_\rho \cdot (T_B + \Delta T_B) - \widetilde{A}_h \cdot \Delta l \cdot U \cdot (T_J - T_B)$$

where U denotes the overall heat-transfer coefficient between the catalytic tubular bed and the jacket and $\widetilde{A}_h = \pi d_t$ is the area of heat transfer per unit length between the reactor and the cooling or heating jacket. We can rewrite this equation in form of the DE

$$q \cdot \rho \cdot C_\rho \frac{dT_B}{dl} = A(1 - \epsilon) \cdot \rho_s \cdot k_0 \cdot e^{-\gamma/y_B} \cdot C_{A_B} \cdot C_{B_B}(-\Delta H) \cdot \eta + \widetilde{A}_h \cdot u \cdot (T_J - T_B) \ .$$

By setting $y_B = T_B/T_f$, $y_J = T_J/T_{J_f}$, $m = T_{J_f}/T_f$, $\alpha = (A(1 - \epsilon) \cdot \rho_s \cdot k_0 \cdot C_{B_f})/q$, $\beta = (-\Delta H \cdot C_{A_f} \cdot C_{B_f})/(\rho \cdot C_\rho \cdot T_f)$, and $\alpha_c = (\widetilde{A}_h \cdot U)/(q \cdot \rho \cdot C_\rho)$, this DE becomes

$$\frac{dy_B}{dl} = \eta \cdot \alpha \cdot \beta \cdot e^{-\gamma/y_B} \cdot x_{AB} \cdot x_{BB} + \alpha_c \cdot (m \cdot y_J - y_B) \qquad (7.3)$$

in dimensionless form. This is an IVP with the initial condition $y_B = 1$ when $l = 0$.

Next we consider the heat balance for the cooling jacket in incremental form as depicted in Figure 7.2. Note first that ours is a case of cocurrent flow in the jacket and reactor. This simplifies the mathematical problem when compared with countercurrent cooling. Countercurrent cooling is physically more efficient, but it transforms the problem mathematically into a more demanding two point boundary value problem which we want to avoid here; see problem 3 of the Exercises.

The heat-balance equation for the jacket is

$$q_J \cdot \rho_J \cdot C_{\rho_J} \cdot T_J - \widetilde{A}_h \cdot U \cdot \Delta l \cdot (T_J - T_B) = q_J \cdot \rho_J \cdot C_{\rho_J} \cdot (T_J + \Delta T_J) .$$

This transforms readily to the DE

$$q_J \cdot \rho_J \cdot C_{\rho_J} \frac{dT_J}{dl} = -\widetilde{A}_h \cdot U \cdot (T_J - T_B) .$$

And in dimensionless form with $b = (q \cdot \rho \cdot C_\rho)/(q_J \cdot \rho_C \cdot C_{\rho_C})$ the DE becomes the IVP

$$\frac{dy_J}{dl} = -\alpha_C \cdot b \cdot (y_J - y_B/m) \tag{7.4}$$

with the initial condition $y_C = y_{C_f} = 1 \ (= T_{C_f}/T_{C_f})$ at $l = 0$.

7.1.3 Reactor Model Summary

In summary we have formulated four IVPs that describe our problem:

1. *The mass-balance design equations for component A*

$$\frac{dx_{AB}}{dl} = -\alpha \cdot \eta \cdot e^{-\gamma/y_B} \cdot x_{AB} \cdot x_{BB} \quad \text{with} \tag{7.5}$$

$$\alpha = (A \cdot (1-\epsilon) \cdot \rho_s \cdot k_0 \cdot C_{B_f})/q, \ \eta = \frac{\text{actual rate of reaction}}{\text{intrinsic rate of reaction}} = \frac{e^{-\gamma/y} \cdot x_A \cdot x_B}{e^{-\gamma/y_B} \cdot x_{AB} \cdot x_{BB}}$$

the pellet efficiency factor, and x_A and x_B the dimensionless concentrations in the catalyst pellet. Finally y is the dimensionless temperature of the catalyst pellet. Equation (7.5) is an IVP with the initial condition $x_{AB} = 1$ at $l = 0$.

2. *The mass-balance design equations for component B*

$$\frac{dx_{BB}}{dl} = -a \cdot \alpha \cdot \eta \cdot e^{-\gamma/y_B} \cdot x_{AB} \cdot x_{BB} \tag{7.6}$$

with $a = C_{A_f}/C_{B_f}$ and the initial condition $x_{BB} = 1$ at $l = 0$.

3. *The heat-balance design equations for the tubular reactor*

$$\frac{dy_B}{dl} = \eta \cdot \alpha \cdot \beta \cdot e^{-\gamma/y_B} \cdot x_{AB} \cdot x_{BB} + \alpha_C \cdot (m \cdot y_J - y_B) \tag{7.7}$$

with the initial condition $y_B = 1$ at $l = 0$.

4. *The heat-balance design equations for the cocurrent cooling jacket*

$$\frac{dy_C}{dl} = -\alpha_C \cdot b \cdot (y_J - y_B/m) \tag{7.8}$$

with $y_C = 1$ at $l = 0$.

These four IVPs need to be solved simultaneously from $l = 0$ to $l = L_t$, the length of the tubular reactor.

Since the cooling jacket has cocurrent flow, the model consists of the set of four coupled initial value differential equations (7.5) to (7.8). Note that the first three DEs (7.5) to (7.7) contain the variable catalyst effectiveness factor η. Thus there are other equations to be solved at each point along the length $0 \le l \le L_t$ of the reactor tube, namely the equations for the catalyst pellet's effectiveness factor η.

7.1.4 The Catalyst Pellet Design Equations, Calculating η along the Length of the Reactor

Here we consider a spherical catalyst pellet with negligible intraparticle mass- and negligible heat-transfer resistances. Such a pellet is nonporous with a high thermal conductivity and with external mass and heat transfer resistances only between the surface of the pellet and the bulk fluid. Thus only the external heat- and mass-transfer resistances are considered in developing the pellet equations that calculate the effectiveness factor η at every point along the length of the reactor.

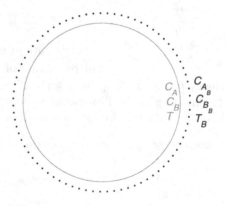

Catalyst pellet diagram
Figure 7.3

Mass balance for component A:

From Figure 7.3 we deduce the following mass-balance equation

$$a_p \cdot K_{g_A} \cdot (C_{A_B} - C_A) = W_p \cdot k_0 \cdot e^{-E/(R \cdot T)} \cdot C_A \cdot C_B$$

where a_p is the surface area of the catalyst pellet, K_{g_A} is the external mass-transfer coefficient for the pellet, and W_p is the mass of the pellet. This equation relies on the intrinsic rate of reaction being given per unit mass of the catalyst as shown on p. 426. In addition, C_A, C_B, and T are the respective component concentrations and temperature of the pellet as indicated in Figure 7.3. After introducing dimensionless parameters and setting $\alpha' = W_p \cdot k_0 \cdot C_{B_f}/(a_p \cdot K_{g_A})$, we obtain

$$x_{AB} - x_A = \alpha' \cdot e^{-\gamma/y} \cdot x_A \cdot x_B .$$

Mass balance for component B:

Similarly we have

$$a_p \cdot K_{g_B} \cdot (C_{B_B} - C_B) = W_p \cdot k_0 \cdot e^{-E/(R \cdot T)} \cdot C_A \cdot C_B ,$$

or

$$x_{BB} - x_B = \alpha' \cdot d \cdot a \cdot e^{-\gamma/y} \cdot x_A \cdot x_B$$

for $d = K_{g_A}/K_{g_D}$ and $a = C_{A_f}/C_{B_f}$.

Heat balance for the pellet:

Here we have

$$a_p \cdot h \cdot (T - T_B) = W_p \cdot k_0 \cdot e^{-E/(R \cdot T)} \cdot C_A \cdot C_B \cdot (-\Delta H) ,$$

or

$$y - y_B = \alpha' \cdot \beta' \cdot e^{-\gamma/y} \cdot x_A \cdot x_B$$

for $\beta' = (-\Delta H) \cdot K_{g_A} \cdot C_{A_f} \cdot C_{B_f}/(h \cdot T_f)$.

7.1.5   Pellet Model Summary

1. *For component A*

$$x_{AB} - x_A = \alpha' \cdot e^{-\gamma/y} \cdot x_A \cdot x_B \tag{7.9}$$

   with $\alpha' = W_p \cdot k_0 \cdot C_{B_f}/(a_p \cdot K_{g_A})$.

2. *For component B*

$$x_{BB} - x_B = \alpha' \cdot d \cdot a \cdot e^{-\gamma/y} \cdot x_A \cdot x_B \tag{7.10}$$

   for $a = C_{A_f}/C_{B_f}$ and $d = K_{g_A}/K_{g_B}$.

3. *For the heat balance*

$$y - y_B = \alpha' \cdot \beta' \cdot e^{-\gamma/y} \cdot x_A \cdot x_B \tag{7.11}$$

for $\beta' = (-\Delta H) \cdot K_{g_A} \cdot C_{A_f} \cdot C_{B_f}/(h \cdot T_f)$.

For given values of the parameters we can compute the effectiveness factor η as follows: By integrating the bulk-phase IVPs (7.5) to (7.8) we find x_{AB}, x_{BB}, and y_B. Then the catalyst pellet equations (7.9) to (7.11) allow us to find y, x_A and x_B from which we can find η by the formula

$$\eta = \frac{e^{-\gamma/y} \cdot x_A \cdot x_B}{e^{-\gamma/y_B} \cdot x_{AB} \cdot x_{BB}} . \tag{7.12}$$

7.1.6   Manipulation and Reduction of the Equations

We now try to reduce the seven coupled model equations, namely the IVP equations (7.5) to (7.8) for the reactor and the transcendental equations (7.9) to (7.11) for the catalyst pellet, if this is possible.

First we work on the four reactor equations (7.5) to (7.8). If we subtract a times equation (7.5) from equation (7.6) we obtain

$$\frac{d(x_{BB} - a \cdot x_{AB})}{dl} = 0 .$$

Thus the function $x_{BB}(l) - a \cdot x_{AB}(l)$ is constant. At the initial value $l = 0$ we know that $x_{BB} = 1 = x_{AB}$. Therefore we have

$$x_{BB} = 1 + a \cdot (x_{AB} - 1) . \tag{7.13}$$

Consequently we can drop the design equation (7.6) for component B from our list and simplify (7.5) to become

$$\frac{dx_{AB}}{dl} = -a \cdot \eta \cdot e^{-\gamma/y_B} \cdot x_{AB} \cdot (1 + a \cdot (x_{AB} - 1)) . \tag{7.14}$$

And whenever we need to know x_{BB}, we can simply find it via x_{AB} obtained by integrating equation (7.14) and then substituting x_{AB} into equation (7.13).

Can we also reduce the heat-balance equations and combine them with the mass-balance design equations in some fashion?

For the nonadiabatic case, the answer is rather complicated, even when the cooling jacket temperature y_J is constant. But for the adiabatic case with $\alpha_C = 0$ we can add equation (7.7) and β times equation (7.5) to obtain

$$\frac{d(y_B + \beta \cdot x_{AB})}{dl} = 0 . \tag{7.15}$$

And thus $y_B(l) + \beta \cdot x_{AB}(l) = constant = y_B(0) + \beta \cdot x_{AB}(0) = 1 + \beta$ according to the initial conditions. Thus

$$x_{AB} = \frac{1 + \beta - y_B}{\beta} . \tag{7.16}$$

Substituting equation (7.16) into equation (7.7) (with $\alpha_C = 0$, as assumed) gives us

$$\frac{dy_B}{dl} = \alpha \cdot \eta \cdot e^{-\gamma/y_B} \cdot (1 + \beta - y_B) \cdot x_{BB} . \qquad (7.17)$$

If we combine the equations (7.16) and (7.13), we obtain the following formula for x_{BB}:

$$x_{BB} = \frac{\beta + a(1 - y_B)}{\beta} . \qquad (7.18)$$

This transforms equation (7.17) into

$$\frac{dy_B}{dl} = \alpha \cdot \eta \cdot e^{-\gamma/y_B} \cdot (1 + \beta - y_B) \cdot \frac{\beta + a(1 - y_B)}{\beta} . \qquad (7.19)$$

This equation is a nonlinear DE that we can integrate in order to find $y_B(l)$. Recall that at each point on the graph of y_B we must also calculate η from the catalyst pellet equations (7.9) to (7.11) and the effectiveness factor equation (7.12). Recall that we can easily calculate x_{AB} and x_{BB} from y_B using the formulas

$$(7.16) \quad x_{AB} = \frac{1 + \beta - y_B}{\beta} \quad \text{and} \quad (7.13) \quad x_{BB} = 1 + a \cdot (x_{AB} - 1) .$$

Note that for the special case of equimolar feed we have $a = 1$ and $x_{AB} = x_{BB}$.

Next we try to simplify the catalyst pellet design equations (7.9) to (7.11) which are used to calculate η for all points along the length of the reactor.
If we subtract $d \cdot a$ times equation (7.9) for component A from equation (7.10) for component B in the pellet, we obtain

$$-d \cdot a \cdot (x_{AB} - x_A) + x_{BB} - x_B = 0 ,$$

i.e., $x_B = x_{BB} + d \cdot a \cdot (x_A - x_{AB})$. Recalling that $x_{BB} = 1 + a \cdot (x_{AB} - 1)$ gives us

$$x_B = (1 - a) + a \cdot (1 - d) \cdot x_{AB} + d \cdot a \cdot x_A . \qquad (7.20)$$

Here we notice that for an equimolar feed of the components A and B and for $K_{g_A} = K_{g_B}$ we have $a = 1$ and $d = 1$ so that $x_A = x_B$, which is a very special case.
If we subtract β' times equation (7.9) from equation (7.11) we obtain $-\beta' \cdot (x_{AB} - x_A) + y - y_B = 0$ or

$$x_A = \frac{y_B + \beta' \cdot x_{AB} - y}{\beta'} , \qquad (7.21)$$

and consequently

$$x_B = (1 - a) + a \cdot (1 - d) \cdot x_{AB} + d \cdot a \cdot \frac{y_B + \beta' \cdot x_{AB} - y}{\beta'} . \qquad (7.22)$$

Equation (7.22) allows us to express x_B linearly in terms of y, x_{AB}, and y_B. Therefore, we can drop equation (7.10) for component B at the pellet from our list. Thus we can substitute the equations (7.21) and (7.20) into equation (7.9) to obtain an equation in y only and in which y_B and x_{AB} can be obtained from the solution of the reactor bulk equations (7.19), (7.16), and (7.13).

In order to find y only one equation must be solved for the computed values of x_{AB} and y_B that were earlier obtained from the bulk-phase reactor DEs (7.14) and (7.19). In fact, we need only integrate equation (7.19) for y_B and then for each computed value of y_B we can compute x_{AB} from equation (7.16) and x_{BB} from equation (7.13).

Once we have obtained y in this way, we can find x_A and x_B from the equations (7.21) and (7.22). Thus we can now compute the effectiveness factor $\eta = \eta(l)$ from its definition

$$\eta = \frac{e^{-\gamma/y} \cdot x_A \cdot x_B}{e^{-\gamma/y_B} \cdot x_{AB} \cdot x_{BB}} . \tag{7.23}$$

The transcendental equation that must be solved for the pellet after substituting (7.21) and (7.22) into equation (7.9) is

$$y - y_B = \alpha' \cdot e^{-\gamma/y} \cdot (y_B + \beta' \cdot x_{AB} - y) \cdot \left\{ (1-a) + a \cdot (1-d) \cdot x_{AB} + d \cdot a \cdot \frac{y_B + \beta' \cdot x_{AB} - y}{\beta'} \right\} . \tag{7.24}$$

Figure 7.4 gives a flow diagram for the overall solution procedure. It starts with solving the reactor initial value differential equation (7.19) for y_B, followed by substituting y_B into the equations (7.16) and (7.18) to obtain x_{AB} and x_{BB}, then solving equation (7.24) to obtain y and then x_A and x_B from (7.21) and (7.22), and finally obtaining η from equation (7.23).

Once the local effectiveness factor η has been found, the new value of η is used in equation (7.19) for the next integration step to find y_B and so forth.

In practical terms, we suggest to calculate a new value of $\eta = \eta^{(n)}$ at every length l that is a small integer multiple of the catalyst pellet diameter d_p. Thus we solve the IVP and the associated transcendental equations $n = L_t/d_p$ times, each time for $i \cdot d_p \leq l \leq (i+1) \cdot d_p$ and $i = 0, 1, ..., n-1$.

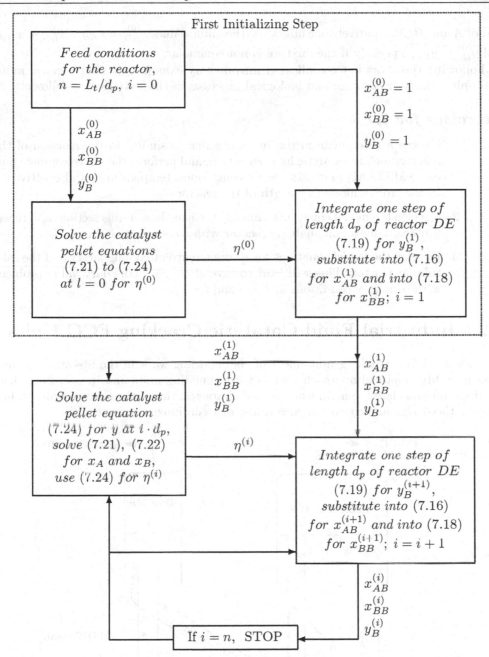

Schematic sequence of computations for the adiabatic case
Figure 7.4

The initial settings $x_{AB}^{(0)} = x_{BB}^{(0)} = y_B^{(0)} = 1$ (in the top right corner of Figure 7.4) use C_{A_f} as the reference concentration and T_f as the reference temperature and assume equimolar feed of A and B. Alternatively one might use the initial values $x_{AB}^{(0)} = x_{AB_f}$, $x_{BB}^{(0)} = x_{BB_f}$, and $y_B^{(0)} = y_{B_f}$, especially if the mixture is not equimolar.

Following this short but complicated introductory example, we study several industrial units and their chemical and biological processes in the sections that follow.

Exercises for 7.1

1. Choose a reaction from the literature that is similar to the reaction of this reactor, such as catalytic hydrogenation, and perform the above computations using MATLAB to obtain the concentrations, temperatures, and effectiveness factor profiles along the length of the reactor.

2. Develop the model equations, simplify them as done in this section, and repeat problem 1 for a nonadiabatic reactor with a cocurrent cooling jacket.

3. Develop the model equations for a countercurrent cooling jacket and the same tubular reactor. This will lead to several coupled boundary value problems with boundary conditions at $l = 0$ and $l = L_t$.

7.2   Industrial Fluid Catalytic Cracking FCC Units

For industrial fluid cracking units most of the modeling work in the literature is based upon a highly empirical approach that helps in building units and in operating them, but does not elucidate the main features and characteristics of the units in order to help improve the design and control of such units, or to optimize their output.

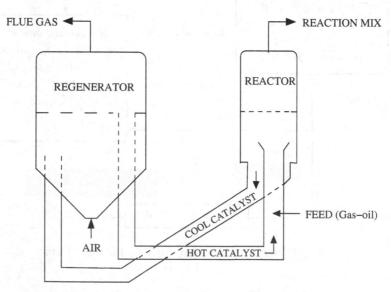

Schematic representation of an FCC unit of type IV
Figure 7.5

A better understanding of the behavior of FCC units can be obtained through mathe-matical models coupled with industrial verification and cross verification of these models. The mathematical model equations need to be solved for both design and simulation pur-poses. Most of the models are nonlinear and therefore they require numerical techniques like the ones described in the previous chapters.

7.2.1   Model Development for Industrial FCC Units

Many models for fluid catalytic cracking have been studied. We introduce a model that seems to have an optimal degree of sophistication and that succeeds to describe many industrial units accurately.

The most simple version of our model considers the two-phase nature of the fluidized beds in the reactor and in the regenerator in a simplified way. The kinetic model that we use considers three pseudocomponents in modeling type IV FCC units. This model is a consecutive-parallel model formed of three lumped components and coke as follows:

$$\text{Gas Oil } (A_1) \xrightarrow{K_1} \text{Gasoline } (A_2) \xrightarrow{K_2} \text{Coke } + \text{Gases } (A_3)$$
$$\overset{K_3}{\longmapsto}$$

The rates of reactions, measured in $g(converted)/(g(catalyst) \cdot s)$ for the consecutive-parallel network model are as follows:

The rate of disappearance of gas oil is

$$R_1 = \frac{\Psi_R \cdot \epsilon}{(1 - \epsilon)\rho_s} (K_1 + K_3) \cdot C_{A_1}^2 ; \tag{7.25}$$

the net rate of appearance of gasoline is

$$R_2 = \frac{\Psi_R \cdot \epsilon}{(1 - \epsilon)\rho_s} (K_1 \cdot C_{A_1}^2 - K_2 \cdot C_{A_2}) , \quad \text{and} \tag{7.26}$$

the rate of appearance of coke is

$$R_3 = \frac{\Psi_R \cdot \epsilon}{(1 - \epsilon)\rho_s} (K_1 \cdot C_{A_1}^2 + K_3 \cdot C_{A_2}) , \tag{7.27}$$

where the temperature dependence of the rate constants K_i $(i = 1, 2, 3)$ is given by

$$K_i = A_i \cdot e^{-E_i/(R_G \cdot T)} .$$

For the coke that burns in the regenerator, the continuous reaction model for solids of unchanging size is used because of its simplicity.

In our simple coke burning model, the gaseous reactant (oxygen) is assumed to be present uniformly throughout the catalyst particle at a constant concentration without diffusional resistance. The oxygen reacts with the solid reactant B, consisting of coke deposited on the catalyst according to the rate equation

$$R_C = A_C \cdot e^{-E_C/(R_G \cdot T_G)} \cdot (1 - x_{CB}) \cdot C_0 . \tag{7.28}$$

Based on these kinetic relations and using the simple two-phase model for bubbling fluidized beds, the FCC model equations for the steady states of the unit can be given in dimensionless form as follows:

The reactor dense-phase equations:

Gas oil balance:

$$B_R \cdot (x_{1f} - x_{1D}) - \Psi_R \cdot x_{1D}^2 \cdot (\alpha_1 \cdot e^{-\gamma_1/Y_{RD}} + \alpha_3 \cdot e^{-\gamma_3/Y_{RD}}) = 0 . \tag{7.29}$$

Gasoline balance:

$$B_R \cdot (x_{2f} - x_{2D}) - \Psi_R \cdot (\alpha_2 \cdot x_{2D} \cdot e^{-\gamma_2/Y_{RD}} - \alpha_1 \cdot x_{1D}^2 \cdot e^{-\gamma_1/Y_{RD}}) = 0 . \tag{7.30}$$

Coke balance:

$$C_R^{'} \cdot (\Psi_R - \Psi_C) + \Psi_R \cdot R_{cf} = 0 . \tag{7.31}$$

Heat balance:

$$a_1 \cdot (Y_{GD} - Y_{RD}) + B_R \cdot (Y_\nu - Y_{RD}) + a_2 \cdot (Y_{GF} - Y_\nu) - (\Delta \overline{H}_\nu + \Delta \overline{H}_{LR}) + \Delta \overline{H}_{cr} = 0 . \tag{7.32}$$

The reactor bubble-phase equations:

Gas oil balance:

$$x_{1B} - x_{1D} = (x_{1f} - x_{1D}) \cdot e^{-a_R \cdot H_R} . \tag{7.33}$$

Gasoline balance:

$$x_{2B} - x_{2D} = (x_{2f} - x_{2D}) \cdot e^{-a_R \cdot H_R} . \tag{7.34}$$

Heat balance:

$$Y_{RB} - Y_{RD} = (Y_\nu - Y_{RD}) \cdot e^{-a_R \cdot H_R} . \tag{7.35}$$

The regenerator dense-phase equations:

Coke balance:

$$C_G^{'} \cdot (\Psi_G - \Psi_R) - R_c = 0 . \tag{7.36}$$

Heat balance:

$$B_G \cdot (Y_{AF} - Y_{GD}) + a_3 \cdot (Y_{RD} - Y_{GD}) - \Delta \overline{H}_{GL} + \beta_c \cdot R_c = 0 . \tag{7.37}$$

Here oxygen is assumed to be available in excess and in constant concentration throughout.

The regenerator bubble-phase equation:

There is no need for an oxygen mass balance in the bubble phase because oxygen is assumed available in excess and the bubble is assumed to be free of solids.

Heat balance:

$$Y_{GB} - Y_{GD} = (Y_{AF} - Y_{GD}) \cdot e^{-a_G \cdot H_G} . \tag{7.38}$$

The parameters in these equations are

$$a_1 = F_C \cdot C_{ps}/(A_{RI} \cdot H_R \cdot C_{pf} \cdot \rho_f) , \qquad a_R = Q_{ER} \cdot A_{RB}/G_{RI} ,$$
$$a_2 = F_{GM} \cdot C_{pl}/(A_{RI} \cdot H_R \cdot C_{pf} \cdot \rho_f) , \qquad a_G = Q_{EG} \cdot A_{GB}/G_{GI} ,$$
$$a_3 = F_C \cdot C_{ps}/(A_{GI} \cdot H_G \cdot C_{pa} \cdot \rho_a) , \qquad \gamma_i = E_i/(R_G \cdot T_{rf}) ,$$
$$\alpha_1 = A_1 \cdot \epsilon \cdot C_{A_1 f} , \quad \alpha_2 = A_2 \cdot \epsilon , \quad \text{and} \quad \alpha_3 = A_3 \cdot \epsilon \cdot C_{A_1 f} ,$$

where A_i denotes the preexponential factor for reaction i and $i = 1, 2, 3$.
The space velocities B_R and B_G in the reactor and the regenerator are given by

$$B_R = \frac{G_{RI} + G_{RB} \cdot (1 - e^{-a_R \cdot H_R})}{A_{RI} \cdot H_R} \quad \text{and} \tag{7.39}$$

$$B_G = \frac{G_{GI} + G_{GB} \cdot (1 - e^{-a_G \cdot H_G})}{A_{GI} \cdot H_G} . \tag{7.40}$$

In dimensionless variables the rate of coke combustion is

$$R_C = \alpha_c \cdot e^{-\gamma_c/Y_G} \cdot (1 - \Psi_G) \tag{7.41}$$

for $\alpha_c = A_C \cdot C_0 \cdot (1 - \epsilon)$ and $\gamma_c = E_C/(R_G \cdot T_{rf})$.
The exothermicity factor for the coke burning reaction is

$$\beta_C = \overline{C}_m \cdot (-\Delta H_c)/(T_{rf} \cdot C_{pa} \cdot \rho_a) . \tag{7.42}$$

The rate of coke formation is

$$R_{cf} = (\alpha_3 \cdot e^{-\gamma_3/Y_{RD}} \cdot x_{1D}^2 + \alpha_2 \cdot e^{-\gamma_2/Y_{RD}} \cdot x_{2D}) . \tag{7.43}$$

The endothermicity factors for the three cracking reactions are given by:

$$\beta_i = C_{A_1 f} \cdot (-\Delta H_i)/(T_{rf} \cdot C_{pf} \cdot \rho_f) \quad \text{for} \quad i = 1, 2, 3 . \tag{7.44}$$

The overall endothermic heat of cracking is described by the equation

$$\Delta \overline{H}_{cr} = \Psi_R \cdot (\alpha_1 \cdot \beta_1 \cdot x_{1D}^2 \cdot e^{-\gamma_1/Y_{RD}} + \alpha_2 \cdot \beta_2 \cdot x_{2D} \cdot e^{-\gamma_2/Y_{RD}} + \alpha_3 \cdot \beta_3 \cdot x_{1D}^2 \cdot e^{-\gamma_3/Y_{RD}}) . \tag{7.45}$$

Finally

$$C_R' = F_C \cdot \overline{C}_m/(A_{RI} \cdot H_R \cdot C_{A_1 f}) \quad \text{and} \quad C_G' = F_C \cdot /(A_{GI} \cdot H_G) .$$

The design or simulation of FCC units involves numerically solving the above 21 equations and relations (7.25) to (7.45). The solution process will be discussed later. For the simulation of industrial units and the verification of this model for industrial data, the majority of these 21 equations are used to calculate various parameters in the 10 equations numbered (7.29) to (7.38). Specifically, equations (7.33) to (7.35) compute the concentration and temperature profiles in the bubble phase of the reactor and equation (7.38) computes the temperature profile in the regenerator. This leaves the main equations (7.29) to (7.32), (7.36), and (7.38) as six coupled equations in the six state variables x_{1D}, x_{2D}, Y_{RD}, Ψ_R, Ψ_G, and Y_{GD}.
The model recognizes the two-phase nature of the fluidized beds in both the reactor and the regenerator in a simple manner and it uses a reaction network that relates the

different steps of the cracking reactions to each other and to the reactor model.

From a hydrodynamics point of view, our simplified two-phase model uses many simplifying assumptions for the system. A sample of results for industrial type IV FCC units is shown in Figure 7.6. This figure shows how the multiplicity of the steady states dominates the behavior of the unit. It also shows the effect of the efficiency of the catalyst regeneration through carbon burning in the regenerator.

With regard to the kinetics of the cracking reaction, our approach based on lumping several components into three pseudocomponents is reasonably sound for modeling the system, but the reaction network has a strong deficiency in lumping coke and light hydrocarbons into one component while the two have completely different roles. A four-component network may improve the reliability of the model. However, a much more detailed network is needed for computing the product distribution and gasoline octane number.

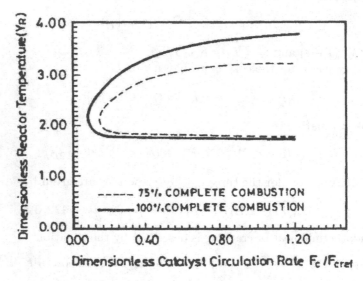

Effect of the degree of coke combustion to CO_2 on the bifurcation diagrams for the reactor temperature versus the catalyst circulation rate

Figure 7.6

Our first simple model will be developed further to account for several effects, namely:

1. *The change in volumetric gas flow rates between inlet and outlet of the reactor and regenerator.*

   The volumetric gas flow rate at the outlet of the reactor is

$$G_R = \left[\frac{x_1}{\overline{M}_{GO}} + \frac{x_2}{\overline{M}_{GS}} + \frac{(1 - x_1 - x_2) \cdot (1 - W_c)}{\overline{M}_C} \right] \cdot \frac{F_{GM} \cdot R_G \cdot T_R}{P_R} . \quad (7.46)$$

   The volumetric gas flow rate at the outlet of the regenerator is

$$G_C = (0.5 \cdot W_H + 0.5 \cdot W_{CO}) \cdot \frac{R_c \cdot R_G \cdot T_G}{100 \cdot P_G} + \frac{F_{AF} \cdot R_G \cdot T_G}{29 \cdot P_G} , \quad (7.47)$$

where F_{GM} and F_{AF} are the feed mass flow rate of gas oil to the reactor and of air to the regenerator, respectively (in kg/sec).

2. *The partial cracking of gasoline and gas oil to lighter hydrocarbons.*
   The heats of cracking that are generally used are based on the complete cracking of hydrocarbons to carbon and hydrogen and thus they are overestimated. We modify the heats of cracking so that they are closer to the industrial values obtained empirically.

3. *The lumping of light hydrocarbon gases with coke.*
   This is not suitable for realistic modeling of commercial FCC units. To remedy this, the amount of light gases is obtained from the weight ratio of coke to (coke + gases) as given in industrial plants data.

4. *The recycle stream.*
   The kinetic scheme does not account for the formation of HCO (heavy cycle oil), which are the fractions produced in addition to gasoline. In refineries, the only fraction that is recycled is HCO and in order to calculate it, the ratio of $HCO/(HCO + LCO + CSO)$ is obtained from plant data.

Both steady-state models for type IV FCC units require only the solution of a set of equations. The solution of our 21 and the 23 equations models formed by the equations (7.25) to (7.45) or (7.25) to (7.47), respectively, can be checked against the industrial data in order to compare the accuracy and to validate the model.

Evaluation of the model:

The model output using the 23 equations model coupled with the above 4 industrial modifications compares favorably with the data of two industrial FCC units. Both units were found to be operating near their middle steady state, see Figure 7.7, and the multiplicity region for each system covers a very wide range of parameters.
The model can help to investigate the effect of the feed composition on the performance of an FCC unit as will be shown in the next section.
Figure 7.7 shows the simulated Van Heerden[1] heat generation-heat removal versus reactor temperature Y_R diagram on top and the unreacted gas oil and gasoline yield profiles below for both industrial units. It is clear that the maximum gasoline yield for both industrial units occurs in the multiplicity region and specifically near the middle unstable saddle-type state. However, the simulation graphs show that neither unit is operating at its optimum condition since their operating points are slightly shifted from the maximal possible gasoline yield. A simple change in the operating variables can shift each unit to its maximum gasoline yield and thereby achieve a considerable improvement in productivity.
   The graphs of Figure 7.7 were obtained from the 23 equations model coupled with our four modifications on p. 440.

---

[1]C. van Heerden, Dutch chemical engineer, 19..-

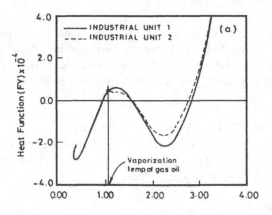

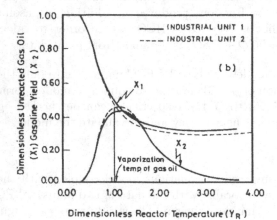

(a)   Heat functions for units 1 and 2
(b)   Conversion and gasoline yield for units 1 and 2
Figure 7.7

It is clear from the graphs that both industrial FCC units have multiple steady states and that the steady state with the most desirable production of gasoline is the middle unstable saddle-type state.

Since FCC units are usually operated at their middle unstable steady state, extensive efforts are needed to analyze the design and dynamic behavior of open loop and closed loop control systems to stabilize the desirable middle steady state.

7.2.2   Static Bifurcation in Industrial FCC Units

The two-phase nature of the reactor and regenerator in Figure 7.5 should always be recognized. The gas entering the reactor bed splits into two parts, one part rising between the individual catalyst particles in a dense-phase gas, and another rising through the bed in form of bubbles that forms the bubble phase. There is an exchange of mass and heat between the two phases.

In addition to the familiar assumptions for modeling bubbling fluidized-bed catalytic reactors, the following specific assumptions for FCC units are used in deriving our model.

1. The change of the number of moles during the reaction is negligible.

2. The heats of reaction and physical properties of the gases and solids are constant.

The above two assumptions will be relaxed in the next section when verifying this model for industrial FCC units.

3. The refractivity parameter W in

$$\frac{k}{k_0} = \left(\frac{C_A}{C_{A_0}}\right)^W \tag{7.48}$$

varies widely between about 0.1 and 0.7, depending on the feed stock. For simplicity we will sometimes set $W = 1.0$. k and k_0 are the feed reactivities at the concentration C_A and C_{A_0}, the fresh feed concentration, respectively.

4. Excess air is used in the regenerator, i.e. the oxygen concentration in the regenerator is kept constant.

The Steady State Model

The mass- and heat-balance equations for the steady-state model are the equations (7.25) to (7.47). In the following, we describe a simple procedure to compute the model parameters.

1. *Computation of two phase parameters:*
   The two phase parameters are computed as follows. The bubble velocity U_b (in cm/sec) is given by

$$U_b = U_{br} + U_2 - U_{mf} \tag{7.49}$$

where

$$U_{br} = 22.26(D_B)^{1/2} \tag{7.50}$$

and D_B is the bubble diameter in cm.

In a bed where the bubbles are fast and large, the net upward velocity of the bubble is

$$U_b = \frac{U_2 - (1 - \delta)U_{mf}}{\delta} . \tag{7.51}$$

Hence the volume fraction of bubbles is

$$\delta = \frac{U_2 - U_{mf}}{U_b - U_{mf}} . \tag{7.52}$$

Therefore, the area occupied by the bubble and cloud phase is

$$A_C = A \cdot \delta \tag{7.53}$$

where A is the cross-sectional area of the bed. Consequently, the area outside the bubble phase, i.e., the dense-phase area is

$$A_I = A - A_c = (1 - \delta) \cdot A \ . \tag{7.54}$$

The bubble flow rate is given by

$$G_c = A(U_2 - U_{mf}) \tag{7.55}$$

and the dense-phase gas flow rate which flows at minimum fluidization velocity is given by

$$G_I = A \cdot U_{mf} \ . \tag{7.56}$$

2. *Kinetics of cracking and coke burning:*
   Our kinetic scheme uses three components:

$$\text{Gas Oil } (A_1) \xrightarrow{K_1} \text{Gasoline } (A_2) \xrightarrow{K_2} \text{Coke } + \text{ Gases } (A_3)$$

$$\underline{\hspace{3cm} K_3 \hspace{3cm}}\uparrow$$

Here A_1 represents gas oil, A_2 gasoline, and A_3 represents coke and dry gases. The rate of disappearance of gas oil, the rate of appearance of gasoline, and the rate of appearance of coke and light gases are given by the equations (7.25), (7.26), and (7.27), respectively. The rate constants can be written in Arrhenius[2] form as follows:

$$K_1 = 0.095 \cdot 10^2 \cdot e^{-21321.664/(R_G \cdot T_R)} \quad m^3/kg \cdot s \ ,$$
$$K_2 = 0.077 \cdot 10^2 \cdot e^{-70466.93/(R_G \cdot T_R)} \quad l/sec \ ,$$
$$K_3 = 473.8 \cdot 10^2 \cdot e^{-109313.28/(R_G \cdot T_R)} \quad m^3/kg \cdot s \ ,$$

where the activation energies are in units of $kJ/kmol \cdot K$.

These starting values are used as initial guesses for fitting the model to industrial data and the preexponential factors are changed to obtain the best fit. This is done because the kinetic parameters depend upon the specific characteristics of the catalyst and of the gas oil feedstock. This complexity is caused by the inherent difficulties with accurate modeling of petroleum refining processes in contradistinction to petrochemical processes. These difficulties will be discussed in more details later. They are clearly related to our use of pseudocomponents. But this is the only realistic approach available to-date for such complex mixtures.

Gas oil and gasoline cracking are assumed to be second order and first order processes, respectively. It was found that W ranges between 0.1 and about 0.7.

The rate of coke burning is given in equation (7.28) and the value of the preexponential factor and activation energy are

$$K_c = A_c \cdot e^{-E_c/(R_G \cdot T)} \quad m^3/(kmol \cdot sec) \ , \qquad \text{and}$$

$$A_c = 1.6821 \cdot 10^8 \ , \quad E_c = 28.444 \cdot 10^3 \ kJ/(kmol \cdot K) \ .$$

---

[2]Svante August Arrhenius, Swedish chemist, 1859 – 1927

3. *Steady state mass- and heat-balance equations:*
   The reactor and regenerator mass and heat-balance equations for the dense phase
   and the bubble phase are given by equations (7.29) to (7.45). The catalyst activities
   in the reactor and regenerator are defined by the following two relations

$$\psi_R = \frac{\overline{C}_m - X_{CR}}{\overline{C}_m} \quad \text{and} \tag{7.57}$$

$$\psi_G = \frac{\overline{C}_m - X_{CG}}{\overline{C}_m}, \tag{7.58}$$

where

$\overline{C}_m$ =   ratio of total amount of coke deposits necessary for complete
deactivation/total amount of catalyst.

X_{CG} =   total amount of coke deposits in regenerator/total amount of catalyst
ratio in regenerator.

X_{CR} =   total amount of coke deposits in reactor/total amount of catalyst
ratio in the reactor.

Solution of the Steady State Equations

The steady-state equations can be manipulated to take the form of heat generation and
heat removal functions, i.e., a modified Van Heerden diagram. This manipulation can be
carried out in different ways, all leading to the same results and here we choose to obtain
the heat generation and removal functions of the regenerator as a function of the reactor
temperature.
The solution proceeds as follows.

1. Choose a value of Y_{RD} within the desired range.

2. Assume a value of Ψ_R as an initial guess.

3. Compute x_{1D} from the following equation which can easily be derived from equation
   (7.29).

$$x_{1D} = \frac{-B_R + \sqrt{B_R^2 + 4\Psi_R \cdot B_R \cdot x_{1f} \cdot (\alpha_1 e^{-\gamma_1/Y_{RD}} + \alpha_3 e^{-\gamma_3/Y_{RD}})}}{2\Psi_R \cdot (\alpha_1 e^{-\gamma_1/Y_{RD}} + \alpha_3 e^{-\gamma_3/Y_{RD}})}$$

$$= f_1(Y_{RD}) \tag{7.59}$$

4. Calculate x_{2D} from equation (7.30) in the following equivalent explicit form:

$$x_{2D} = \frac{\Psi_R \cdot \alpha_1 \cdot f_1^2(Y_{RD}) e^{-\gamma_1/Y_{RD}} + B_R \cdot x_{2f}}{\Psi_R \cdot \alpha_2 e^{-\gamma_2/Y_{RD}} + B_R}$$

$$= f_2(Y_{RD}) \tag{7.60}$$

5. Calculate the rate of coke formation R_{cf} from equation (7.43) and Ψ_G from equation
   (7.31).

6. Calculate $\Delta \overline{H}_{cr}$ from equation (7.45) and Y_{GD} from equation (7.32).

7. Calculate the rate of coke burning from equation (7.41).

8. Check the correctness of the assumed value Ψ_R by computing its value from equation (7.36).

9. If the residual in step 8 is too large, correct the value of Ψ_R using interpolation. Then repeat steps 3 to 8. Once the residual is small enough ($\leq 10^{-6}$), substitute Ψ_R in the heat-balance equation of the regenerator (7.37), written in the form

$$\overline{R}(Y_{RD}) \;=\; B_G \cdot Y_{Af} + a_3 \cdot Y_{RD} \;=\; (B_G + a_3) \cdot Y_{GD} - \beta_c R_c + \Delta \overline{H}_{LG} \;=\; \overline{G}(Y_{RD}) \,.$$
$$(7.61)$$

This is a nonlinear equation due to R_C on the right-hand side because the definition of R_C in equation (7.28) involves the exponentiation of variables.

10. Change the value of Y_{RD} and repeat steps 2 to 9 until the required range of the reactor temperature has been covered.

Steady State Simulation Results for an Industrial Scale FCC Unit

For the results in this section, all system parameters that are not explicitly mentioned in the figures below are as specified in the following table.

Data for one fluid catalytic cracking unit	
Height and diameter of the regenerator bed	25.05, 13.50 m
Height and diameter of the reactor bed, respectively	25.50, 8.25 m
Pressure in both vessels	1.5 atm
Voidage in both vessels	0.4
U_{mf} in regenerator and reactor, respectively	16×10^{-4}, 4×10^{-4} m/sec
Density and average size of catalyst, respectively	450 kg/m^3, 60 μm
Average molecular weight of gas oil and gasoline, respectively	180, 110
Boiling point of gas oil and gasoline, respectively	539 K, 411 K
Heat of reaction of gas oil to coke ΔH_3	8.126×10^3 kJ/kg
Heat of reaction of gas oil to coke ΔH_1	6.036×10^2 kJ/kg
Heat of reaction of gasoline to coke ΔH_2 (by difference)	7520 kJ/kg
Latent heat of vaporization of gas oil	264.7 kJ/kg
Unless otherwise stated	$Q_{ER} = Q_{EG} = \infty$
Heat of combustion of coke	30.19×10^3 kJ/kg
Reference temperature (T_{rf})	500 K

Determining the steady state for given sets of parameters:

Below we present and discuss our simulation results obtained by numerically solving the model equations (7.29) to (7.47) as described in the ten step adaptive procedure above. The results are based on manipulating equation (7.61), so that the heat removal line becomes independent of the feed temperature and its slope independent of the reactor temperature and the other variables. Once this has been achieved, we can assume that the slope of the heat removal line is constant over time.

The gas flow rate in the reactor and the regenerator are assumed constant as well and are calculated – according to the ideal gas law, see Section 3.4 – from their feed mass flow rates F_{GM} and F_{AF} to the reactor at vaporization conditions for the gas oil and at feed conditions for the air to the regenerator, respectively.

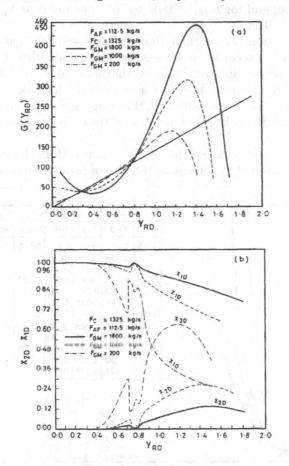

Effect of gas oil flow rate:
(a)   Heat generation function and heat removal line vs. reactor temperature
(b)   Yield and conversion vs. reactor temperature
Figure 7.8

Figure 7.8 shows the effect of the gas oil flow rate on the modified heat generation function $\overline{G}(Y_{RD})$ in Figure 7.8(a), as well as the conversion $1 - x_{1D}$ and the gasoline yield x_{2D} in Figure 7.8(b). We notice that the gasoline yield functions in terms of the reactor temperature have several maxima. This is caused by differences in the activation energies of the reactions between the reactor and the regenerator. When $F_{GM} = 200\ kg/sec$, three such maxima occur in the $-\ .\ -\ .\ -$ graph of x_{2D} in Figure 7.8(b). The largest of these gives the maximum gasoline yield at a reactor dimensionless temperature of around 1.2. For $F_{GM} = 1000$ and $1800\ kg/sec$, two maxima of x_{2D} occur and the maximal gasoline yield occurs at the larger maximum, namely at the reactor dimensionless temperatures of approximately 1.45 and 1.5, respectively. Under the given system conditions three steady states are obtained for $F_{GM} = 200 \ kg/sec$ at $Y_{RD} = 0.19$, 0.62, 1.26 with $x_{2D} = 0.0$, 0.18, 0.63. For $F_{GM} = 1000 \ kg/sec$ the steady states are at $Y_{RD} = 0.28$, 0.77, 1.5 with $x_{2D} = 0.0$, 0.02, 0.26, and for $F_{GM} = 1800 \ kg/sec$ they occur at $Y_{RD} = 0.27$, 0.78, 1.66 with $x_{2D} = 0.0$, 0.01, 0.11.

From these simulations we conclude that the increase in the gas oil flow rate at constant operating conditions causes an increase in the temperature of the system together with a decrease in the gasoline yield. Notice that as F_{GM} increases, the corresponding yield Y_{GD} for the same value of Y_{RD} increases since the increase in F_{GM} causes an increase in the amount of coke that is formed, thus causing a larger amount of coke to enter the regenerator where the coke gets burned, and thus this leads to an increased heat of combustion.

The increase in Y_{GD} may exceed the allowable limits. If this happens, the regenerator temperature can be lowered by increasing the rate of catalyst circulation F_C as depicted in Figure 7.9.

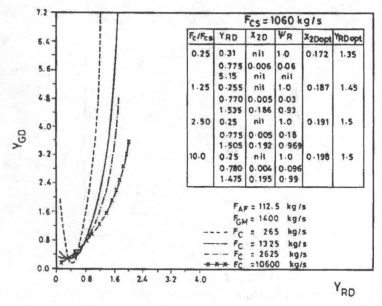

Regenerator temperature versus reactor temperature and the effect of the catalyst circulation rate. (The embedded table shows the effect of F_C on the steady-state temperature, the yield and the catalyst activity.)

Figure 7.9

By changing F_C the system's steady states are changed and consequently the gasoline yield changes, too. This shows the complexity of the interaction between the variables of the system and a need for mathematical models to optimize the operation of FCC units. The most useful models must be able to account for all varying parameters simultaneously.

We notice in Figure 7.9 that an increase of the catalyst circulation rate increases the yield by reducing the overheating of the system that ultimately destroys the gasoline to coke and dry gases. Therefore, increasing the catalyst circulation rate to control the regenerator temperature has a favorable effect on gasoline yield, too.

The plots in Figures 7.8 and 7.9 make both Q_{ER} and Q_{EG} infinite and therefore the dense phase and bubble phase conditions are identical and are equal to the output conditions of the reactor and the regenerator in this example. In case of finite exchange rates between the bubble and dense phases in reactor and regenerator, the output conditions from the reactor and the regenerator can be obtained by mass and heat balances for the concentration and the temperature of both phases and these expressions use the same symbols as before, but without the subscript D (used to signify the "dense phase" before).

The effect of the exchange rate between the bubble and dense phases is discussed further in the next section for one industrial FCC unit. There the simulation results are verified and cross-verified against the corresponding industrial data.

Determining the air feed temperature for maximum gasoline yield:

Here equation (7.61) is used without modification and the air feed temperature Y_{RD} is used as the manipulated variable to obtain the maximum gasoline yield. The term $a_3 \cdot Y_{RD}$ on the left-hand side of (7.61) can be moved so that the chosen variable Y_{RD} appears only on the nonlinear right-hand side of equation (7.61), while the left-hand side of the equation (7.61) still contains Y_{Af}. This manipulation allows us to solve the equation without having to use numerical optimization techniques.

In this setting, the maximum gasoline yield x_{2D} and the corresponding reactor temperature $Y_{RD_{max}}$ are determined from a plot of x_{2D} vs. Y_{RD}. This optimal value for Y_{RD} is then plugged into the function $G(Y_{RD_{max}}) = \overline{G}(Y_{RD_{max}}) - a_3 \cdot Y_{RD_{max}} = B_G \cdot Y_{Af_{max}}$. Since B_G is known, this allows us to find $Y_{Af_{max}}$ easily.

Optimum air feed temperature at different values of the catalyst circulation rate with $F_{cs} = 1060 \ kg/sec$

F_c/F_{cs}	$T_{FA} \ K$	F_c/F_{cs}	$T_{FA} \ K$
1.25	1958.5	1.875	665.0
2.5	308.6	5.0	282.6

The table above gives the optimum air feed temperature obtained graphically as described above. It uses different values of F_C for the given parameters $F_{AF} = 112.5 \ kg/sec$, $F_{GM} = 1400 \ kg/sec$ and $Q_{ER} = Q_{EG} = \infty$.

It is important to notice that the optimum feed temperature T_{FA} decreases sharply as F_C increases. This is due to the fact that increased quantities of heat are removed from the regenerator as the catalyst circulation rate increases.

Hysteresis loops:

The hysteresis phenomenon, i.e., static bifurcation as described earlier for the nonisothermal CSTR that is associated with multiple steady states plays an important role in start-up, control and in the stability of chemical reactors. The desired optimum steady state may be within the multiplicity region, and hence special start-up procedures must be followed in order to attain the desired steady state.

As discussed in Chapters 3 and 4 of this book, when such conditions prevail, a large disturbance may shift the reactor outside the region of stability of the optimum steady state and towards the stability region of another, but undesirable steady states. When the disturbance is removed, the reactor does not generally restore itself to its original steady state and an irreversible drop of productivity will generally occur.

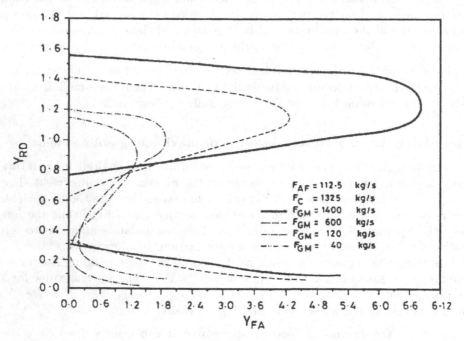

Hysteresis loops of reactor temperature versus air feed temperature
(Effect of gas oil flow rate)
Figure 7.10

Figure 7.10 shows the hysteresis loops of Y_{RD} vs. Y_{FA} for various values of F_G. It is clear that the multiplicity range is in general quite large and increases as the gas oil flow rate F_{GM} increases.

The various graphs in Figure 7.10 also show that the behavior of the FCC unit is governed by the multiplicity of the steady states.

Summary of the results:

A relatively simple mathematical model composed of 21 or 23 transcendental and rational equations numbered (7.25) to (7.47) was presented to describe the steady-state behavior of type IV FCC units. The model lumps the reactants and products into only three groups. It accounts for the two-phase nature of the reactor and of the regenerator using hydrodynamics principles. It also takes into account the complex interaction between the reactor and the regenerator due to catalyst circulation. The effect of some parameters on the steady-state behavior has been presented in this section and a simple graphical technique given for determining the optimum air feed temperature for each given value of the catalyst circulation rate. The optimal air feed temperature may, however, exceed allowable limits, hence the catalyst circulation rate may have to be adjusted to bring down both the optimum air feed temperature and the regenerator temperature.

The results obtained from this model provide a sufficiently good understanding of the complex interactions between the reactor and the regenerator and their effect on the overall behavior of the system. This model is more realistic and reliable than earlier empirical models. The model can be applied to optimize and control the steady state of FCC units. One further advantage of this model is the ease with which the model equations can be solved graphically using a modified van Heerden diagram.

This model involves some assumptions which can be relaxed in order to become more effective and more for accurate in simulating industrial units.

Industrial verification of our steady-state model and further static-bifurcation studies of industrial FCC units follow.

7.2.3 Industrial Verification of the Steady State Model and Static Bifurcation of Industrial Units

We relax some of our previous FCC system assumptions now and restate our model assumptions first. Namely:

1. There is no reaction in the bubble phase.

2. The dense phase is perfectly mixed and the bubble phase is in plug flow with respect to mass and heat.

3. The interchange between bubble-phase gas and dense-phase gas is by bulk flow and diffusion.

4. An average mean bubble diameter is used throughout the bed.

5. The heats of reaction and the physical properties are constant except for diffusivities, the density of vaporized gas oil, and the density of air, which are functions of temperature.

6. Excess air is used in the regenerator.

The starting values of the kinetic rate constants and the heats of the three reactions are as follows.

$$
\begin{array}{llll}
K_1 &=& 4.468 \cdot e^{-20883.3/(R_G \cdot T)} \quad m^3/(kg \cdot sec); & \Delta H_1 &=& 603.9 \; kJ/kg \\
K_2 &=& 10773.5 \cdot e^{-75240/(R_G \cdot T)} \quad 1/sec; & \Delta H_2 &=& 7520.0 \; kJ/kg \\
K_3 &=& 17.092 \cdot e^{-37620/(R_G \cdot T)} \quad m^3/(kg \cdot sec); & \Delta H_3 &=& 8126.0 \; kJ/kg
\end{array}
$$

The heat of combustion of coke is calculated from the relation

$$(-\Delta H_c) = \left[4100 + 10100 \cdot \left(\frac{CO_2}{CO_2 + CO} \right) + 3370 \cdot \left(\frac{H}{C} \right) \right] \cdot 2.326 \ kJ/kg \qquad (7.62)$$

where $CO_2/(CO_2 + CO)$ refers to the relative volume of these gases in the flue gases from the regenerator and H/C is the atomic ratio of hydrogen to carbon in the coke that is formed.

The heat losses in the reactor are assumed to be 0.5% of the heat that enters with the catalyst, and the heat losses in the regenerator are assumed to be 0.5% of the heat supplied by coke burning. The rest of the data is given in the following table.

**Basic data for the two industrial units and
the characteristics of their feedstock, products and catalyst**

Average molecular weight of gas oil and gasoline, respectively	180, 110
Boiling point of gas oil and gasoline, respectively	539, 411 K
Latent heat of vaporization of gas oil	265.15 kJ/kg
Reference temperature T_{rf}	500 K
Concentration of gasoline in gas oil feed C_{A2f}	0
Voidage in both vessels	0.4
Height of expanded bed in reactor and regenerator, for the operating conditions of unit 1.	7, 7.2 m

The two industrial fluid catalytic cracking units that we consider are of type IV with U-bends. The two units vary in their input parameters which lead to different outputs. The following table contains the plant data for the two commercial FCC units under consideration.

Plant data for both units

	Unit 1	Unit 2
Fresh feed flow rate, kg/sec	16.782	13.476
Recycle HCO flow rate, kg/sec	2.108	2.111
Combined feed ratio CFR	1.1256	1.1566
Air flow rate, kg/sec	10.670	10.670
Catalyst circulation rate, kg/sec	88.605	117.113
Combined feed temperature, K	527.	538.
Air feed temperature, K	436.	433.
Hydrogen in coke, wt %	4.17	6.79
$HCO/(HCO + CSO + LCO)$	0.286	0.419
Coke/(Coke+gases)	0.221	0.207
$CO_2/(CO_2 + CO)$, m^3/m^3	0.75	0.613
$(-\Delta H)_c$, kJ/kg	31.24×10^3	30.78×10^3

Additional plant data that is common to both units is given below.

Plant data common to both units

Regenerator dimensions	$5.334\ m\ ID \times 14.859\ m\ TT$
Reactor dimensions	$3.048\ m\ ID \times 12.760\ m\ TT$
Catalyst retention in reactor	$17.5\ metric\ tons$
Catalyst retention in generator	$50\ metric\ tons$
API of raw oil feed	28.7
Reactor pressure	$225.4938\ kPa$
Regenerator pressure	$254.8708\ kPa$
Average particle size	$0.00072\ m$
Pore volume of catalyst	$0.00031\ m^3/kg$
Apparent bulk density	$800\ kg/m^3$
Catalyst surface area	$215\ m^2/g$

The two industrial units are relatively small compared to more modern large capacity riser-reactor units.

Simulation Procedure; Verification and Cross Verification

The model was fitted to the first FCC unit using corrections for the frequency factors. The modified frequency factors that were used are: $K'_{1_0} = 1.5K_{1_0}$, $K'_{2_0} = 0.5K_{2_0}$, $K'_{3_0} = 1.5K_{3_0}$. We note that in practice these correction factors vary from one catalyst to the other, depending upon the type of catalyst, its activity and age, as well as the type of feedstock used.

To insure that the correction factors are not just empirical numbers that fit only one specific FCC unit at one set of operating conditions, the model should be cross-verified. For this, the same model with these correction factors is used to simulate the second FCC unit. This unit uses a very similar catalyst with almost the same activity and age and a very similar feedstock. Success of the second simulation acts as a cross-verification here and ensures that the corrected model gives a reasonably good general representation of FCC type IV units. However, in practice the catalyst activity should be regularly checked to ensure that the preexponential factors used in the simulation model are still valid.

Simulation and Bifurcation Results; Discussion for two Industrial FCC Units

Comparison of the model results with plant data:

Figure 7.11 (a) shows the heat function vs. the reactor temperature for both units 1 and 2 and Figure 7.11 (b) shows the corresponding unreacted gas oil and gasoline yield.

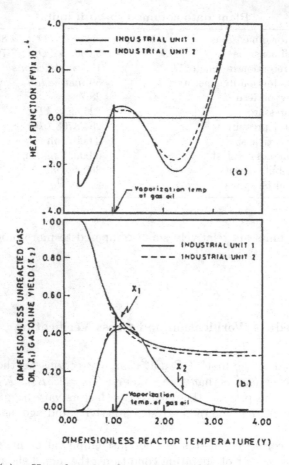

(a)   Heat function for units 1 and 2
(b)   Conversion and gasoline yield for units 1 and 2
Figure 7.11

The following table contains the simulated and measured output variables for both units 1 and 2.

Simulation results of the FCC units 1 and 2

Parameters	Model results		Plant data		% error	
	Unit 1	Unit 2	Unit 1	Unit 2	Unit 1	Unit 2
Y	1.55	1.57	1.55	1.57	0	0
YG	1.98	1.85	1.9	1.88	+4.21	−1.6
x1	0.382	0.355	0.39	0.322	−2.05	+10.25
x2	0.39	0.39	0.415	0.41	−6.02	−4.9
Coke kg/sec	4.2	4.083	3.681	4.073	+14.1	+0.25
HCO kg/sec	2.057	2.354	2.109	2.111	−2.45	−11.6

It is obvious that our model agrees very well with the plant output data. The simulation could be improved by using more accurate kinetic data, and even further improved if the kinetic scheme were extended to include separate routes for the formation of coke and light gases. It is clear from the above table and Figure 7.11 that both units are operating at their middle unstable steady state and that neither is operating at its maximum gasoline yield point.

Effect of some operating parameters:

For units operating at the middle unstable steady state, the steady-state response of the system is generally very sensitive to variations in the operating conditions. Therefore, a detailed parametric investigation of such a system is highly advised. We now do so for the present model to give our readers a deeper insight into the behavior of these rather complex FCC systems.

The optimization problem for the system in a neighborhood of its unstable middle steady state is quite different from that in a neighborhood of the stable high temperature steady state. Clearly there are trade-offs between the gasoline yield of the system and its stability, i.e. in some cases, higher yields correspond to an operating point closer to the critical bifurcation point. The parametric presentation is simplified by the fact that the dense gas oil x_{1D} and gasoline dimensionless concentrations x_{2D} versus Y diagrams do not change with variations in F_C, Y_{Af}, or Y_{Gf}. The only changes come from the operating temperatures of the reactor and regenerator.

For any given operating temperature the corresponding x_{1D} and x_{2D} values can be obtained from the same curve in Figure 7.11(b). Our parametric investigation is confined to the FCC unit 1 with its $x_{1D} - x_{2D}$ versus Y diagram shown in Figure 7.11(b).

Effect of catalyst circulation rate:

Figure 7.12(a) shows that for the middle steady state the reactor temperature Y_R decreases from 1.75 to 1.56 and to 1.5 as the catalyst circulation rate F_C increases from 44.302 to 88.605 and to 177.21 kg/sec.

The corresponding conversion $1 - x_{1D}$ decreases while the corresponding gasoline yield x_{2D} increases from 0.314 to 0.385 and to 0.41. It is clear from the Figures 7.12(a) and (b) that the response of the middle steady state to a change in any parameter reacts in the opposite way to the response of the other steady states. This is important for optimizing and controlling FCC units and it will be discussed in more detail in the sequel.

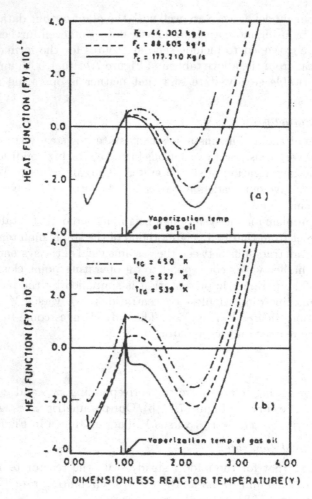

(a)   Effect of catalyst circulation rate on heat function for unit 1
(b)   Effect of gas oil feed temperature on heat function for unit 1
Figure 7.12

Effect of gas oil feed temperature:

For the middle steady state, Figure 7.12(b) shows that the increase in the gas oil feed temperature Y_{Gf} from 450 K to 527 K and to 539 K causes the dimensionless reactor temperature to decrease from 1.76 to 1.56 and to 1.07. This increase in Y_{Gf} in turn causes a decrease in conversion and an increase in gasoline yield from 0.314 to 0.3875 and to 0.3971, see Figure 7.11(b). The opposite response directions for these middle steady state parameters is again evident.

Bifurcation diagrams:

Obviously the reactor is not operating at the optimal operating conditions. The optimal operating conditions cannot be estimated unless the multiplicity regions are constructed for different parameters, because as the system approaches maximum yield it gets closer and closer to the bifurcation point where ignition of the reactor can occur. In this region the reactor temperatures at differing steady states can be calculated by solving the heat-balance equation (7.61) with a bisection method such as MATLAB's `fzero`.

Effect of FC on the bifurcation behavior:

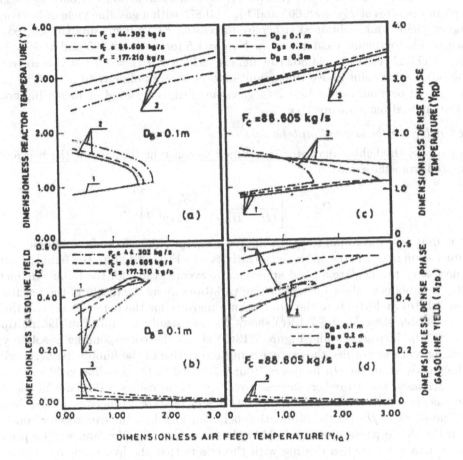

(a) Effect of catalyst circulation rate on the bifurcation diagram of $Y(=Y_{RD})$ vs. Y_{Af}
(b) Effect of catalyst circulation rate on the bifurcation diagram of x_2 vs. Y_{Af}
(c) Effect of DB on the bifurcation diagram for Y_{RD} vs. Y_{Af}
(d) Effect of DB on the bifurcation diagram for x_2 vs. Y_{Af}
    [1 = low-temperature steady state; 2 = middle steady state;
    3 = high-temperature steady state]
Figure 7.13

Figure 7.13(a) shows the bifurcation diagram for three different values of the catalyst circulation rate $F_C = 177.21$, 88.605, and 44.303 kg/sec. The multiplicity region increases as F_C decreases. As the air temperature increases, the gasoline yield corresponding to the middle steady state increases also, see Figure 7.13(b). However, the system is approaching the bifurcation point and at an air temperature higher than 1.7 (for the case of $F_C = 44.302$), ignition occurs i.e. the reactor operates at the high-temperature steady state. This steady state yields a negligible amount of gasoline as represented by the lower branch of the bifurcation diagram of Figure 7.13(b). At higher circulation rates, the drop in gasoline yield occurs at $Y_{fa} = 1.65$ for $F_C = 88.605$ and at $Y_{fa} = 1.6$ for $F_C = 177.21$. The plant operates at $F_C = 88.605$ and $Y_{fa} = 0.875$ with a gasoline yield of 0.39 or 39%. A higher yield can be obtained either by increasing F_C or increasing the air feed temperature. The optimum yield occurs at $Y_{AF} = 1.5$ for $F_C = 88.605$ and at $Y_{AF} = 1.46$ for $F_C = 177.21$. For $F_C = 44.302$ the highest yield is at $Y_{AF} = 1.7$ but this corresponds to the bifurcation point above where ignition occurs. A gasoline yield of 0.47, i.e., a 20% improvement over our earlier best 39% gasoline yield, is achieved by increasing both F_C and the air feed temperature Y_{AF}.

Effect of D_B on the bifurcation behavior:

In our model the bubble diameter is assumed as constant throughout the bed and it is calculated from the equation

$$D_B = \left[\frac{(U - U_{mf})H_{mf}}{(H - H_{mf}) \cdot 0.711 g^{1/2}} \right]^2 . \tag{7.63}$$

For the unit, D_B was found to be 0.1 m. We now study the effect of the bubble diameter on bifurcation in order to exhibit the sensitivity of a FCC system to the bubble diameter. It is necessary to calculate a good approximate average value for the bubble diameter or to alter the model to allow for bubble size variations along the height of the reactor. The figures 7.13(c) and (d) show the bifurcation diagrams for the dense phase and differing average bubble sizes. Figure 7.13(c) shows the increase in the range of the multiplicity region as D_B increases, while Figure 7.13(d) shows the corresponding gasoline yields. According to these graphs, the system is quite sensitive to the bubble diameter. For the middle steady state, as D_B increases from 0.1 to 0.2 m the gasoline yield in the dense phase increases, but a further increase to $D_B = 0.3$ m causes a decrease in the dense-phase gasoline yield. At the high temperature steady state an increase in D_B decreases x_{2D}. The effect of D_B on the bifurcation behavior based on the exit reactor conditions, i.e., on the dense phase plus the bubble phase, is quite complex and will be presented in more details later when dealing with the effects that the hydrodynamic parameters have on the performance of FCC units. We conclude that better model simulation of industrial units can be achieved by obtaining more accurate experimental estimates for the average bubble size. However, taking the bubble size variations along the height of the bed into account complicates the model considerably, without affecting its prediction quality much.

Summary of this section's results:

Our model simulates the industrial data for two FCC units quite well. It shows that the two reactors are operated at the middle unstable steady state. The model has been verified and used in a parametric investigation in which the response directions at the middle steady state of the input parameters is opposite to that of the high temperature steady state. The bifurcation results show a rather large multiplicity region and the sensitive regions near the bifurcation points have been studied. We have learnt that the two commercial FCC units are not operating at their optimum gasoline yield conditions and the model was used to suggest changes for improving the gasoline yield without a formal optimization study.

7.2.4 Preliminary Dynamic Modeling and Characteristics of Industrial FCC Units

The Dynamic Model

In this section we develop a dynamic model from the same basis and assumptions as the steady-state model developed earlier. The model will include the necessarily unsteady-state dynamic terms, giving a set of initial value differential equations that describe the dynamic behavior of the system. Both the heat and coke capacitances are taken into consideration, while the vapor phase capacitances in both the dense and bubble phase are assumed negligible and therefore the corresponding mass-balance equations are assumed to be at pseudosteady state. This last assumption will be relaxed in the next subsection where the chemisorption capacities of gas oil and gasoline on the surface of the catalyst will be accounted for, albeit in a simple manner. In addition, the heat and mass capacities of the bubble phases are assumed to be negligible and thus the bubble phases of both the reactor and regenerator are assumed to be in a pseudosteady state. Based on these assumptions, the dynamics of the system are controlled by the thermal and coke dynamics in the dense phases of the reactor and of the regenerator.

Unsteady-state heat balance for the reactor:

After simple manipulations the unsteady-state heat-balance equations for the reactor can be written as follows.

For the dense phase:

The unsteady thermal behavior of the dense phase is described by the nonlinear ordinary differential equation

$$\epsilon_{HR}\frac{dY_{RD}}{dt} = B_R(Y_\nu - Y_{RD}) + a_1(Y_{GD} - Y_{RD}) + a_2(Y_{GF} - Y_\nu) - (\Delta\overline{H}_\nu + \Delta\overline{H}_{LR}) + \Delta\overline{H}_{cr} \tag{7.64}$$

with the initial condition $Y_{RD} = Y_{RD}(0)$ at time $t = 0$.

For the bubble phase (in pseudosteady state):

The assumption of pseudosteady state in the bubble phase and the general assumptions allow us to describe the bubble-phase temperature profile using the algebraic equation (7.35).

Unsteady-state heat balance for the regenerator:

Simple manipulations similar to those used for the reactor give the following heat-balance equations for the regenerator.

For the dense phase:

The thermal behavior of the regenerator is described by the following nonlinear ordinary differential equation.

$$\epsilon_{HG}\frac{dY_{GD}}{dt} = B_G(Y_{AF} - Y_{GD}) + \beta_c \cdot R_c + a_3(Y_{RD} - Y_{GD}) - \Delta\overline{H}_{LG} \tag{7.65}$$

with the initial condition $Y_{GD} = Y_{GD}(0)$ at $t = 0$.

For the bubble phase:

The assumption of pseudosteady state in the bubble phase and the general assumptions allow us to write the bubble-phase temperature profile in the simple algebraic form given by equation (7.38).

Unsteady-state carbon balance in the reactor:

The dynamics of the carbon inventory in the reactor, expressed in terms of the catalyst activity Ψ_R, are described by the following nonlinear ordinary differential equation

$$\epsilon_{MR}\frac{d\Psi_R}{dt} = C'_R(\Psi_G - \Psi_R) - \Psi_R \cdot R_{cf} \tag{7.66}$$

for the initial condition $\Psi_R = \Psi_R(0)$ at $t = 0$. Here R_{cf} is given by equation (7.43).

Unsteady-state carbon balance in the regenerator:

The dynamics of the carbon inventory in the regenerator, expressed in terms of catalyst activity Ψ_G, are described by the following nonlinear ordinary differential equation

$$\epsilon_{MG}\frac{d\Psi_G}{dt} = R_c - C'_R(\Psi_G - \Psi_R) \tag{7.67}$$

for the initial condition $\Psi_G = \Psi_G(0)$ at $t = 0$. Here R_c is given by equation (7.41).

Pseudosteady-state gas oil and gasoline mass balances in the reactor:

Equations (7.29) and (7.33) represent the pseudosteady-state gas oil balances for the dense and bubble phases, respectively, while equations (7.30) and (7.34) do the same for gasoline. Equations (7.32) and (7.35) are used for the steady-state dense- and bubble-phase temperatures of the reactor, and equations (7.37) and (7.38) for the dense and bubble-phase temperatures in the regenerator.

The exit concentrations and temperatures from the reactor and the regenerator are the result of the mixing of gas from the dense phase and from the bubble phase at the exit of the two subunits.

The exit gas oil dimensionless concentration is

$$x_{1o} = (G_{IR} \cdot x_{1D} + G_{CR} \cdot x_{1B})/(G_{IR} + G_{CR}) \equiv X_1 . \tag{7.68}$$

The exit gasoline dimensionless concentration is

$$x_{2o} = (G_{IR} \cdot x_{2D} + G_{CR} \cdot x_{2B})/(G_{IR} + G_{CR}) \equiv X_2 . \tag{7.69}$$

The exit reactor dimensionless temperature is

$$Y_{Ro} = \frac{G_{IR} \cdot \rho_{IR} \cdot C_{P_{IR}} \cdot Y_{RD} + G_{CR} \cdot \rho_{CR} \cdot C_{P_{CR}} \cdot Y_{RB}}{G_{IR} \cdot \rho_{IR} \cdot C_{P_{IR}} + G_{CR} \cdot \rho_{CR} \cdot C_{P_{CR}}} \equiv Y_R , \tag{7.70}$$

and the exit regenerator dimensionless temperature is

$$Y_{G_o} = \frac{G_{IG} \cdot \rho_{IG} \cdot C_{P_{IG}} \cdot Y_{GD} + G_{CG} \cdot \rho_{CG} \cdot C_{P_{CG}} \cdot Y_{GB}}{G_{IG} \cdot \rho_{IG} \cdot C_{P_{IG}} + G_{CG} \cdot \rho_{CG} \cdot C_{P_{CG}}} \equiv Y_G . \qquad (7.71)$$

The above dynamic model equations are defined in terms of the same state variables as the steady-state model was. The parameters used are also the same as those of the steady-state model except for the additional four dynamic parameters ϵ_{HR}, ϵ_{HG}, ϵ_{MR}, and ϵ_{MG}.

Results for the Dynamic Behavior of FCC Units and their Relation to the Static Bifurcation Characteristics

The dynamic behavior of industrial unit 1 is studied in this section. The steady-state behavior of unit 1 has been studied previously. Both open-loop and closed-loop feedback controlled configurations will be presented.

Figure 7.14(a) shows the gasoline yield x_{2D} vs. the dimensionless dense phase temperature Y_{RD}. The heat generation function for this case with $Y_{fa} = 0.872$ and $FCD = 1.0$ is shown in Figure 7.19 when $K_c = 0$. The bifurcation diagrams are shown in Figure 7.14(b) and (c). The bifurcation diagram with FCD as the bifurcation parameter has very different characteristics than the diagram with Y_{fa} as the bifurcation parameter. For the latter, at constant FCD values Y_{fa} can be varied within the physically reasonable range to give a unique ignited (high temperature) steady state, but it is not possible to obtain a unique quenched (low temperature) state, see Figure 7.14(b). When FCD is the bifurcation parameter at a constant value of Y_{fa}, then FCD can be varied to give a quenched unique state, but it is not possible to obtain a unique ignited state within the physically reasonable range of the parameters (Figure 7.14(c)). There is also a narrow region of five steady states in Figure 7.14(b).

We have chosen the steady state with $Y_{fa} = 0.872$ and $FCD = 1.0$ giving a dense phase reactor temperature of $Y_{RD} = 1.5627$ (Figure 7.14(b) and (c)) and a dense-phase gasoline yield of $x_{2D} = 0.387$ (Figure 7.14(a)). This is the steady state around which we will concentrate most of our dynamic analysis for both the open-loop and closed-loop control system. We first discuss the effect of numerical sensitivity on the results. Then we address the problem of stabilizing the middle (desirable, but unstable) steady state using a switching policy, as well as a simple proportional feedback control.

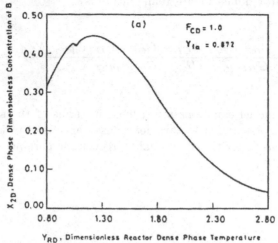

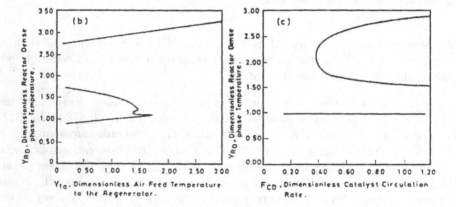

(a)   Gasoline yield vs. dimensionless reactor dense-phase temperature
(b)   Bifurcation diagram of the dimensionless reactor dense-phase temperature vs. the dimensionless air feed temperature of the regenerator
(c)   Bifurcation diagram of the dimensionless reactor dense-phase temperature vs. the dimensionless catalyst circulation rate

Figure 7.14

Numerical sensitivity:

The set of four ordinary differential equations (7.64) to (7.67) for the dynamical system are quite sensitive numerically. Extreme care should be exercised in order to obtain reliable results. We advise our students to experiment with the standard IVP integrators ode... in MATLAB as we have done previously in the book. In particular, the stiff integrator ode15s should be tried if ode45 turns out to converge too slowly and the system is thus found to be stiff by numerical experimentation.

In many cases a very small accuracy limit $(TOL < 10^{-8})$ must be specified in order to obtain accurate results. Using a value of TOL as small as 10^{-6} often does not give accurate results here.

Actually this is a very tricky point, for with the advent of computing in chaotic situations, it is quite difficult to specify what it means to have an "accurate result".

In this IVP example with $TOL = 10^{-6}$ we see a chaotic like response of the system in Figure 7.15.

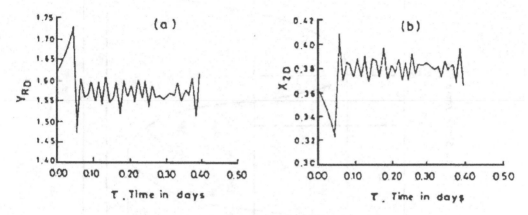

Numerically chaotic results
(a)   Dimensionless reactor dense-phase temperature Y_{RD}
(b)   Dimensionless gasoline dense phase concentration X_{2D}
Figure 7.15

This chaotic behavior of the solution curves persists over a range of TOL values, such as for $TOL = 10^{-7}$ and $TOL = 10^{-8}$. The chaotic looking trajectories may suggest chaotic behavior of the system. However, with a more stringent control over the allowed accuracy of the solution, such as by using $TOL = 10^{-10}$ for example, this chaotic behavior of the solution trajectories disappears.

Thus we must conclude that the results presented in Figure 7.15 are clearly due to numerical inaccuracies and not to system chaos. The reader should be cautious.

This no-chaos conclusion can be validated further by trying different methods of integration with different complexities in MATLAB. Each integrator $\texttt{ode..}$ of MATLAB shows that a different tolerance bound is needed in order to obtain the correct unchaotic results without numerical instability. Moreover, when each method is executed with the appropriately tight tolerance, then all methods give the same correct and unchaotic result.

Simple control strategies to stabilize the middle unstable steady state:

In order to avoid the continuous drift of the system away from the unstable steady state with its high gasoline yield, we discuss two simple control strategies in this section: we can operate the system dynamically around this unstable steady state or we can use a suitable feedback control that stabilizes the unstable steady state.

Dynamic operation around the middle unstable steady state:

Switching the feed conditions around the values needed to operate at the desired middle steady state has been suggested by many investigators.

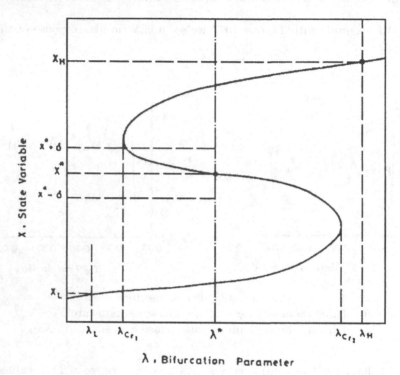

Schematic for a simple switching policy around the middle unstable steady state
Figure 7.16

The underlying principle is quite simple and we illustrate it in Figure 7.16. Figure 7.16 shows the desired state x^* that corresponds to the bifurcation parameter feed variable λ^*. The desired state x^* is unstable. In theory it is possible to operate the system as closely as possible to the unstable state x^* by alternating the feed condition parameter between the value λ_H and λ_L in order to keep the system operating near x^*. More specifically when the state variable x drops below x^* to a value of $x^* - \delta$ for a given small tolerance δ then λ is switched to the higher value λ_H that corresponds to a unique state x_H. Thus the system is forced to increase x. When the state variable exceeds x^* in this dynamic process and it reaches the value $x^* + \delta$ which still lies on the middle branch of the bifurcation curve, then λ is switched to λ_L that corresponds to a much lower state parameter x_L and the system is forced to decrease x and so on. Our feed condition switching law is:

$$\text{Set } \lambda = \lambda_H \quad \text{if} \quad x \leq x^* - \delta \text{ or if } dx/dt > 0 \text{ for } x^* - \delta < x < x^* + \delta \; ;$$

$$\text{set } \lambda = \lambda_L \quad \text{if} \quad x \geq x^* + \delta \text{ or if } dx/dt < 0 \text{ for } x^* - \delta < x < x^* + \delta \; .$$

The specific dynamic behavior of the system depends upon the chosen values of λ_H, λ_L, and δ. These can be varied to get the best desirable response. Notice that in Figure 7.16 we are taking $\lambda_L < \lambda_{cr1}$, the first static limit point and $\lambda_L > \lambda_{cr2}$, the second static limit point. In principle λ_L and λ_H can be chosen anywhere between λ_{cr1} and λ_{cr2} or even beyond. However, the actual switching may lead to oscillation and excite a chaotic behavior of the system. Taking λ_H and λ_L both outside the region between λ_{cr1} and λ_{cr2} seems to be the least troublesome choice regarding the dynamic behavior of industrial FCC units.

However, differing from the simple switching policy used in Figure 7.16, the FCC problem has the added complication that changing one input parameter is not sufficient to control the system because of the specific shape of the bifurcation diagrams shown in Figures 7.14(b) and (c). When the dense-phase reactor temperature goes above the desired middle steady-state temperature it can be forced down by switching the FCD value to a lower value that corresponds to a unique (low temperature) steady state. But when the dense phase reactor temperature goes below the middle steady state temperature it often cannot be brought up by only switching FCD, because to do so would require an unrealistically high value of FCD that corresponds to a unique (high temperature) steady state. The exact opposite applies when trying to use Y_{fa} as the switching parameter. Therefore, we suggest a dual switching policy that employs simultaneous FCD and Y_{fa} switching and uses Y_{RD} as the measured variable.
The dual switching control law is as follows:

$$
\begin{aligned}
\text{Set } Y_{fa} = 0.872 \quad &\text{and} \quad FCD = FCD_L \qquad &\text{if} \qquad &Y_{RD} \geq Y_{RD}(mss) + \delta \ ; \\
\text{Set } Y_{fa} = Y_{faH} \quad &\text{and} \quad FCD = 1.0 \qquad &\text{if} \qquad &Y_{RD} \leq Y_{RD}(mss) - \delta \ ,
\end{aligned}
$$

where $Y_{RD}(mss)$ denotes the system's middle steady state, $FCD_L < FCD_{rc}$, and $Y_{faH} > Y_{farc}$.

Figure 7.17(a) to (d) shows the behavior of the system using a dual switching policy with $\delta = 0.1$, $FCD_L = 0.2$ and $Y_{faH} = 2.0$.

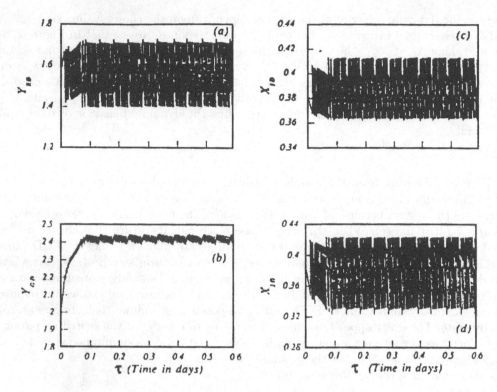

Dynamic response of the system under the switching policy $\delta = 0.1$, $Y_{faH} = 2.0$ and
$$FCD_L = 0.2 \ .$$
(a)   Dimensionless reactor dense-phase temperature
(b)   Dimensionless regenerator dense-phase temperature
(c)   Dimensionless gas oil dense-phase concentration
(d)   Dimensionless gasoline dense-phase concentration
Figure 7.17

Although the reactor dense-phase temperature is allowed to vary by the dimensionless value of $\delta = \pm 0.1$ that corresponds to temperature changes of $\pm 50^0\ C$, the dense-phase dimensionles temperature of the reactor in Figure 7.17(a) oscillates between 1.745 and 1.345. This represents a temperature oscillation with an amplitude of about 0.4 units or $200^\circ\ C$.

The regenerator dense-phase temperature is depicted in Figure 7.17(b). Its oscillation amplitude is settling at a relatively low value, but the center line of these oscillations is drifting upwards, away from the steady-state regenerator temperature to a very high value above 2.4 or $1200^0\ K$. This shows that our dual switching policy has a pathological effect on the regenerator temperature which drifts to high values.

The unreacted gas oil in the dense phase oscillates as shown in Figure 7.17(c). The dense-phase gasoline yield x_{2D} also oscillates, namely between the values 0.32 and 0.42, see Figure 7.17(d). The steady-state gasoline yield for this case is $x_{2D} = 0.387$ or about

39%.

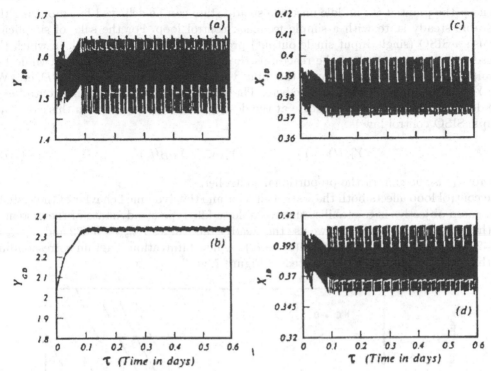

Dynamic response of the system under the dual switching policy with
$\delta = 0.03$, $Y_{faH} = 1.85$ and $FCD_L = 0.28$
(a)   Dimensionless reactor dense-phase temperature
(b)   Dimensionless regenerator dense-phase temperature
(c)   Dimensionless gas oil dense-phase concentration
(d)   Dimensionless gasoline dense-phase concentration
Figure 7.18

Figure 7.18(a) to (d) shows the behavior of the system when using a dual switching policy with another set of parameters, namely $\delta = 0.03$, $FCD_L = 0.28$ and $Y_{faH} = 1.85$.

Although δ is much smaller here ($\delta = 0.03$, or around $15^0\ C$) than in Figure 7.17, the dense-phase reactor temperature still has a large amplitude. It varies between 1.65 and 1.46. This is an amplitude of 0.19 or $95^0\ C$.

For the regenerator dense-phase temperature drawn in Figure 7.18(b), the amplitude again settles to a low value, but the center line of the oscillations also drifts away from the steady-state dense-phase temperature of the regenerator to a very high value above 2.3 or $1150^0\ K$.

The unreacted gas oil in the dense phase oscillates as shown in Figure 7.18(c). The dense-phase gasoline yield x_{2D} oscillates strongly between the values 0.36 and 0.41 in Figure 7.18(d).

Feedback control of the system:

To keep the plant at its middle unstable steady state can be achieved by stabilizing the unstable steady state with a simple feedback control loop. For the sake of simplicity, we use a SISO (single input single output) proportional feedback control, in which the dense-phase temperature of the reactor is the controlled measured variable, while the manipulated variable can be any of the input variables of the system Y_{fa}, FCD, etc. We use Y_{fa} as the manipulated variable here. The set-point of the proportional controller is the dense-phase reactor temperature at the desired middle steady state in this case. Our simple SISO control law is

$$Y_{fa}(t) = Y_{fass} + K_C(Y_{RDss} - Y_{RD}(t)) , \tag{7.72}$$

where K_C is the gain of the proportional controller.

The control loop affects both the static behavior and the dynamic behavior of the system. Our main objective is to stabilize the unstable saddle-type steady state of the system. In the SISO control law (7.72) we use the steady-state values $Y_{fass} = 0.872$ and $Y_{RDss} = 1.5627$ as was done in Figures 7.14(a) to (c). A new bifurcation diagram corresponding to this closed-loop case is constructed in Figure 7.20.

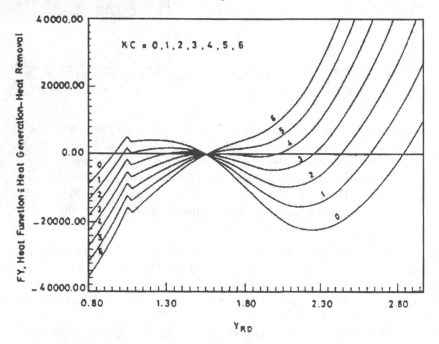

Effect of K_C on the heat function versus Y_{RD} diagrams
Figure 7.19

The heat function is obviously altered by the control as seen in Figure 7.19. The high- and low-temperature steady states change with the change of the proportional gain K_C of the controller. The middle steady state does not change with K_C since it is the set point of the proportional controller. The change of the heat function in dependence of K_C is rather complex and a better explanation can be given from the bifurcation diagram in Figure 7.20.

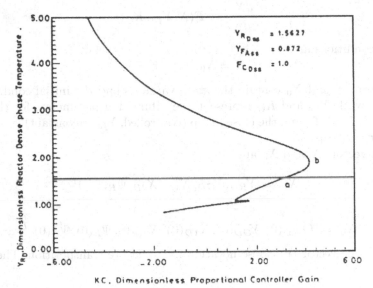

Pitchfork bifurcation of Y_{RD} versus K_C for the closed-loop SISO system with simple proportional controller

Figure 7.20

The bifurcation diagram represents an imperfect pitchfork diagram, where the middle steady state persists over the entire range of K_C, even for negative values of K_C, i.e., even for positive feed back control, which would destabilize the system.

For negative values of K_C of magnitude greater than about 5.6, the steady-state temperatures (other than at the middle steady state) tend to infinity. Further discussion will be restricted to the more practical range of $K_C > 0$.

7.2.5   Combined Static and Dynamic Bifurcation Behavior of Industrial FCC Units

The steady-state version of the model that we use in this section is unchanged from the previous sections.

We have studied the dynamic behavior of FCC units in Section 7.2.3. Here we explain the dynamic bifurcation behavior of FCC type IV units. The dynamic model that we use will be more general than the earlier one. Specifically, we will relax the assumption of negligible mass capacity of gas oil and gasoline in the dense catalyst phase. This relaxation is based upon considering the catalyst chemisorption capacities of the components.

The Dynamic Model

The dynamic model can be written in the following vector differential equation form

$$\frac{dX}{d\hat{\tau}} = F(X, Y_{fa}, K_C) . \tag{7.73}$$

The initial conditions are

$$X = X_0 \quad \text{at} \quad \hat{\tau} = 0 . \tag{7.74}$$

The vectors X and X_0 contain the state variables and the initial conditions vector, respectively, while Y_{fa} and K_C represent the bifurcation parameters for the open loop (uncontrolled, $K_C = 0$) and the closed loop (controlled, $Y_{fa} = $ constant) systems, respectively, and $\hat{\tau} = t/\epsilon_{HR}$.

The column vectors X and X_0 are

$$X = (Y_{RD}, Y_{GD}, X_{1D}, X_{2D}, \Psi_R, \Psi_G)^T \tag{7.75}$$

and

$$X_0 = (Y_{RD}(0), Y_{GD}(0), X_{1D}(0), X_{2D}(0), \Psi_R(0), \Psi_G(0))^T , \tag{7.76}$$

where we use the vector transpose notation $..^T$ to denote transposition. The vector F is given by

$$F = \begin{pmatrix} B_R(Y_\nu - Y_{RD}) + a_2(Y_{GF} - Y_\nu) - \Delta\overline{H}_\nu + \Delta\overline{H}_{cr} + a_1(Y_{GD} - Y_{RD}) - \Delta\overline{H}_{LR} \\ N_1[B_G(Y_{fas} + K_C(Y_{sp} - Y_{RD}) - Y_{GD}) + \beta_C \cdot R_c + a_3(Y_{RD} - Y_{GD}) - \Delta\overline{H}_{LG}] \\ N_2[B_R(X_{3f} - X_{1D}) - (\Psi_R(\alpha_1 \cdot e^{-\gamma_1/Y_{RD}} + \alpha_3 \cdot e^{-\gamma_3/Y_{RD}})) \cdot X_{1D}^2] \\ N_3[B_R(X_{4f} - X_{2D}) - \Psi_R(\alpha_2 \cdot X_{2D} \cdot e^{-\gamma_2/Y_{RD}} - \alpha_1 \cdot X_{1D}^2 \cdot e^{-\gamma_1/Y_{RD}})] \\ N_4[C'_R(\Psi_G - \Psi_R) - \Psi_R \cdot R_{cf}] \\ N_5[R_c - C'_G(\Psi_G - \Psi_R)] \end{pmatrix} \tag{7.77}$$

where $\Delta\overline{H}_{LR} = 0.005a_1(Y_{GD} - Y_{RD})$, $\Delta\overline{H}_{LG} = 0.005\beta_C \cdot R_C$, $N_1 = \epsilon_{HR}/\epsilon_{HG}$, $N_2 = \epsilon_{HR}/\phi_{m3}$, $N_3 = \epsilon_{HR}/\phi_{m4}$, $N_4 = \epsilon_{HR}/\epsilon_{MR}$, and $N_5 = \epsilon_{HR}/\epsilon_{MG}$.

Here ϕ_{m3} and ϕ_{m4} are the mass capacitances due to the chemisorption of gas oil and gasoline, respectively, on the catalyst which has a high surface area. These parameters can have very appreciable and important dynamic implications on the system. The values of N_1, N_2, and N_5 are calculated from the operating and physical parameters of the industrial unit 1 under study as given in the tables on p. 452 and p 453. The parameters ϕ_{m3} and ϕ_{m4} are taken equal to ϵ_{HR}, i.e., we use $N_2 = N_3 = 1.0$. The effect of values differing from 1 for N_2 and N_3 on the dynamic behavior of FCC systems has not been investigated, i.e., the effect of the chemisorption capacities of the catalyst on the dynamic behavior of the whole system is not well understood.

Some static and dynamic bifurcation results:

The dimensionless air feed temperature used for the FCC unit 1 is $Y_{fas} = 0.872$ and the dimensionless operating temperature of the reactor dense phase is $Y_{RD} = 1.5627$. These operating conditions correspond to a steady state on the intermediate branch B of Figure 7.21(a) at the point labeled a.

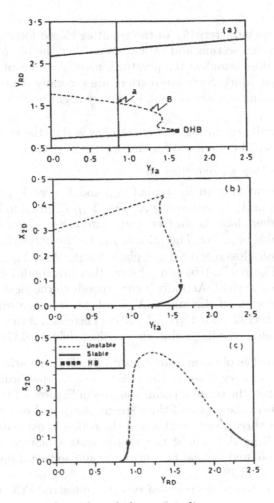

Open loop bifurcation diagram
(a)   Dimensionless reactor dense-phase temperature vs.
      dimensionless air feed temperature to regenerator
(b)   Dimensionless reactor dense-phase gasoline concentration vs.
      dimensionless air feed temperature to regenerator
(c)   Dimensionless reactor dense-phase gasoline concentration vs.
      dimensionless reactor dense-phase temperature
Figure 7.21

Point a lies in the middle of the multiplicity region with three steady states. Therefore, this point is an unstable saddle-type steady state. It is clear from Figures 7.21(b) and (c) that this operating point does not correspond to the maximum gasoline yield X_{2D}. How to alter the operating conditions so that the FCC unit operates at the maximum gasoline yield has been discussed in the previous sections. As explained there, a simple way to stabilize such unstable steady states is to use a negative feedback proportional control.

The static bifurcation characteristics of the resulting closed loop system have also been discussed in the previous section and we have seen that the bifurcation diagram of the reactor dense-phase dimensionless temperature, namely a plot of Y_{RD} versus the controller gain K_C is a pitchfork. Such bifurcations are generally structurally unstable when any of the system parameters are altered, even very slightly.

Next, the industrially operating steady state, as well as the steady state giving maximum gasoline yield will be discussed for the open loop system.

Behavior of the open loop uncontrolled unit

Figure 7.21 shows the bifurcation diagram of Y_{RD} and X_{2D} vs. Y_{fa} in Figures 7.21(a) and (b) for the open loop unit, as well as the X_{2D} vs. Y_{RD} diagram in Figure 7.21(c), where all the parameters other than the air feed temperature Y_{fa}, are assigned their industrial values given in the table of p. 452. The value of Y_{fa} for the industrial unit is 0.872 and the corresponding dimensionless reactor dense phase temperature Y_{RD} is 1.5627 as shown in Figure 7.21(a). The Figures 7.21(b) and (c) show that this condition does not correspond to the maximum gasoline yield. Actually it corresponds to the yield $X_{2D} = 0.38088$ while the maximal gasoline yield of the unit is $X_{2D} = 0.437885$. According to Figure 7.21(b) this occurs at $Y_{fa} = 1.419213$ for $Y_{RD} = 1.19314$. Therefore, if the system were operated at this maximum gasoline yielding point, the gasoline yield would increase by about 15%.

The following caveat is obvious from Figure 7.21: The operating conditions for maximum gasoline yield are very close to the critical points, specifically to the static limit points (sometimes called the turning points) as seen in Figures 7.21(a) and (b). This optimal operating point is in the region of three steady states and very close to the boundary between the region of three steady states and the region of five steady states. It is to be expected that controlling the unit at this steady state is not easy. Again proportional feedback control can be used instead to control the unit so that it operates at its maximal gasoline yield of $x_{2D} = 43.7885\%$.

This concludes our condensed overview of type IV industrial FCC units.

Exercises for 7.2

1. Revisit Section 4.4.6 on the numerics for the neurocycle enzyme system in light of our comments on the numerical sensitivity and chaos of IVPs on p. 462.

   Specifically, verify that the data used for Figure 4.55 on p. 238 leads to chaotic behavior for the neurocycle enzyme system. Do so by changing the MATLAB integrator ode... and the tolerances used by them (as explained on p. 462) in the MATLAB program neurocycle.m of Chapter 4.

   Repeat for the data leading to Figures 4.54 and 4.56 of Section 4.4.6 and note the different numerical behavior for varying tolerances.

2. Develop an algorithm to construct the static bifurcation diagram for an industrial type IV FCC unit.

3. Develop the mathematical model of a modern riser reactor FCC unit and construct its static bifurcation diagram numerically.

Conclusions

In this section we have presented modeling results for industrial type IV FCC units that produce high octane number gasoline from gas oil. Such units consist of two connected bubbling fluidized beds with continuous circulation of the catalyst between the two vessels, the reactor and the regenerator. The steady-state design equations are nonlinear transcendental equations which can be solved using the techniques described in the earlier chapters of the book.

The unsteady state is described by ordinary differential equations or IVPs that can be solved routinely via MATLAB's ODE suit of programs.

The reader should try and reproduce the graphical results of this section by developing her or his own MATLAB algorithms and verifying her or his results versus the results given graphically in this section.

7.3 The UNIPOL® process for the Production of Polyethylene and Polypropylene

The bubbling fluidized-bed reactor technology (UNIPOL®) of Union Carbide[3] employs the Ziegler-Natta[4] catalyst to produce polyethylene. This process can produce HDPE (high density polyethylene) and LDPE (low density polyethylene) in the same reactor through a relatively simple change in the operating conditions. The reaction zone of the reactor is a bubbling bed of polymer particles that grow around very small catalyst particles while both are fluidized by a continuous flow of gaseous components consisting of the make-up feed and recycled gas. See Figure 7.22. To succeed the gas flow rate must exceed the minimum fluidization flow velocity by a factor of 3 to 6. It is essential that the bed always contains particles to prevent local hot spots. For this purpose the unit has to entrap and distribute the particulate catalyst throughout the reaction zone. The main reactant is ethylene. It is used to fluidize the solid particles and also for cooling. The reaction temperature is kept between 100^0 and 130^0 C and the pressure at 300 psi while the catalyst productivity is between 2 and 20 $g \cdot PE/g$ of catalyst. The polymer is formed in small granules of $0.5 - 1$ mm diameter. The granular product is discharged intermittently into a tank from which unreacted gases are removed. After blending and stabilization, the product is ready for processing and pelletizing.

---

[3]Union Carbide, USA company, 1917-
[4]Karl Ziegler, German chemist, 1898-1973
  Guilio Natta, Italian chemist, 1903-1979

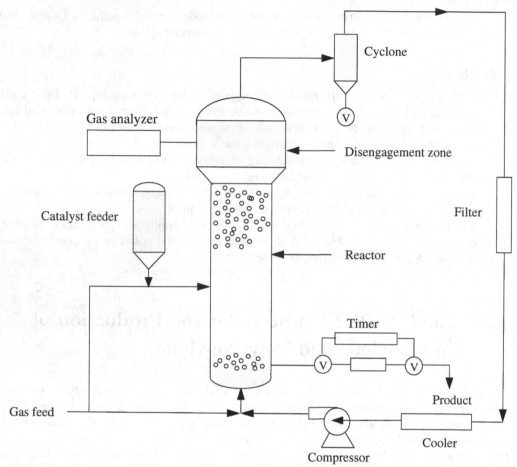

Simplified flow diagram of the Union Carbide UNIPOL® process (V = valve)
Figure 7.22

The molecular weight distribution in the process is controlled by the choice of catalyst. The density (between 0.915 and 0.97 g/cm^3) is controlled by the amount of comonomer added. This versatile process avoids using hydrocarbon dilutants or solvents. In comparison to conventional high pressure LDPE units, Union Carbide has reported significant savings in plant investment and energy costs with this gas-phase technology.

The polymerization of ethylene is a highly exothermic reaction and when highly exothermic reactions occur in fluidized-bed reactors, unusual steady state and dynamic behavior may occur.

7.3.1   A Dynamic Mathematical Model

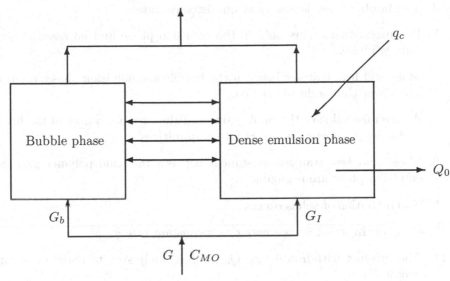

Schematic diagram for the polyethylene reactor model
Figure 7.23

Figure 7.23 shows a schematic diagram for the freely bubbling fluidized bed that produces polyethylene. Here q_c denotes the catalyst injection rate, Q_0 is the product withdrawal rate, G represents the gas input flow rate with the monomer concentration C_{MO}. The monomer is introduced at the bottom of the reactor through a distributor that is very important for the smooth operation of the reactor. In the distributor the total input gas G is split into G_b for the bubble phase and G_I for the dense phase. Since the polymer particles are hot and possibly active, they must be prevented from settling to avoid agglomeration. It is essential that the fluidized bed always contains particles that prevent localized hot spots. During startup the reaction zone is therefore usually charged with a bed of polyethylene particles before the gas flow is initiated. The catalyst is stored in a feeder tank under a nitrogen blanket and it is injected into the bed at a rate of mass equal to its consumption rate. At the same time, makeup gas is fed to the bed at a rate equal to the mass rate of polymer removal. The polymer product is continuously withdrawn from the reactor at a rate such that the bed height remains constant.

General Assumptions

We make the following simplifying and reasonable assumptions in developing the model.

1. The dense emulsion phase is perfectly mixed and is at incipient fluidization conditions with constant voidage.

2. The flow of gas in excess of the minimum fluidization requirement passes through the bed in the form of bubbles.

3. The bubbles are spherical, of uniform size, and in plug flow.

4. The bubble phase is always at quasisteady state.

5. Polymerization occurs only in the emulsion phase and no reaction occurs in the bubble phase.

6. Mass and heat transfer between the bubble and emulsion phases occur at uniform rates over the height of the bed.

7. An average value of the bubble size and thus average values of the heat and mass exchange parameters are used for computations.

8. Mass- and heat-transfer resistances between the solid polymer particles and the emulsion phase are negligible.

9. No elutriation of solids occurs.

10. Catalyst injection is continuous at a constant rate q_c.

11. The product withdrawal rate Q_0 is always adjusted to maintain a constant bed height H.

12. There is no catalyst deactivation.

Hydrodynamic Relations

Based on the two-phase model of fluidization, the following basic relations hold:

$$G_I = U_{mf} \cdot A = U_I \cdot A_I \,, \tag{7.78}$$

$$A_I = A \cdot (1 - \delta^*) \,, \tag{7.79}$$

$$A_B = A \cdot \delta^* \,, \tag{7.80}$$

$$A = A_I + A_B \,, \tag{7.81}$$

$$U_I = U_{mf}/(1 - \delta^*) \,, \quad \text{and} \tag{7.82}$$

$$U_e = U_I/\epsilon_{mf} \,. \tag{7.83}$$

Here δ^* is the relative fraction of the bubble phase.
The superficial velocity at minimum fluidizing conditions is given by the correlation

$$\frac{1.75 \cdot Re_{mf}^2}{\epsilon_{mf}^3} + \frac{150 \cdot (1 - \epsilon_{mf}) \cdot Re_{mf}}{\epsilon_{mf}^3} - \frac{d_p^3 \cdot \rho_g (\rho_s - \rho_g) \cdot g}{\mu^2} = 0 \tag{7.84}$$

where $Re_{mf} = U_{mf} \cdot d_p/((1 - \epsilon_{mf}) \cdot \mu)$ is the Reynolds[5] number at minimum fluidization conditions, d_p is the particle diameter, ϵ_{mf} is the voidage at minimum fluidization, and μ is the viscosity.

---

[5]Osbourne Reynolds, English engineer and physicist, 1842-1912

The bubble rising velocity U_B is given by

$$U_B = U_0 - U_{mf} + 0.711\sqrt{g \cdot d_B} \, . \tag{7.85}$$

The relative fraction δ^* of the bubble phase is

$$\delta^* = (U_0 - U_{mf})/U_B \, . \tag{7.86}$$

The velocity U_e of the emulsion gas is

$$U_e = U_{mf}/(\epsilon_{mf} \cdot (1 - \delta^*)) \, . \tag{7.87}$$

The bed voidage ϵ_{mf} at minimum fluidization conditions is

$$\epsilon_{mf} = 0.586 \cdot \Phi^{-0.72} \cdot \left(\frac{\mu^2}{\rho_g \cdot g \cdot (\rho_s - \rho_g) \cdot d_p^3}\right)^{0.029} \left(\frac{\rho_g}{\rho_s}\right)^{0.021} \, . \tag{7.88}$$

The bubble diameter D_B can be found from the relation

$$D_B = D_{BM} - (D_{BM} - D_{B0}) \cdot e^{-0.15 \cdot H/D} \, , \tag{7.89}$$

where $D_{BM} = 0.652(A \cdot (U_0 - U_{mf}))^{0.4}$ and $D_{B0} = 0.00376(U_0 - U_{mf})^2$.
The heat transfer Nu_W with the wall is given by

$$Nu_W = 0.16 \cdot Pr^{0.4} \cdot Re_p^{0.76} \left(\frac{\rho_s \cdot C_{ps}}{\rho_g \cdot C_{pg}}\right)^{0.4} \cdot Fr^{-0.2} \cdot \left(\Phi_W \frac{L_{mf}}{L_f}\right)^{0.36} \, , \tag{7.90}$$

with the Prandtl[6] number

$$Pr = \frac{\mu \cdot C_{pg}}{K_g}, \quad Fr = \frac{U_0^2}{g \cdot d_p}, \quad \frac{L_{mf}}{L_f} = \frac{U_B}{U_B - U_0 + U_{mf}} \, ,$$

and Φ_W is assumed equal to 0.6.

The mass- and heat-transfer coefficients between the bubble and dense phases are connected via the following relations:
For the mass transfer:

$$\frac{1}{K_{be}} = \frac{1}{K_{bc}} + \frac{1}{K_{ce}} \, , \tag{7.91}$$

where

$$K_{bc} = 4.5 \cdot \frac{U_{mf}}{D_B} + 5.85 \cdot \left(\frac{g^{0.5} \cdot D_G}{D_B^{2.5}}\right)^{0.5} \quad \text{and} \tag{7.92}$$

$$K_{ce} = 6.78 \cdot \left(\frac{\epsilon_{mf} \cdot D_G \cdot U_B}{D_B^3}\right)^{0.5} \, . \tag{7.93}$$

For the heat-transfer coefficient:

$$\frac{1}{H_{be}} = \frac{1}{H_{bc}} + \frac{1}{H_{ce}} \, , \tag{7.94}$$

---

[6]Ludwig Prandtl, German physicist, 1875-1953

where

$$H_{bc} = 4.5 \cdot \frac{U_{mf} \cdot \rho_g \cdot C_{pg}}{D_B} + 5.85 \cdot \left(\frac{g^{0.5} \cdot K_p \cdot \rho_g \cdot C_{pg}}{D_B^{2.5}} \right)^{0.5} \quad \text{and} \qquad (7.95)$$

$$H_{ce} = 6.78 \cdot (\rho_g \cdot C_{pg} \cdot K_g)^{0.5} \left(\frac{\epsilon_{mf} \cdot U_B}{D_B^3} \right) . \qquad (7.96)$$

The above relations and parameter values are all taken from the literature.

The Model Equations

Bubble-phase monomer mass-balances:

The mass balance on the inactive bubble phase for the monomer is given by

$$G_B \frac{dC_{Mb}}{dz^*} = -K_{be} \cdot A_B \cdot (C_{Mb} - C_{Me}) \qquad (7.97)$$

with $C_{Mb} = C_{Me0} = C_{M0}$ at $z^* = 0$. This equation can be rearranged as

$$\frac{dC_{Mb}}{dz^*} = -\frac{K_{be} \cdot A_B}{G_B} \cdot (C_{Mb} - C_{Me}) . \qquad (7.98)$$

For the dimensionless height $z = z^*/H$ and for $K_B = K_{be} \cdot H/U_B$ and $U_B = G_B/A_B$, the integration of the linear differential equation (7.98) yields

$$(C_{Mb} - C_{Me}) = (C_{M0} - C_{Me}) \cdot e^{-K_B \cdot z} . \qquad (7.99)$$

With the dimensionless parameters $X_{1B} = C_{Mb}/C_0$, $X_{10} = C_{M0}/C_0$, and $X_1 = C_{Me}/C_0$, equation (7.99) takes on the dimensionless form

$$X_{1B} = X_1 + (X_{10} - X_1) \cdot e^{-K_B \cdot z} . \qquad (7.100)$$

Dense emulsion-phase monomer mass-balances:

Here the mass-balance equation is

$$A_I \cdot H \cdot \epsilon_{mf} \frac{dC_{Me}}{dt} = G_I \cdot (C_{M0} - C_{Me}) + K_{be} \cdot A_B \cdot \int_0^H (C_{Mb} - C_{Me}) \, dz^*$$
$$-K_P(T_e) \cdot C_{Me} \cdot A_I \cdot H \cdot (1 - \epsilon_{mf}) \cdot \rho_s \cdot X_{cat} - Q_0 \cdot C_{Me} \cdot \epsilon_{mf} . \quad (7.101)$$

The integral in (7.101) can be evaluated by solving the DE (7.98), namely

$$\int_0^H (C_{Mb} - C_{Me}) \, dz^* = \frac{(C_{M0} - C_{Me}) \cdot H}{K_B} \cdot (1 - e^{-K_B}) . \qquad (7.102)$$

By substituting (7.102) into equation (7.101) and using the known previous hydrodynamic relations, equation (7.101) becomes

$$\frac{dC_{Me}}{dt} = \frac{U_e}{H}(C_{M0} - C_{Me}) + \frac{K_{be} \cdot \delta^*}{(1 - \delta^*) \cdot \epsilon_{mf}} \cdot \frac{(C_{M0} - C_{Me})}{K_B} \cdot (1 - e^{-K_B})$$
$$-\frac{1}{\epsilon_{mf}} \cdot K_P(T_e) \cdot C_{Me} \cdot (1 - \epsilon_{mf}) \cdot \rho_s \cdot X_{cat} - \frac{Q_0 \cdot C_{Me}}{A \cdot H \cdot (1 - \delta^*)} . \quad (7.103)$$

Put into dimensionless form, this equation becomes

$$\frac{dX_1}{d\bar{t}} = \bar{\bar{B}}_R \cdot (X_{10} - X_1) + \bar{U} \cdot \Gamma \cdot (X_{10} - X_1) \cdot (1 - e^{-K_B})$$
$$- \zeta \cdot X_1 \cdot X_2 \cdot e^{-\gamma_a/Y_D} - \frac{\zeta \cdot X_1^2 \cdot X_2 \cdot e^{-\delta/Y_D}}{\hat{a} + X_1},$$ (7.104)

where

$$\bar{\bar{B}}_R = \frac{U_e \cdot t_0}{H}; \qquad K_B = \frac{K_{be} \cdot H}{U_B}; \qquad \bar{t} = t/t_0;$$

$$\hat{a} = \frac{\rho_s \cdot (1 - \epsilon_{mf})}{C_0 \cdot \epsilon_{mf}}; \qquad \gamma_a = \frac{E_a}{R_G \cdot T_{ref}}; \qquad X_2 = X_{cat};$$

$$K_P = K_{P0} \cdot e^{-\gamma_a/Y_D}; \qquad \zeta = \frac{t_0 \cdot K_{P0} \cdot \rho_s \cdot (1 - \epsilon_{mf})}{\epsilon_{mf}}; \qquad \Gamma = \frac{\delta^*}{1 - \delta^*};$$

$$\bar{U} = \frac{U_B \cdot t_0}{H \cdot \epsilon_{mf}}; \quad \text{and} \quad Y_D = T_e/T_{ref}.$$

Catalyst mass balance in the dense emulsion phase:
Here the mass-balance equation is

$$A_I \cdot H \cdot (1 - \epsilon_{mf}) \cdot \rho_s \frac{dX_{cat}}{dt} = q_c - Q_0 \cdot X_{cat} \cdot (1 - \epsilon_{mf}) \cdot \rho_s.$$ (7.105)

This can be reduced to

$$\frac{dX_{cat}}{dt} = \frac{q_c}{A \cdot H \cdot (1 - \delta^*) \cdot (1 - \epsilon_{mf}) \cdot \rho_s} - \frac{Q_0 \cdot X_{cat}}{A \cdot H \cdot (1 - \delta^*)}.$$ (7.106)

Next we use the dimensionless parameter $f_0 = Q_0 \cdot t_0/(A \cdot H \cdot (1 - \delta^*))$ and assume a constant bed height H so that the volumetric flow rate Q_0 of product removal equals the rate of increase in bed volume due to polymerization. This gives us

$$Q_0 \cdot (\rho_s(1 - \epsilon_{mf}) + \epsilon_{mf} \cdot C_{Me}) = A \cdot H \cdot (1 - \delta^*) \cdot (1 - \epsilon_{mf}) \cdot \rho_s \cdot K_P(T_e) \cdot C_{Me} \cdot X_{cat}.$$ (7.107)

By rearrangement of (7.107) we obtain

$$\frac{Q_0 \cdot t_0}{A \cdot H \cdot (1 - \delta^*)} = \frac{t_0 \cdot (1 - \epsilon_{mf}) \cdot \rho_s \cdot K_P(T_e) \cdot C_{Me} \cdot X_{cat}}{\rho_s \cdot (1 - \epsilon_{mf}) + \epsilon_{mf} \cdot C_{Me}} = f_0,$$ (7.108)

or in dimensionless form:

$$f_0 = \frac{t_0 \cdot (1 - \epsilon_{mf}) \cdot \rho_s \cdot K_P(T_e) \cdot C_{Me} \cdot X_{cat}}{\rho_s(1 - \epsilon_{mf}) + \epsilon_{mf} \cdot C_{Me}} = \frac{\zeta \cdot X_1 \cdot X_2 \cdot e^{-\gamma_a/Y_D}}{\hat{a} + X_1}.$$ (7.109)

Finally the dimensionless form of the catalyst mass-balance equation (7.105) is

$$\frac{dX_2}{d\bar{t}} = \frac{t_0 \cdot q_c}{A \cdot H \cdot (1 - \delta^*) \cdot (1 - \epsilon_{mf}) \cdot \rho_s} - \frac{\zeta \cdot X_1 \cdot X_2^2 \cdot e^{-\gamma_a/Y_D}}{\hat{a} + X_1}.$$ (7.110)

Energy balance on the bubble phase:

Here the energy-balance equation is

$$\frac{d}{dz^*}(C_{Mb} \cdot (T_b - T_{ref})) = \frac{H_{be}}{U_B \cdot C_{pg}} \cdot (T_e - T_b) \tag{7.111}$$

with the initial condition $T(0) = T_{b0}$.

Since in the emulsion phase the bubbles rise much faster than the gas, we use the average value $\bar{C}_{Mb}$ for C_{Mb} and simplify equation (7.111) to become

$$\frac{dT_b}{dz^*} = \frac{H_{be}}{U_B \cdot C_{pg} \cdot \bar{C}_{Mb}} \cdot (T_e - T_b) . \tag{7.112}$$

When put into dimensionless form, this equation becomes

$$\frac{dY_B}{dz} = \frac{K_H}{\bar{X}_{1B}} \cdot (Y_D - Y_B) . \tag{7.113}$$

Integration of (7.113) gives us

$$Y_D - Y_B = (Y_{B0} - Y_D) \cdot e^{-K_H \cdot z / \bar{X}_{1B}} \tag{7.114}$$

with $Y_B = T_b/T_{ref}$, $K_H = H_{be} \cdot H/(U_B \cdot C_{pg} \cdot C_0)$, and $Y_{B0} = T_{b0}/T_{ref} = Y_F$. The average $\bar{Y}_B$ of Y_B is

$$\bar{Y}_B = \int_0^1 Y_B \, dz = Y_D + \frac{Y_{B0} - Y_D}{K_H} \cdot \bar{X}_{1B} \cdot (1 - e^{-K_H / \bar{X}_{1B}}) . \tag{7.115}$$

And the average value $\bar{X}_{1B}$ of X_{1B} can be derived from the bubble-phase monomer-balance equation (7.100) as

$$\bar{X}_{1B} = \int_0^1 X_{1B} \, dz = X_1 + \frac{X_{10} - X_1}{K_B} \cdot (1 - e^{-K_B}) . \tag{7.116}$$

Energy balance on the emulsion phase:

The energy-balance equation on the emulsion phase is

$$A_I \cdot H \cdot ((1 - \epsilon_{mf}) \cdot \rho_s \cdot C_{ps} + \epsilon_{mf} \cdot C_{Me} \cdot C_{pg}) \frac{dT_e}{dt} + A_I \cdot H \cdot (T_e - T_{ref}) \cdot \epsilon_{mf} \cdot C_{pg} \frac{dC_{Me}}{dt} =$$

$$= -G_I \cdot C_{Me} \cdot C_{pg} \cdot (T_e - T_f) + A_B \cdot H_{be} \int_0^H (T_b - T_e) dz^* +$$

$$+ A_I \cdot H \cdot (-\Delta H_r) \cdot K_p(T_e) \cdot \rho_s \cdot (1 - \epsilon_{mf}) \cdot X_{cat} \cdot C_{Me}$$

$$- Q_0 \cdot (1 - \epsilon_{mf}) \cdot \rho_s \cdot C_{ps} \cdot (T_e - T_f) - Q_0 \cdot \epsilon_{mf} \cdot C_{Me} \cdot C_{pg} \cdot (T_e - T_f)$$

$$- \pi \cdot D \cdot H \cdot h_w \cdot (T_e - T_w) . \tag{7.117}$$

This can be rewritten as

$$((1 - \epsilon_{mf}) \cdot \rho_s \cdot C_{ps} + \epsilon_{mf} \cdot C_0 \cdot X_1 \cdot C_{pg}) \frac{d(Y_D \cdot T_{ref})}{dt} + T_{ref} \cdot (Y_D - 1) \cdot \epsilon_{mf} \cdot C_{pg} \frac{C_0 \cdot dX_1}{dt} =$$

$$= -\frac{U_e \cdot \epsilon_{mf} \cdot C_{pg} \cdot C_0 \cdot X_1}{H} \cdot (Y_D - Y_F) \cdot T_{ref} + \frac{\delta^*}{1 - \delta^*} \cdot H_{be} \cdot T_{ref} \int_0^1 (Y_B - Y_D) \, dz +$$

$$+ (-\Delta H_r) \cdot K_{P0} \cdot e^{-\gamma_a/Y_D} \cdot (1 - \epsilon_{mf}) \cdot \rho_s \cdot X_1 \cdot C_0 \cdot X_2$$

$$- \frac{Q_0 \cdot (1 - \epsilon_{mf}) \cdot \rho_s \cdot C_{ps} \cdot (Y_D - 1) \cdot T_{ref}}{A \cdot H \cdot (1 - \delta^*)}$$

$$- \frac{Q_0}{A \cdot H \cdot (1 - \delta^*)} \cdot \epsilon_{mf} \cdot X_1 \cdot C_0 \cdot C_{pg} \cdot (Y_D - 1) \cdot T_{ref}$$

$$- \frac{\pi \cdot D \cdot h_w}{A \cdot (1 - \delta^*)} \cdot (Y_D - Y_w) \cdot T_{ref} \qquad (7.118)$$

with $Y_F = T_F/T_{ref}$, $Y_w = T_w/T_{ref}$, $Y_D = T_e/T_{ref}$, and $Y_B = T_b/T_{ref}$.
The integral inside equation (7.118) can be evaluated with the help of the inert bubble-phase energy-balance equation (7.114) to yield

$$\int_0^1 (Y_B - Y_D) \, dz = \int_0^1 (Y_{B0} - Y_D) \cdot e^{-K_H \cdot z/\bar{X}_{1B}} \, dz$$

$$= (Y_{B0} - Y_D) \cdot \frac{\bar{X}_{1B}}{K_H} \cdot (1 - e^{-K_H/\bar{X}_{1B}}) . \qquad (7.119)$$

To obtain our final dimensionless dense-phase heat-balance equation, we substitute (7.119) into (7.118) and use the following parameters:

$$b' = \frac{(1 - \epsilon_{mf}) \cdot \rho_s \cdot C_{ps}}{\epsilon_{mf} \cdot C_0 \cdot C_{pg}} ; \qquad K_H \cdot \bar{U} \cdot \Gamma = \frac{\delta^*}{1 - \delta^*} \cdot \frac{H_{be} \cdot t_0}{C_0 \cdot \epsilon_{mf} \cdot C_{pg}} ;$$

$$\gamma = \frac{-\Delta H_r}{C_{pg} \cdot T_{ref}} ; \quad \text{and} \quad \bar{K} = \frac{t_0 \cdot h_w}{D \cdot \epsilon_{mf} \cdot C_0 \cdot C_{pg} \cdot (1 - \delta^*)} .$$

Then

$$\frac{dY_D}{d\bar{t}} + \frac{Y_D - 1}{b' + X_1} \cdot \frac{dX_1}{d\bar{t}} = \bar{\bar{B}}_R \cdot X_1 \cdot \frac{Y_F - Y_D}{b' + X_1} +$$

$$+ K_H \cdot \bar{U} \cdot \Gamma \cdot \frac{(Y_{B0} - Y_D) \cdot (\bar{X}_{1B}/K_H) \cdot (1 - e^{-K_H/\bar{X}_{1B}})}{b' + X_1} +$$

$$+ \frac{\gamma \cdot \zeta \cdot X_1 \cdot X_2 \cdot e^{-\gamma_a/Y_D}}{b' + X_1} - \frac{\zeta \cdot X_1 \cdot X_2 \cdot (Y_D - 1) \cdot e^{-\gamma_a/Y_D}}{\hat{a} + X_1}$$

$$- 4 \cdot \bar{K} \cdot \frac{Y_D - Y_w}{b' + X_1} . \qquad (7.120)$$

In equation (7.120) the derivative $dX_1/d\bar{t}$ was given earlier in (7.104), and this term should be used here to give us one DE in $Y_D(\bar{t})$.

For specific values of the reactor height H and the reactor diameter D, the operating parameters are the feed temperature T_f, the wall temperature T_w, the catalyst injection rate q_c, and the fluidizing gas velocity U_0. The following table gives a base set of design parameters, operating variables, and chemical and physical properties for this UNIPOL® reactor.

C_{ps}	0.456	$cal/(g \cdot K)$	T_f	300	K	d_p	0.05		cm
C_{pg}	0.44	$cal/(g \cdot K)$	T_{ref}	300	K	D_G	$6 \cdot 10^{-3}$		cm^2/sec
ρ_s	0.95	g/cm^3	T_{B0}	330	K	Φ_w	0.6		
ρ_g	0.029	g/cm^3	$T_e(0)$	300	K	C_0	0.029		g/cm^3
ρ_{cat}	2.37	g/cm^3	T_w	340	K	q_c	150		g/h
K_g	$7.6 \cdot 10^{-5}$	$cal/(cm \cdot s \cdot K)$	U_{mf}	5.2	cm/sec	d_B	50		cm
K_{p0}	$4.167 \cdot 10^6$	$cm^3/(g(cat) \cdot sec)$	U_0	26	cm/sec	ϵ_{mf}	0.38		
E_a	9000	cal/mol	U_B	80	cm/sec	δ^*	0.115		
μ	$1.16 \cdot 10^{-4}$	$g/(cm \cdot sec)$	U_e	15.4	cm/sec	D	250		cm
$(-\Delta H_r)$	916	cal/g	t_0	$1.08 \cdot 10^5$	s	H	600		cm

7.3.2   Numerical Treatment

The dynamic model consists of the three differential equations (7.104), (7.110), and (7.120). These define an initial value problem with initial conditions at $\bar{t} = 0$. The dynamic, unsteady state of this system is described by these highly nonlinear DEs, while the steady states are defined by nonlinear transcendental equations, obtained by setting all derivatives in the system of differential equations (7.104), (7.110), and (7.120) equal to zero.

The reader can use the MATLAB ODE solvers that were presented earlier to solve this IVP. As a guide we include two parameter continuation diagrams for this problem in Figure 7.24.

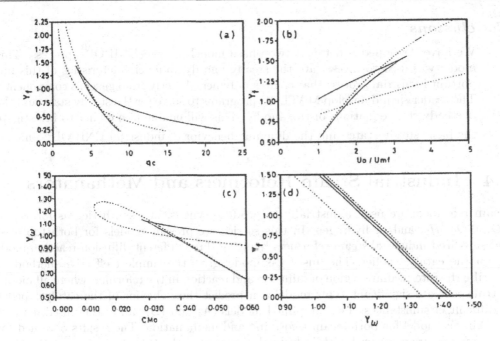

Two parameter continuation diagrams
Figure 7.24

The subplots of Figure 7.24 are drawn for the following base values:

Parameter	subplot (a)	subplot (b)	subplot (c)	subplot (d)
$Y_f = T_f/T_{ref}$	variable	variable	1.175	variable
q_c	variable	250	250	250
U_0/U_{mf}	2.8	variable	2.8	2.8
C_{M0}	0.025	0,03	variable	0.03
$Y_w = T_w/T_{ref}$	1.05	1.05	variable	variable

Exercises for 7.3

1. The fluidization conditions of the UNIPOL[R] process are more complicated due to the change in particle size during the "growth" of the polymer particles.
   Develop a more sophisticated model that takes this particle growth into account.

2. Using the simple model of this section, find the static bifurcation characteristics for a typical industrial UNIPOL[R] unit.

3. Compute the basic dynamic characteristics for an industrial UNIPOL[R] reactor.

Conclusions

We have developed a detailed two-phase model for the UNIPOL® process. This model was used to investigate the steady and dynamic characteristics of this important industrial process that creates polymers directly from gaseous components. The reader should develop MATLAB programs to solve for the steady state and the unsteady-state equations of this model. This will enable him or her to investigate the basic steady states and the dynamic behavior of industrial UNIPOL® units.

7.4   Industrial Steam Reformers and Methanators

Steam reformers are used industrially to produce syngas, i.e., synthetic gas formed of CO, CO_2, H_2, and/or hydrogen. In this section we present models for both top-fired and side-fired industrial steam reformers by using three different diffusion-reaction models for the catalyst pellet. The dusty gas model gives the simplest effective method to describe the intermediate region of diffusion and reaction in the reformer, where all modes of transport are significant. This model can predict the behavior of the catalyst pellet in difficult circumstances. Two simplified models (A) and (B) can also be used, as well as a kinetic model for both steam reforming and methanation. The results obtained for these models are compared with industrial results near the thermodynamic equilibrium as well as far from it.

Methanators are usually used in the ammonia production line to guard the catalyst of the ammonia converters from the ill effect of carbon monoxide and carbon dioxide. This section includes precise models for different types of methanators using the dusty gas model with reliable kinetic expression and the results are compared with those of the simplified models (A) and (B).

7.4.1   Rate Expressions

An intrinsic rate equation for steam reforming of methane uses an integral flow reactor and the industrial catalyst $Ni/MgAl_2O_4$ spinel of Haldor Topsøe[7]. In it the Langmuir[8]-Hinshelwood[9] and Hougen[10]-Watson[11] principles are used with a mechanism of thirteen steps, three of which are rate determining steps, and ten occur at equilibrium.

The three reactions are:

$$
\begin{array}{lll}
CH_4 + H_2O & \leftrightarrow \quad 3H_2 + CO & \qquad I \\
CO + H_2O & \leftrightarrow \quad H_2 + CO_2 & \qquad II \\
CH_4 + 2H_2O & \leftrightarrow \quad 4H_2 + CO_2 & \qquad III
\end{array}
$$

---

[7]Haldor Topsøe A/S, Danish company, 1940-
[8]Irvine Langmuir, USA chemist and engineer, 1881-1957
[9]Cyril Norman Hinshelwood, British chemist, 1897-1967
[10]Olav Andreas Hougen,USA chemical engineer, 1893-1986
[11]Kenneth Merle Watson, USA chemical engineer, 1903-1989

The corresponding rate equations are:

$$r_I = \frac{k_1}{p_{H_2}^{2.5}} \cdot \left(p_{CH_4} \cdot p_{H_2O} - \frac{p_{H_2}^3 \cdot p_{CO}}{K_1} \right) / DEN^2 , \qquad (7.121)$$

$$r_{II} = \frac{k_2}{p_{H_2}} \cdot \left(p_{CO} \cdot p_{H_2O} - \frac{p_{H_2} \cdot p_{CO_2}}{K_2} \right) / DEN^2 , \qquad (7.122)$$

$$r_{III} = \frac{k_3}{p_{H_2}^{3.5}} \cdot \left(p_{CH_4} \cdot p_{H_2O}^2 - \frac{p_{H_2}^4 \cdot p_{CO_2}}{K_1 \cdot K_2} \right) / DEN^2 , \qquad (7.123)$$

where

$$DEN = 1 + K_{CO} \cdot p_{CO} + K_{H_2} \cdot p_{H_2} + K_{CH_4} \cdot p_{CH_4} + K_{H_2O} \cdot p_{H_2O}/p_{H_2} . \qquad (7.124)$$

The reactions I and III are endothermic while the water-gas shift reaction II is weakly exothermic. The above rate equations cannot be used when the concentration of hydrogen is zero because then the rate expressions become infinite. Hence the presence of hydrogen is necessary for these equations to be applicable. From a technical point of view, the feed cannot be hydrogen free in order to reduce any Ni oxide with the packed catalyst and therefore the division by the partial pressure of hydrogen p_{H_2} in equation (7.124) does not cause any practical problems.

For the catalyst tubes of a fixed-bed catalytic reactor, a large amount of physico-chemical data is needed for precise modeling and simulation. The heart of this data is related to the catalyst pellets, namely the intrinsic kinetics and diffusional resistances. However, other data is also needed, including physical properties of the gas mixture and the heat-transfer coefficients. Important for modeling is the fact that the composition and temperature of the mixture changes with the progress of the reaction. Therefore, the physical data should be updated every so often along the length of the catalyst tubes in order to achieve an accurate simulation of the reactor. The kinetic rate coefficients, the adsorption equilibrium constants and other physico-chemical data for industrial steam reformers and other catalytic reactors are given in the tables and descriptions of the next three pages.

Kinetic coefficients and heats of reaction

Rate coefficients:	
$k_1 = 9.490 \cdot 10^{16} \cdot e^{-28879/T}$	$kmol \cdot kPa^{0.5}/(kg \cdot h)$
$k_2 = 4.390 \cdot 10^{4} \cdot e^{-8074.3/T}$	$kmol \cdot kPa^{-1}/(kg \cdot h)$
$k_3 = 2.290 \cdot 10^{16} \cdot e^{-29336/T}$	$kmol \cdot kPa^{0.5}/(kg \cdot h)$
Adsorption equilibrium constants:	
$K_{CH4} = 6.65 \cdot 10^{-6} \cdot e^{4604.28/T}$	kPa^{-1}
$K_{H2O} = 1.77 \cdot 10^{3} \cdot e^{-10666.35/T}$	kPa^{-1}
$K_{H2} = 6.12 \cdot 10^{-11} \cdot e^{9971.13/T}$	kPa^{-1}
$K_{CO} = 8.23 \cdot 10^{-7} \cdot e^{8497.71/T}$	kPa^{-1}

(continued)

Equilibrium constants:

$$K_1 = 10266.76 \cdot e^{(-2630/T+30.11)} \qquad\qquad kPa^2$$
$$K_2 = e^{(4400/T-4.063)} \qquad\qquad \text{(dimensionless)}$$
$$K_3 = K_1 \cdot K_2 \qquad\qquad kPa^2$$

Heats of reaction:

$$\Delta H_I = 206.31 \cdot 10^3 + 4.186 \int_{25}^{T} (3 \cdot C_{PH2} + C_{PCO} - C_{PCH4} - C_{PH2O})\, dT \qquad kJ/kmol$$

$$\Delta H_{III} = 165.11 \cdot 10^3 + 4.186 \int_{25}^{T} (4 \cdot C_{PH2} - C_{PCO2} - C_{PCH4} - 2 \cdot C_{PH2O})\, dT \quad kJ/kmol$$

$$\Delta H_{II} = \Delta H_{III} - \Delta H_I \qquad\qquad kJ/kmol$$

Diffusivities

The *Knudsen*[12] *diffusivity* is

$$D_{Ki} = 34.92 \cdot 10^{-6} \cdot \bar{\bar{r}} \cdot \sqrt{T/M_i} \qquad m^2/h ,$$

where $\bar{\bar{r}}$ is the average pore radius in m, M_i is the molecular weight of component i, and T is the temperature measured in K.

The *binary molecular diffusivities* are

$$D_{ij} = 0.00214 - 0.000492 \cdot \sqrt{\frac{M_i + M_j}{M_i \cdot M_j}} \cdot \frac{0.36 \cdot T^{1.5}}{P_T \cdot \bar{\sigma}_{ij}^2 \cdot \Omega_D} ,$$

where the binary molar diffusivity D_{ij} is measured in m^2/h, M_i denotes the molecular weight of component i, T is the temperature in K, P_T is the total pressure in atm, $\bar{\sigma}_{ij}$ is the collision diameter in *Angstrom*[13], given by

$$\bar{\sigma}_{ij} = \frac{\bar{\sigma}_i + \bar{\sigma}_j}{2} ,$$

Ω_D is the diffusion collision integral

$$\Omega_D = 0.42541 + \frac{0.82133 - 6.8314 \cdot 10^{-2}/\bar{e}_{ij}}{\bar{e}_{ij}} + 0.26628 \cdot e^{-0.012733 \cdot \bar{e}_{ij}} ,$$

and $\bar{e}_{ij}$ is the characteristic energy

$$\bar{e}_{ij} = \frac{T}{\sqrt{\bar{e}_i \cdot \bar{e}_j}} .$$

The values for $\bar{\sigma}_i$ and $\bar{e}_i$ are as follows

[12]Martin Hans Christian Knudsen, Danish physicist, 1871-1949
[13]Anders Jonas Ångström, Swedish physicist, 1814-1874

Force constants and collision diameters

Component	$\bar{\sigma}_i$ in $Angstrom$	$\bar{e}_i$
CH_4	3.882	136.5
H_2O	2.665	363.0
H_2	3.305	33.3
CO	3.590	110.3
CO_2	3.996	190.0
N_2	3.681	91.5

The *viscosity correlations* and *specific heats* are readily available in the literature.

Heat-transfer parameters

The overall *heat-transfer coefficient* U between the catalyst tubes and their surroundings is given by the equation

$$\frac{1}{U} = \frac{d_i}{2 \cdot \lambda_{st}} \ln(d_{t,0}/d_i) + \frac{1}{\alpha_\omega} + \frac{d_i}{8 \cdot \lambda_{er}} \quad \text{in } m^2 \cdot h \cdot C/kJ ,$$

where d_i and $d_{t,0}$ are the internal and external tube dimensions in m, respectively, λ_{st} is the thermal conductivity of the tube metal in $kJ/(m^2 \cdot h \cdot C)$, and α_ω and λ_{er} are the wall heat-transfer coefficient and effective conductivity of the catalyst bed, respectively, given by

$$\alpha_\omega = \frac{8.694 \cdot \lambda_{er}^o}{d_i^{4/3}} + \frac{0.512 \cdot \lambda_g \cdot d_i \cdot Re_P \cdot Pr^{1/3}}{D_P} \quad \text{in } kJ/(m^2 \cdot h \cdot C)$$

and

$$\lambda_{er} = \lambda_{er}^o + \frac{0.111 \cdot \lambda_g \cdot Re_P \cdot Pr^{1/3}}{1 + 46 \cdot (D_P/d_i)^2} \quad \text{in } kJ/(m \cdot h \cdot C) .$$

Here λ_g denotes the process gas conductivity in $kJ/(m \cdot h \cdot C)$, Re_P is the Reynolds[14] number based on the equivalent particle diameter. It is equal to $Re_P = g \cdot D_P/\mu$, where Pr is the Prandtl[15] number $Pr = C_P \cdot \mu/\lambda_g$, λ_{er} is the effective conductivity of the catalyst bed measured in $kJ/(m \cdot h \cdot C)$, and λ_{er}^o is the static contribution of the effective conductivity of the catalyst bed with the value

$$\lambda_{er}^o = \epsilon \cdot (\lambda_g + 0.95 \cdot hru \cdot D_P) + \frac{0,95 \cdot (1 - \epsilon)}{2/(3 \cdot 0.2988) + 1/(10 \cdot \lambda_g + hrs \cdot D_P)}$$

for $hru = (0.1952 \cdot (T/100)^3)/(1 + 0.125 \cdot \epsilon/(1 - \epsilon))$ and $hrs = 0.0651 \cdot (T/100)^3$.

The *thermal conductivities* and the *boiling points* of all components are readily available in the literature.

---

[14]Osbourne Reynolds, English engineer and physicist, 1842-1912
[15]Ludwig Prandtl, German physicist, 1875-1953

Catalyst parameters

The *characteristic length* is the thickness of the equivalent slab used in the single catalyst pellet equation and it is defined as the thickness l_c of the catalyst slab that gives the same external surface to volume ratio as the original pellet. For Raschig[16] rings this is given by

$$l_c = \frac{(D_{P,o} - D_{P,i}) \cdot h}{2 \cdot (D_{P,o} - D_{P,i}) + h}$$

where $D_{P,o}$ and $D_{P,i}$ are the outside and inside diameters of the ring shaped catalyst pellet and h is its height.

The *shape factor* for Raschig rings is given by

$$\Phi_s = \frac{D_P^2}{0.5 \cdot (D_{P,o}^2 - D_{P,i}^2) + D_{P,o} + D_{P,i}} \, .$$

7.4.2   Model Development for Steam Reformers

Industrial steam reformers are usually fixed-bed reactors. Their performance is strongly affected by the heat transfer from the furnace to the catalyst tubes. We will model both top-fired and side-fired configurations.

Modeling of steam reformer tubes:

The tubes are loaded with industrial catalyst pellets which usually are nickel catalysts. In the mass-balance equations, methane and carbon dioxide are the key components and their conversion and yield are defined respectively as

$$x_{CH_4} = \frac{F_{CH_4}^0 - F_{CH_4}}{F_{CH_4}^0} \quad \text{and} \tag{7.125}$$

$$x_{CO_2} = \frac{F_{CO_2} - F_{CO_2}^0}{F_{CH_4}^0} \, . \tag{7.126}$$

Our model assumptions are as follows:

1. The reaction mixture behaves as an ideal gas.

2. The single reactor tube performance is representative of any other tube in the furnace.

3. The system operates at a steady state.

4. We assume a one dimensional model for mass, heat, and momentum transfer. This means that the composition, heat, and pressure is uniform at any cross section of the catalyst bed in the reformer tubes.

---

[16]Friedrich August (Fritz) Raschig, German chemist, 1863-1928

5. All hydrocarbons in the feed higher than methane are assumed to be instantly cracked into CH_4, CO_2, H_2, and CO. Consequently the reaction system inside a reformer tube is described by the rate expressions of the kinetics of steam reforming in the methane reactions I, II, and III.

6. The axial diffusion of mass and heat is negligible.

7. The system is heterogeneous, external mass and heat transfer between the pellet and the bulk gas is negligible, but the intraparticle diffusional resistance is considerable.

The mass-balance equations:

The differential mass-balance equations for methane and carbon dioxide are

$$\frac{dx_{CH_4}}{dl} = \frac{A_t \cdot \rho_b \cdot \eta_{CH_4} \cdot R_{CH_4}}{F_{CH_4}^0} \quad \text{and} \tag{7.127}$$

$$\frac{dx_{CO_2}}{dl} = \frac{A_t \cdot \rho_b \cdot \eta_{CO_2} \cdot R_{CO_2}}{F_{CH_4}^0} \,, \tag{7.128}$$

where η_{CH_4} is the catalyst effectiveness factor for methane and η_{CO_2} is the catalyst effectiveness factor for carbon dioxide, while $R_{CH_4} = r_I + r_{III}$ and $R_{CO_2} = r_{II} + r_{III}$. The mole fraction of each component at any point along the reformer can be computed in terms of x_{CH_4} and x_{CO_2} and the feed mole fractions as follows.

$$Y_{CH_4} = \frac{Y_{CH_4}^0 \cdot (1 - x_{CH_4})}{1 + 2Y_{CH_4}^0 \cdot x_{CH_4}} \tag{7.129}$$

$$Y_{CO_2} = \frac{Y_{CH_4}^0 \cdot (F_{CO_2}^0 / F_{CH_4}^0 + x_{CO_2})}{1 + 2Y_{CH_4}^0 \cdot x_{CH_4}} \,, \tag{7.130}$$

$$Y_{H_2O} = \frac{Y_{CH_4}^0 \cdot (F_{H_2O}^0 / F_{CH_4}^0 - x_{CH_4} - x_{CO_2})}{1 + 2Y_{CH_4}^0 \cdot x_{CH_4}} \,, \tag{7.131}$$

$$Y_{H_2} = \frac{Y_{CH_4}^0 \cdot (F_{H_2}^0 / F_{CH_4}^0 + 3 \cdot x_{CH_4} + x_{CO_2})}{1 + 2Y_{CH_4}^0 \cdot x_{CH_4}} \,, \quad \text{and} \tag{7.132}$$

$$Y_{CO} = \frac{Y_{CH_4}^0 \cdot (F_{CO}^0 / F_{CH_4}^0 + x_{CH_4} - x_{CO_2})}{1 + 2Y_{CH_4}^0 \cdot x_{CH_4}} \,. \tag{7.133}$$

The mole fractions of the inerts are obtained by subtraction.

Energy-balance equations:

An energy balance on a differential element gives us

$$\frac{dT}{dl} = \frac{1}{G \cdot \bar{C}_p} \left(\frac{4 \cdot U \cdot (T_{t,0} - T)}{d_i} + \sum_{j=1}^{3} (-\Delta H_j) \cdot \rho_b \cdot \eta_j \cdot r_j \right) \,, \tag{7.134}$$

where G is the mass flow rate of the mixture, $\bar{C}_p$ is the average specific heat of the mixture, and $T_{t,0}$ is the temperature of the furnace that surrounds the catalyst tubes.

Momentum-balance equation:

The differential equation for the pressure drop across the differential element is

$$\frac{dP}{dl} = \frac{-f_p \cdot G^2}{g_c \cdot \rho_g \cdot \Phi_s \cdot D_p} . \tag{7.135}$$

7.4.3   Modeling of Side-Fired Furnaces

The total amount of heat that is transferred to the reformer tubes is obtained from the flames and the furnace gas. For this heat transfer, we use the following assumptions.

1. The furnace is a well stirred enclosure with a mean temperature different from its exit temperature.

2. Radiation is the main mode of heat transfer; heat transfer by convection is negligible.

3. The flames and the furnace gas are distinct sources of heat. The radiation from these two sources do not interfere with each other.

4. 2% of the released heat is lost to the surroundings.

5. There is no net radiative exchange between the enclosing walls (fire-box), and the radiating gas and tubes. Heat is lost from the enclosing wall to the surroundings only.

6. The furnace gas is assumed to be grey to radiation.

Radiative transfer from the furnace gases to the reformer tubes:

The radiation heat-transfer equation from the furnace gas to the reformer tubes has the form

$$Q_R = (\bar{G} \cdot \bar{S}_t)_R \cdot (\bar{E}_g - \bar{E}_t) , \tag{7.136}$$

where

$$\frac{1}{(\bar{G} \cdot \bar{S}_t)_R} = \frac{1}{\alpha \cdot A_{CP}} \left(\frac{1}{e_t} - 1 \right) + \frac{1}{(\bar{G} \cdot \bar{S}_t)_{R,black}} ,$$

$$(\bar{G} \cdot \bar{S}_t)_{R,black} = e_g \cdot \left(\alpha \cdot A_{CP} + \frac{A_R}{(1 + e_g/((1 - e_g) \cdot F_{Rt})} \right) ,$$

$$\bar{E}_g = \sigma \cdot T_g^4 , \quad \bar{E}_t = \sigma \cdot T_{t,0}^4 , \quad F_{Rt} = \frac{\alpha \cdot A_{CP}}{A_R + \alpha \cdot A_{CP}} ,$$

and $0 \le A_R \le 0.5 \cdot \alpha \cdot A_{CP}$.

The heat released by the furnace gas is transferred to the reformer tubes, the flue gas and 2% is lost to the surroundings. This heat balance can be written as

$$\int_0^L Q_R \, dl = 0.98 \cdot Q_n - Q_G . \tag{7.137}$$

7.4.4 Model for a Top-Fired Furnace

The Roesler[17] model gives rise to the two equations

$$\frac{d^2\bar{E}_1}{dl^2} = \alpha_1(\zeta \cdot (\bar{E}_1 - \sigma \cdot \xi \cdot T_g^4) + e_t \cdot a_t \cdot (\bar{E}_1 - \sigma \cdot \xi \cdot T_{t,0}^4) +$$

$$+ e_r \cdot a_r \cdot ((1 - \xi)\bar{E}_1 - \xi\bar{E}_2)) \quad \text{and} \tag{7.138}$$

$$\frac{d^2\bar{E}_2}{dl^2} = \alpha_2(e_t \cdot a_t \cdot (\bar{E}_2 - \sigma \cdot (1 - \xi) \cdot T_{t,0}^4) + e_r \cdot a_r(\xi \cdot \bar{E}_2 - (1 - \xi) \cdot \bar{E}_1)) \tag{7.139}$$

with the boundary conditions:
at $l = 0$

$$\frac{1}{\alpha_1} \cdot \frac{d\bar{E}_1}{dl} = -\frac{1}{\alpha_2} \cdot \frac{d\bar{E}_2}{dl} = \frac{e_r}{2 - e_r} \cdot ((1 - \xi)\bar{E}_1 - \xi \cdot \bar{E}_2) ; \tag{7.140}$$

at $l = L$

$$-\frac{1}{\alpha_1} \cdot \frac{d\bar{E}_1}{dl} = \frac{1}{\alpha_2} \cdot \frac{d\bar{E}_2}{dl} = \frac{e_r}{2 - e_r} \cdot ((1 - \xi)\bar{E}_1 - \xi \cdot \bar{E}_2) \tag{7.141}$$

for $\alpha_1 = \zeta + a_t + a_r$ and $\alpha_2 = a_t - a_r$.
The heat balance on the differential element of the flue gas stream yields

$$G_g \cdot C_g \frac{d(T_g - F \cdot (T^* - T_{g,o}))}{dl} = 2\zeta \cdot (\bar{E}_1 - \xi \cdot \bar{E}_g) \tag{7.142}$$

for $T_{g,0} = (\bar{E}_{1,0}/(\sigma \cdot \xi))^{1/4}$.
The heat transfer by radiation to the reformer tubes has the form

$$Q_g = 2e_t \cdot a_t \cdot (\bar{E}_r - \bar{E}_t) \cdot V \tag{7.143}$$

for $E_r = \sigma T_r^4$ and $T_r = ((\bar{E}_1 - \bar{E}_2)/\sigma)^{1/4}$.

7.4.5 Modeling of Methanators

Methanators are usually adiabatic fixed-bed reactors. The kinetics of methanization are described by the reverse $\hat{r}_I$, $\hat{r}_{II}$ and $\hat{r}_{III}$ of the reactions I, II, and III. Hence there are two exothermic reactions I and II and one endothermic one, III. Carbon dioxide and methane are the key components here.

Mass-balance equations:

The conversion and yield of carbon dioxide and methane are defined in the opposite sense of the yield and conversion used in steam reformers in equations (7.125) and (7.126) and the same differential mass balance equations (7.127) and (7.128) are used. The mole fractions of all components are computed from similar relations as given before.

---

[17]F. C. Roesler, USA chemical engineer, 19..

Energy-balance equation:

The energy-balance equation for a differential element is

$$\frac{dT}{dl} = \frac{\rho_b}{G \cdot \bar{C}_p} \cdot \sum_{j=1}^{3} (-\Delta H_j) \cdot \eta_j \cdot \hat{r}_j \, . \tag{7.144}$$

Momentum-balance equation:

The momentum-balance equation for the methanator is the same as for the steam reformer, namely equation (7.135).

7.4.6 Modeling the Catalyst Pellets for Steam Reformers and Methanators using the Dusty Gas Model

The profiles of the effectiveness factor η along the length of a steam reformer as well as a methanator can be computed in three different ways, namely by the dusty gas model, as well as by our simplified models (A) and (B).

Dusty gas model:

When using the dusty gas model we obtain the following gradient equation for the total diffusive flux

$$-grad \, C_i = \frac{N_i}{D_{ki}^e} + \sum_{j \neq i}^{n} \frac{Y_j \cdot N_i - Y_i \cdot N_j}{D_{ij}^e} \, . \tag{7.145}$$

Simplified model (A):

Equation (7.145) results from the rigorous dusty gas model, but unfortunately, it is not easy to implement for a multi-component system. Therefore, we will use simplified equations for the flux relations (7.145). The ordinary diffusion term in formula (7.145) can be approximated by

$$\frac{N_i}{D_{im}^e} = \sum_{j \neq i}^{n} \frac{Y_j \cdot N_i - Y_i \cdot N_j}{D_{ij}^e} \, . \tag{7.146}$$

Thus equation (7.145) takes the simplified form

$$N_i = -D_i^e \cdot grad \, C_i \, , \tag{7.147}$$

where $1/D_i^e = 1/D_{ki}^e + 1/D_{im}^e$ and D_{im}^e is obtained from the diffusion equation

$$\frac{1}{D_{im}} = \frac{1}{1 - Y_i} \cdot \sum_{j \neq i}^{n} \frac{Y_i}{D_{ij}} \tag{7.148}$$

for component i through the stagnant components of the mixture. Moreover, the effective diffusion coefficients are set equal to $D_{im}^e = D_{im} \cdot \epsilon/\tau$ with $\epsilon = 0.625$ and $\tau = 2.74$. These values were obtained experimentally.

Note that equation (7.148) is only valid for dilute systems, but it can be used as an approximation for multi-component systems.

Simplified model (B):

Here we use equation (7.146) to find D_{im}^e from the relation

$$\frac{1}{D_{im}^e} = \sum_{j \neq i}^{n} \frac{Y_j - Y_i \cdot (N_j/N_i)}{D_{ij}^e} \tag{7.149}$$

and approximate the flux ratios by the ratio of rates of reaction to get

$$\frac{1}{D_{im}^e} = \sum_{j \neq i}^{n} \frac{Y_j - Y_i \cdot (R_j/R_i)}{D_{ij}^e} . \tag{7.150}$$

7.4.7 Numerical Considerations

The differential equations (7.127), (7.128), (7.134), and (7.135) for the material, energy, and momentum balances for a steam reformer and the corresponding DEs for a methanator are initial value problems that can be solved in MATLAB using one of the familiar MATLAB functions ode.... Care should be taken to select appropriate integrators, i.e., ode23 or ode45, in case the simple Runge-Kutta embedding formulas converge quickly, or ode15s if they do not and the IVP system is found to be stiff. The initial conditions are the feed conditions.

For the reformer we assume that the outer wall temperature profile of the reformer tubes decouples the heat-transfer equations of the furnace from those for the reformer tubes themselves. The profile is correct when the heat flux from the furnace to the reformer tube walls equals the heat flux from the tube walls to the reacting mixture. We must carry out sequential approximating iterations to find the outer wall temperature profile $T_{t,0}$ that converges to the specific conditions by using the difference of fluxes to obtain a new temperature profile $T_{t,0}$ for the outer wall and the sequence of calculations is then repeated. In other words, a $T_{t,0}$ profile is assumed to be known and the flux Q from the furnace is computed from the equations (7.136) and (7.137), giving rise to a new $T_{t,0}$. This profile is compared with the old temperature profile. We iterate until the temperature profiles become stationary, i.e., until convergence.

The furnace equation (7.136) that describes the heat transfer in the side fired furnace is a single transcendental equation used to compute and iterate $T_{t,0}$ and Q_R. It can be solved using fzero of MATLAB.

The differential equations (7.138) and (7.139) for the top-fired furnace, however, describe a boundary value problem with the boundary conditions (7.140) and (7.141) that can be solved via MATLAB's built-in BVP solver bvp4c that uses the collocation method, or via our modified BVP solver bvp4cfsinghouseqr.m which can deal with singular Jacobian search matrices; referred to in Chapter 5. On the other hand, the differential equation (7.142) is a simple first-order IVP.

Notice that the methanator is working under adiabatic conditions so that its design equations, i.e., its material, energy, and momentum equations can be integrated straightforward as IVPs in MATLAB.

The effectiveness factors at each point along the length of the reactor are calculated for the key components methane and carbon dioxide by using either the dusty gas model or one of our simplified models (A) and (B). The dusty gas model gives rise to more complicated two point boundary value differential equations (BVPs) for the catalyst pellets. These are solved in MATLAB via `bvp4c` or `bvp4cfsinghouseqr.m` as done in Chapter 5.

7.4.8 Some Computed Simulation Results for Steam Reformers

In this section we collect some computed results when simulating industrial steam reformers. We compare the actual plant outputs with those obtained by simulation using the three models that we have developed earlier. We investigate both the close to thermodynamic equilibrium case and the far from thermodynamic equilibrium case.

For the industrial steam reformer called Plant (1):

Plant (1) is a top-fired reformer and its construction and operating conditions are summarized below.

Top-fired reformer (1) construction data and operating conditions

Reformer tubes		Catalyst pellet	
Heated length of reformer tubes	13.72 m	Shape	Raschig[18] rings
Inside diameter of reformer tube	0.0978 m	Dimensions	$1.6 \times 0.6 \times 1.6\ cm$
Outside diameter of tubes	0.1154 m	Porosity	0.51963
Ratio of tube pitch and diameter	3.3	Tortuosity	2.74
Number of tubes	897	Bulk density	1362 Kg/m^3
Furnace dimensions	$31.834 \times 35.49 \times 13.72\ m$	Mean pore radius	80$\mathring{A}$
Number of burners	204	Catalyst character-istic length	0.001948 cm
Process gas flow rate and composition		*Fuel*	
Flowrate (natural gas)	4430.7 $Kg \cdot mol/h$	Flowrate	4430.7 $Kg \cdot mol/h$ + 9 to 11% excess air at 576 K
Composition (in mol %, dry basis)		Composition (in mol %, wet basis)	
CH_4	73.10	CH_4	16.59
C_2H_6	19.90	C_2H_6	1.41
C_3H_8	2.00	C_3H_8	0.03
C_4H_{10}	0.10	CH_3OH	0.16
H_2	1.80	H_2	71.54
CO_2	0.90	H_2O	0.06
N_2	2.20	CO	1.68
		CO_2	2.80
		N_2	5.73

(continued)

	Inlet conditions per tube		Feed composition in catalyst tubes (in mol %)	
Process gas flow (methane equivalent)	3.953 $kg \cdot mol/h$		CH_4	20.22
Temperature	760 K		H_2O	72.00
Pressure	2837.1 kPa		H_2	4.92
Steam to methane ratio	3.5610		CO_2	2.44
Hydrogen to methane ratio	0.2432		N_2	0.42
Carbon dioxide to methane ratio	0.1209			
Nitrogen to methane ratio	0.0204			

The data below shows an almost identical performance of the plant and the simulated computations for temperature, pressure, and composition of the process gas obtained by all three models, the dusty gas model (abbreviated as DGM) and the simplified models (A) and (B).

We note that this industrial unit (1) is operating quite close to its thermodynamic equilibrium.

Comparison between Plant (1)'s actual output and simulated results with the dusty gas model (DGM) and the simplified models (A) and (B)

Variable	Actual Plant (1)	Calculated via:		
		DGM	model (A)	model (B)
Process gas temp. K	1130.0	1130.9	1131.1	1130.7
Process gas press. kPa	2200.0	2197.8	2197.8	2197.1
Conversion of CH_4	0.8527	0.8588	0.8600	0.8600
Equilibrium conversion of CH_4	0.8616	0.8634	0.8641	0.8631
Conversion of CO_2	0.2862	0.2787	0.2763	0.2792
Equilibrium conversion of CO_2	0.2891	0.2804	0.2800	0.2796
Process gas composition (in mol %)				
H_2O	35.71	36.35	36.37	36.35
CH_4	2.24	2.12	2.10	2.10
H_2	46.46	46.49	46.50	46.47
CO_2	6.19	6.00	5.96	6.00
CO	9.07	8.75	8.76	8.76
N_2	0.33	0.29	0.31	0.32

For the industrial steam reformer called Plant (2):

Plant (2) is a side-fired steam reformer with the following construction and operating data.

---

[18] Friedrich August (Fritz) Raschig, German chemist, 1863-1928

Side-fired reformer (2) construction data and operating conditions

Reformer tubes and furnace		*Catalyst pellet*	
Heated length of reformer tubes	11.95 m	Shape	Raschig rings
Inside diameter of reformer tube	0.0795 m	Dimensions	$1.6 \times 0.6 \times 1.6$ cm
Outside diameter of tubes	0.102 m	Porosity	0.51963
Ratio of tube pitch and diameter	2.4	Tortuosity	2.74
Number of tubes	176	Bulk density	1362 Kg/m^3
Furnace width	2 m	Mean pore radius	80$\mathring{A}$
Number of burners	112	Catalyst character-	
		-istic length	0.001948 cm

Process gas flow rate and composition		*Fuel*	
Flowrate	4085.7 $Kg \cdot mol/h$	Flowrate	313.6 $Kg \cdot mol/h$
(natural gas)		(at 319.1 K)	+ 15% excess air
			at ambient temp.
Composition (in mol %, dry basis)		Composition (in mol %, wet basis)	
CH_4	79.6	CH_4	12.4
C_2H_6	13.2	H_2	67.6
C_3H_8	2.6	CO	5.8
C_4H_{10}	0.4	$C)_2$	13.6
H_2	2.1	N_2	0.6
N_2	2.1		

Inlet conditions per tube		*Feed gas composition for the catalyst tubes*	
		(in mol %)	
Process gas flow	4.03 $kg \cdot mol/h$	CH_4	16.78
(methane equivalent)			
Temperature	733 K	H_2O	77.17
Pressure	2452.1 kPa	H_2	4.19
Steam to methane ratio	4.6	CO_2	1.53
Hydrogen to methane ratio	0.25	N_2	0.33
Carbon dioxide to			
methane ratio	0.091		
Nitrogen to methane ratio	0.02		

The data below shows that the measured output and the simulated data for Plant (2) again agree very closely for each of our three models.

**Comparison between Plant (2) actual output and simulated
results with the dusty gas model (DGM) and
the simplified models (A) and (B)**

Variable	Actual Plant (2)	Calculated via: DGM	model (A)	model (B)
Process gas temp. K	981.1	985.2	985.5	985.2
Process gas press. kPa	2087.3	2076.3	2076.0	2076.1
Conversion of CH_4	0.5720	0.5741	0.5735	0.5742
Equilibrium conversion of CH_4	0.5863	0.5989	0.5997	0.5990
Conversion of CO_2	0.3790	0.3717	0.3710	0.3718
Equilibrium conversion of CO_2	0.3820	0.3850	0.3852	0.3850

Process gas composition (in mol %)

H_2O	52.12	51.41	51.43	51.39
CH_4	5.93	5.99	6.00	5.99
H_2	32.35	32.96	32.94	32.98
CO_2	6.51	6.50	6.50	6.51
CO	2.78	2.86	2.85	2.85
N_2	0.30	0.28	0.28	0.28

Note that all three models give almost the same exit conversion and yield for methane and carbon dioxide and that the second unit (2) is also operating relatively closely to its thermodynamic equilibrium, though further away from it when compared to Plant (1). The close agreement between the industrial performance data and the simulated data for the reformers (1) and (2) that was obtained by three different diffusion-reaction models validates the models that we have used, at least for plants operating near their thermodynamic equilibria.

However, a simple comparison of the stated simulation results does not favor any particular model. To differentiate the models, we propose run comparison tests of the three models for a steam reformer, called Plant (3), that runs far from its thermodynamic equilibrium.

Plant (3):

A short $1\ m$ isothermal steam reformer tube was used for this test. The reformer (3) was run under the same operating conditions as reformer (2), but with a relatively large catalyst pellet of characteristic length $0.007619\ m$. For Plant (3) the exit methane conversion X, the CO_2 yield $\bar{X}$, and the equilibrium values X_e and $\bar{X}_e$ for methane and carbon dioxide, respectively, are as follows.

Exit conversions and yields of methane and carbon dioxide simulation results with the dusty gas model (DGM) and the simplified models (A) and (B) for Plant (3)

	Methane			Carbon dioxide		
	Exit conversion X	Equilibrium conversion X_e	% diff.	Exit yield $\bar{X}$	Equil. yield $\bar{X}_e$	% diff.
DGM	0.2756	0.4710	−41.49	0.1925	0.3414	−43.61
Model (A)	0.2298	0.4710	−51.21	0.1689	0.3414	−50.53
Model (B)	0.2104	0.4710	−55.33	0.1575	0.3414	−53.87

For Plant (3) the exit conversions and yields of methane and carbon dioxide obtained by all three models are much lower than the equilibrium values. Therefore, Plant (3) is run far from its thermodynamic equilibrium. Large differences between the predictions of the three models exist in the data: the exit conversion simulations of methane differ by 16 to 23%; that of the carbon dioxide yield by 12 to 18%. Since the dusty gas model is the more rigorous one, we can use its simulation output as a base for comparison in place of experimental or industrial data which is unavailable in this case.

It is reasonable to conclude from our simulations that for steam reformers that operate close to their thermodynamic equilibrium it is sufficient to use a relatively simple model to compute the catalyst pellet effectiveness factor η. In particular, model (B) is the simplest and it should be used under these conditions. However, for more efficient steam reformers that operate far from their thermodynamic equilibrium, the more complicated and more rigorous dusty gas model for the diffusion and reaction inside the catalyst pellets should be used to obtain more precise results for the effectiveness factor η since both simplified models (A) and (B) give erroneous results here. We reiterate that the closeness of the three model results for systems operating near their thermodynamic equilibrium does not mean that all three models can be used equally reliably in other regions of operation. This is so because in the region of operation far from thermodynamic equilibrium the effectiveness factors are rather small and the simplified pseudohomogeneous models (A) and (B) will greatly overestimate the rates of reaction, thereby overestimating the amount of heat absorbed by the reaction which in turn will alter the prediction of the wall and furnace temperatures and therefore give erroneous predictions.

7.4.9   Simulation Results for Methanators

In this section we check our models against two industrial adiabatic fixed-bed methanators. Along the length of the methanators we calculate the effectiveness factor from the dusty gas model (DGM) and our simplified models (A) and (B).

A methanator, called Plant M-one:

Here are the design and actual operating conditions of Plant M-one:

Methanator Plant M-one: construction, design, and operating conditions

Dimensions			
Length of methanator	4.80 m		
Inside diameter of methanator	2.686 m		
Catalyst pellet			
Shape	spherical particle		
Diameter	0.476 to 0.794 cm		
Porosity	0.625		
Tortuosity	2.74		
Bulk density	1014 Kg/m^3		
Mean pore radius	80$\mathring{A}$		
Characteristic length	0.10583 cm		
Inlet conditions:	*Design*	*Actual*	
Feed flow rate	3392.3 $kmol/h$	3398.3 $kmol/h$	
Methane molar flow	16.557 $kmol/h$	13.523 $kmol/h$	
Temperature	586.3 K	577.8 K	
Pressure	2074.75 kPa	1823.85 kPa	
Feed composition (in mol %)			
CO	0.428	0.160	
	(4280 ppm)	(1600 ppm)	
CO_2	0.050	0.0	
	(500 ppm)		
H_2	74.052	73.410	
CH_4	0.448	0.410	
H_2O	0.483	0.120	
Inerts	24.5	25.9	

The design and actual output data from Plant M-one and the simulated results from the three models are given in the following two tables.

Comparison between methanator Plant M-one design output and simulated results with the dusty gas model (DGM) and the simplified models (A) and (B)

Variable	Design	Calculated via: DGM	model (A)	model (B)
Temperature K	616.3	622.1	622.0	622.0
Pressure kPa	2036.8	2071.2	2071.2	2073.5

(continued)

Composition (in mol %)

CO	0.0	0.157	0.1267	0.1539
		(1570 *ppm*)	(1267 *ppm*)	(1539 *ppm*)
CO_2	0.0	0.530	15.513	17.062
		(5300 *ppm*)	(155,130 *ppm*)	(170,620 *ppm*)
H_2	73.27	73.27	73.27	73.28
CH_4	0.975	0.975	0.973	0.970
H_2O	1.02	1.02	1.02	1.02
Inerts	24.735	24.735	24.735	24.735

**Comparison between methanator Plant M-one <u>actual</u> output and
simulated results with the dusty gas model (DGM) and
the simplified models (A) and (B)**

Variable	Actual	Calculated via:		
		DGM	model (A)	model (B)
Temperature K	585.2	589.6	589.5	
Pressure kPa	1787.1	1820.1	1820.2	1820.2

Composition (in mol %)

CO	0.0	0.609	0.067	0.068
		(6090 *ppm*)	(670 *ppm*)	(680 *ppm*)
CO_2	0.0	$5.6 \cdot 10^{-3}$	$1.7 \cdot 10^{-6}$	$2.0 \cdot 10^{-6}$
		(56 *ppm*)	(0.017 *ppm*)	(0.02 *ppm*)
H_2	73.42	73.16	73.16	73.16
CH_4	0.570	0.572	0.572	0.730
H_2O	0.280	0.281	0.281	0.282
Inerts	25.730	25.983	25.983	25.983

These results show good agreement between the design and actual output temperature,
pressure and composition with the calculated results for all models at the exit of methanator Plant M-one. Note that the pressure drop along methanator Plant M-one is small for
the design and actual conditions and the simplified models (A) and (B) give almost the
same exit pressure.

Methanator Plant M-two:

The design conditions for methanator Plant M-two are as follows.

Methanator Plant M-two: construction and design conditions

Dimensions		Inlet conditions	
Length of methanator	3.048 m	Feed flow rate	77444 m^3/h
Inside diameter of methanator	2.667 m	Methane molar flow	1420.1 $kmol/h$
Catalyst pellet		Temperature	613.6 K
Shape	spherical particle	Pressure	2199.16 kPa
Diameter	0.476 to 0.794 cm	*Feed composition (in mol %)*	
Porosity	0.625	CO	0.350
Tortuosity	2.74		(3500 ppm)
Bulk density	1040 Kg/m^3	CO_2	0.098
Mean pore radius	80$\mathring{A}$		(980 ppm)
Char. length	0.10583 cm	H_2	93.550
		CH_4	4.399
		H_2O	1.60

The output data from methanator Plant M-two and the simulated results from the three models are given in the following table.

Comparison between methanator Plant M-two design output and simulated results with the dusty gas model (DGM) and the simplified models (A) and (B)

Variable	Design	Calculated via:		
		DGM	model (A)	model (B)
Temperature K	644.10	644.18	645.63	646.0
Pressure kPa	2184.88	2115.46	2116.47	2117.01
Composition (in mol %)				
CO	0.0	0.005	0.006	0.007
		(50 ppm)	(60 ppm)	(70 ppm)
CO_2	0.0	0.033	0.048	0.054
		(330 ppm)	(480 ppm)	(540 ppm)
H_2	92.94	93.07	93.10	93.11
CH_4	4.892	4.822	4.811	4.817
H_2O	2.16	2.07	2.04	2.02

The calculated temperature, pressure, and composition from the models are again in good agreement with the design output data of methanator Plant M-two, indicating the validity of our models.

Exercises for 7.4

1. **Project**

   Create a MATLAB program to simulate steam reformers and methanators of sections 7.4.8 and 7.4.9 as indicated in Section 7.4.7 and test your program against the given data of Section 7.4.8 and 7.4.9.

2. Find other sets of industrial data and compare the model results with the industrial ones.

Conclusions

A rigorous dusty gas model and two simplified models have been used to simulate industrial steam reformers and methanators. The basic principles for the solution of both the nonadiabatic steam reformer and the adiabatic methanator are given. The details of developing solution algorithms from the models are left to the reader as a serious and extensive project.

The predictions from the three models for steam reforming are equally reliable for industrial reactors that operate near their thermodynamic equilibrium. Significant deviations occur when the system is operating far from thermodynamic equilibrium, though. In this case the dusty gas model is still quite accurate while the simplified diffusion-reaction models are no longer accurate enough.

The industrial data that is supplied in this section serves to ensure the accuracy of the students' programs. Once established and verified against industrial data, these programs can be used for design, simulation, optimization, scaling-up, and parametric investigations.

7.5   Production of Styrene for the Polystyrene Industry

This section deals with industrial fixed-bed reactors for the dehydrogenation of ethylbenzene to styrene. Styrene is an important intermediate product in the petrochemical industry that is mainly used to produce the polymer polystyrene. The intrinsic kinetics for this reaction are not available in the literature and all known kinetic expressions have been extracted from industrial data by using a pseudohomogeneous model for the reactor. The models developed on this basis are only valid for the particle size and operating conditions that were used during the acquisition of the kinetic data. Thus these models are not reliable over a range of operating and design conditions wider than that under which they were gathered.

As a general caveat with published data, we recommend that the reader check for printing mistakes by looking up multiple sources and comparing them. Great care with using outside data must be exercised at the beginning of any simulation work.

Here we develop a pseudohomogeneous model first and investigate how to solve this model numerically. This is later followed by a more rigorous heterogeneous model.

7.5.1   The Pseudohomogeneous Model

The Rate Equations

In the process, ethylbenzene is dehydrogenated to styrene in a fixed-bed catalytic reactor. The feed stream is preheated and mixed with superheated steam before being injected into the reactor at a temperature above 490^0C. The steam serves as a dilutant and decokes the catalyst, thereby extending its life. The steam also supplies the necessary heat for the endothermic dehydrogenation reaction. For our model we have chosen six reactions to represent the plant data.

Reaction (1):

$$C_6H_5CH_2CH_3 \longleftrightarrow C_6H_5CHCH_2 + H_2 \qquad (7.151)$$
$$ethylbenzene \longleftrightarrow styrene + hydrogen$$

Reaction (2):

$$C_6H_5CH_2CH_3 \longrightarrow C_6H_6 + C_2H_4 \qquad (7.152)$$
$$ethylbenzene \longrightarrow benzene + ethylene$$

Reaction (3):

$$C_6H_5CH_2CH_3 + H_2 \longrightarrow C_6H_5CH_3 + CH_4 \qquad (7.153)$$
$$ethylbenzene + hydrogen \longrightarrow toluene + methane$$

Reaction (4):

$$2H_2O + C_2H_4 \longrightarrow 2CO + 4H_2 \qquad (7.154)$$
$$steam + ethylene \longrightarrow carbon\ monoxide + hydrogen$$

Reaction (5):

$$H_2O + CH_4 \longrightarrow CO + 3H_2 \qquad (7.155)$$
$$steam + methane \longrightarrow carbon\ monoxide + hydrogen$$

Reaction (6):

$$H_2O + CO \longrightarrow CO_2 + H_2 \qquad (7.156)$$
$$steam + carbon\ monoxide \longrightarrow carbon\ dioxide + hydrogen$$

Each of the six reactions is assumed to occur catalytically and all but reaction (1) are irreversible.

The respective rate equations for the reactions are

$$r_1 = k_1 \cdot (p_{EB} - p_{ST} \cdot p_{H_2}/K_{EB}) , \qquad (7.157)$$
$$r_2 = k_2 \cdot p_{EB} , \qquad (7.158)$$
$$r_3 = k_3 \cdot p_{EB} \cdot p_{H_2} , \qquad (7.159)$$
$$r_4 = k_4 \cdot p_{H_2O} \cdot p_{ETH}^{0.5} , \qquad (7.160)$$
$$r_5 = k_5 \cdot p_{H_2O} \cdot p_{MET} , \quad and \qquad (7.161)$$
$$r_6 = k_6 \cdot (P_T/T^3) \cdot p_{H_2O} \cdot p_{CO} . \qquad (7.162)$$

Here the rate constants k_i are expressed in the form

$$k_i = e^{A_i - E_i/(R_G \cdot T)} .$$ (7.163)

The A_i and E_i are the apparent frequency factors and the activation energies of reaction i as given below for each of the six reactions $i = 1, ..., 6$.

Initial values of frequency factors and activation energies of the reactions

Reaction number	Frequency factors A_i (dimensionless) according to equation (7.163)	Activation energy E_i (in kJ/kmol)
(1)	−0.0854	90891.40
(2)	13.2392	207989.23
(3)	0.2961	91515.26
(4)	−0.0724	103996.71
(5)	−2.9344	65723.34
(6)	21.2402	73628.40

The rates of reactions depend on the partial pressures of the components, the equilibrium constants of the ethylbenzene, the total pressure, and the temperature as expressed in equations (7.157) to (7.162). Here the total pressure and temperature of the reactor is known, the equilibrium constant of ethylbenzene is given in the table on p. 508, and the partial pressures can be computed in terms of the conversions.

The reactions (1), (2), and (3) are expressed in terms of ethylbenzene conversion as follows.

$$x_{EB_{ST}} = \frac{F_{ST} - F_{ST}^0}{F_{EB}^0} \quad \text{from reaction (1) ,}$$

where $x_{EB_{ST}}$ is the conversion of ethylbenzene to styrene, or the yield of styrene;

$$x_{EB_{BZ}} = \frac{F_{BZ} - F_{BZ}^0}{F_{BZ}^0} \quad \text{from reaction (2) ,}$$

where $x_{EB_{BZ}}$ is the conversion of ethylbenzene to benzene, or the yield of benzene; and

$$x_{EB_{TOl}} = \frac{F_{TOl} - F_{TOl}^0}{F_{TOl}^0} \quad \text{from reaction (3)}$$

with $x_{EB_{TOl}}$ denoting the conversion of ethylbenzene to toluene, or the yield of toluene. This gives us the total conversion of ethylbenzene x_{EB} as the sum of these three conversions, or

$$x_{EB} = \frac{F_{EB}^0 - F_{EB}}{F_{EB}^0} = x_{EB_{ST}} + x_{EB_{BZ}} + x_{EB_{TOl}} .$$

The reactions (4), (5), and (6) are expressed in terms of steam conversion as follows.

$$x_{H_2O_{COI}} = \frac{F_{COI} - F_{CO}^0}{F_{H_2O}^0} \quad \text{from reaction (4) ;}$$

$$x_{H_2O_{COII}} = \frac{F_{COII} - F_{CO}^0}{F_{H_2O}^0} \quad \text{from reaction (5) ;}$$

$$x_{H_2O_{CO_2}} = \frac{F_{CO_2} - F_{CO_2}^0}{F_{H_2O}^0} \quad \text{from reaction (6) ;}$$

giving us the total conversion of steam as the sum of these conversions

$$x_{H_2O} = \frac{F_{H_2O}^0 - F_{H_2O}^0}{F_{H_2O}^0} = x_{H_2O_{COI}} + x_{H_2O_{COII}} + x_{H_2O_{CO_2}} .$$

The number of moles of the various components in the reactor are as follows:

For ethylbenzene $F_1 = F_{EB} = F_{EB}^0 - x_{EB} \cdot F_{EB}^0$;

For styrene $F_2 = F_{ST} = F_{ST}^0 + x_{EB_{ST}} \cdot F_{EB}^0$;

For benzene $F_3 = F_{BZ} = F_{BZ}^0 + x_{EB_{BZ}} \cdot F_{EB}^0$;

For toluene $F_4 = F_{TOL} = F_{TOL}^0 + x_{EB_{TOL}} \cdot F_{EB}^0$;

For ethylene $F_5 = F_{ETH} = F_{ETH}^0 + x_{EB_{BZ}} \cdot F_{EB}^0 - 0.5 \cdot x_{H_2O_{COI}} \cdot F_{H_2O}^0$;

For methane $F_6 = F_{MET} = F_{MET}^0 + x_{EB_{TOL}} \cdot F_{EB}^0 - 0.5 \cdot x_{H_2O_{COI}} \cdot F_{H_2O}^0$;

For steam $F_7 = F_{H_2O} = F_{H_2O}^0 - x_{H_2O} \cdot F_{H_2O}^0$;

For hydrogen $F_8 = F_{H_2} = F_{H_2}^0 + (x_{EB_{ST}} - x_{EB_{TOL}}) \cdot F_{EB}^0 +$
$\qquad\qquad\qquad + (2 \cdot x_{H_2O_{COI}} + 3 \cdot x_{H_2O_{COII}} + x_{H_2O_{CO_2}}) \cdot F_{H_2O_F}^0$;

For carbon monoxide $F_9 = F_{CO} = F_{CO}^0 + (x_{H_2O_{COI}} + x_{H_2O_{COII}} - x_{H_2O_{CO_2}}) \cdot F_{H_2O}^0$;

For carbon dioxide $F_{10} = F_{CO_2} = F_{CO_2}^0 + x_{H_2O_{CO_2}} \cdot F_{H_2O}$

The total mole flow in the reactor is given by

$$F_T = \sum_{i=1}^{10} F_i$$

and the mole fraction Y_i of component i is $Y_i = F_i/F_T$. Finally the partial pressures p_i are $p_i = Y_i \cdot P_T$ for each of the above listed ten components, ethylbenzene to carbon dioxide.

The rates of formation for the ten different components are

For ethylbenzene	$R_{EB} = -r_1 - r_2 - r_3$;
For styrene	$R_{ST} = r_1$;
For benzene	$R_{BZ} = r_2$;
For toluene	$R_{TOL} = r_3$;
For ethylene	$R_{ETH} = r_2 - 0.5 \cdot r_4$;
For methane	$R_{MET} = r_3 - r_5$;
For steam	$R_{H_2O} = -r_4 - r_5 - r_6$;
For hydrogen	$R_{H_2} = r_1 - r_3 + 2r_4 + 3r_6 + r_6$;
For carbon monoxide	$R_{CO} = r_4 + r_5 - r_6$;
And for carbon dioxide	$R_{CO_2} = r_6$.

Model Equations

We now formulate six material-balance equations for the six reactions (1) to (6) of the equations (7.151) to (7.156), as well as an energy balance and a pressure-drop equation.

The material-balance equations for the first three reactions (1), (2), and (3) have the form

$$\frac{dx_i}{dl} = \frac{\rho_b \cdot A_t \cdot r_i}{F_{EB}^0} , \qquad (7.164)$$

where the index $i = 1, 2, 3$ stands for the respective reaction number (1) to (3) and x_i is the fractional conversion of ethylbenzene in each of the three reactions.
The material-balance equations for the last three reactions (4), (5), and (6) have the form

$$\frac{dx_i}{dl} = \frac{\rho_b \cdot A_t \cdot r_i}{F_{H_2O}^0} , \qquad (7.165)$$

where the index $i = 4, 5, 6$ stands for the respective reaction number (4) to (6) and x_i is the fractional conversion of steam in each of these reactions.

The Ergun[19] equation can be used to compute the pressure profile along the length of the catalyst bed

$$\frac{dP}{dl} = 9.807 \cdot 10^{-4} \cdot \frac{(1 - \epsilon) \cdot G_0}{D_P \cdot \epsilon^3 \cdot \rho_G} \left(\frac{150 \cdot (1 - \epsilon) \cdot \mu_G}{D_P} + \frac{1.75 \cdot G_0}{g_c} \right) . \qquad (7.166)$$

The energy-balance differential equation can be derived from the equation

$$\sum_{i=1}^{10} F_i \cdot Cp_i \cdot dT + \sum_{j=1}^{6} \Delta H_j \cdot A_t \cdot \rho_b \cdot r_j \cdot dl = U \cdot A_H \cdot dl \cdot (T - T_e) . \qquad (7.167)$$

---

[19]Sabri Ergun, Turkish chemical engineer, 19..–

Since the reactor is adiabatic and no heat is transferred to the surroundings, we have $U \cdot A_H \cdot dl \cdot (T - T_e) = 0$, and therefore the energy-balance equation is given by

$$\frac{dT}{dl} = \frac{\sum_{j=1}^{6}(-\Delta H_j) \cdot A_t \cdot \rho_b \cdot r_j}{\sum_{i=1}^{10} F_i \cdot Cp_i} . \qquad (7.168)$$

The heats of reactions can be computed as functions of temperature from the relation

$$\Delta H_j = a_j + b_j \cdot T . \qquad (7.169)$$

The set of values a_j and b_j for each of the reactions is as follows.

Values of a_j and b_j of the heat of reactions $\Delta H_j = a_j + b_j \cdot T$

Reaction number	a_j (in $kJ/kmol$)	b_j (in $kJ/(kmol \cdot K)$)
(1)	120759.11	4.56
(2)	108818.11	−7.96
(3)	−53178.20	−13.19
(4)	82065.74	8.84
(5)	211255.19	16.58
(6)	−45223.66	10.47

The molar heat capacitances Cp_i and Cp_j of the components are given as functions of temperature by the following two equations:

$$Cp_i = \alpha_i + \beta_i \cdot T + \gamma_i \cdot T^2 \qquad (7.170)$$

for the organic components, and

$$Cp_j = d_j + e_j \cdot T + f_j/T^2 . \qquad (7.171)$$

for the inorganic ones. The sets of constants in equation (7.170) for the molar heat capacitances of the organic components are as follows.

Constants of the molar heat capacitances Cp_i for the organic components

Component	α_i (in $kJ/(kmol \cdot K)$)	β_i (in $kJ/(kmol \cdot K^2)$)	$-\gamma_i \cdot 10^5$ (in $kJ/(kmol \cdot K^3)$)
Ethylbenzene	−57.2835	0.5155	30.928
Styrene	−11.02	0.075	5.69
Benzene	−1.71	0.325	11.10
Toluene	2.41	0.392	13.10
Ethylene	11.85	0.120	3.65
Methane	14.16	0.076	1.80

And the sets of constants for the inorganic components are:

Constants of the molar heat capacitances Cp_j for the inorganic components

Component	d_j $kJ/(kmol \cdot K)$	e_j $kJ/(kmol \cdot K^2)$	$f_j \cdot 10^{-5}$ $kJ/(kmol \cdot K^{-3})$
Steam	30.57	0.0375	–
Hydrogen	27.30	$3.27 \cdot 10^{-3}$	5.02
Carbon monoxide	28.43	$4.10 \cdot 10^{-3}$	4.61
Carbon dioxide	44.26	$8.79 \cdot 10^{-3}$	86.30

Numerical Solution of the Model Equations

The differential equations (7.164), (7.165), (7.166), and (7.168) form a pseudohomogeneous model of the fixed-bed catalytic reactor. More accurately, in this pseudohomogeneous model, the effectiveness factors η_i are assumed to be constantly equal to 1 and thus they can be included within the rates of reaction k_i. Such a model is not very rigorous. Because it includes the effects of diffusion and conduction empirically in the catalyst pellet, it cannot be used reliably for other units.

We need to solve the IVP from $l = 0$ to $l = L_t$, the total length of the bed. This can be achieved readily by any of the MATLAB IVP solvers `ode..`, once the initial conditions at $l = 0$ are specified.

Simulation of an Industrial Reactor Using the Pseudohomogeneous Model

The above pseudohomogeneous model was used to simulate an industrial reactor with the following data.

Molar feed rates and industrial reactor data

Component	Value and dimension	Reactor data	Value
Ethylbenzene (EB)	36.87 $kmol/h$	Inlet pressure	2.4 bar
Styrene (ST)	0.67 $kmol/h$	Inlet temperature	922.59 K
Benzene (BZ)	0.11 $kmol/h$	Catalyst bed diameter	1.95 m
Toluene (TOL)	0.88 $kmol/h$	Catalyst bed length L_t	1.70 m
Steam (H_2O)	453.10 $kmol/h$	Catalyst bulk density	2146.27 $kg(cat)/m^3$
Total molar feed	491.88 $kmol/h$	Catalyst particle	
Mass flow rate	12238.67 kg/h	diameter	$4.7 \cdot 10^{-3}$ m

Equilibrium relations data for reaction (1)

Data	Value and dimension	Constants	Value
Equilibrium constant	$K_{EB} = e^{-\Delta F_0/(R_G \cdot T)}$ bar	a $(kJ/kmol)$	122725.157
ΔF_0	$a + b \cdot T + c \cdot T^2$ $kJ/kmol$	b $(kJ/(kmol \cdot K))$	−126.267
		c $(kJ/(kmol \cdot K^2))$	$-2.194 \cdot 10^{-3}$

Here is a comparison chart of the reactor output data and the results of our simulation for the same plant.

**Actual and pseudohomogeneous model results
for an industrial reactor**

Component	Conversion		Molar flow rate $kmol/h$		Molar fraction %	
	plant	model	plant	model	plant	model
Ethylbenzene	47.25	47.30	19.45	19.39	-	3.78
Steam	-	1.20	-	447.62	-	87.36
	Yield %					
Styrene	40.41	40.34	15.57	15.57	-	3.04
Benzene	3.77	3.72	1.50	1.48	-	0.29
Toluene	3.12	3.24	2.03	2.08	-	0.41
Ethylene	-	3.49	-	0.33	-	0.07
Methane	-	3.09	-	0.53	-	0.10
Hydrogen	-	39.06	-	22.62	-	4.42
Carbon monoxide	-	0.0067	-	0.03	-	0.006
Carbon dioxide	-	0.60	-	2.73	-	0.53
Total molar flow rate	-		-	512.37	-	100

	actual	model
Exit temperature	850 K	851.98 K
Exit pressure	2.32 bar	2.36 bar

Note that the results of our simulation via the pseudohomogeneous model tracks the actual plant very closely. However, since the effectiveness factors η_i were included in a lumped empirical fashion in the kinetic parameters, this model is not suitable for other reactors. A heterogeneous model, using intrinsic kinetics and a rigorous description of the diffusion and conduction, as well as the reactions in the catalyst pellet will be more reliable in general and can be used to extract intrinsic kinetic parameters from the industrial data.

7.5.2 Simulation of Industrial Units Using the more Rigorous Heterogeneous Model

Here we develop a heterogeneous model that is based on the more rigorous dusty gas model for diffusion and reaction in porous catalyst pellets.

The Catalyst Pellet Equations

The six reactions described in (7.151) to (7.156) with their rate equations (7.157) to (7.162) take place in the catalyst pellet. Reactants and products are diffusing simultaneously through the pores of the catalyst. In our development of the diffusion-reaction model equations we make the following assumptions.

1. The porous structure of the pellet is homogeneous.

2. Mass transfer through the catalyst pellet occurs by diffusion and only ordinary molecular and Knudsen[20] diffusion are considered.

3. The gases diffusing through the pellet obey the ideal gas law.

4. Viscous flow is negligible and the reactor conditions are isobaric.

5. The concentration profiles are symmetric around the center of the pellet.

6. The pellets are isothermal.

7. External mass- and heat-transfer resistances are negligible.

8. The system is at steady state.

Mass Balance Equations for the Catalyst Pellet:

The catalyst packing of the reactor consists of an iron oxide Fe_2O_3, promoted with potassium carbonate K_2CO_3 and chromium oxide Cr_2O_3. The catalyst pellets are extrudates of a cylindrical shape. Since at steady state the problem of simultaneous diffusion and reaction are independent of the particle shape, an equivalent slab geometry is used for the catalyst pellet, with a characteristic length making the surface to volume ratio of the slab equal to that of the original shape of the pellet.

The mass-balance equations for component i can be expressed as

$$\frac{dN_i}{dz} = \rho_S \cdot R_i . \tag{7.172}$$

Equation (7.172) is the generalized equation for the flux of component i diffusing through the catalyst pellet, subject to the boundary condition $N_i(0) = 0$ at the center $z = 0$.
The mass-balance equations and the dusty gas model equations provide necessary equations for predicting the diffusion through the catalyst pellet. The dusty gas model equations are

$$-\frac{dC_i}{dz} = \frac{N_i}{D_{ki}^e} + \sum_{\substack{i,j=1 \\ i \neq j}}^{n} \frac{Y_i \cdot N_j - Y_j \cdot N_i}{D_{ij}^e} \tag{7.173}$$

with the boundary condition $C_i(r_p) = C_i$ at the surface when $z = r_p$. Using the dusty gas model requires us to find the fluxes by solving the material balance differential equation (7.172) simultaneously with the dusty gas differential equation (7.173). The coupled boundary value problems (7.172) and (7.173) can be solved via MATLAB using `bvp4c` or `bvp4cfsinghouseqr` as detailed in chapter 5.

Algebraic Manipulation of the Pellet Mass Balance Equations:

The ten flux equations and ten dusty gas model equations have to be solved simultaneously. By counting the number of stoichiometric equations and the number of reactants and products, it becomes clear that the flux values of four components can be expressed

---

[20]Martin Hans Christian Knudsen, Danish physicist, 1871-1949

by the flux values of the remaining six components. Simple algebraic manipulations give us these relations:

$$N_{EB} = -N_{ST} - N_{BZ} - N_{TOL} ; \qquad (7.174)$$

$$N_{H_2O} = -N_{CO} - 2 \cdot N_{CO_2} ; \qquad (7.175)$$

$$N_{MET} = 2 \cdot N_{BZ} + N_{TOL} - 2 \cdot N_{ETH} - N_{CO} - N_{CO_2} ; \qquad (7.176)$$

$$N_{H_2} = 4 \cdot N_{CO_2} + 3 \cdot N_{CO} + 2 \cdot N_{ETH} + N_{ST} - 2 \cdot N_{BZ} - N_{TOL} . \qquad (7.177)$$

Therefore, only six flux equations and six dusty gas model equations need to be solved simultaneously to obtain the concentration profiles. The performance of the pellet is expressed in terms of the effectiveness factor of each reaction and of each component. The effectiveness factor for component j is

$$\eta_j = \frac{1}{r_P \cdot R_{j_B}} \int_0^{r_P} R_j \, dz \qquad (7.178)$$

and for reaction i it is

$$\eta_i = \frac{1}{r_P \cdot r_{i_B}} \int_0^{r_P} r_i \, dz . \qquad (7.179)$$

These definite integrals define η_j for both reactions and components. They can be easily evaluated in MATLAB by calling quad... once the BVPs have been solved. Use >> help quad to learn how to evaluate definite integrals in MATLAB.

Model Equations of the Reactor

We now formulate six mass-balance DEs for the six reactions inside the reactor, as well as the energy balance and pressure-drop equations.

The mass-balance equations for the bulk phase of the reactor are given by

$$\frac{dx_i}{dl} = \frac{\eta_i \cdot \rho_b \cdot A_t \cdot r_i}{F_{EB}^0} \qquad (7.180)$$

where the index $i = 1, 2, 3$ distinguishes between the first three reactions (1), (2), and (3), and x_i is the fractional conversion of ethylbenzene in each of the three reactions. And

$$\frac{dx_j}{dl} = \frac{\eta_j \cdot \rho_b \cdot A_t \cdot r_j}{F_{H_2O}^0} \qquad (7.181)$$

where the index $j = 4, 5, 6$ refers to the reactions numbered (4), (5), and (6), and x_j is the fractional conversion of steam in each of the last three reactions. The molar flow rates of all ten components can be computed in terms of these six conversions and the feed molar flow rates using a simple mass balance equation.

The pressure-drop equation is the same as equation (7.166).

The energy-balance differential equation is

$$\frac{dT}{dl} = \frac{\sum_{j=1}^{6} \left((-\Delta H_j) \cdot \eta_j \cdot A_t \cdot \rho_b \cdot r_j \right)}{\sum_{i=1}^{10} (F_i \cdot Cp_i)} . \qquad (7.182)$$

Here the index $j = 1, ..., 6$ refers to the reactions (1) through (6) defined at the start of this section and the index $i = 1, ..., 10$ refers to the 10 components of the system.

The solution to the reactor model differential equations (7.166) and (7.180) to (7.182) simulates the molar flow rates and the pressure drop and energy balance of the reactor. The solution of the catalyst pellet boundary value differential equations (7.172) and (7.173) provides the effectiveness factors η_j for each reaction labeled $j = 1, ..., 6$ for use inside the differential equations (7.180) to (7.182).

The main objective of the pellet model is to calculate the effectiveness factors for the six reactions that take place inside the reactor. These factors are defined as the ratios of the actual rates occurring inside the pellet and the rates occurring in the bulk phase, i.e., inside the pellet when diffusional resistances are negligible.

The catalyst pellet boundary value differential equations (7.172) and (7.173) can be solved via MATLAB using `bvp4c` or `bvp4cfsinghouseqr` as practiced in Chapter 5. The reactor model DEs (7.166) and (7.180) to (7.182) can be solved via MATLAB's standard IVP solvers `ode...` The reactor model equations and the catalyst pellet equations used to compute the effectiveness factors η_j are all coupled.

Extracting Intrinsic Rate Constants

The industrial rates obtained earlier from the pseudohomogeneous model actually include diffusional limits and are suitable for the specific reactor with the specific catalyst particle size for which the data was extracted. Such pseudohomogeneous models do not account explicitly for the catalyst packing of the reactor. On the other hand, heterogeneous models account for the catalyst explicitly by considering the diffusion of reactants and of products through the pores of the catalyst pellet.

In this section we refer to the same industrial reactor as in 7.5.1, with its data given on p. 508. Further reactor specifications and catalyst-bed properties of this plant are as follows.

Reactor specifications and catalyst-bed properties

Reactor diameter	D_R	$1.95\ m$
Catalyst bed depth	L_B	$1.70\ m$
Catalyst bulk density	ρ_C	$2.146.27\ kg(cat)/m^3$
Catalyst particle diameter	D_P	$4.7 \cdot 10^{-3}\ m$
Catalyst mean pore radius	r_P	$2400\ \mathring{A}$
Catalyst porosity	ϵ	0.35 dimensionless
Catalyst tortuosity factor	τ	4.0 dimensionless
Reactor inlet pressure	P_T	$2.4\ bar$
Reactor inlet temperature	T	$922.59\ K$

The apparent kinetic data as discussed in the previous section is given in terms of the dimensionless frequency factors A_i and the activation energy E_i, measured in $kJ/kmol$ with

$$k_i = e^{A_i - E_i/(R_G \cdot T)} \tag{7.183}$$

for each of the six reactions in the plant as follows.

Frequency factors and activation energies of the apparent rates of reactions and the frequency factors of the intrinsic rates in equation (7.183)

Reaction	A_i intrinsic	A_i apparent	E_i, apparent and intrinsic
(1)	0.8510	−0.0854	90891.40
(2)	14.0047	13.2392	207989.23
(3)	0.5589	0.2961	91515.71
(4)	0.1183	−0.0724	103996.71
(5)	−3.2100	−2.9344	65723.34
(6)	21.2423	21.2402	73628.40

The set of constants in columns 3 and 4 above are not intrinsic rate constants. They can, however, be used as starting values or as the initial guess in an iterative scheme to obtain kinetics that are suitable for the heterogeneous model.

The heterogeneous model of this section is used to extract the intrinsic kinetic constants from the industrial data. The iteration scheme to find these constants is as follows.

1. The rate constants of the pseudohomogeneous model are used as starting values or the initial guess in the heterogeneous model. The results obtained from the heterogeneous model with these settings will show lower conversion (as is to be expected) when compared with the results of the actual industrial plant and of the pseudohomogeneous model simulation.

2. For components resulting from a single reaction, such as styrene, benzene, toluene, and carbon dioxide, the molar flow rates from the industrial reactor are divided by the corresponding rates obtained from the heterogeneous model and multiplying factors are established for the respective four reactions.

3. The remaining two rate constants are multiplied by suitable factors to give the best match between the heterogeneous model and the industrial data, or equivalently, the data obtained from the pseudohomogeneous model.

4. The rate equations are then multiplied by the respective factors and the program is restarted.

5. The new results are again compared with the industrial ones and new multiplying factors are set.

6. The above five steps are repeated until the results for the heterogeneous model are very closely agreeing with the industrial data.

The intrinsic frequency factors A_i from this iterative procedure are listed in the second column of the previous table as 'A_i intrinsic'. The results from the heterogeneous model with these intrinsic rates and the industrial data is as follows where MFR denotes the Molar Flow Rate, P_E is the exit pressure in *bar* and T_E the exit temperature in K.

Results of the heterogeneous model with the intrinsic rate constants (found iteratively) and the output data of the industrial reactor

| | Industrial reactor | | Heterogeneous model | |
	MFR	Conversion %	MFR	Conversion %
Ethylbenzene	19.45	47.25	19.29	47.45
Steam	-	-	447.32	1.26

	MFR	Yield %	MFR	Yield %
Styrene	15.57	40.41	15.53	40.71
Benzene	1.50	3.77	1.64	4.12
Toluene	1.03	3.12	2.07	3.17

P_E	2.32 *bar*	2.382 *bar*
T_E	850.0 *K*	851.1 *K*

The final table of this section shows the corresponding exit effectiveness factors, the apparent rates of reactions of the components at the exit, and the intrinsic rates.

Values of the effectiveness factors and the apparent and intrinsic rates of consumption of components by individual and overall reactions

	Effectiveness factor	Apparent rate ($\times 10^4$)	Intrinsic rate ($\times 10^4$)
Reaction (1)	0.37	2.89	7.81
Reaction (2)	0.85	0.55	0.65
Reaction (3)	0.87	1.23	1.41
Reaction (4)	0.95	1.25	1.32
Reaction (5)	1.17	0.81	0.695
Reaction (6)	1.06	2.07	1.95
Ethylbenzene	0.47	4.67	9.88
Styrene	0.37	−2.89	−7.81
Benzene	0.84	−0.55	−0.65
Toluene	0.87	−1.23	−1.41
Methane	0.58	−0.42	−0.72
Ethylene	1.01	0.079	0.078
Hydrogen	0.29	−8.66	-30.08
Steam	0.197	4.13	20.964
Carbon monoxide	0.000488	0.0083	17.01
Carbon dioxide	1.06	−2.07	−1,95

The effectiveness factors for the reactions and the components are changing along the length of the reactor. The data above shows the effectiveness factors and rates at the exit. Here the effectiveness factor of a component is computed from the rate of its consumption at the exit. For components which are formed in a single reaction and are not involved in any other reactions, such as styrene, benzene, toluene and carbon dioxide, the effectiveness factor is determined by the effectiveness factor of their reaction, i.e., by the reactions (1) to (3) and (6). Note that for most of the reactions (except for reactions (5) and (6) with η slightly greater than 1), the intrinsic rates of reactions are higher than the apparent rates. The results of our table show that the η values of the six main reactions are not very far from unity. This justifies our procedure that is based on correcting the frequency factors of diffusional limitations while keeping the activation energies unaltered. Thus the heterogeneous model helps to separate the effectiveness factors from the actual rate equations, thereby allowing us to extract intrinsic kinetics. This technique is helpful when no intrinsic kinetics data is available, but it requires sufficient industrial data to start the numerical calculations.

Numerically the heterogeneous model involves IVPs for the reactor and BVPs for the catalyst pellets. These problems can be solved as before via MATLAB.

Exercises for 7.5

1. Develop the heterogeneous model of this section and a MATLAB algorithm for its numerical solution. Verify your work for the supplied industrial data.

2. Use your MATLAB code of problem 1 to investigate the effect of

   (a) the feed flow rate;

   (b) the feed concentration of ethylbenzene;

   (c) the feed temperature;   and

   (d) the feed pressure, respectively,

   on the styrene yield and productivity.

Conclusions

In this section we have applied the modeling and numerical techniques of this book to simulate and extract intrinsic kinetic parameters from industrial data for an industrial reactor that produces styrene.

7.6   Production of Bioethanol

We describe the biochemical process of fermentation used to produce ethanol.

In this fermentation process, sustained oscillations have been reported frequently in experimental fermentors and several mathematical models have been proposed. Our approach in this section shows the rich static and dynamic bifurcation behavior of fermentation systems by solving and analyzing the corresponding nonlinear mathematical models. The results of this section show that these oscillations can be complex leading to chaotic behavior and that the periodic and chaotic attractors of the system can be exploited for increasing the yield and productivity of ethanol. The readers are advised to investigate the system further.

7.6.1   Model Development

In microbial fermentation processes, biomass acts as the catalyst for substrate conversion and at the same time it is produced by the process itself. This is a biochemical example of autocatalysis. One can classify the biofermentation models according to their description or assumptions for the biomass.

Segregated or corpuscular models regard biomass as a population of individual cells. Consequently, the corresponding mathematical model is based on statistical equations. Such models are valuable for describing the variations in a given populations such as the age distribution amongst the cells. This approach is also useful for describing stochastic events, in which case probability and statistics are applied.

If there is no need for describing the variations of the biomass population or the stochastic events and provided that the number of cells under consideration is large, then continuum or unsegregated models are more convenient. Continuum models regard biomass as a chemical complex in solution or as a multiphase system without concentration gradients within the separate phases. The description of nonstochastic interactions between the biomass and its environment by continuum models equates to describing them by segregated models in the case of large populations with a normal variance distribution.

Continuum models can be subdivided into unstructured and structured models. Unstructured models regard the biomass as one compound which does not vary in composition under environmental changes. Structured models regard the biomass as consisting of at least two different compounds and they describe the interactions between the various constituting compounds, the biomass and the environment. It is interesting to study the occurrence of oscillations in anaerobic cultures experimentally and theoretically. Besides, these oscillations give us a perfect tool for in-depth studies of the microorganism physiology.

An unsegregated-structured two-compartment representation considers biomass as being divided into two compartments, the K-compartment and the G-compartment. These two compartments contain specific groupings of macromolecules, namely the K-compartment is identified with RNA, carbohydrates and monomers of macromolecules, while the G-compartment is identified with proteins, DNA and lipids.

The oscillatory behavior of fermentors can be investigated utilizing such a model. Here the synthesis of a cellular component that is essential for both growth and product formation depends nonlinearly on the ethanol concentration. Hence the inhibition by ethanol does not directly influence the specific growth rate of the culture, but it affects it indirectly instead.

We use a two-compartment model in this section. One of the most widely used models for fermentation processes is the maintenance model in which the substrate S consumption r_S is expressed in the form

$$r_S = \frac{1}{Y_{SX}} \cdot r_X + m_S \cdot C_X . \tag{7.184}$$

In (7.184) the first term on the right-hand side accounts for the growth rate and the second term accounts for the maintenance. Here the growth term and the maintenance factor are used in their classical definitions.

The rate of growth of the biomass is given by

$$r_X = \mu \cdot C_X . \tag{7.185}$$

The relatively simple unsegregated-structured model is based on an internal key compound e of the biomass. The activity of this compound is expressed in terms of the concentrations of the substrate C_S, of the product C_P and of the compound C_e of the biomass itself. Thus the rate of formation of the key compound e is given by

$$r_e = g_1(C_S) \cdot g_2(C_P) \cdot C_e \qquad (7.186)$$

where the substrate dependence function $g_1(C_S)$ is given by a Monod[21] type relation of the form

$$g_1(C_S) = \frac{C_S}{K_S + C_S} . \qquad (7.187)$$

By checking the model against experimental data, the relation between the ethanol concentration C_P and the ethanol dependence function $g_2(C_P)$ is found to be a second order polynomial

$$g_2(C_P) = k_1 - k_2 \cdot C_P + k_3 \cdot C_P^2 . \qquad (7.188)$$

This four dimensional model accounts for four concentrations, namely C_S of the substrate S, C_P of the product ethanol P, C_X of the microorganisms or the biomass X, and C_e of the internal key compound e. We can modify the dynamic model that represents the concentrations of the components X, S, and P together with the mass ratio of the components e and X. To do so we define the dimensionless fraction $E = C_e/C_X$ of the biomass that belongs to component e and the maximum possible specific growth rate μ_{max} that would be obtained if $E = 1$, i.e., if the whole biomass were active. The specific growth rate can be written as $\mu = C_S \cdot E \cdot \mu_{max}/(K_S + C_S)$. With these settings the dynamic model is described by the following set of four ODEs:

$$\frac{dE}{dt} = \left(\frac{k_1}{\mu_{max}} - \frac{k_2}{\mu_{max}} \cdot C_P + \frac{k_3}{\mu_{max}} \cdot C_P^2 \right) \cdot \mu - \mu \cdot E , \qquad (7.189)$$

$$\frac{dC_X}{dt} = \mu \cdot C_X + D \cdot (C_{X0} - C_X) , \qquad (7.190)$$

$$\frac{dC_S}{dt} = -\left(\frac{1}{Y_{SX}} \cdot \mu + m_S \right) \cdot C_X + D \cdot (C_{S0} - C_S) , \qquad (7.191)$$

$$\frac{dC_P}{dt} = \left(\frac{1}{Y_{PX}} \cdot \mu + m_P \right) \cdot C_X + D \cdot (C_{P0} - C_P) . \qquad (7.192)$$

We point out that the balance equation (7.189) for the mass ratio of component e and X expressed via E is independent of the type of reactor used. It states that the rate of formation of E represented by the term

$$\left(\frac{k_1}{\mu_{max}} - \frac{k_2}{\mu_{max}} \cdot C_P + \frac{k_3}{\mu_{max}} \cdot C_P^2 \right) \cdot \mu$$

must be at least the same as the dilution rate of E represented by the term $\mu \cdot E$ in equation (7.189). In the equations (7.189) to (7.192) we set $\mu_{max} = 1 \ hr^{-1}$ for simplicity.

---

[21] Jaques Monod, French biochemist, 1910-1976

If needed, equation (7.189) can be replaced by a differential equation for the component e concentration, namely

$$\frac{dC_e}{dt} = (k_1 - k_2 \cdot C_P + k_3 \cdot C_P^2) \cdot \frac{C_S \cdot C_e}{K_S + C_S} + D \cdot (C_{e0} - C_e) \tag{7.193}$$

to get the same results. The dilution rate D is equal to q/V where q denotes the constant flow rate into the fermentor and V its active volume. In our examples we assume that both q and V are constant.

For steady-state solutions, the set of the four differential equations (7.189) to (7.192) (or equivalently the DEs (7.190) to (7.193)) reduces to a set of four coupled rational equations in the unknown variables E (or e), C_X, C_S, and C_P. To solve the corresponding steady-state equations, we interpret the equations (7.189) to (7.192) as a system of four coupled scalar homogeneous equations for the right-hand sides of the DE system in the form $F(E, C_X, C_S, C_P) = 0$. The resulting coupled system of four scalar equations is best solved via Newton's[22] method after finding the Jacobian[23] DF by partial differentiation of the right-hand-side functions f_i of the equations (7.189) to (7.192). I.e.,

$$DF = \begin{pmatrix} grad(f_1(E, C_X, C_S, C_P)) \\ \vdots \\ grad(f_4(E, C_X, C_S, C_P)) \end{pmatrix}$$

for the gradients

$$grad(f_i(E, C_X, C_S, C_P)) = \begin{pmatrix} df_i/dE, & df_i/dC_X, & df_i/dC_S, & df_i/dC_P \end{pmatrix}$$

of the four component functions f_i of $F = (f_1, \ldots, f_4)^T$ on the right-hand sides of the DEs (7.189) to (7.192).

Note that in case of multiplicity different starting values for E, C_X, C_S, and C_P will lead to different stable steady states. MATLAB itself does not include a built-in Newton method solver since the main work is to find the Jacobian DF by partially differentiating the component functions f_i explicitly by hand for each separate nonlinear system of equations.

The dynamic behavior of the model consisting of the four ODEs in (7.189) to (7.192) can be found using a suitable MATLAB IVP solver `ode...` as previously outlined.
The four-dimensional model simulates the oscillatory behavior of an experimental continuous fermentor quite successfully in the high feed sugar concentration region.
We will use the model to explore the complex static/dynamic bifurcation behavior of this system in the two-dimensional $D - C_{S0}$ parameter space and show the implications of bifurcation phenomena on substrate conversion and ethanol yield and productivity.
The system parameters for the specific fermentation unit under consideration are given below.

---

[22]Isaac Newton, British physicist and mathematician, 1643-1727
[23]Carl Jacobi, German mathematician, 1801-1851

Parameter	Value	Parameter	Value	Parameter	Value
k_1 hr^{-1}	16.0	k_2 $m^3/(kg \cdot hr)$	0.497	k_3 $m^6/(kg^2 \cdot hr)$	0.00383
m_S hr^{-1}	2.166	m_P hr^{-1}	1.1		
Y_{SX} (dim.less)	0.0244498	Y_{PX} (dim.less)	0.0526315	K_S kg/m^3	0.5
C_{XO} kg/m^3	0	C_{PO} kg/m^3	0	C_{eO} kg/m^3	0

7.6.2   Discussion of the Model and Numerical Solution

The parameters of this model offer a physiologically adequate description of the growth and fermentation of *Zymomonas mobilis*. Furthermore, this model is highly consistent with experimental fermentor data. Specifically, it predicts the response of the steady state RNA content of the biomass to elevated ethanol concentrations qualitatively. The effect of an elevated ethanol concentration on the fermentation kinetics resembles the effect of elevating the temperature of the fermentation broth.

The oscillatory behavior of product-inhibited cultures cannot simply be described by a common inhibition term in the equation for the biomass growth. A better description must include an indirect or delayed effect of the product ethanol on the biomass growth rate as indicated in experiments. The decay rate μ_{max} was introduced to account for the accumulation of the inhibitory product pyruvic acid. Other more mechanistic, structured models can be formed that relate to the internal key-compound e. In these, the inhibitory action of ethanol is accounted for in the inhibition of the key-compound e formation. Mathematically, however, these two model descriptions are equivalent, except that the key-compound e is washed out as a part of the biomass in continuous cultures and the rate constant μ_{max} does not vary. Our proposed indirect inhibition model provides a good qualitative description of the experimental results shown in Figure 7.25.

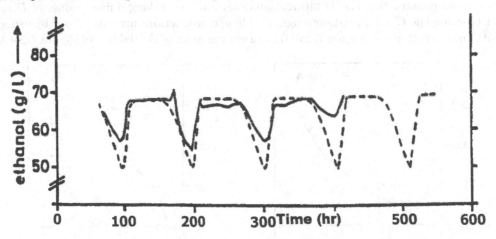

Comparison of experimental (—) and simulated (– –) results

Figure 7.25

The quantitative description, however, is not optimal, as it was necessary to adapt some parameters values for describing the oscillations at different dilution rates. A quantitatively more adequate model must probably also account for the inhibition of the total fermentation including the growth rate independent metabolisms and the dying-off of the biomass at long contact times and high ethanol concentrations.

7.6.3 Graphical Presentation

Our model consists of the four ordinary differential equations (7.189) to (7.192) in the dynamics case and of the corresponding set of coupled rational equations in the static case. These two sets of equations can be solved and studied via MATLAB in order to find the system's steady states, the fermentor's dynamic behavior and to control it.

In the design problem, the dilution rate $D = q/V$ is generally unknown and all other input and output variables are known. In simulation, usually D is known and we want to find the output numerically from the steady-state equations. For this we can use the dynamic model to simulate the dynamic behavior of the system output. Specifically, in this section we use the model for simulation purposes to find the static and dynamic output characteristics, i.e., static and dynamic bifurcation diagrams, as well as dynamic time traces.

As usual in this industrial problems chapter, we do not include actual MATLAB programs for this model. The readers should, however, be well prepared to create such programs and try to verify our included graphical results.

The bifurcation analysis, i.e., the analysis of the steady states and the dynamic solutions is carried out for the dilution rate D as the bifurcation parameter. We have chosen D as the bifurcation parameter since the flow rate $q = V/D$ is directly related to D and q is most easily manipulated during the operation of a fermentor.

Figure 7.26 is a two-parameter continuation diagram of D versus C_{S0}, showing the static limit points. The one parameter bifurcation diagram is constructed by taking a fixed value of C_{S0} and constructing the D bifurcation diagram, then taking a fixed values of D and constructing the C_{S0} bifurcation diagram. These constructions are carried out by using a continuation technique. Figure 7.26(B) is an enlargement of dotted box of Figure 7.26(A).

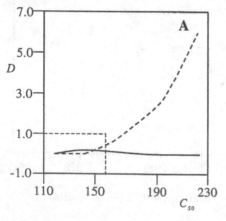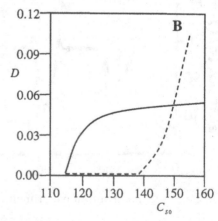

Bifurcation diagram
Figure 7.26

In order to evaluate the performance of the fermentor as an ethanol producer, we calculate the conversion of substrate, the product yield of ethanol and its productivity according to the following simple relations:

Substrate (sugar) conversion: $\quad X_S = \dfrac{C_{S0} - C_S}{C_{C0}}$;

Ethanol yield: $\quad Y_P = \dfrac{C_P - C_{P0}}{C_{S0}}$;

Ethanol productivity of the fermentor: $\quad P_P = C_P \cdot D$.

The latter measures the production rate per unit volume in $kg/(m^3 \cdot hr)$. For oscillatory and chaotic cases the average conversion $\overline{X}_S$, the average yield $\overline{Y}_P$, and the average production rate $\overline{P}_P$, as well as the average ethanol concentration $\overline{C}_P$ are also computed. They are defined as

$$\overline{X}_S = \int_0^{\tau} X_S \, \frac{dt}{\tau}, \; \overline{Y}_P = \int_0^{\tau} Y_P \, \frac{dt}{\tau}, \; \overline{P}_P = \int_0^{\tau} P_P \, \frac{dt}{\tau}, \; \text{and} \; \overline{C}_P = \int_0^{\tau} C_P \, \frac{dt}{\tau},$$

respectively. Here the value of τ is chosen as the length of one period of oscillation in the periodic case, and in the chaotic case, τ is taken large enough to reasonably represent the "average" behavior of the chaotic attractor.

As before we use the dilution rate D as the bifurcation parameter for varying values of C_{S0}. The role of these two variables can easily be exchanged to achieve similar results with C_{S0} as the bifurcation parameter for varying dilution rates D.

Our first case involves $C_{S0} = 140 \; kg/m^3$:

Figure 7.25 compares the experimental results for this data, drawn in a solid curve, with the simulated results in dashed form. Further details of the static and dynamic bifurcation behavior of this system are shown in Figure 7.27.

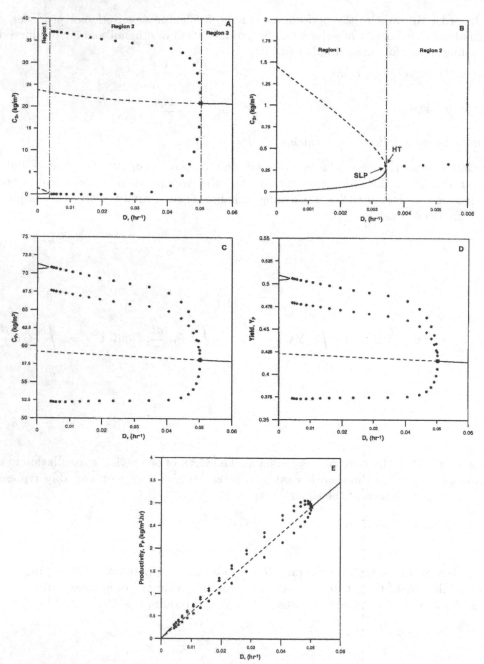

Bifurcation diagram at $C_{S0} = 140 \ kg/m^3$ with D as the bifurcation parameter
Steady state branch: — stable; – – unstable
Periodic branch: • • • stable; ♦ ♦ ♦ average of oscillations
Figure 7.27

Figure 7.27(A) shows the bifurcation diagram for the substrate concentration C_S with vertical demarcations between three different regions. The static bifurcation diagram is an incomplete S shape hysteresis type curve with a static limit point (SLP) at the very low value of $D = 0.0035\ hr^{-1}$, see Figure 7.27(B). The dynamic bifurcation shows a Hopf[24] bifurcation point (HB), i.e., the value of D at which the fermentor starts to oscillate for $D_{HB} = 0.05\ hr^{-1}$ with a periodic branch emanating from it as shown in Figures 7.27(C) and (D). The region in the neighborhood of the static limit point is enlarged in Figure 7.27(B). The periodic branch emanating from HB terminates homoclinically with infinite period when it touches the saddle point very close to the SLP at $D_{HT} = 0.0035\ hr^{-1}$. Figure 7.27(C) is the bifurcation diagram for the ethanol concentration . Figure 7.27(C) shows that the average ethanol concentrations for the periodic attractors are higher than those corresponding to the unstable steady states. Figures 7.27(D) and (E) show the bifurcation diagrams for the ethanol yield Y_P and the ethanol production rate P_P and the average yield and production rate for the periodic branch are depicted by diamonds ◆.

Figure 7.28 shows the period of oscillations as the periodic branch approaches the homoclinical bifurcation point; the period tends to infinity indicating homoclinical termination of the periodic attractor at $D_{HT} = 0.0035\ hr^{-1}$.

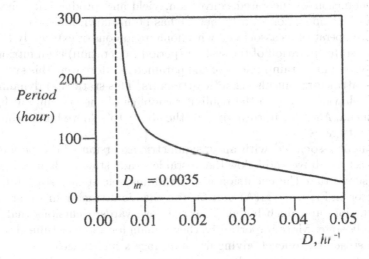

Period changes with D at $C_{S0} = 140\ kg/m^3$
Figure 7.28

Region 1 in Figures 7.27(A) and (B) with $D < D_{HT}$ contains three point attractors. It is characterized by the fact that two of the attractors are unstable and only the steady state with the lowest conversion is stable. The highest (almost complete) conversion occurs in this region for the upper stable steady state, with low C_S concentrations depicted in Figure 7.27(A) and likewise with high C_P concentrations in Figure 7.27(C). This steady state also gives the highest ethanol yield at around 0.51, see Figure 7.27(D). On the other

---

[24]Eberhard Frederich Ferdinand Hopf, Austrian mathematician, 1902-1983

hand, this region has the lowest ethanol production rate according to Figure 7.27(E). This is due to the low values of the dilution rate D. For a given fermentor with active volume V, this corresponds to a very low flow rate.

Region 2 of Figure 7.27 with $D_{HT} < D < D_{HB}$ is characterized by a unique periodic attractor surrounding the unstable steady middle state. The periodic attractor starts at the HB point and terminates homoclinically at a point very close to SLP as shown in Figures 7.27(A) and 7.27(B). As shown in Figures 7.27(C) to (E), in this region the average of the oscillations for the periodic attractor give us higher values for $\overline{C}_P$, $\overline{Y}_P$ and $\overline{P}_P$ than the corresponding steady states do. This makes the operation of the fermentor under periodic conditions more productive because the fermentor will give higher ethanol concentrations by achieving a higher sugar conversion. Comparing the values of the static branch and the average of the periodic branch in this region at $D = 0.045\ hr^{-1}$ for example shows the following percentage improvements: For $\overline{C}_P$: 9.34%, for $\overline{X}_S$: 9.66%, for $\overline{Y}_P$: 8.67%, and for $\overline{P}_P$ the improvement amounts to 9.84%. Therefore, the best production policy for ethanol concentration, yield and productivity in this case is to operate the fermentor at the periodic attractor. In general, there is a trade-off between concentration and productivity, which requires an economic optimization study to determine the optimum value for D.

This phenomenon of increased conversion, yield and productivity through deliberate unsteady-state operation of a fermentor has been known for some time. Deliberate unsteady-state operation is associated with nonautonomous or externally forced systems. The unsteady-state operation of the system (periodic operation) is an intrinsic characteristic of this system in certain regions of the parameters. Moreover, this system shows not only periodic attractors but also chaotic attractors. This static and dynamic bifurcation and chaotic behavior is due to the nonlinear coupling of the system which causes all of these phenomena. And this in turn gives us the ability to achieve higher conversion, yield and productivity rates.

Physically this is associated with an unsymmetric excursion of the periodic or chaotic dynamic trajectory above and below the unstable steady state as depicted in Figure 7.29. Figure 7.29 shows that the excursion above the unstable steady state for both the periodic attractor in Figure 7.29(A) and for the chaotic attractor in Figure 7.29(B). The upwards excursions are much higher than the downwards excursions that go below the unstable steady state. More importantly, they remain for a longer time above than below the unstable steady-state level, giving us on average a higher yield.

We notice that the conversion, yield and productivity are very sensitive to changes in the neighborhood of a HB point. This sensitivity is not only qualitative regarding the birth of oscillations for $D < D_{HB}$, but also quantitative as seen by comparing the conversion, yield and productivity for $D > D_{HB}$ and their average values for $D < D_{HB}$. A further decrease for values of D beyond D_{HB} causes the average values of conversion, yield and productivity to increase, but not as sharply as in the neighborhood of D_{HB}.

Region 3 of Figure 7.27 is characterized by having a unique stable steady state with conversion, yield and productivity characteristics very close to those of the unstable steady state of region 2.

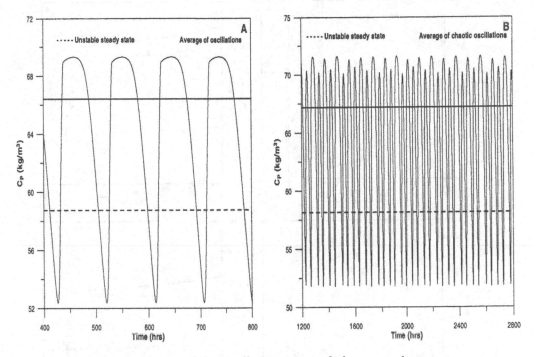

Unequal excursions of the oscillations around the unsteady state
(A)   Periodic attractor at $C_{S0} = 140 \ kg/m^3$ and $D = 0.02 \ hr^{-1}$
(B)   Chaotic attractor at $C_{S0} = 140 \ kg/m^3$ and $D = 0.045842 \ hr^{-1}$
Figure 7.29

Next we study the system when $C_{SO} = 149 \ kg/m^3$.

Figure 7.30 shows bifurcation diagrams with D as the bifurcation parameter.

The bifurcation diagram of Figure 7.30 again is an incomplete S shaped hysteresis type
curve with the static limit point SLP shifted to the much higher value of $D_{SLP} = 0.051 \ hr^{-1}$ when compared with the previous case with $C_{SO} = 140 \ kg/m^3$. A unique
periodic attractor exists between the Hopf bifurcation point at $D_{HB} = 0.0515 \ hr^{-1}$ and
D_{SLP}, followed by a region of bistability characterized by stable periodic and point attractors between D_{SLP} and the first period-doubling point at $D_{PD} = 0.041415 hr^{-1}$. In
this region each of the two attractors has its own domain of attraction. This has important practical implications, not only with regards to start-up, but also with respect to
feasible control policies. The amplitudes of the oscillations increase as D increases, see
Figure 7.30(A). However, differing from the previous case, prior to HT a complex period
doubling PD scenario starts.

Period doubling occurs at D_{PD}, as shown in the Poincaré diagram in Figure 7.31(A)
and the period versus D diagram in Figure 7.31(B).

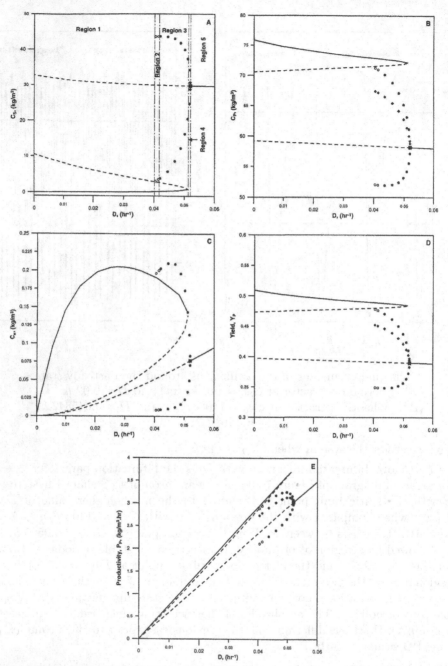

Bifurcation diagram at $C_{S0} = 149\ kg/m^3$ with D as the bifurcation parameter
Steady state branch: — stable; – – unstable
Periodic branch: • • • stable; ○ ○ ○ unstable; ◆ ◆ ◆ average of oscillations
Figure 7.30

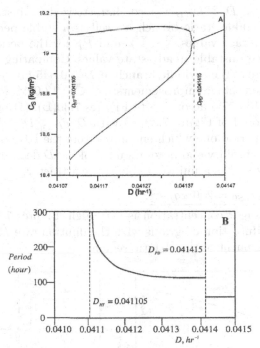

(A)   One dimensional Poincaré diagram at $C_{S0} = 149\ kg/m^3$
(B)   Period change with D at $C_{S0} = 149\ kg/m^3$
Figure 7.31

At D_{PD} the periodicity of the system changes form period one (P1) to period two (P2). Figure 7.31(A) shows that as D decreases further, the periodic attractor P2 grows in size till it touches the middle unstable saddle-type steady state and the oscillations disappear homoclinically at $D_{HT} = 0.041105\ hr^{-1}$ without completing the Feigenbaum[25] period-doubling sequence to chaos.

Region 1 of Figure 7.30 is characterized by having three steady states. Two of these are unstable and only one steady state is stable. It has a very high conversion and is depicted in the lowest branch of Figure 7.30(A) and in the topmost branch of Figure 7.30(B). The conversion rate C_e shows a nonmonotonic behavior. It initially increases as D decreases until it reaches a maximum value of $0.2\ kg/m^3$ at $D = 0.025\ hr^{-1}$. For smaller values of D it decreases monotonically towards zero. This nonmonotonic behavior of C_e with respect to D is due to the nonlinear term $g_2(C_P) = k_1 - k_2 \cdot C_P + k_3 \cdot C_P^2$, for in this region g_2 is nonmonotonic in C_P. This can be seen for the data $k_1 = 16$, $k_2 = 0.497$, and $k_3 = 0.00383$, given on p. 519, by differentiating $g_2(C_P) = 16 - 0.497 \cdot C_P + 0.00383 \cdot C_P^2$ with respect to C_P and finding that $g_2'(64.88) = 0$, i.e., that g_2 has its minimum in region 1 and is therefore nonmonotonic here.

Region 2 of Figure 7.30(A) with $D_{PD} < D < D_{HB}$ is characterized by bistability which calls for special considerations during start-up and in control. In this region there is a very high conversion stable static branch together with a stable period two branch.

---

[25]Mitchell Jay Feigenbaum, USA physicist, 1944-

Region 3 with $D_{SLP} < D < D_{HB}$ is also characterized by bistability. Here there is a very high conversion stable static branch as well as a stable periodic branch. Again it is obvious that the average values $\overline{X}_S$, $\overline{Y}_P$ and $\overline{P}_P$ for the periodic branch are higher than the corresponding unstable steady-state values. Comparing the values of the static branch and the average of the periodic branch at $D = 0.045\ hr^{-1}$ in region 3 for example shows the following percentage improvements: For $\overline{C}_P$: 13.02%, for $\overline{X}_S$: 13.33%, for $\overline{Y}_P$: 13.02%, and for $\overline{P}_P$: 13.577% according to Figures 7.30(B), (D) and (E).

The very narrow region 4 of Figure 7.30(A) with $D_{SLP} < D < DHB$ has a unique periodic attractor with period one which emanates from the HB point at which the stable static branch loses its stability and becomes unstable as D decreases, see Figure 7.30(A). Region 5 with $D > D_{HB}$ has a unique stable static attractor.

Our last case uses $\underline{C_{S0} = 200\ kg/m^3}$.

In this case the feed sugar concentration is very high. Figures 7.32(A) to (D) show the static and dynamic bifurcation diagrams with the dilution rate D as the bifurcation parameter and an enlargement of the chaotic region.

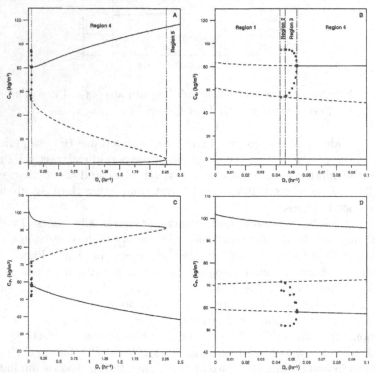

Bifurcation diagram at $C_{S0} = 200\ kg/m^3$ with D as the bifurcation parameter
Steady state branch: — stable; – – unstable
Periodic branch: • • • stable; ○ ○ ○ unstable; ♦ ♦ ♦ average of oscillations
Figure 7.32

This case is characterized by fully developed chaos in region 2. This region is the characteristic region for this case due to the presence of period doubling to the fully developed chaos into two bands. The developmental sequence is from P1 → P2 → P4 → P8 → • • • → fully developed chaos. This terminates homoclinically at $D_{HT} = 0.045835 \ hr^{-1}$ according to Figure 7.32(B). The one dimensional Poincaré[26] diagram in Figure 7.33(A) is enlarged in Figure 7.33(B), where the two bands of chaos and the period-doubling sequence are clearly shown.

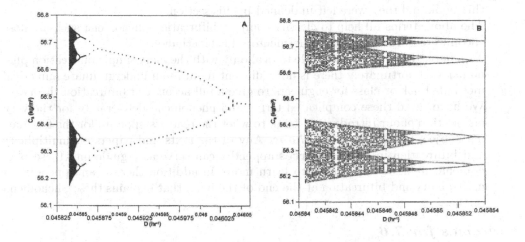

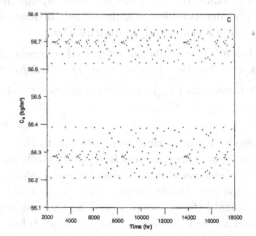

Dynamic characteristics at $C_{S0} = 200 \ kg/m^3$ and $D = 0.04584 \ hr^{-1}$

(A)   One dimensional Poincaré diagram

(B)   Enlarged chaos region from plot (A)

(C)   Return point histogram

Figure 7.33

---

[26]Jules Henri Poincaré, French mathematician, 1854-1912

Figure 7.33(C) is the return point histogram with the Poincaré surface drawn at $C_X = 1.55 \ kg/m^3$ for variable C_S values at $D = 0.04584 \ hr^{-1}$.

Remark

The reader will have noticed by now that we have introduced many new terms such as *Hopf bifurcation point, static limit point, Feigenbaum sequence, Poincaré diagram* and so forth in this section. These terms have never occurred before in this book and they were left undefined in this section.

The above terms all help to classify complex bifurcation phenomena such as those encountered in this section on fermentors for bioethanol.

This book thus far has been able to deal only with the most simple bifurcation phenomena. Unfortunately there is not sufficient room in an undergraduate numerical methods book or class for engineers to give a full account of bifurcation theory.

We have saved these complicated terms and phenomena deliberately for the very last section of our introductory book to whet the reader's appetite for the fascinating subject of mathematical **chaos**. Any of the texts and papers on multiplicity and bifurcation in our **Resources** appendix can serve as a guide into these phenomena, if the reader desires to learn more. In addition there is an appendix on multiplicity and bifurcation at the end of the book that explains these phenomena further.

Exercises for 7.6

1. Find the steady states of the fermentor given by the model (7.189) to (7.192) using Newton's method with a wide variety of starting values for the data from the given plots that exhibits multiplicity.

2. Perform a bifurcation analysis using the feed concentration C_{S0} as the bifurcation parameter for the two dilution rates $D = 0.05 \ hr^{-1}$ and $D = 0.045 \ hr^{-1}$.

3. By using our model, find regions of the parameters with chaotic behavior.

Conclusions

We have used a model for anaerobic fermentation in this section to simulate the oscillatory behavior of an experimental fermentor. Both the steady state and the dynamic behavior of the fermentor with *Zymomonas mobilis* were investigated. The four ODE model simulates the fermentor quite well. Further studies have shown that this model is suitable for scaling-up and for the design of commercial fermentors. Our model has shown the rich static and dynamic bifurcation characteristics of the system, as well as its chaotic ones. All these characteristics have been confirmed experimentally and the oscillatory/chaotic fermentor model is highly suitable for design, optimization and control purposes.

Additional Problems for Chapter 7

1. Develop a model for an industrial riser reactor FCC unit. Collect the necessary data and develop the MATLAB code needed to design the unit from your model.

2. Develop a model for an authothermic circulating fluidized-bed (CFB) reformer for the production of hydrogen from heptane and a MATLAB code for the design of the unit.

3. Develop a model for a packed-bed immobilized fermentor and the MATLAB code for the design of the unit.

Appendix 1:
Linear Algebra and Matrices

Why this appendix ?

When we glance across the main subjects areas, methods, algorithms and codes of this book, we see the following key elements:

We have encountered several scalar equations, some of which are transcendental (Chapter 3), some involve matrices (Chapter 6); but most equations and models are differential in nature (Chapters 4, 5, 6, 7), involving time or location dependent functions for dynamic models of various chemical and biological engineering plants and apparatus.

Why then is there no appendix on differential equations?

A first course on DEs typically studies how to solve equations of the form $y'(t) = f(t, y(t))$. Such a course develops methods to find explicit solutions for specific (theoretically solvable) classes of differential equations and besides, it studies the behavior of solutions of DEs for which there are or are not any known explicit solution methods. In a nutshell, such a course looks at DEs and their solutions both quantitatively and qualitatively.

Unfortunately, the number of classes of DEs with known explicit solutions is rather small. Worse, these theoretically solvable DEs do not often appear, if at all, in practical models such as are needed by the chemical and biological engineer, except for linear DEs which we will treat later in this appendix.

How does one handle DEs that have no known explicit algebraic nor formulistic solution such as the ones of this book? How can they even be solved?

The fundamental difficulty in solving DEs explicitly via finite formulas is tied to the fact that antiderivatives are known for only very few functions $f : \mathbb{R} \to \mathbb{R}$. One can always differentiate (via the product, quotient, or chain rule) an explicitly given function $f(x)$ quite easily, but finding an antiderivative function F with $F'(x) = f(x)$ is impossible for all except very few functions f. Numerical approximations of antiderivatives can, however, be found in the form of a table of values (rather than a functional expression) numerically by a multitude of integration methods such as collected in the **ode...** m file suite inside MATLAB. Some of these numerical methods have been used for several centuries, while the algorithms for stiff DEs are just a few decades old. These codes are very reliable and easy to use in MATLAB.

Therefore, solving a DE by obtaining a table of approximate values for the antiderivative F of a given function f is a cinch nowadays. While theoretical studies of DEs helps us to understand them, it does not help with actually solving DEs for applied problems.

Why do we need to understand Linear Algebra and Matrices for solving DEs.
To be successful in solving applied and mostly differential problems numerically, we must know how to implement our physico-chemical based differential equations models inside standard numerical ODE solvers. The numerical ODE solvers that we use in this book are integrators that work only for first-order differential equations and first-order systems of differential equations. [Other DE solvers, for which we have no need in this book, are discretization methods, finite element methods, multigrid methods etc.]

We have shown in Section 1.2, p. 35ff and Section 5.2, p. 304ff and p. 314ff, for example, how to transform and reduce high-order systems of several coupled DEs to first-order systems of DEs so that we can apply the standard numerical integrators. The reader should be very familiar with this reduction process to first-order systems.

Note that systems of first-order DEs have the matrix/vector form

$$y'(t) = A(t, y(t))$$

with a vector valued function $y : \mathbb{R} \to \mathbb{R}^n$, such as the equations (1.8), (5.66), and (5.71) on the above mentioned pages. For example, the right-hand-side function $A(\omega, y(\omega))$ of the first-order differential equation that is equivalent to equation (5.71) on p. 315 is derived from two coupled second-order DEs. This is implemented inside our corresponding m-file `pellet4runfwd.m` by the following lines of code involving vectors in $\mathbb{R}^4$.

```
function dydx = frhs4(x,y,phi,bt,ga,Sh,Nu)
dydx = [ y(2,:);
         phi^2*exp(ga*(1-1./y(3,:))).*y(1,:);
         y(4,:);
         -bt*phi^2*exp(ga*(1-1./y(3,:))).*y(1,:) ];
```

Further complications of the DE (5.71) arise due to the boundary conditions which require us to define a "singular term matrix" at $\omega = 0$, see p. 315 for more details.

To use numerical integrators for solving ODEs thus requires us to handle vectors and matrices throughout the implementation process.

More generally, our readers need to think vectorially and to envision matrices, linear concepts, and matrix and vector notation throughout this book and, we believe, in any other project that involves numerical computations.

Specifically, in Chapter 3 we create a surface for a transcendental function $f(\alpha, y)$ as an elevation matrix whose zero contour, expressed numerically as a two row matrix table of values, solves the nonlinear CSTR bifurcation problem. In Chapter 6 we investigate multi-tray processes via matrix realizations; in Chapter 5 we benefit from the least squares matrix solution to find search directions for the collocation method that helps us solve BVPs and so on. Matrices and vectors are everywhere when we compute numerically.

That is, after the laws of physics and chemistry and differential equations have helped us find valid models for the physico-chemical processes.

Therefore, we include this short and modern appendix on matrices and linear algebra. Modern in the sense that determinants play no role.

Our list of topics is as follows:

(A) Basic Notions of Linear Algebra

(B) Row Reduction and Systems of Linear Equations

(C) Subspaces, Linear (In)dependence, Matrix Inverse and Bases

(D) Basis Change and Matrix Similarity

(E) Eigenvalues and Eigenvectors, Diagonalizable Matrices

(F) Orthonormal Bases, Normal Matrices and the Schur Normal Form

(G) The Singular Value Decomposition and Least Squares Problems

(H) Linear Differential Equations

(A) Basic Notions of Linear Algebra

Linear algebra deals with finite dimensional real or complex spaces, called $\mathbb{R}^n$ or $\mathbb{C}^n$ for any positive integer n. A typical n-**vector** $x \in \mathbb{R}^n$ or $\mathbb{C}^n$ has the form of a row

$$x = \begin{pmatrix} x_1 & x_2 & \cdots & x_n \end{pmatrix} \text{ where each } x_i \text{ is a real or complex number,}$$

or the form of a column

$$x = \begin{pmatrix} x_1 \\ x_2 \\ \vdots \\ x_n \end{pmatrix} = \begin{pmatrix} x_1 & x_2 & \cdots & x_n \end{pmatrix}^T \text{ where each } x_i \text{ is a real or complex number .}$$

Vectors of the same shape (same dimension n), both rows or both columns, can be added if they belong to the same space by adding the corresponding entries such as $(1,\ 3,\ -4) + (3,\ 0,\ 17) = (4,\ 3,\ 13) \in \mathbb{R}^3$. Vectors can be stretched, i.e., multiplied by a constant such as in $-2 \cdot (1,\ 7,\ -3.1,\ 2-i,\ 1.4)^T = (-2,\ -14,\ 6.2,\ -4+2i,\ -2.8)^T \in \mathbb{C}^5$.

Linear algebra studies *linear functions* $f : \mathbb{R}^n \to \mathbb{R}^m$ defined by the property that

$$f(\alpha x + \beta y) = \alpha f(x) + \beta f(y) \text{ for all vectors } x \text{ and } y \text{ and all scalars } \alpha \text{ and } \beta ,$$

or equivalently that

$$f(x+y) = f(x) + f(y) \text{ and } f(\alpha x) = \alpha f(x) \text{ for all vectors } x \text{ and } y \text{ and all scalars } \alpha .$$

The **dot product** of two row or column vectors x and $y \in \mathbb{R}^n$ with components x_i and y_i for $i = 1, ..., n$ is defined as

$$x \cdot y = \sum_{i=1}^{n} x_i y_i .$$

With x^T denoting the **transposed vector** of x, i.e., the row vector for x if x is a column vector and vice versa, the dot product can be visualized in the form of a matrix product as a row times a column vector

$$x \cdot y = x^T y = \begin{pmatrix} x_1 & x_2 & ... & x_n \end{pmatrix} \begin{pmatrix} y_1 \\ y_2 \\ \vdots \\ y_n \end{pmatrix} = x_1 y_1 + x_2 y_2 + ... + x_{n-1} y_{n-1} + x_n y_n .$$

This concept allows us to express every linear function $f : \mathbb{R}^n \to \mathbb{R}^m$ as a constant matrix times vector product. Here a **matrix** $A \in \mathbb{R}^{m,n}$ is an m by n rectangular array of numbers a_{ij} in $\mathbb{R}$ or $\mathbb{C}$ where m counts the number of rows in A, while n is the number of A's columns.

$$A = \begin{pmatrix} a_{11} & a_{12} & ... & a_{1n} \\ a_{21} & a_{22} & ... & a_{2n} \\ \vdots & \vdots & & \vdots \\ a_{m-1,1} & a_{m-1,2} & ... & a_{m-1,n} \\ a_{m1} & a_{m2} & ... & a_{mn} \end{pmatrix} = \begin{pmatrix} \vdots & \vdots & & \vdots \\ c_1 & c_2 & ... & c_n \\ \vdots & \vdots & & \vdots \end{pmatrix} = \begin{pmatrix} ... & r_1 & ... \\ ... & r_2 & ... \\ & \vdots & \\ ... & r_m & ... \end{pmatrix} .$$

An m by n matrix A can be expressed three ways: in terms of its entries a_{ij}, where the first index i denotes the row that a_{ij} appears in and the second index j denotes the column; or a matrix $A = A_{m,n}$ can be denoted by its n columns $c_1, ..., c_n \in \mathbb{R}^m$; or by its m rows $r_1, ..., r_m \in \mathbb{R}^n$ as depicted above.

We define the **standard unit vectors** e_i of $\mathbb{R}^n$ to be the n-vector of all zero entries, except for position i which is one, such as $e_1 = (1, 0, 0, 0)^T \in \mathbb{R}^4$ or $e_{n-1} = (0, 0, ..., 1, 0)^T \in \mathbb{R}^n$ in column notation. With this notation every linear function $f : \mathbb{R}^n \to \mathbb{R}^m$ can be represented as a constant matrix times vector product $f(x) = Ax$ for

$$A = \begin{pmatrix} \vdots & \vdots & & \vdots \\ f(e_1) & f(e_2) & ... & f(e_n) \\ \vdots & \vdots & & \vdots \end{pmatrix}_{m,n}$$

where the matrix times vector product Ax is either evaluated one dot product at a time as

$$Ax = \begin{pmatrix} ... & r_1 & ... \\ ... & r_2 & ... \\ & \vdots & \\ ... & r_m & ... \end{pmatrix} \begin{pmatrix} x_1 \\ \vdots \\ x_n \end{pmatrix} = \begin{pmatrix} r_1 \cdot x \\ \vdots \\ r_m \cdot x \end{pmatrix}$$

or as a *linear combination* of the n columns $f(e_1) \in \mathbb{R}^m$ of A

$$Ax = \begin{pmatrix} \vdots & \vdots & & \vdots \\ f(e_1) & f(e_2) & \cdots & f(e_n) \\ \vdots & \vdots & & \vdots \end{pmatrix} \begin{pmatrix} x_1 \\ \vdots \\ x_n \end{pmatrix} = x_1 \begin{pmatrix} \vdots \\ f(e_1) \\ \vdots \end{pmatrix} + \ldots + x_n \begin{pmatrix} \vdots \\ f(e_n) \\ \vdots \end{pmatrix} \in \mathbb{R}^m .$$

(B) Row Reduction and Systems of Linear Equations

One main benefit of matrices and vectors is the notational ease which their theory offers. Explicitly written out systems of m *linear equations* in n unknowns x_i have the form

$$
\begin{aligned}
a_{11}x_1 + a_{12}x_2 + \cdots + a_{1n}x_n &= b_1 \\
a_{21}x_1 + a_{22}x_2 + \cdots + a_{2n}x_n &= b_2 \\
&\vdots \\
a_{m1}x_1 + a_{m2}x_2 + \cdots + a_{mn}x_n &= b_m
\end{aligned}
$$

for given coefficients a_{ij}, $i = 1, ..., m$, $j = 1, ..., n$ and right-hand-side entries b_k, $k = 1, ..., m$. This readily translates into a matrix times vector linear equation, namely

$$Ax = \begin{pmatrix} a_{11} & a_{12} & \cdots & a_{1n} \\ a_{21} & a_{22} & \cdots & a_{2n} \\ \vdots & \vdots & & \vdots \\ a_{m1} & a_{m2} & \cdots & a_{mn} \end{pmatrix} \begin{pmatrix} x_1 \\ \vdots \\ x_n \end{pmatrix} = \begin{pmatrix} \cdots & r_1 & \cdots \\ \cdots & r_2 & \cdots \\ & \vdots & \\ \cdots & r_m & \cdots \end{pmatrix} \begin{pmatrix} x_1 \\ \vdots \\ x_n \end{pmatrix} = \begin{pmatrix} b_1 \\ \vdots \\ b_m \end{pmatrix} .$$

Systems of linear equations can be simplified without affecting their solution by

(a) adding multiples of one equation to any other,

(b) multiplying any equation by an arbitrary nonzero constant, or

(c) interchanging any two equations.

These three legitimate operations relate to row operations performed on the *augmented matrix*

$$(A \mid b) = \begin{pmatrix} \cdots & r_1 & \cdots & \vline & b_1 \\ \cdots & r_2 & \cdots & \vline & b_2 \\ & \vdots & & \vline & \vdots \\ \cdots & r_m & \cdots & \vline & b_m \end{pmatrix}$$

that involve

(a) the addition of any multiple of an extended row ($\ldots r_i \ldots \mid b_i$) to any other,

(b) multiplying any extended row by a nonzero constant, and

(c) interchanging any two extended rows in the augmented matrix $(A \mid b)$.

The row operations (a) to (c) are performed on $(A \mid b)$ until the front m by n matrix A achieves **row echelon form**. In a row echelon form R of A each row has a first nonzero entry, called a **pivot**, that is further to the right than the leading nonzero entry (pivot) of any previous row, or it is the zero row.

The particular upper triangular shape of R makes the equation corresponding to its last nonzero row have the least number of variables so that it can be solved most easily. The remaining equations are then solved from the REF via **backsubstitution** from the bottom row on up.

Columns with pivots in a REF are called **pivot columns**, those without pivots are **free columns**. The number of pivots in a REF of a matrix A is called the **rank** of A.

A system of linear equations $Ax = b$ is **solvable** if and only if $\text{rank}(A) = \text{rank}(A \mid b)$ or alternatively if and only if b is contained in the set of linear combinations of the column vectors in A.
A system of linear equations $Ax = b$ is uniquely solvable if it is solvable and if there are as many pivots as A has columns. Otherwise there are infinitely many solutions if the system is solvable.

In MATLAB the command $A\backslash b$ solves $Ax = b$ if A is square $(m = n)$ and the REF of A has n pivots. We have used this command in several of the m files of Section 6.2.

(C) Subspaces, Linear (In)dependence, Matrix Inverse and Bases

A subset of $\mathbb{R}^n$ or $\mathbb{C}^n$ is called a **subspace** if it is closed under vector addition and vector scaling. Typical subspaces are (1) the set of all linear combinations of a number of vectors $c_1, \ldots, c_k \in \mathbb{R}^n$. For

$$A = \begin{pmatrix} \vdots & \vdots & & \vdots \\ c_1 & c_2 & \cdots & c_k \\ \vdots & \vdots & & \vdots \end{pmatrix}$$

this set of linear combinations of the c_i is equal to the **image** $\text{im}(A) = \{y \in \mathbb{R}^m \mid y = Ax \text{ for } x \in \mathbb{R}^k\}$, i.e, to the space spanned linearly by the columns c_i of A.
Another way to describe a subspace is as (2) the **kernel** or **nullspace** of a matrix B_{mn}, formally defined as $\ker(B) = \{x \in \mathbb{R}^n \mid Bx = 0 \in \mathbb{R}^m\}$.

An important application of the theory of linear equations is to decide which vectors among a given set of subspace generating vectors are essential for the given subspace, and which can replicate the action of others and therefore are not needed.
Such minimal spanning sets of vectors for a given subspace are called a **basis** for the subspace. The vectors of a basis for a subspace are also a maximally linearly independent set of vectors. Here we call a set of k column vectors c_k in $\mathbb{R}^n$ **linearly independent** if

the row echelon form of the matrix

$$A = \begin{pmatrix} \vdots & \vdots & & \vdots \\ c_1 & c_2 & \cdots & c_k \\ \vdots & \vdots & & \vdots \end{pmatrix}$$

has k pivots. If there are less than k pivots in the REF of $A_{n,k}$ and there is no pivot in one column j with $1 \le j \le k$, then the linear system

$$\begin{pmatrix} \vdots & \vdots & & \vdots \\ c_1 & c_2 & \cdots & c_{j-1} \\ \vdots & \vdots & & \vdots \end{pmatrix} x = \begin{pmatrix} \vdots \\ c_j \\ \vdots \end{pmatrix}$$

is solvable according to the earlier rank solvability condition. Thus c_j is a linear combination of the c_i for $i < j$, i.e., it is not needed to generate the subspace span$\{c_1, c_2, ..., c_k\}$.

As matrices represent linear mappings between finite dimensional vector spaces we are interested to find out which linear mappings can be inverted, i.e., for which matrices $A_{m,n} : \mathbb{R}^n \to \mathbb{R}^m$ does there exist an ***inverse matrix*** $A_{n,m}^{-1} : \mathbb{R}^m \to \mathbb{R}^n$ with $A^{-1}(Ax) = x$ for all $x \in \mathbb{R}^n$, and how can we find A^{-1} from A if possible.

Functions $f : M \to N$ between two sets M and N are invertible if

    (i) $f(M) = N$ and

    (ii) $f(a) = f(b)$ implies $a = b$.

The first property (i) is essential for f^{-1} to have N as its domain, i.e., for f^{-1} to be defined on the complete range set N of f.

The second property (ii) is essential for f^{-1} to be a ***function*** in the unique assignment sense.

Specifically for matrices $A_{m,n}$ and their induced linear mappings $A : \mathbb{R}^n \to \mathbb{R}^m$, property (i) requires that $Ax = b$ is solvable for every $b \in \mathbb{R}^m$, i.e., that every row of a REF of A must contain a pivot, while property (ii) requires that the kernel of A is zero., i.e., that $Ax = b$ is uniquely solvable for every $b \in \mathbb{R}^m$. Both (i) and (ii) can thus only be true for a matrix $A_{m,n}$ if $m = n = \text{rank}(A)$.

A square and full rank matrix $A_{n,n}$ can be inverted by row reduction of the multiple augmented matrix $(A \mid I_n)$ where I_n is the n by n identity matrix comprised of the n standard unit vectors $e_i \in \mathbb{R}^n$ as its columns. If this row reduction is carried out until the identity matrix I_n appears in the left half of $(I_n \mid A^{-1})$, the right half gives us the matrix inverse A^{-1} with $A^{-1}A = I_n = AA^{-1}$.

(D) Basis Change and Matrix Similarity

Every subspace $\mathcal{U} \subset \mathbb{R}^n$ that differs from the trivial subspace $\{0\}$ has many bases.

One basis of $\mathbb{R}^n$ is the **standard unit vector basis** $\mathcal{E} = \{e_1,\ e_2,\ ...,\ e_n\}$. Since

$$x = \begin{pmatrix} x_1 \\ x_2 \\ \vdots \\ x_n \end{pmatrix} = x_1 e_1 + x_2 e_2 + ... + x_n e_n$$

we call $x = x_{\mathcal{E}}$ the standard **coordinate vector** of the point $x \in \mathbb{R}^n$. If $\mathcal{U} = \{u_1,\ u_2,\ ...,\ u_n\}$ is another basis of $\mathbb{R}^n$, then by the unique spanning property of a basis there is a vector

$$x_{\mathcal{U}} = \begin{pmatrix} \alpha_1 \\ \alpha_2 \\ \vdots \\ \alpha_n \end{pmatrix}$$

with $x = \alpha_1 u_1 + \alpha_2 u_2 + ... + \alpha_n u_n$. We call the coefficient vector $(\alpha_1, \ ..., \ \alpha_n)^T \in \mathbb{R}^n$ the $\mathcal{U}$ coordinate vector $x_{\mathcal{U}}$ of x if

$$x_{\mathcal{E}} = x = \begin{pmatrix} \vdots & & \vdots \\ u_1 & \cdots & u_n \\ \vdots & & \vdots \end{pmatrix} \begin{pmatrix} \alpha_1 \\ \vdots \\ \alpha_n \end{pmatrix} = U x_{\mathcal{U}}$$

where $U_{n,n}$ is the matrix with columns u_1 to u_n.

If $\mathcal{V} = \{v_1,\ v_2,\ ...,\ v_n\}$ is a third basis of $\mathbb{R}^n$, then the coordinate vectors $x_{\mathcal{U}}$ and $x_{\mathcal{V}}$ for the same point $x \in \mathbb{R}^n$ are related by the equation

$$x = x_{\mathcal{E}} = U x_{\mathcal{U}} = V x_{\mathcal{V}}\ .$$

Note that both U and V are invertible, or **nonsingular** since the u_i and the v_j are both linearly independent as bases. Thus

$$x_{\mathcal{U}} = U^{-1} V x_{\mathcal{V}} \quad \text{and} \quad x_{\mathcal{V}} = V^{-1} U x_{\mathcal{U}}\ ,$$

i.e., the matrix product $U^{-1} V$ transforms $\mathcal{V}$ coordinate vectors $x_{\mathcal{V}}$ into $\mathcal{U}$ coordinate vectors $x_{\mathcal{U}}$ and $V^{-1} U$ translates in the reverse direction.

How does a **basis change** in $\mathbb{R}^n$ alter the matrix representation of a linear map $f : \mathbb{R}^n \to \mathbb{R}^m$?

In subsection (A) we have expressed a given linear transformation f as a matrix A with respect to the standard unit vector basis $\mathcal{E} = \{e_1,\ e_2,\ ...,\ e_n\}$, namely

$$A = \begin{pmatrix} \vdots & \vdots & & \vdots \\ f(e_1) & f(e_2) & \cdots & f(e_n) \\ \vdots & \vdots & & \vdots \end{pmatrix} = A_{\mathcal{E}}\ .$$

If we want to represent the linear transformation f with respect to the basis $\mathcal{U}$ instead, we can try to use $A_\mathcal{E}$, but only for transforming $\mathcal{E}$-vectors to $\mathcal{E}$-vectors. In order to find $A_\mathcal{U}$ that maps $\mathcal{U}$-vectors to $\mathcal{U}$-vectors as f does, we first transform the $\mathcal{U}$ coordinate vector $x_\mathcal{U}$ by multiplication from the left by the matrix U with columns u_i into the $\mathcal{E}$ coordinate vector $x = x_\mathcal{E} = Ux_\mathcal{U}$. This vector is then mapped by $A_\mathcal{E}$ to the $\mathcal{E}$-vector $A_\mathcal{E}Ux_\mathcal{U}$. Now we translate this vector back into $\mathcal{U}$ coordinates via U^{-1} to obtain the $\mathcal{U}$-vector $(Ax)_\mathcal{U} = U^{-1}A_\mathcal{E}U \, x_\mathcal{U}$.

Thus the matrix $U^{-1}A_\mathcal{E}U$ is the *matrix representation* $A_\mathcal{U}$ of the linear mapping f with respect to the basis $\mathcal{U}$.

Square matrices A and B, both of size n, with $A = X^{-1}BX$ for a nonsingular n by n matrix X are called *similar matrices*. Similar matrices thus represent the same linear transformation, but with respect to different bases of $\mathbb{R}^n$ if $X \neq I_n$, the n by n *identity matrix*.

(E) Eigenvalues and Eigenvectors, Diagonalizable Matrices

For a given square matrix A_{nn} our aim is to find and explore bases $\mathcal{U}$ of $\mathbb{R}^n$ or $\mathbb{C}^n$ in which the matrix representation $A_\mathcal{U} = U^{-1}A_\mathcal{E}U$ reveals much of the properties of the underlying linear transformation f that the generally dense standard matrix representation $A = A_\mathcal{E}$ of f often hides.

For example, if

$$X^{-1}AX = D = \begin{pmatrix} \lambda_1 & & 0 \\ & \ddots & \\ 0 & & \lambda_n \end{pmatrix}$$

is diagonal, then $AX = XD$. Or written out column by column for X we have $Ax_i = \lambda_i x_i$ for each $i = 1, ..., n$, i.e., the action of A on $\mathbb{R}^n$ is completely separated into n individual actions on each of the basis vectors x_i inside the matrix X. This gives a phenomenally clear picture of the linear transformation $y \mapsto Ay$. The numbers λ and the corresponding vectors $x \neq 0$ with

$$A\,x = \lambda\,x$$

are called *eigenvalues* and *eigenvectors* of A, respectively.

Every square matrix $A \in \mathbb{R}^{n,n}$ or $\mathbb{C}^{n,n}$ has n eigenvalues in $\mathbb{C}$, where we count repeated eigenvalues repeatedly. Eigenvalues of a matrix are numbers $\lambda \in \mathbb{C}$ for which $A - \lambda I$ is singular for the n by n *identity matrix* I that has ones on its main diagonal and zeros everywhere else. Once the eigenvalues λ_j of $A_{n,n}$ have been found, the corresponding eigenvectors x_j are the nonzero solutions x_j of the homogeneous linear system $(A - \lambda_j I)x_j = 0$.

Note that when solving linear systems $Ax = b$ with real coefficient matrices A and real right-hand-side vectors b that are solvable, the solution vector x is real according to the row reduction process of subsection (B).

This is not so for the matrix eigenvalue problem: the eigenvalues (and eigenvectors) of real matrices $A_{n,n}$ generally can only be found in the complex plane $\mathbb{C}$ (and in $\mathbb{C}^n$). The matrix

$$A = \begin{pmatrix} 0 & -1 \\ 1 & 0 \end{pmatrix}$$

for example has only complex eigenvalues, namely $i = \sqrt{-1}$ and $-i \in \mathbb{C}$.

Some real matrix classes, studied in subsection (F) below, however, have only real eigenvalues and corresponding real eigenvectors. The complication with complex eigenvalues and eigenvectors is caused by the Fundamental Theorem of Algebra which states that all the roots of both real and complex polynomials can only be found in the complex plane $\mathbb{C}$.

Some matrices are ***diagonalizable*** over $\mathbb{C}$ in the sense that they have a basis of - possibly complex - eigenvectors and these matrices can therefore be diagonalized by matrix similarity. For example, this is so for all n by n matrices with n distinct eigenvalues since eigenvectors for different eigenvalues are linearly independent. Other matrices such as

$$\begin{pmatrix} 2 & 3 \\ 0 & 2 \end{pmatrix}$$

are not diagonalizable.

(F) Orthonormal Bases, Normal Matrices and the Schur Normal Form

The standard unit vector basis $\mathcal{E} = \{e_1, e_2, ..., e_n\}$ has two particular properties, namely its vectors e_i are mutually orthogonal and have length 1. Such a basis is called ***orthonormal***. Every basis of a subspace can be orthonormalized by the Gram-Schmidt process to an ***ONB***, short for "orthonormal basis".

An orthonormal basis $\mathcal{U} = \{u_1, u_2, ..., u_k\}$ of a subspace of $\mathbb{R}^n$ or $\mathbb{C}^n$ gives rise to a matrix $U_{n,k}$ with the columns u_i that satisfies $U^T U = I_k$ in the real case and $U^* U = I$ in the complex case. A matrix $U_{n,n} \in \mathbb{R}^{n,n}$ with $U^T U = I_n$ is called an ***orthogonal matrix*** and $U_{n,n} \in \mathbb{C}^{n,n}$ with $U^* U = I_n$ is a ***unitary matrix***.
Here U^T denotes the ***transpose matrix***, namely if $U = (u_{i,j})$ is given by its entries, then $U^T = (u_{j,i})$ with rows and columns exchanged, and $U^* = (\overline{u_{j,i}})$ is the ***complex conjugate matrix*** with complex conjugate transposed entries.
If $U = U^T \in \mathbb{R}^{n,n}$ we call U ***symmetric***, and if $U = U^* \in \mathbb{C}^{n,n}$, U is called ***hermitian***.

Orthogonal (and unitary) matrices preserve ***lengths*** of vectors and ***angles*** between vectors in the sense that $\|Ux\| = \|x\|$ for all vectors x and the ***euclidean vector norm*** $\|x\| = \sqrt{x^T x} = \sqrt{x_1^2 + ... + x_n^2}$, while $\cos(\angle(Ux, Uy)) = \cos(\angle(x, y))$ where the cosine of an angle is defined as $\cos(\angle(x, y)) = (x^T y)/(\|x\| \|y\|)$.

Special orthogonal matrices such as ***Householder matrices*** $H = I_m - 2vv^*$ for a unit column vector $v \in \mathbb{C}^m$ with $\|v\| = v^* v = 1$ can be used repeatedly to zero out the lower triangle of a matrix A_{mn} much like the row reduction process that finds a REF of A in subsection (B). The result of this elimination process is the ***QR factorization*** of $A_{m,n}$ as $A = QR$ for an upper triangular matrix $R_{m,n}$ and a unitary matrix $Q_{m,m}$ that is the product of $n - 1$ Householder elimination matrices H_i.

With respect to some orthonormal basis of $\mathbb{C}^n$, every matrix $A_{n,n} \in \mathbb{R}^{n,n}$ or $\mathcal{C}^{n,n}$ can be triangularized, and some matrices can even be diagonalized orthogonally or unitarily.

This follows from the **Schur Normal Form Theorem**: for every $A_{n,n}$ there exists a **unitary matrix** U (with $U^*U = I_n$) so that $U^*AU = T = (t_{i,j})$ is upper triangular with $t_{i,j} = 0$ for all indices $i > j$.

Specifically for **normal matrices**, defined by the matrix equation $A^*A = AA^*$, this implies orthogonal diagonalizability for all normal matrices, such as **symmetric** (with $A^T = A \in \mathbb{R}^{n,n}$), **hermitian** ($A^* = A \in \mathbb{C}^{n,n}$), **orthogonal** ($A^TA = I$), **unitary** ($A^*A = I$), and **skew-symmetric** ($A^T = -A$) matrices.

(G) The Singular Value Decomposition and Least Squares Problems

From subsection (B) it is clear that some systems of linear equations $Ax = b$ have no solution. This may occur for systems with the same number of unknowns as there are equations or for systems $Ax = b$ with nonsquare system matrices A. There is a remedy for this, namely to try to find the **least squares solution** x_{LS} that minimizes the error $\|Ax - b\|$.

To do so, we introduce the **singular value decomposition** of a real matrix $A_{m,n}$, abbreviated by its acronym **SVD**.

For every real matrix $A_{m,n}$ there are two real orthogonal matrices $V \in \mathbb{R}^{n,n}$ and $U \in \mathbb{R}^{m,m}$ and real nonnegative numbers $\sigma_1 \geq \sigma_2 \geq ... \geq \sigma_r > 0$ with $r \leq \min\{m, n\}$ so that

$$\Sigma_{m,n} = \begin{pmatrix} \sigma_1 & & & & & 0 \\ & \ddots & & & & \\ & & \sigma_r & & & \\ & & & 0 & & \\ & & & & \ddots & \\ 0 & & & & & 0 \end{pmatrix} = U^T A_{m,n} V$$

is a diagonal matrix. Clearly $\mathrm{rank}(A) = \mathrm{rank}(\Sigma) = r$.

Since the length of a vector is not altered under an orthogonal transformation (see subsection (F)), the SVD of A yields

$$\|Ax - b\|^2 = \|U^T Ax - U^T b\|^2 = \|U^T AV(V^T x) - U^T b\|^2 =$$

$$= \|\Sigma V^T x - U^T b\|^2 = \sum_{i=1}^{r} (\sigma_i \alpha_i - u_i^T b)^2 + \sum_{i=r+1}^{\min\{m,n\}} (u_i^T b)^2 ,$$

where α_i denotes the ith component of $V^T x \in \mathbb{R}^n$ for $i = 1, ..., n$.

Note that no variation occurs in the second term $\sum_{i=r+1}^{\min\{m,n\}} (u_i^T b)^2$ in the above formula when we vary x. Its value is the **unavoidable error** of the least squares problem $\min_x \|Ax - b\|$. And the first summed term $\sum_{i=1}^{r} (\sigma_i \alpha_i - u_i^T b)^2$ above becomes zero, i.e., minimal, if we choose

$$x_{LS} = \sum_{i=1}^{r} ((u_i^T b)/\sigma_i) v_i \in \mathbb{R}^n .$$

This is the least squares solution for $Ax = b$ because the ith component of $V^T x_{LS}$ is $\alpha_i = (u_i^T b)/\sigma_i$ by the orthonormality of the columns v_i of V so that the first sum term above becomes zero for x_{LS}.

If A is not a square matrix and we command A\b in MATLAB, then the SVD is invoked and finds the least squares solution to the minimization problem $\min_x \|Ax - b\|$. A slight variant that uses only the **QR factorization** mentioned in subsection (F) for a singular but square system matrix $A \in \mathbb{R}^{n,n}$ is used inside our modified boundary value solver bvp4cfsinghouseqr.m in Chapter 5 in order to deal successfully with singular Jacobian matrices inside its embedded Newton iteration.

(H) Linear Differential Equations

Systems of ordinary *linear differential equations* have the form

$$y'(t) = A \, y(t)$$

for a constant matrix $A \in \mathbb{R}^{n,n}$ and a vector valued function $y : \mathbb{R} \to \mathbb{R}^n$. Such systems in one variable t can always be solved theoretically via matrix theory. In our subject area linear DEs can help us decide stability issues of a system's steady states and they allow us to better understand stiffness problems theoretically.

A single real linear differential equation ($n = 1$) has the form $y'(t) = a\, y(t)$ for $A = a \in \mathbb{R}$. Its general solution is $y(t) = c \cdot e^{a \cdot t}$. If an initial condition such as $y(t_0) = y_0 \in \mathbb{R}$ is prescribed then the solution becomes $y(t) = y_0 \cdot e^{a \cdot (t - t_0)}$.

If we allow a to be complex in the single linear DE $y' = ay$, then we can immediately infer that the solution function $y(t)$ will grow without bound if the real part α of $a = \alpha + i \cdot \beta$ ($\alpha, \beta \in \mathbb{R}$) is positive. The solution $y(t)$ will decay to zero as $t \to \infty$ if $\alpha < 0$, and the solution will oscillate if $\beta \neq 0$, possibly growing infinitely large or decaying to zero depending on the sign of α. This is so since for complex $a = \alpha + i \cdot \beta$ we have

$$y(t) = y(0) \cdot e^{a \cdot t} = y(0) \cdot e^{(\alpha + i\beta) \cdot t} = y(0) \cdot e^{\alpha \cdot t} \cdot e^{i \cdot \beta t} = y(0) \cdot e^{\alpha \cdot t} \cdot (\cos(\beta t) + i \cdot \sin(\beta t)) \, .$$

For a system of linear DEs $y'(t) = A \, y(t)$ ($n > 1$) we assume for simplicity that the matrix A is diagonalizable in the sense of subsection (E), i.e., there is a nonsingular matrix X, containing a basis of eigenvectors for A in its columns so that $X^{-1}AX = D$ is a diagonal matrix with the diagonal entries $\lambda_1, ..., \lambda_n \in \mathbb{C}$.

Then $A = XDX^{-1}$ and $y'(t) = XDX^{-1} \, y(t)$. Multiplying on the left by X^{-1} we obtain the equivalent system $X^{-1}y'(t) = DX^{-1}y(t)$. Hence $z(t) = X^{-1}y(t)$ satisfies the decoupled linear system of DEs $z'(t) = Dz(t)$. This system can be solved analogously as a single linear DE, namely one DE at a time by $z(t) = (c_1 e^{\lambda_1 t}, c_2 e^{\lambda_2 t}, \ldots, c_n e^{\lambda_n t})^T$ using the n eigenvalues λ_i of A that appear on the diagonal of D. Finally the solution of $y'(t) = A \, y(t)$ is $y(t) = Xz(t)$.

The situation is more complex if A is not diagonalizable. But in either case the location of the eigenvalues $\lambda_i \in \mathbb{C}$ of A determines the behavior of the solution $y(t) \in \mathbb{R}^n$ as t grows. In particular, if all eigenvalues λ_i have negative real parts, i.e., if they lie in the left halfplane of $\mathbb{C}$, then the solution $y(t)$ will decay to zero over time from any initial condition $y(t_0) \in \mathbb{C}^n$. If at least one eigenvalue has a positive real part, the solution will grow without bound in $\mathbb{C}^n$. And the solution will oscillate or spiral in or out if A has nonreal eigenvalues.

Appendix 2 on bifurcation describes a *linearization* process for nonlinear systems of DEs at a steady state. Linearization forms a locally equivalent linear DE system $y'(t) = Ay(t)$ at a steady state. The eigenvalues of the linearized system matrix A determine the dynamic stability or instability of the particular steady state from their lay with respect to the imaginary axis in $\mathbb{C}$.

Moreover, the eigenvalues of the matrix A in the linearization of a nonlinear system determine the system's *stiffness*. We have encountered stiffness with DEs in Chapters 4 and 5 for both initial value problems (*IVPs*) and boundary value problems (*BVPs*). A formal definition of stiffness for systems of DEs is given in Section 5.1 on p. 276.

Stiff IVPs are best solved by the integrators of the MATLAB `ode...` suite of functions that end in the letter s. In practical terms, one need not construct a linearization to check for stiffness, but rather compare the run times for ordinary integrators (without an "s" in their MATLAB name) and for stiff integrators (with an "s"). If the ordinary ones take too long, the IVP problem is most likely stiff and a MATLAB integrator with a name ending in s should be used to solve the problem more successfully.

For *BVPs*, long run times due to stiffness can often be ameliorated by reversing the direction of integration. A reversal of the direction of integration means a switch to a differential equation with the negative of the previous right-hand side, i.e., to one with the role of the previous apparently stiff pair of eigenvalues reversed in sign to become a nonstiff pair. This follows from the definition of stiffness on p. 276, namely that the smaller magnitude eigenvalue of a stiff pair must be negative for stiffness. However, when reversing the direction of integration, all matrix eigenvalues of the linearized model matrix A change sign to become the eigenvalues of $-A$ upon reversal. But another, formerly nonstiff pair of eigenvalues may now become stiff, leading to further complications. We have encountered such problems and dealt with them repeatedly in the MATLAB codes of Chapter 5. Again linearization gives us only a theoretical tool to understand BVP solvers, but it is not recommended in practice since this is a very time consuming "pencil and paper" process.

Appendix 2
Bifurcation, Instability, and Chaos in Chemical and Biological Systems

This appendix gives an introduction to multiplicity of steady states and to bifurcation and chaos in Chemical and Biological Engineering.
[The numbers in square brackets [..] refer to the extended bibliography contained on the accompanying CD.]

This book differs in many aspects from all other books on this subject.

One of the differences is our treatment of the multiple steady states case as the general case and considering the unique steady-state case as a special case.
This is more logical, more general in scope and easier to comprehend. Besides, it forces our students to become aware of an important phenomenon which has been long neglected in chemical and biological engineering education despite its great importance and ubiquitous occurrence in chemical and biological engineering systems.

The multiplicity phenomenon in chemically reactive systems was first observed in 1918 by Liljenroth [1] for ammonia oxidation and it later appeared in the Russian literature of the 1940s [2]. However, it was not until the 1950s that major investigations of this phenomenon began. This development was inspired by the Minnesota school of Amundson and Aris and their students [3-5]. The Prague school also had a notable contribution in expanding the field [13-16].
For the last 50 years multiplicity phenomena have been studied in general in chemical and biological engineering research and in chemical/biological reaction engineering research in particular. The fascination with this phenomenon has widely spread into the biological engineering literature where it is referred to as "short-term memory" [17, 18]. Part of the applied math literature has also studied these phenomena, but in more general and more abstract terms under the name of "bifurcation theory" [19].

A major breakthrough in understanding these phenomena in chemical reaction engineering was achieved by Ray and co-workers [20, 21] in the 1970s. They uncovered a large variety of bifurcation behavior in nonadiabatic continuous stirred tank reactors. In addition to the usual hysteresis type bifurcation, they uncovered different types of bifurcation diagrams [21], the most important of which is the "isola". This is a closed curve that is disconnected from the rest of the continuum of steady states.

Isolas were also found by Elnashaie et al. [22-24] in the 1980s for enzyme systems where the rate of reaction depends nonmonotonically upon two of the system's state variables, the substrate concentration and pH. This was later shown to be applicable to the acetylcholinesterase enzyme system [52-54].

Later development in singularity theory, especially the pioneering work of Golubitsky and Schaeffer [19], has provided a powerful tool for analyzing the bifurcation behavior of chemically reactive systems. These techniques have been used extensively, elegantly and successfully by Luss and his co-workers [6-11] to uncover a large number of possible types of bifurcation. They were also able to apply the technique successfully to complex reaction networks as well as to distributed parameter systems.

Many laboratory experiments have been successfully designed to confirm the existence of bifurcation behavior in chemically reactive systems [25-29], as well as in enzyme systems [18].

Multiplicity or bifurcation behavior was found to occur in many other systems such as distillation [30], absorption with chemical reaction [31], polymerization of olefins in fluidized beds [32], char combustion [33, 34], the heating of wires [35] and in a number of processes used for manufacturing and processing electronic components [36, 37].

Although the literature is rich in theoretical investigations of bifurcation and in experimental laboratory work for verification, it is rather poor with regard to this phenomenon in industrial systems. In fact, only a few papers have been published that address the question whether the multiplicity of steady states is important industrially or whether it is mainly of theoretical and intellectual interest. The fundamental research in this area has certainly raised the intellectual level of industrial chemical and biological engineering and it has helped us tremendously to develop a more advanced and rigorous approach to modeling chemical and biological engineering processes. Moreover, besides the intellectual benefits of bifurcation research for the engineering level-of-thinking and for the simulation of difficult processes, the multiplicity phenomenon is very important and highly relevant to certain industrial units. For example, using a heterogeneous model an industrial TVA ammonia converter was found to be operating in the multiplicity region [38]. The source of multiplicity in this case was found to be the countercurrent flow of the reactants in the catalyst bed with respect to the flow in the cooling tubes and heat exchanger. The multiplicity of steady states in TVA ammonia converters had been established much earlier from a simple pseudo homogeneous model [39, 40]. In fluid catalytic cracking (FCC) units for the conversion of heavy gas oil to gasoline and light hydrocarbons complex bifurcation behavior was also found, as well as their usual operation at a middle, unstable saddle-type state [42-44]. For these systems, the two stable high- and low-temperature steady states give very low gasoline yields, while the optimum operating state for maximum gasoline yield is almost always the middle, saddle-type unstable state. Autothermic circulating fluidized-bed (CFB) reformers for the production of hydrogen from methane and from heptane have also been shown by Elnashaie and co-workers to exhibit multiplicity of the steady states [45-50]. Elnashaie and co-workers have shown the rich static/dynamic bifurcation and possibly chaotic behavior associated with a diffusion reaction model for the brain acetylcholine neurocycle [51-54].

Despite the broad interest in bifurcation behavior of chemical and biological systems as manifested in the chemical and biological engineering literature, the industrial interest in this phenomenon has been extremely limited. It seems that the industrial philosophy is to avoid the troublesome regions of operating and design parameters which are characterized by instabilities, quenching, ignition etc., where design and control can be quite difficult. This conservative philosophy was quite justifiable before the great advances of the last decades in developing rigorous models for chemical/biological systems and the concurrent advances in computing power and digital computer control. The present state of affairs suggests that the industrial philosophy should change in the direction of exploring the advantages of increased conversion, higher yield and selectivity in the multiplicity region of the operating and design parameters of many systems.

In the last two decades many academic researchers have enabled us to exploit these possible instabilities for higher conversions, selectivities and yields [56-67]. In the Russian literature, Matros and co-workers [67-69] have demonstrated the advantage of operating catalytic reactors deliberately under unsteady-state conditions to achieve improved performance. Industrial enterprises in the west are showing great interest in benefiting from this multiplicity phenomenon. And although a deliberate unsteady-state operation is not directly related to the bifurcation phenomenon and its associated instabilities, it nevertheless demonstrates a definite change of the present conservative industrial attitude.

A 2.1 Sources of Multiplicity

Catalytic and biocatalytic reactors are exceptionally rich in bifurcation and instability problems. These can come from many sources, of which we outline a few below.

A-2.1.1 Isothermal or Concentration Multiplicity

This source of multiplicity is probably the most intrinsic one in catalytic and biocatalytic reactors, for it occurs due to the nonmonotonic dependence of the intrinsic rate of reaction upon the concentration of reactants and products. Although a decade ago, nonmonotonic kinetics of catalytic reactions were considered the exceptional case, today it is clear that nonmonotonic kinetics in catalytic reactions are much more widespread than previously thought. The reader can learn more about various examples from the long list of catalytic reactions exhibiting nonmonotonic kinetics [71-76].

However, nonmonotonic kinetics alone will not produce multiplicity. Such kinetics have to be coupled with a diffusion process, either in the form of a mass-transfer resistance between the catalyst pellet surface and the bulk gas, or within the pores of the pellets. If the flow conditions and the catalyst pellets size are such that diffusional resistances between the bulk gas phase and the catalytic active centers are negligible and the system can be described accurately by a pseudohomogeneous model and consequently the bulk gas phase is in plug flow, then multiplicity is not possible. However, if for such a pseudohomogeneous reactor the bulk-phase flow conditions are not in plug flow then multiplicity of the steady states is possible [77, 78]. The range of deviations from plug flow which gives rise to multiple steady states corresponds to shallow beds with small gas-flow rates, a situation which is not applicable to most industrial reactors. Therefore, multiplicity due to axial dispersion in fixed beds is not very widespread; however, multiplicity due to mass- and heat-transfer resistances associated with the catalyst pellet is. Moreover, multiplicity is also widespread in CSTRs and fluidized beds.

A word of caution is necessary here. Isothermal multiplicity that is the result of nonmonotonic kinetics occurs only when the nonmonotonic kinetic dependence of the rate of reaction upon species concentration is sharp enough. For flat nonmonotonic behavior, multiplicity can occur only for bulk-phase concentrations which are too high to have much practical relevance.

A-2.1.2 Thermal Multiplicity

This is the most widespread and extensively investigated type of multiplicity. It is associated with exothermic reactions. In fact, this type of multiplicity results from nonmonotonic behavior associated with the change of the rate of reaction under the simultaneous variation of the reactants concentration and the temperature accompanying the exothermic reactions that take place within the boundaries of the system, i.e., inside the reactor. For an exothermic reaction, as the reaction proceeds the reactants deplete. This tends to cause a decrease in the rate of reaction, while at the same time the heat release increases the temperature. This causes the rate of reaction to increase according to the Arrhenius dependence of the rate of reaction upon temperature. These two conflicting effects on the rate of reaction lead to a nonmonotonic dependence of the rate of reaction upon the reactant concentration. And in turn this leads to possibly multiple steady states. However, similar to the isothermal case, multiplicity will not occur in fixed-bed catalytic reactors caused by this effect alone. There again has to be some diffusional mechanism coupled with the reaction in order to give rise to multiplicity. Therefore, for a fixed-bed catalytic reactor in which the flow conditions are such that external mass- and heat-transfer resistance between the surface of the catalyst pellets and the bulk gas phase are negligible, where moreover the catalyst pellet size, pore structure and conductivity are such that intraparticle heat and mass transfer resistances are also negligible so that the system can be described accurately by a pseudohomogeneous model, and where the bulk gas phase is in plug flow while the system is adiabatic or cooled co-currently, we note that multiplicity of the steady states is not possible. If the bulk-flow condition is not in plug flow, however, then multiplicity is possible. However, as indicated earlier, this situation is very unlikely in industrial fixed-bed catalytic reactors.

Therefore, the main source of multiplicity in fixed-bed catalytic reactors is through the coupling between the exothermic reaction and the catalyst pellet mass- and heat-transfer resistances.

Isothermal or concentration multiplicity and thermal multiplicity may co-exist in certain systems, when the kinetics are nonmonotonic, the reaction is exothermic and the reactor system is nonisothermal.

A-2.1.3 Multiplicity due to Reactor Configuration

Industrial fixed-bed catalytic reactors have a wide range of different configurations. The configuration of the reactor itself may give rise to multiplicity of the steady states when other sources alone are not sufficient to produce the phenomenon. Most well known is the case of catalytic reactors where the gas phase is in plug flow and all diffusional resistances are negligible, while the reaction is exothermic and is countercurrently cooled. One typical example for this is the TVA type ammonia converter [38-40].

A-2.2 Simple Quantitative Explanation of the Multiplicity Phenomenon

We consider three very simple lumped parameter systems, described by similar algebraic equations, namely

1. a CSTR (continuous stirred tank reactor);

2. a nonporous catalyst pellet; and

3. a cell with a permeable membrane containing an enzyme.

A reaction is taking place in each system with a rate of reaction $r = f(C_A)$, where $f(C_A)$ is a nonmonotonic function.

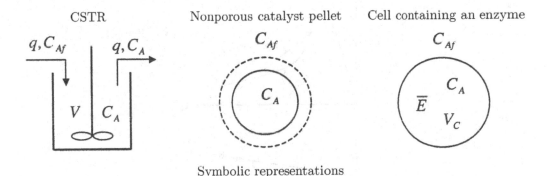

Symbolic representations
Figure 1 (A-2)

In the three examples of Figure 1 (A-2), the rate of reaction equations are as follows: For the CSTR it is

$$q \cdot (C_{Af} - C_A) = V \cdot f(C_A)$$

where q is the volumetric flow rate, V is the reactor volume, C_{Af} is the feed concentration, C_A is the effluent concentration of the reactant, and $f(C_A)$ is the reaction rate, based on the unit volume of the reactant mix.

For the nonporous catalyst pellet, the rate equation is

$$a_p \cdot K_g \cdot (C_{Af} - C_A) = W_p \cdot f(C_A)$$

where a_p is the surface area of the pellet, W_p is the mass of the pellet, K_g is the mass-transfer coefficient, C_{Af} is the concentration of the reactant in bulk fluid, C_A is the concentration of the reactant on the pellet surface, and $f(C_A)$ is the reaction rate based on the unit volume of the reaction mixture.

Finally for the cell containing an enzyme we have the equation

$$a_C \cdot p_A \cdot (C_{Af} - C_A) = \overline{E} \cdot V_C \cdot f(C_A)$$

where a_C is the surface area of the cell, V_C is the volume of the cell, p_A is the permeability of the cell membrane, $\overline{E}$ is the weight of enzyme per unit volume of the cell, C_{Af} is the concentration of the reactant in bulk fluid, C_A is the concentration of the reactant inside the cell, and $f(C_A)$ is the reaction rate based on the unit mass of enzyme.

Note that the rate of reaction equations have the same form for all three cases, namely

$$\alpha \cdot (C_{Af} - C_A) = f(C_A) ,$$

where $\alpha = q/V$ for the CSTR, $\alpha = (a_p \cdot K_g)/W_p$ for the catalyst pellet and $\alpha = (a_c \cdot p_A)/(\overline{E} \cdot V_C)$ for the biocell.

Therefore, the three systems behave similarly. Their typical behavior is depicted in Figure 2 (A-2).

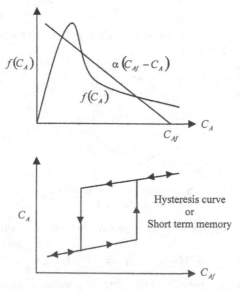

Typical hysteresis curve for concentration multiplicity
Figure 2 (A-2)

Figure 2 (A-2) shows that for a certain combination of parameters, the supply line $\alpha \cdot (C_{Af} - C_A)$ intersects the consumption curve $f(C_A)$ in three points giving rise to multiple steady states.

A-2.3 Bifurcation and Stability

In the multiplicity region of the operating and design parameters, the stability characteristics of a system are quite different from those in the unique steady state region of parameters for the same system. It is important to realize that for systems with multiplicity behavior for certain parameter regions, the mere existence of parameter region with multiplicity has implications on the behavior of the system under external disturbances that may move the system into its multiplicity region. This is so even when the system is operating in the region of a unique steady state

A-2.3.1 Steady State Analysis

A full appreciation of the stability characteristics of any system requires dynamic modeling and analysis of the system. Detailed dynamic modeling and analysis are beyond the scope of this Appendix. However, broad insights into the stability and dynamic characteristics of a system can be extracted from a steady-state analysis. In subsection A-2.3.2 we give a simple and brief introduction to the dynamical side of the picture.

To illustrate, a steady-state analysis suffices to determine whether a system is operating in or near the multiplicity region. In the multiplicity case, global stability is not possible for any of the steady states. Instead, each steady state has its own region of asymptotic stability (RAS), or expressed in the more modern terminology of dynamical systems theory, each steady state has a "basin" or "domain" of attraction. This has important implications on the dynamic behavior, the stability and control of the system. Moreover, the initial direction of a dynamic system trajectory can be predicted by steady-state arguments. Of course qualitative and quantitative dynamic questions may need to be answered separately. These questions can only be answered through dynamic modeling and analysis of the system.

To clarify the above points we consider a simple homogeneous continuous stirred tank reactor (CSTR), in which consecutive exothermic reactions

$$A \xrightarrow{k_1} B \xrightarrow{k_2} C$$

take place. We assume that the reactor is at steady-state conditions, see Figure 3 (A-2).

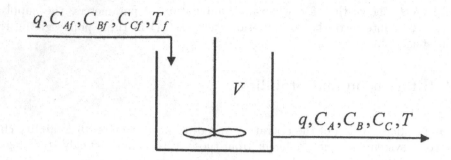

$$q, C_{Af}, C_{Bf}, C_{Cf}, T_f$$

$$V$$

$$q, C_A, C_B, C_C, T$$

Schematic diagram of a CSTR
Figure 3 (A-2)

The mass- and heat-balance equations for the model are given in dimensionless form as follows.

$$X_{Af} = X_A + \alpha_1 \cdot e^{-\gamma_1/y} \cdot X_A \tag{7.194}$$

$$X_{Bf} = X_B + \alpha_2 \cdot e^{-\gamma_2/y} \cdot X_B - \alpha_1 \cdot e^{-\gamma_1/y} \cdot X_A \tag{7.195}$$

$$y - y_f = \alpha_1 \cdot \beta_1 \cdot e^{-\gamma_1/y} \cdot X_A + \alpha_2 \cdot \beta_2 \cdot e^{-\gamma_2/y} \cdot X_B + \overline{K}_c \cdot (y_c - y) \tag{7.196}$$

Here we use the dimensionless variables $X_A = C_A/C_{ref}$, $X_{Af} = C_{Af}/C_{ref}$, $X_B = C_B/C_{ref}$, $X_{Bf} = C_{Bf}/C_{ref}$, $y = T/T_{ref}$, and $y_c = T_c/T_{ref}$, together with the dimensionless parameters $\alpha_i = (V \cdot k_{i0})/q$, $\gamma_i = E_i/(R_G \cdot T_{ref})$, $\beta_i = (\Delta H_i) \cdot C_{ref}/(\rho \cdot C_p \cdot T_{ref})$, and $\overline{K}_c = U \cdot A_H/(q \cdot \rho \cdot C_p)$ for the two consecutive reactions numbered $i = 1, 2$.
In the adiabatic case we set $\overline{K}_c = 0$ in equation (7.196). The above equations with $\overline{K}_c = 0$ also represent the nonporous catalyst pellet with external mass- and heat-transfer resistances and negligible intraparticle heat-transfer resistance but the parameters have a different physical meaning as explained earlier on p. 552.

For a given set of parameters the equations (7.194) to (7.196) can be solved simultaneously to obtain the concentrations X_A and X_B and the temperature at the exit of the reactor for the CSTR and at the surface of the catalyst pellet for the nonporous catalyst pellet.

However, it is also possible to reduce the three equations (7.194) to (7.196) to a single nonlinear equation in y, together with two explicit linear equations for computing X_A and X_B, once y has been determined. This single nonlinear equation is

$$R(y) = (y - y_f) + \overline{K}_v \cdot (y - y_c) =$$
$$= \frac{\alpha_1 \cdot \beta_1 \cdot e^{-\gamma_1/y} \cdot X_{Af}}{1 + \alpha_1 \cdot e^{-\gamma_1/y}} + \frac{\alpha_2 \cdot \beta_2 \cdot e^{-\gamma_2/y} \cdot \alpha_1 \cdot e^{-\gamma_1/y} \cdot X_{Af}}{(1 + \alpha_1 \cdot e^{-\gamma_1/y}) \cdot (1 + \alpha_2 \cdot e^{-\gamma_2/y})} = G(y) \tag{7.197}$$

The right-hand side of equation (7.197) is proportional to the heat generation and it will be called the "heat generation function" $G(y)$, while the left-hand side is proportional to the heat removal due to the flow and the cooling jacket. Therefore,

$$R(y) = (y - y_f) + \overline{K}_v \cdot (y - y_c)$$

is called the "heat removal function".

Equation (7.197) can be solved using any of the standard methods such as bisection or Newton's method. However it is more instructive to solve it graphically by plotting $G(y)$ and $R(y)$ versus y as shown in Figure 4 (A-2) for a parameter case with maximally three steady states.

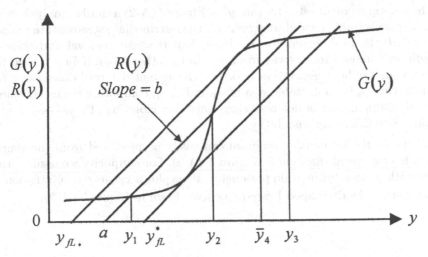

Diagram for the heat generation function $G(y)$ and the heat removal function $R(y)$
for the case of maximally three steady states
Figure 4 (A-2)

The slope of the heat removal line is $b = 1 + \overline{K}_c$ and it intersects the horizontal axis at $a = (y_f + \overline{K}_c \cdot y_c)/(1 + \overline{K}_c)$.

Specifically, for the adiabatic case with $\overline{K}_c = 0$ we get $b_{ad} = 1$ and $a_{ad} = y_f$.

For the middle line $R(y)$ with fixed slope equal to b and horizontal axis intercept equal to a in Figure 4 (A-2), there are maximally three steady states y_1, y_2 and y_3. Stability information of the three steady states and a qualitative analysis of the dynamic behavior of the system can be obtained from the static diagram. The steady-state temperatures y_1, y_2 and y_3 correspond to points where the heat generation and heat removal are equal. That is the defining property of a steady state.

If, for example, the reactor temperature is disturbed from y_3, the high temperature steady state, to $\overline{y}_4$ that is not a steady-state temperature, then the static diagram helps us to determine the direction of temperature change towards a steady state of the system as follows. At $\overline{y}_4$ the heat generation exceeds the heat removal since $G(\overline{y}_4) > R(\overline{y}_4)$ from the graph. Therefore, the temperature will increase. A quantitative analysis of the temperature-time trajectory can of course only be determined from a dynamic model of the system.

Regarding stability, let us examine y_3, for example, using the above kind of argument. Suppose we disturb y_3 infinitesimally by $\delta y > 0$. Then the diagram of Figure 4 (A-2) indicates that the heat removal is larger than the heat generation and therefore the system will cool back down towards y_3. If the disturbance $\delta y < 0$, then the heat generation is larger than the heat removal and therefore the system will heat back up towards y_3. This shows that the steady state at y_3 is stable for the given parameters. However this is only a necessary condition for stability but it is not sufficient. Other dynamic stability conditions should be checked by, for example, computating the eigenvalues of the linearized dynamic model, see [80, 81].

For the low-temperature steady state y_1 in Figure 4 (A-2) a similar analysis shows that this steady state is stable as well. However, for the intermediate steady-state temperature y_2 and $\delta y > 0$ the heat generation is larger than the heat removal and therefore the system will heat up and move away from y_2. On the other hand, if $\delta y < 0$ then the heat removal exceeds the heat generation and thus the system will cool down away from y_2. We conclude that y_2 is an unstable steady state. For y_2, computing the eigenvalues of the linearized dynamic model is not necessary since any violation of a necessary condition for stability is sufficient for instability.

The shape of the bifurcation diagram can easily be predicted from the simple heat generation-heat removal diagram in Figure 4 (A-2). For simplicity we assume the adiabatic case with y_f as a bifurcation parameter. If we plot y versus the bifurcation parameter y_f we obtain the $\mathcal{S}$-shaped hysteresis curve shown in Figure 5 (A-2).

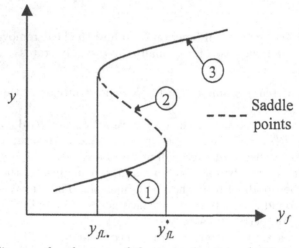

Bifurcation diagram for the case of three steady states (with two limit points)
Figure 5 (A-2)

The points y_{fl_*} and y_{fl}^* are the limit points (bifurcation points) of the multiplicity region. Multiple steady states exist for all y_f values of with $y_{fl_*} \leq y_f \leq y_{fl}^*$.

The curves shown in Figures 4 (A-2) and 5 (A-2) are generic for systems with maximally three steady states. It is possible to have maximally five, seven etc, i.e., any odd number of steady states for different sets of parameters. Figure 6 (A-2) depicts the heat generation and heat removal functions for a system with maximally five steady states.

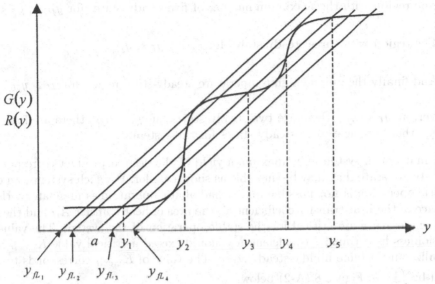

Heat generation and heat removal functions for a system with five steady states
Figure 6 (A-2)

Regarding stability, the five steady states $y_1, \ldots, y_5$ depicted in Figure 6 (A-2) alternate in their stability behavior: the low temperature y_1 is stable, y_2 is unstable, y_3 is stable, y_4 is unstable and the high temperature steady state y_5 is stable. This can be deduced as before from the graph. The bifurcation multiple S-curve diagram for this case is shown in Figure 7 (A-2).

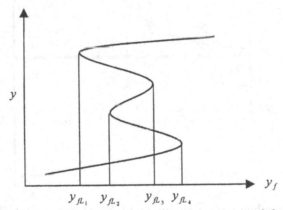

Schematic bifurcation diagram with five steady states and four limit points
Figure 7 (A-2)

The above diagram naturally divides into five regions with respect to y_f.

1. The unique low-temperature steady-state region with $y_f < y_{fl_1}$.

2. The adjacent region with three steady states for $y_{fl_1} < y_f < y_{fl_2}$.

3. The region with the maximum number of five steady states for $y_{fl_2} < y_f < y_{fl_3}$.

4. The region with three steady states for $y_{fl_3} < y_f < y_{fl_4}$.

5. And finally the unique high-temperature steady-state region for $y_f > y_{fl_4}$.

Moreover, at $y_f = y_{fl_1}$ there are two steady states, at $y_f = y_{fl_3}$ there are four, and at $y_f = y_{fl_4}$ there are again two steady states for the system.

For many such systems the maximum yield of the desired product corresponds to a middle steady state that may be unstable as shown earlier. For such systems, an efficient adiabatic operation is not possible and nonadiabatic operation is mandatory. However, the choice of the heat-transfer coefficient U, the area of heat transfer A_H and the cooling jacket temperature are critical for the stable operation of the system. The value of the dimensionless heat-transfer coefficient $\overline{K}_c$ should exceed a critical value $\overline{K}_{c,crit}$ in order to stabilize an unstable middle steady state. The value of $\overline{K}_{c,crit}$ corresponds to the line marked by ④ in Figure 8 (A-2) below.

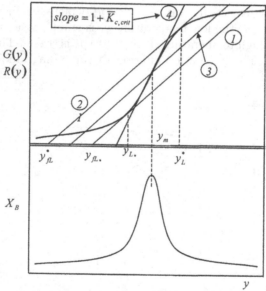

Schematic diagram when the maximum yield X_B of component B corresponds to a
middle unstable steady state
Figure 8 (A-2)

Changing $\overline{K}_c$ increases the slope of the heat removal line because its slope is $1 + \overline{K}_c$. If a bifurcation diagram is drawn for this nonadiabatic case with $\overline{K}_c$ as the bifurcation parameter and the jacket cooling temperature is the temperature of the middle steady state y_m, we obtain a pitchfork type bifurcation diagram as shown in Figure 9 (A-2).

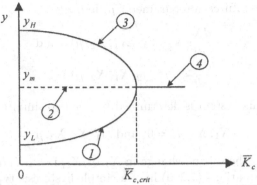

Pitchfork bifurcation diagram of y versus $\overline{K}_c$
Figure 9 (A-2)

If $\overline{K}_c = 0$ in Figure 9 (A-2), the three steady states y_H, y_m and y_L are those of the adiabatic case. For $\overline{K}_c < \overline{K}_{c,crit}$, y_H and y_L change but y_m remains the same because it is the jacket temperature. However, y_m is a middle, an unstable steady state when $\overline{K}_c < \overline{K}_{c,crit}$. At $\overline{K}_{c,crit}$ the multiplicity disappears, giving rise to a unique steady state y_m. From a static point of view, this steady state is stable. However, dynamically it may be unstable for a certain range of $\overline{K}_c > \overline{K}_{c,crit}$ giving rise to limit cycle behavior in the form of a periodic attractor in this region, see [80], for example.

For a CSTR system, Figure 4 (A-2) shows maximally three steady states and Figure 6 (A-2) maximally five, with the possibility of fewer steady state for other specific y parameter values. This maximal steady-state number depends on the shape of the heat generating curve $G(y)$. For different heat generation functions there may possibly be seven, nine, or any odd maximal number of steady states for certain values of y.

A-2.3.2 Dynamic Analysis

The steady states found to be unstable by static analysis are always unstable. However, steady states that are stable from a static point of view may prove to be unstable when a full dynamic analysis is performed. For example, the branch ② in Figure 9 (A-2) is always unstable, while the branches marked by ① , ③ and ④ in Figure 9 (A-2) may be stable or unstable depending upon the dynamic characteristics of the system. This ultimately can only be determined through a proper dynamic stability analysis of the system. As mentioned earlier, our stability analysis for the CSTR is mathematically equivalent to that for a catalyst pellet when using a lumped parameter model and to a distributed parameter model made discrete by techniques such as orthogonal collocation. However, in the latter technique, the system's dimensionality increases considerably, with n dimensions for each state variable, where n is the number of internal collocation points.

As illustration for a dynamic stability analysis we consider a simple two-dimensional

ODE system with one bifurcation parameter μ, namely

$$\frac{dX_1}{dt} \;=\; f_1(X_1, X_2, \mu) \quad \text{and} \tag{7.198}$$

$$\frac{dX_2}{dt} \;=\; f_2(X_1, X_2, \mu) \,. \tag{7.199}$$

The steady state of this system is determined by the two simultaneous equations

$$f_1(X_1, X_2, \mu) = 0 \quad \text{and} \quad f_1(X_1, X_2, \mu) = 0 \tag{7.200}$$

since at a steady state there is no change in X_1 or X_2, i.e., $dX_1/dt = 0 = dX_2/dt$.

We assume that the equations (7.200) have a simple hysteresis type static bifurcation as depicted by the solid curves in Figures 10 to 12 (A-2). The intermediate static dashed branch is always unstable (saddle points), while the upper and lower branches can be stable or unstable depending on the position of eigenvalues in the complex plane for the right-hand-side matrix of the linearized form of equations (7.198) and (7.199). The static bifurcation diagrams in Figures 10 to 12 (A-2) have two static limit points which are usually called saddle-node bifurcation points.

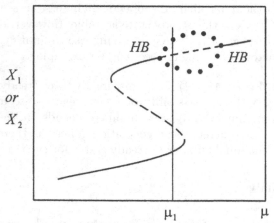

Bifurcation diagram for equations (7.198) and (7.199) with two Hopf bifurcation points

Figure 10 (A-2)

The stability characteristics of the steady-state points is then determined via an eigenvalue analysis of the linearized version of the two DE (7.198) and (7.199). The linearized form of equations (7.198) and (7.199) is as follows:

$$\frac{d\hat{x}_1}{dt} \;=\; g_{11} \cdot \hat{x}_1 + g_{12} \cdot \hat{x}_2 \quad \text{and} \tag{7.201}$$

$$\frac{d\hat{x}_2}{dt} \;=\; g_{21} \cdot \hat{x}_1 + g_{22} \cdot \hat{x}_2 \,, \tag{7.202}$$

where $\hat{x}_1 = x_i - x_{iss}$ with $x_{iss} = x_i$ at the steady state, and $g_{ij} = \partial f_i/\partial x_j|_{ss} \in \mathbb{R}$ for $i, j = 1, 2$.

From the linearized equations (7.201) and (7.202) we form the right-hand side matrix

$$A = \begin{pmatrix} g_{11} & g_{12} \\ g_{21} & g_{22} \end{pmatrix}.$$

With the trace $\mathrm{tr}(A) = g_{11} + g_{22}$ and the determinant $\det(A) = g_{11} \cdot g_{22} - g_{12} \cdot g_{21}$ as customarily defined, we find the eigenvalues of A as the roots of the quadratic equation $\lambda^2 - \mathrm{tr}(A) \cdot \lambda + \det(A) = 0$. In particular, the eigenvalues of A are

$$\lambda_{1,2} = \frac{\mathrm{tr}(A) \pm \sqrt{(\mathrm{tr}(A))^2 - 4 \cdot \det(A)}}{2}.$$

The most important dynamic bifurcation is Hopf bifurcation. This occurs when λ_1 and λ_2 cross the imaginary axis into the right half-plane of $\mathbb{C}$ as the bifurcation parameter μ changes. At the crossing point both roots are purely imaginary with $\det(A) > 0$ and $\mathrm{tr}(A) = 0$, making $\lambda_{1,2} = \pm i \cdot \sqrt{\det(A)}$. At this value of μ, periodic solutions (stable limit cycles) start to exist as depicted in Figures 10 and 11 (A-2).

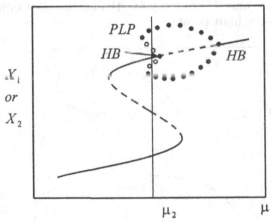

Bifurcation diagram for equations (7.198) and (7.199) with two Hopf bifurcation points
and one periodic limit point
Figure 11 (A-2)

This is called Hopf bifurcation. Figure 10 (A-2) shows two Hopf bifurcation points with a branch of stable limit cycles connecting them. Figure 13 (A-2) shows a schematic diagram of the phase plane for this case when $\mu = \mu_1$. In this case a stable limit cycle surrounds an unstable focus and the behavior of the typical trajectories are as shown. Figure 11 (A-2) shows two Hopf bifurcation points in addition to a periodic limit point (PLP) and a branch of unstable limit cycles in addition to the stable limit cycles branch.

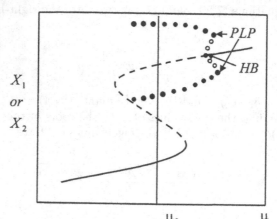

X_1
or
X_2

μ_3 μ

Bifurcation diagram for equations (7.198) and (7.199) with one Hopf bifurcation point, two periodic limit points and one homoclinical orbit (infinite period bifurcation point)

Figure 12 (A-2)

In Figures 10 - 12 (A-2) the following symbols are used:
a solid curve —— for the stable branch of the bifurcation diagram, a dashed curve – – – for the saddle points, a solidly dotted curve • • • for the stable limit cycles, a curve made of circles o o o for unstable limit cycles, HB for the Hopf bifurcation points, and PLP to indicate a periodic limit point.

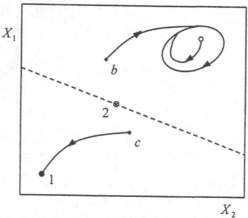

X_1

X_2

Phase plane for $\mu = \mu_1$ and Figure 10 (A-2). —— indicates stable limit cycle trajectories, ⊗ the unstable saddle, • the stable steady state (node or focus), o the unstable steady state (node or focus), and – – – the separatrix

Figure 13 (A-2)

Figure 12 (A-2) shows one Hopf bifurcation point, one periodic limit point and the stable limit cycle terminates at a homoclinical orbit (infinite period bifurcation).

For $\mu = \mu_3$ we observe an unstable steady state surrounded by a stable limit cycle similar to the case in Figure 13 (A-2). However in this case, as μ decreases below μ_3, the limit cycle grows until we reach a limit cycle that passes through the static saddle point as shown in Figure 14 (A-2).

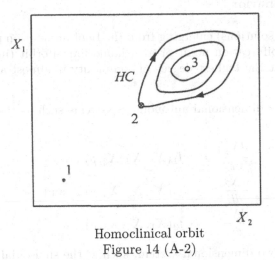

Homoclinical orbit
Figure 14 (A-2)

This limit cycle represents a trajectory that starts at the static saddle point and ends after "one period" at the same saddle point. This trajectory is called the homoclinical orbit and will occur at some critical value μ_{HC}. It has an infinite period and therefore this bifurcation point is called "infinite period bifurcation". For $\mu < \mu_{HC}$ the limit cycle disappears. This is the second most important type of dynamic bifurcation after Hopf bifurcation.

Figure 14 (A-2) shows the phase plane for this case when $\mu = \mu_2$.

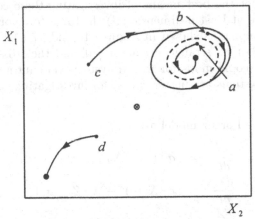

Phase plane for $\mu = \mu_2$ and Figure 11 (A-2). —— indicates stable limit cycle trajectories, − − − the unstable limit cycle, $\otimes$ the unstable saddle, and • the stable steady state (node or focus)
Figure 15 (A-2)

In Figure 15 there is an unstable limit cycle surrounding a stable focus and the unstable limit cycle is surrounded by a stable limit cycle.

A-2.3.3 Chaotic Behavior

Limit cycles (periodic solutions) emerging from the Hopf bifurcation point and terminating at another Hopf bifurcation point or at a homoclinical orbit (infinite period bifurcation point) represent the highest degree of complexity in almost all two- dimensional autonomous systems.

However, for higher dimensional autonomous systems such as the three dimensional system

$$\frac{dX_1}{dt} = f_1(X_1, X_2, X_3, \mu) \ ,$$
$$\frac{dX_2}{dt} = f_2(X_1, X_2, X_3, \mu) \ , \quad \text{and} \tag{7.203}$$
$$\frac{dX_3}{dt} = f_2(X_1, X_2, X_3, \mu) \ ,$$

or a nonautonomous two dimensional system, such as the sinusoidal forced two dimensional system

$$\frac{dX_1}{dt} = f_1(X_1, X_2, \mu) + A \cdot \sin(\omega \cdot t) \tag{7.204}$$
$$\frac{dX_2}{dt} = f_2(X_1, X_2, \mu) \ , \tag{7.205}$$

higher degrees of dynamic complexity are possible, including period doubling, quasi periodicity (torus) and chaos.

Phase plane plots are not the best means of investigating these complex dynamics, for in such cases (which are at least 3-dimensional) the three (or more) dimensional phase planes can be quite complex as shown in Figures 16 and 17 (A-2) for two of the best known attractors, the Lorenz strange attractor [80] and the Rössler strange attractor [82, 83]. Instead stroboscopic maps for forced systems (nonautonomous) and Poincaré maps for autonomous systems are better suited for investigating these types of complex dynamic behavior.

The equations for the Lorenz model are

$$\frac{dX}{dt} = \sigma \cdot (Y - X) \ ,$$
$$\frac{dY}{dt} = r \cdot X - Y - X \cdot Z \ , \quad \text{and} \tag{7.206}$$
$$\frac{dZ}{dt} = -b \cdot Z + X \cdot Y \ .$$

Stroboscopic and Poincaré maps are different from phase plane plots in that they plot the variables on the trajectory at specifically chosen and repeated time intervals. For example, for the forced two dimensional system (7.204) and (7.205) these points are taken at every forcing period. For the Poincaré map, the interval of strobing is not as simple and many different techniques can be applied. Different phase planes can be used for a deeper insight into the nature of strange attractors in these cases. A periodic solution (limit cycle) on the phase plane will appear as one point on the Stroboscopic or the Poincaré map. When period doubling takes place, period two appears as two points on the map, period four as four points and so on. Quasi periodicity (torus) looks very complicated on the phase plane. It will show up as an invariant closed circle with discrete points on a stroboscopic or Poincaré map. When chaos takes place, a complicated collection of points appears on the stroboscopic map. Their shapes have fractal dimensions and are usually called strange attractors.

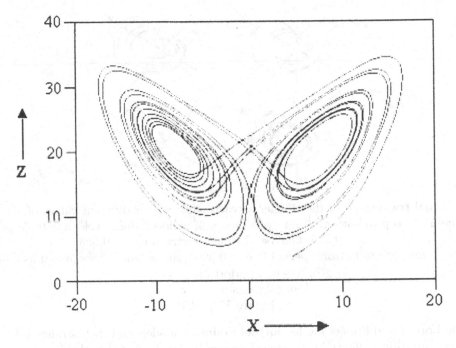

Lorenz strange attractor projected onto the x-z plane for $\sigma = 10$, $b = 8/3$ and $r = 28$

Figure 16 (A-2)

The equations for the Rössler model are

$$\frac{dX}{dt} = Z - Y\,,$$
$$\frac{dY}{dt} = X + a \cdot Y\,, \text{ and} \qquad (7.207)$$
$$\frac{dZ}{dt} = b + Z \cdot (X - c)\,.$$

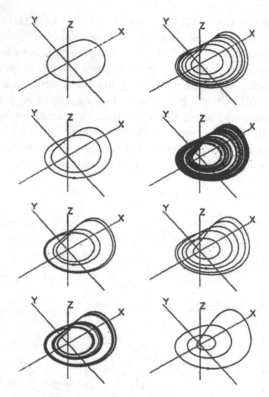

Final trajectories of the Rössler attractor [82, 83] for different values of a.
Left column, top to bottom: limit cycle, $a = 0.3$; period 2 limit cycle, $a = 0.35$; period 4, $a = 0.375$; four-band chaotic attractor, $a = 0.386$.
Right column, top to bottom: period 6, $a = 0.3904$; single-band chaos, $a = 0.398$; period 5, $a = 0.4$; period 3, $a = 0.411$.
[For all cases $b = 2$, $c = 4$.]
Figure 17 (A-2)

The Lorenz and Rössler models are deterministic models and their strange attractors are therefore called "deterministic chaos" to emphasize the fact that this is not a random or stochastic behavior.

Conclusions

The previous notes give our readers the minimal information that is necessary to appreciate the possibly very complex bifurcation, instability and chaos behavior of chemical and biological engineering systems. While the current industrial practice in petrochemical petroleum refining and in biological systems does not heed the importance of these phenomena and their implications on the design, optimization and control of catalytic and biocatalytic reactors, it is more than obvious that these phenomena are extremely important. Bifurcation, instability and chaos in these systems are generally due to nonlinearity and specifically to nonmonotonicity which is widespread in catalytic and biocatalytic reactors. Such nonlinearity and nonmonotonicity is often the result of exothermicity or of the nonmonotonic dependence of the rate of reaction upon the concentration of the reactant species or the bell shaped pH curve for enzyme activities.

Through healthy scientific interaction between industry and academia we expect in the near future that these important phenomena will form the basis of industrial plant design and operations and that more academicians will turn their attention to investigate these phenomena in industrial systems.

For further useful reading on bifurcation, instability and chaos we recommend:

1. The work of Uppal et al. [20, 21] and that of Ray [56] on the bifurcation behavior of continuous stirred tank reactors.

2. The work of Teymour and Ray [84-86] on bifurcation, instability and chaos of polymerization reactors.

3. The work of Elnashaie and co-workers on the bifurcation and instability of industrial fluid catalytic cracking (FCC units) [42-44].

4. The book by Elnashaie and Elshishni on the chaotic behavior of gas-solid catalytic systems [101].

5. The review paper by Razon and Schmitz [87] on bifurcation, instability and chaos for the oxidation of carbon monoxide on platinum.

6. The papers of Wicke and Onken [88] and Razon et al. [89] which give different views on the almost similar dynamics observed during the oxidation of carbon monoxide on platinum. Wike and Onken [88] analyze it as statistical fluctuations while Razon et al. [89] analyze it as chaos. Later, Wicke changes his view and analyzes the phenomenon as chaos [90].

7. The work of the Minnesota group on the chaotic behavior of sinusoidally forced two dimensional reacting systems is useful from the points of view of the analysis of the system as well as the development of suitable efficient numerical techniques for investigating these systems [91-93].

8. The book by Marek and Schreiber [94] is important for chemical/biological engineers who want to enter this exciting field. The book covers a wide spectrum of subjects including differential equations, maps and asymptotic behavior; transition from order to chaos; numerical methods for studies of parametric dependences, bifurcations and chaos; chaotic dynamics in experiments; forced and coupled chemical oscillators; chaos in distributed systems. It contains two appendices, the first dealing with normal forms and their bifurcation diagrams and the second, a computer program for constructing solution and bifurcation diagrams.

9. The software package "Auto 97" [95] is useful for computing bifurcation diagrams. It uses an efficient continuation technique for both static and periodic branches of bifurcation diagrams.

10. The software package "Dynamics" [96] is useful in computing Floquette multipliers and Lyapunov exponents.

11. Elnashaie et al. [97] presents a detailed investigation of bifurcation, instability and chaos in fluidized-bed catalytic reactor for both the unforced (autonomous) and forced (nonautonomous) cases.

The compilation of papers on chaos in different fields published by Cvitanovic [98] is also useful as an introduction.

The dynamic behavior of fixed-bed reactors has not been extensively investigated in the literature. Apparently the only reaction which has received close attention is the oxidation over platinum catalysts. The investigations reveal interesting and complex dynamic behavior and show the occurrence of oscillatory and chaotic behavior [88-90]. It is easy to speculate that further studies of the dynamic behavior of catalytic/biocatalytic reactions will reveal similar complex dynamics like those discovered for the CO oxidation over a Pt catalyst, since most of these phenomena are due to nonmonotonicity of the rate process which is widespread in catalytic systems.

There are many other interesting and complex dynamic phenomena besides oscillation and chaos which have been observed but not followed in depth both theoretically and experimentally. One example is the wrong directional behavior of catalytic fixed-bed reactors, for which the dynamic response to input disturbances is opposite of that suggested by the steady-state response [99, 100]. This behavior is most probably connected to the instability problems in these catalytic reactors as shown crudely by Elnashaie and Cresswell [99]. Recently Elnashaie and co-workers [102-105] have also shown rich bifurcation and chaotic behavior of an anaerobic fermentor for producing ethanol. They have shown that the periodic and chaotic attractors may give higher ethanol yield and productivity than the optimal steady states. These results have been confirmed experimentally [105].

It is obvious that these briefly discussed phenomena (bifurcation, instability, chaos, and wrong directional responses) may discover unexpected shortcomings in the current design, operation and control of industrial catalytic and biocatalytic reactors. This merits extensive theoretical and experimental research as well as specific research of industrial units.

Static and dynamic bifurcation and chaotic behavior are fundamental and very important phenomena which should be familiar to every chemical and biological engineering graduate. These phenomena should form an essential part of their undergraduate training [106] since these phenomena are widespread in chemical and biological engineering systems and have important practical, economic, and environmental implications on the behavior and yield of these systems. Understanding and analyzing these phenomena is essential for the rational design, optimization, operation and control of such systems. The phenomena themselves can be both helpful and dangerous. Not knowing about them precludes an engineer from proper safety precautions and from exploiting them for economic and environmental gains. The modern field of chaos control is very important to know, for example. In some cases knowing the bifurcation implications for a system may be neither harmful nor beneficial. But the system behavior needs to be analyzed and understood, especially in biological systems. The sources of these phenomena are nonlinearity and synergetic nonlinear coupling. However, nonlinearity alone does not create bifurcating systems. It seems that for bifurcation to occur, at least one of the processes' dependence on one of the state variables needs to be nonmonotonic. A coupling between reaction and diffusion also seems to be necessary for the occurrence of bifurcation phenomena. For example, nonmonotonic processes that take place in a homogeneous plug flow reactor do not show any of these phenomena. However, as soon as diffusion comes into the picture, either through axial dispersion in a homogeneous system or through solid-gas interaction in a heterogeneous system, these phenomena start to become a part of the characteristics of the system.

Another important point is what may be called "feedback of information", such as countercurrent operation, recycle or continuous circulation (e.g., the novel autothermic circulating fluidized presented by Elnashaie and co-workers). This feedback of information can give rise to bifurcation and chaos as well [46-50].

These simple and mostly intuitive arguments apply mainly to chemical and biological reaction engineering systems. For other systems such as fluid flow systems, the sources and causes of bifurcation and chaos can be quite different. It is well established that the transition from laminar flow to turbulent flow is a transition from nonchaotic to chaotic behavior. The synergetic interaction between hydrodynamically induced bifurcation and chaos and that resulting from chemical and biological reaction/diffusion is not well studied and calls for an extensive multidisciplinary research effort.

Recall:

In this appendix, the numbers in square brackets [..] refer to the extended bibliography contained on the accompanying CD.

Appendix 3
Contents of the CD and
How to Use it

The CD included with this book contains a copyright and disclaimer notice and:

1. the folders, subfolders, and individual m files that have been developed and tested for Chapters 1, 3, 4, 5 and 6 of this book, as well as

2. an overview `README` file,

3. a short explanation of the 95 supplied m files in `main.txt` in ASCII mode,

4. and in a clickable version in `main.html`. These two `main` files give an overview of each m file; `main.html` can be used to display the text of each m file in a browser window.

5. a file `resources.txt` with further references beyond those contained in the Resources section that immediately follows this appendix.

6. Moreover, we have included a zip file `CBE-book.zip` on the CD with the contents of the CD itself in compressed form. This zip file is used for easy installation of our m file library on any hard drive. Local installation of the m files will facilitate handling of our codes under MATLAB.

We suggest to read the README file on the CD first.

When working with MATLAB we advise our readers to first create a separate folder or directory on his/her hard drive that is used exclusively for MATLAB m files, such as one called `matlabin`, for example.

As the next step, copy the zip file `CBE-book.zip` from the CD onto this MATLAB specific folder `matlabin` and then extract or "unzip" it there. A subfolder called `CBE-book` with several subfolders for our m files, as well as the above mentioned text and html files will then appear inside `matlabin`. The MATLAB desktop should be directed towards a specific subfolder such as `chap5.1.3m` if, for example, one wants to explore or use the m files for Section 5.1.3.

The clickable `main.html` file is included for convenience in navigating between the various m files and the varying chapter folders using a browser.

Each m file subfolder `chap*.*.*m` of the CD is self-contained in the sense that it includes all auxiliary m files for any of the codes of that subfolder.

The file and folder structure of the CD is as follows:

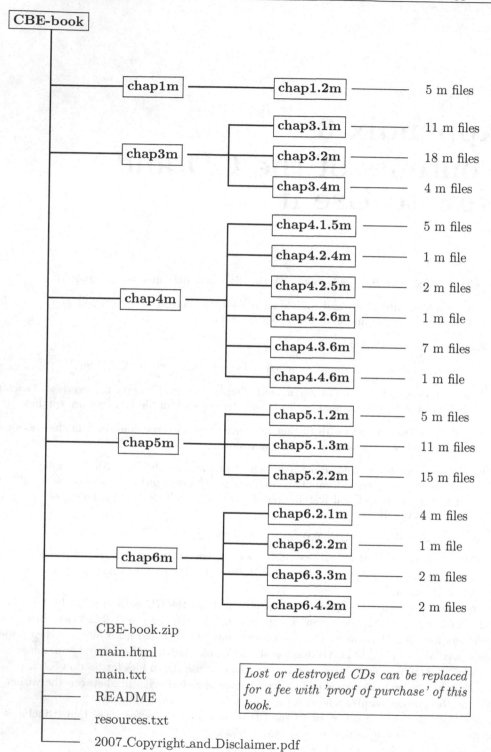

CBE-book.zip

main.html

main.txt

README

resources.txt

2007_Copyright_and_Disclaimer.pdf

Lost or destroyed CDs can be replaced for a fee with 'proof of purchase' of this book.

Resources

This appendix lists the most relevant literature resources for this book. First we give general resources for our subjects, followed by concise chapter specific literature lists. *An extended list of additional papers and books is contained on the accompanying CD.*

General Resources

APPLIED MATHEMATICS :

Linear Algebra and Matrices

H. Anton, C. Rorres, *Elementary Linear Algebra with Applications*, John Wiley, 2005

D. Lay, *Linear Algebra and its Applications*, 3rd updated ed., Addison-Wesley 2005

S. Leon, *Linear Algebra with Applications*, 7th ed, Prentice-Hall, 2005

F. Uhlig, *Transform Linear Algebra*, Prentice-Hall, 2002, ISBN 0-13-041535-9, 502 + xx p

Numerical Analysis

K. Atkinson, W. Han, *Elementary Numerical Analysis*, Wiley 2003

G. Engeln–Müllges, F. Uhlig, *Numerical Algorithms with Fortran, with CD-ROM*, Springer Verlag, 1996, 602 p. (MR 97g:65013) (Zbl 857.65002)

G. Engeln–Müllges, F. Uhlig, *Numerical Algorithms with C, with CD-ROM*, Springer Verlag, 1996, 596 p. (MR 97i:65001) (Zbl 857.65003)

G. F. Gerald, P. O. Wheatley, *Applied Numerical Analysis*, 4th ed, Addison-Wesley, 1989

S. Pizer, V. Wallace, *To Compute Numerically: Concepts and Strategies*, Little, Brown, 1983

A. Quarteroni, F. Saleri, *Scientific Compuying with MATLAB*, Springer, 2001

A. Quarteroni, F. Sacco, F. Saleri, *Numerical Mathematics*, Springer, 2000

J. Stoer, R. Burlirsch, *Introduction to Numerical Analysis*, 3rd ed, Springer, 2002

Polynomial Root Finders

S. Goedecker, *Remark on algorithms to find roots of polynomials*, SIAM J. Sci. Comput., **15** (1994), 1059-1063

V. Pan, *Solving the polynomial equation: some history and recent progress*, SIAM Review, **39** (1997), 187-220

F. Uhlig, *General polynomial roots and their multiplicities in $O(n)$ memory and $O(n^2)$ time*, Linear and Multilinear Algebra, **46** (1999), 327-359. (MR 2001i:12010)

Initial Value Problems

Chapter 17 in Engeln–Müllges and Uhlig; see 'Numerical Analysis' section

A. Iserles, *A First Course in the Numerical Analysis of Differential Equations*, Cambridge U. Press, 1996

L. F. Shampine, M. W. Reichelt, *The MATLAB ODE Suite*, SIAM J. Sci. Comput., **18** (1997), p. 1–22

D. Zwillinger, *Handbook of Differential Equations*, 2nd ed, Academic Press, 1992, 787 p

Boundary Value Problems

W.E. Boyce, R.C. DiPrima, *Elementary Differential Equations and Boundary Value Problems*, 8th ed, John Wiley, 2004, 800 p, ISBN: 0471433381

L.F. Shampine, M.W. Reichelt, J. Kierzenka, *Solving boundary value problems for ordinary differential equations in MATLAB with* bvp4c , ftp://ftp.mathworks.com/pub/doc/papers/bvp

Chemical/Biological Engineering Education

S. Elnashaie, F. Alhabdin, S. Elshishini, *The vital role of mathematical modelling in chemical engineering education*, Math Comput Modelling, 17 (1993), p. 3-11

J. Wei, *Educating chemical engineers for the future*, in Chemical Engineering in a Changing Environment, Engineering Foundation Conference, January 17-22, Santa Barbara, California, (Edited by S. L. Sandler and B.A. Finlayson), 1-12, 1988

Chapter Specific Resources and Literature

For Chapter 1:

D.J. Higham, N.J. Higham, *MATLAB Guide*, 2nd ed, SIAM, 2005

C. B. Moler, *Numerical Computing with MATLAB*, SIAM, 2004, 336 p

W.J. Palm, *Introduction to MATLAB 6 for Engineers*, McGraw-Hill, 2001 (ISBN: 0-07-234983-2)

W.Y. Yang, W. Cao, T-S. Chung, J Morris, *Applied Numerical Methods Using MATLAB*, Wiley-Interscience, 2005 , 528 pages (ISBN: 0471698334)

MATLAB Tutorial. Computer Methods in Chemical Engineering, http://www.glue.umd.edu/ nsw/ench250/matlab.htm

For Chapter 2:

R. Aris, *Manners makyth modellers*, Chem. Eng. Res. Des., 69 (A2) 1991, 165-174

S. Elnashaie, S. Elshishini, *Dynamic Modeling, Bifurcation and Chaotic Behaviour of Gas–Solid Catalytic Reactors*, Gordon and Breach, 1996, 646 p

S. Elnashaie, P. Garhyan, *Conservation Equations and Modeling of Chemical and Biochemical Processes*, Marcel Dekker, 2003, 636 pages (ISBN: 0-8247-0957-8)

What is Systems Theory: http://pespmc1.vub.ac.be/SYSTHEOR.html

For Chapter 3:

M.M. Denn, *Process Modeling*, Longman Science and Technology, 1986

W.F. Ramirez, *Computational Methods in Process Simulation*, 2nd ed, Butterworth-Heinemann, 1997, 512 p (ISBN: 0-7506-3541-X)

G.V. Reklaitis, *Introduction to Material and Energy Balances*, John Wiley, 1983

L. Von Bertalanffy, *General Systems Theory*, George Braziller, New York, 1968

For Chapter 4:

S. Elnashaie, T. Moustafa, T. Alsoudani, S.S. Elshishini, *Modeling and basic characteristics of novel integrated dehydrogenation-hydrogenation membrane catalytic reactors*, Computers & Chemical Engineering, 24(2-7) 2000, 1293-1300

G. F. Froment, K. B. Bischoff, *Chemical Reactor Analysis and Design*, John Wiley, 1990

A. Iserles, *A First Course in the Numerical Analysis of Differential Equations*, Cambridge U. Press, 1996

J.E. Robert Shaw, S.J. Mecca, M.N. Rerick, *Problem Solving. A System Approach*, Academic Press, 1964

J.M. Smith, *Models in Ecology*, Cambridge University Press, Cambridge, 1974

For Chapter 5:

G. R. Cysewski, C.R. Wilke, *Process design and economic studies of alternative fermentation methods for the production of ethanol*, Biotechnology Bioengineering, 20(9), 1421-1444, 1978

S. Elnashaie, S. Elshishini, *Modelling, Simulation and Optimization of Fixed Bed Catalytic Reactors*, Gordon and Breach, 1993, 481 p, ISBN: 2-88124-883-7

S. Elnashaie, S. Elshishini, *Dynamic Modelling, Bifuraction and Chaotic Behaviour of Gas-Solid Catalytic Reactors*, Gordon and Breach, 1996, 646 p, ISBN: 0277-5883

S. Elnashaie, G. Ibrahim, *Heterogeneous Modeling for the Alcoholic Fermentation Process*, Applied Biochemistry and Biotechnology, 19(1), 71-101, 1988

J. Villadsen, W.E. Stewart, *Solution of boundary-value problems by orthogonal collocation*, Chemical Engineering Science, 22, 1483-1501, 1967

For Chapter 6:

N.R. Amundson, *Mathematical Method in Chemical Engineering*, Prentice-Hall, 1966

E. Kreysig, *Advanced Engineering Mathematics*, 2nd ed, John Wiley, 1967

For Chapter 7:

A.E. Abasaeed, S. Elnashaie, *A novel configuration for packed bed membrane fermentors for maximizing ethanol productivity and concentration*, J. Membrane Science, 82(1-2), 75-82, 1993

A. Adris, J. Grace, C. Lim, S. Elnashaie, *Fluidized Bed Reaction System for Steam/ Hydrocarbon Gas Reforming to Produce Hydrogen*, US Patent 5,326,550, Date: June 5, 1994

F.M. Alhabdan, M.A. Abashar, S. Elnashaie, *A flexible software package for industrial steam reformers and methanators based on rigorous heterogeneous models*, Mathematical and Computer Modeling, 16, 77-86, 1992

P.N. Dyer, C.M. Chen, *Engineering development of ceramic membrane reactor system for converting natural gas to H2 and syngas for liquid transportation fuel*, Proceedings of the 2000 Hydrogen Program Review, DOE, 2000

S. Elnashaie, A. Adris, *Fluidized bed steam reformer for methane*, Proceedings of the IV International Fluidization Conference, Banff, Canada, May 1989

S. Elnashaie, A. Adris, M.A. Soliman, A.S. Al-Ubaid, *Digital simulation of industrial steam reformers*, Canadian Journal Chemical Engineering, 70, 786-793, 1992

A.M Gadalla, M.E. Sommer, *Carbon dioxide reforming of methane on nickel catalysts*, Chemical Engineering Science, 44, 2825, 1989

P. Garhyan, S. S. E. H. Elnashaie, S. A. Haddad, G. Ibrahim, S. S. Elshishini, *Exploration and exploitation of bifurcation/chaotic behavior of a continuous fermentor for the production of ethanol*, Chemical Engineering Science, **58** (8), 1479-1496, 2003

I. Jobses, G. Egberts, K. Luyben, J.A. Roels, *Fermentation kinetics of zymomonas mobilis at high ethanol concentrations: oscillations in continuous cultures*, Biotechnology and Bioengineering, 28, 868-877, 1986

A. Mahecha-Botero, P. Garhyan, S. Elnashaie, *Bifurcation, stabilization, and ethanol productivity enhancement for a membrane fermentor*, Mathematical and Computer Modelling, **41** (2005), 391-406

A. Mahecha-Botero, P. Garhyan, S. Elnashaie, *Non-linear characteristics of a membrane fermentor for ethanol production and their implications*, Nonlinear Analysis: Real World Applications, in press.

D. Makel, *Low cost microchannel reformer for hydrogen production from natural gas*, California Energy Commission (CEG), Energy Innovations Small Grant (EISG) Program, 1999

S. Naser, R. Fournier, *A numerical evaluation of a hollow fiber extractive fermentor process for the production of ethanol*, Biotechnology Bioengineering, 32(5), 628-638, 1988

Further references and resources are included on the accompanying CD.

Epilogue

This book deals with an important ingredient of the undergraduate and graduate education in chemical and biological engineering.

A first major part of educating future engineers is to teach how to transform physical, chemical, and biological problems into mathematical equations, called modeling. The next step is to teach how to solve these equations or models numerically.
As a matter of fact, in chemical and biological engineering, the majority of these equations is nonlinear since the reactions are intrinsically nonlinear.
In the "old days", one had to simplify the models and the modeling or design equations greatly in order to make them simple enough for an analytical solution. This was done at the expense of rigor, accuracy, and reliability of the "results".

Thanks to the ever wider use of computers over the last 40 years, the engineering community has devoted serious energy to develop more rigorous, more accurate and more reliable models and design equations for our field. Unfortunately, the resulting more complicated and more representative equations generally do not have closed form analytical solutions.
Since the 1960s we have gone through a long journey with regard to computer power and efficiency, as well as with regard to available numerical codes and algorithms. The first appearance, such as through the IMSL Library, of reliable subroutines that can solve nonlinear equations numerically has exempted us and our students from the tedious and lengthy task of writing programs from scratch for solving detailed problems. Nowadays we can simply call an appropriate subroutine to solve our problem and thereby concentrate more on developing models and implementing the model numerically for the given input data. Then we can quickly analyze the computed results. Resources such as the matrix computation library LAPACK, the development of reliable software for adaptive integrators and stiff ODEs, or the orthogonal collocation method for efficiently solving BVPs have revolutionized our work. Software has given us more detailed and more accurate model solutions than ever possible before.

MATLAB provides a uniformly excellent environment in which engineers can use a wide variety of numerical "engines" that solve almost any equation, algebraic, transcendental, differential, integral etc., without having to "program" from scratch. It also has rich facilities for graphical result representations. MATLAB uses matrices and vectors as its basic building blocks. Therefore it is very economical when dealing with high-dimensional modern engineering systems and offers easy and intuitive programming.

One main aim of this book is to teach current and future engineers how to use MAT-LAB as a "black box" for algorithms and codes in the most efficient and smooth manner in order to solve important chemical and biological engineering problems through realistic nonlinear models.

An added gift of our modern approach is that it will enable future generations of chemical and biological engineers to understand, model, and use the near ubiquitous appearance of multiplicity and bifurcation in chemical and biological systems to their advantage. This book shows how to find multiplicity and how to deal with it for economical and environmental gains.

Putting this advance in engineering at our readers' fingertips makes our efforts worthwhile.

Index

N

List of Photographs

Photos by *Frank Uhlig*, © 2000 – 2005

Printed in the United States
by Bookmasters

Printed in the United States
By Bookmasters